Parasitology Research Monographs

Volume 12

Series Editor
Heinz Mehlhorn
Department of Parasitology
Heinrich Heine University
Düsseldorf
Germany

This book series "Parasitology Research Monographs" presents carefully refereed volumes on selected parasitological topics. Parasites have an increasing impact on animal and human health in the present times of globalization and global warming. Parasites may be agents of diseases and- often at the same time- vectors of other agents of disease such as viruses, bacteria, fungi, protozoa and/or worms. The growth in knowledge of parasitic physiology, cell structure, biotechnological and genetic approaches, ecology, therapeutic capabilities, vaccination, immunology, diagnosis, transmission pathways and many other aspects of parasitology is increasing dramatically, even in the face of the breakthroughs that have already been made. Reflecting these most recent achievements and the importance of parasites as a threat to human and animal health, the series' broad scope concentrates on particularly hot topics that were recently covered by review articles in the journal "Parasitology Research" or in other journals. These reviews offer compact but intense insights into the ongoing research and into the methods and technologies used to control parasites. The volumes in the series build on these topics, and the volume editors are well-known experts in their respective fields. Each volume offers 10 to 20 comprehensive reviews covering all relevant aspects of the topic in focus.

More information about this series at http://www.springer.com/series/8816

Heinz Mehlhorn • Sven Klimpel
Editors

Parasite and Disease Spread by Major Rivers on Earth

Past and Future Perspectives

Springer

Editors
Heinz Mehlhorn
Department of Parasitology
Heinrich Heine University
Düsseldorf, Germany

Sven Klimpel
Institute of Ecology,
Evolution and Diversity,
Senckenberg Biodiversity and
Climate Research Centre (SBiK-F),
Senckenberg Gesellschaft für
Naturforschung (SGN)
Goethe University (GU)
Frankfurt am Main, Germany

ISSN 2192-3671 ISSN 2192-368X (electronic)
Parasitology Research Monographs
ISBN 978-3-030-29060-3 ISBN 978-3-030-29061-0 (eBook)
https://doi.org/10.1007/978-3-030-29061-0

Cover illustration: Deblik, Berlin

This Springer imprint is published by the registered company Springer Nature Switzerland AG.
The registered company address is: Gewerbestrasse 11, 6330 Cham, Switzerland

Preface

The rivers of the globe are the "blood vessels" of the whole life on earth, since all living organisms—bacteria, plants, animals, and of course humans—need water to run their physiological processes successfully. In addition, rivers are the most important pathways for humans and their precursors during their spreading on earth, since they got sufficient drinking water and there they could catch animals, which either live permanently in the water such as fish and mussels or come there daily to drink water. Riversides have been and are still today main places for terrestrial animals to find their sexual partners easily.

Therefore, it is not astonishing that the most important human settlements had been founded along the rivers and that "streets" have been built along these rivers long before ships had been invented and used in constantly increasing amounts.

However, these rivers and their water not only had positive effects on the growth of humans on earth but also enabled the spreading of various agents of diseases such as viruses, bacteria, or worms, which were transmitted either directly by human-to-human contacts, within drinking water (e.g., bacteria, protozoans, or worm eggs), or indirectly by mosquitoes, which breed in water and may transmit viruses, bacteria, protozoans, and even worms during their bloodsucking (acts). All this together occurred permanently in former times leading to the rise and fall of human cultures. Even today, when humans have developed a broad spectrum of drugs and other means to control diseases and thus human and animal populations are widely protected, rivers play an important role in human connections and welfare.

This book considers some important rivers selected from all continents that had been and/or are apparently still today important pathways for the spreading of agents of diseases of humans, animals, and plants endangering the health and (in worst cases) also the survival of humans in some regions especially when multiresistant agents of diseases have been developed or new variations, e.g., of plague or cholera have been developed. Thus, the authors of the chapters of this book contribute insights on diseases along rivers and their distribution in the past and today and give hints how dangers can be avoided or at least minimized.

Düsseldorf, Germany
June 2019

Heinz Mehlhorn

Contents

List of Contributors

Fathy Abdel-Ghaffar Faculty of Sciences, Department of Zoology, Cairo University, Giza, Egypt

Sayed AbdEl-Monem Faculty of Sciences, Department of Zoology, Cairo University, Giza, Egypt

K. G. Alt Institute of Ecology, Evolution and Diversity, Senckenberg Biodiversity and Climate Research Centre (SBiK-F), Senckenberg Gesellschaft für Naturforschung (SGN), Goethe University (GU), Frankfurt am Main, Germany

Sharmênia Araújo Soares Nuto Oswaldo Cruz Foundation (FIOCRUZ), Eusébio, Ceará, Brazil
University of Fortaleza (UNIFOR), Fortaleza, Ceará, Brazil

T. G. Barnard Water and Health Research Centre, Faculty of Health Sciences, University of Johannesburg, Doornfontein, South Africa

C. Bartie Water and Health Research Centre, Faculty of Health Sciences, University of Johannesburg, Doornfontein, South Africa

D. D. Doerge Institute of Ecology, Evolution and Diversity, Senckenberg Biodiversity and Climate Research Centre (SBiK-F), Senckenberg Gesellschaft für Naturforschung (SGN), Goethe University (GU), Frankfurt am Main, Germany

S. Emde Department 26: Fisheries Ecology, North Rhine Westphalian State Agency for Nature, Environment and Consumer Protection, Kirchhundem, Germany

Edenilo Baltazar Barreira Filho Faculty of Education and Culture of Ceará (FAECE), Law School, Fortaleza, Ceará, Brazil

Daniel Grabner Aquatic Ecology and Centre for Water and Environmental Research, University of Duisburg-Essen, Essen, Germany

Jorg Heukelbach School of Medicine, Department of Community Health, Federal University of Ceará (UFC), Fortaleza, Ceará, Brazil

Pewpan M. Intapan Department of Parasitology, Faculty of Medicine and Research and Diagnostic Center for Emerging Infectious Disease, Khon Kaen University, Khon Kaen, Thailand

S. Klimpel Institute of Ecology, Evolution and Diversity, Senckenberg Biodiversity and Climate Research Centre (SBiK-F), Senckenberg Gesellschaft für Naturforschung (SGN), Goethe University (GU), Frankfurt am Main, Germany

J. Kochmann Institute of Ecology, Evolution and Diversity, Senckenberg Biodiversity and Climate Research Centre (SBiK-F), Senckenberg Gesellschaft für Naturforschung (SGN), Goethe University (GU), Frankfurt am Main, Germany

Wanchai Maleewong Department of Parasitology, Faculty of Medicine and Research and Diagnostic Center for Emerging Infectious Disease, Khon Kaen University, Khon Kaen, Thailand

Heinz Mehlhorn Department of Parasitology, Heinrich Heine University, Düsseldorf, Germany

Karim S. Metwaly Faculty of African Postgraduate Studies, Cairo University, Giza, Egypt

P. Muchesa Water and Health Research Centre, Faculty of Health Sciences, University of Johannesburg, Doornfontein, South Africa

Milen Nachev Aquatic Ecology and Centre for Water and Environmental Research, University of Duisburg-Essen, Essen, Germany

Felipe Guhl Nannetti Faculty of Sciences, University of los Andes, Bogotá, Colombia

Monika Niehaus Department of Biology, Düsseldorf University, Düsseldorf, Germany

Marta Cristhiany Cunha Pinheiro School of Medicine, Department of Community Health, Federal University of Ceará (UFC), Fortaleza, Ceará, Brazil

Alberto Novaes Ramos Jr School of Medicine, Department of Community Health, Federal University of Ceará (UFC), Fortaleza, Ceará, Brazil

Juan F. Guhl Samudio Universidad Nacional de Colombia, Bogotá, Colombia

Oranuch Sanpool Department of Parasitology, Faculty of Medicine and Research and Diagnostic Center for Emerging Infectious Disease, Khon Kaen University, Khon Kaen, Thailand

Patrick Scheid Central Institute of the Bundeswehr Medical Service Koblenz, Koblenz, Germany
WG Parasitology and Infection Biology, University Koblenz-Landau, Koblenz, Germany

Hassan M. Sobhy Faculty of African Postgraduate Studies, Cairo University, Giza, Egypt

Martin Stetzkowski Konna e.V., Essen, Germany

Bernd Sures Aquatic Ecology and Centre for Water and Environmental Research, University of Duisburg-Essen, Essen, Germany
Department of Zoology, University of Johannesburg, Johannesburg, South Africa

Tongjit Thanchomnang Faculty of Medicine, Mahasarakham University, Mahasarakham, Thailand

Wei Wang Key Laboratory of National Health Commission on Parasitic Disease Control and Prevention, Wuxi City, Jiangsu Province, China
Jiangsu Provincial Key Laboratory on Parasites and Vector Control Technology, Wuxi City, Jiangsu Province, China
Jiangsu Institute of Parasitic Diseases, Wuxi City, Jiangsu Province, China

Klaus-Christian Withöft Konna e.V., Essen, Germany

Zhongdao Wu Department of Parasitology, Zhongshan School of Medicine, Sun Yat-sen University, Guangzhou, China

Valerie V. Yakimenko Laboratory of Arbovirus Infections, Research Institute of Nature-Foci Infections, Omsk, Russia

Chapter 1
Important Rivers and Their Worldwide Contributions to Epidemics on the Continents

Heinz Mehlhorn

Abstract This chapter offers insights on the importance of rivers selected from all continents. Selection of rivers was done not only according to their length but mainly with respect to their involvement in outbreaks of diseases having spread along their borders in former and recent times and under consideration of the cultural input of the people living along the shores of these "arteries" of civilizations.

In order to get comparable information, seven main topics have been selected. They offer a short comparable overview:

1. *Region*: Placement of the river on earth.
2. *Name*: Names of rivers offer looks in the past.
3. *Length*: This topic gives an idea how many people are involved and are potentially hit by water-borne effects.
4. *Climate*: This topic offers information on potential diseases and the situation of food production along the river.
5. *Early discoveries and history*: This topic explains why diseases are distributed around earth by people spreading along rivers.
6. *Big towns*: This topic selects big towns, which might have contributed as centers of spreading of diseases during growth and in present times.
7. *Literature*: Some selected papers offer deeper insights into the different topics of this book.

Keywords River length · Estuary water · Countries

H. Mehlhorn (✉)
Parasitology, Heinrich Heine University, Düsseldorf, Germany
e-mail: mehlhorn@uni-duesseldorf.de

H. Mehlhorn, S. Klimpel (eds.), *Parasite and Disease Spread by Major Rivers on Earth*, Parasitology Research Monographs 12,
https://doi.org/10.1007/978-3-030-29061-0_1

1.1 Table with a Selection of Important Rivers on the Continents Africa, Asia, Australia, Europe, and North and South America

The different rivers on the different continents on earth are listed according to their length (Table 1.1), which does not correspond to their past or present importance for humans with respect to health, welfare, cultural importance, respectively financial or potential military use. The most important ones are described in a separate part in Sect. 1.2.

1.2 List and Description of Important Rivers of the World in Alphabetical Order

1.2.1 Amazon River (syn. Rio Amazona, Río Marañón, Rio Solimões*)*

1. *Region*: Northern region of the South American continent passing the following countries: Peru, Colombia, Ecuador, Bolivia, Venezuela, and Brazil, where it finally enters the Atlantic Ocean.
2. *Name*: The name was given by the Spanish officer *Francisco de Orellana* (1501–1550) in the year 1541 after fights against tribes including considerable amounts of female warriors who reminded him of the tough amazons in the Greek mythology fighting under the leadership of the famous queen *Penthesilea* in the conquest of Troja (today in Turkey) as had been reported by the Greek poet *Homer* (eighth century B.C.).
3. *Length*: The Amazon river has its origin in the Ucayali-Apurimac river system in Southern Peru (only about 160 km away from the Pacific Ocean) and in the Marañón river. Finally, it enters the Atlantic Ocean after passing a western-to-east route of about 6890 km in Northeastern Brazil. Thus, this river is as long as the Nile river in Africa. Thus, both are considered as the longest rivers on earth. The mouth of the Amazon river is giant reaching a diameter of about 64 km (40 miles) when entering the Atlantic Ocean. The Rio Negro is the largest of all Amazon tributaries followed by the Madeira river.
4. *Climate*: The Amazon river passes very warm, rainy, and highly humid regions and not only gives rise to a splendid green impervious region of high importance for the world climate but also offers excellent conditions for a huge and species-rich animal life spectrum. However, due to high amounts of bloodsucking arthropods, health of humans and farmed animals is endangered by transmission of a high number of agents of diseases.
5. *Early discoveries and history*: In the year 1500 AD, the Portuguese navigator Pedro Alvares Cabral discovered Brazil landing in the Bahia region of our days. The first discovery of the Amazon estuary occurred also in the year 1500 by

Table 1.1 The longest and most important rivers on the different continents

Name	Length in km	Estuary water	Main countries passed
Europe			
Volga river	3688	Caspian Sea	Russia
Danube river	2850	Black Sea	Eight countries
Dnieper river	2285	Black Sea	Russia
Don river	1967	Sea of Azov	Russia
Rhine river	1320	North Sea	Germany
Elbe river (Labe river)	1144	North Sea	Germany
Loire river	1020	Atlantic Ocean	France
Donez river	1016	Don	Russia
Ebro river	930	Mediterranean Sea	Spain
Maas river	925	North Sea	Belgium
Memel river	879	Baltic Sea	Lithuania
Oder river	860	Baltic Sea	Germany, Poland
Rhone river	812	Mediterranean Sea	France
Douro river	776	Atlantic Ocean	Spain
Seine river	776	English Channel	France
Drau river	749	Danube river	Hungary
Po river	682	Adriatic Sea	Italy
Garonne river	650	Gulf of Biscaya	France
Guadalquivir river	579	Gulf of Cádiz	Spain
Mosel river	545	Rhine river	Germany
Main river	524	Rhine river	Germany
Inn river	510	Danube river	Austria
Moldova river (Vltava)	435	Elbe river	Czech Republic
Etsch river	415	Adriatic Sea	Italy
Tiber river	406	Tyrrhenian Sea	Italy
Neckar river	367	Rhine river	Germany
Eger river	350	Elbe river	Czech Republic
Thames river	346	North Sea	England
Africa			
Nile river	6671	Mediterranean Sea	Egypt
Congo river	4377	Gulf of Guinea	Democratic Republic of the Congo
Niger river	4160	Gulf of Guinea	Seven African countries
Zambesi river	2660	Indian Ocean	Zambia
Oranje river	1860	Atlantic Ocean	South Africa
Okavango river	1800	Lake Ngami	Botswana
Limpopo river	1600	Indian Ocean	South Africa
Jubba river	1500	Indian Ocean	Somalia
Senegal river	1430	Atlantic Ocean	Senegal

(continued)

Table 1.1 (continued)

Name	Length in km	Estuary water	Main countries passed
North America			
Mississippi (with Missouri) river	6051	Gulf of Mexico	USA
Mackenzie river	3780	Lake Beaufort	Canada
Yukon river	3185	Bering Sea	Alaska
Saint Lawrence river	3138	Atlantic Ocean	USA
Colorado river	2900	Gulf of California	USA
Rio Grande	2800	Gulf of Mexico	USA, Mexico
Columbia river	2250	Pacific Ocean	USA
Ohio river	1579	Mississippi river	USA
South America			
Amazon river	6518	Atlantic Ocean	Brazil
Rio de la Plata	4700	Atlantic Ocean	Argentina, Uruguay
Rio Madeira	3250	Amazon river	Brazil
Rio Sao Francisco	2897	Atlantic Ocean	Brazil
Orinoco river	2736	Atlantic Ocean	Venezuela
Rio Tocantins	2640	Atlantic Ocean	Brazil
Paraguay river	2621	Paraná	Paraguay
Rio Xingu	1980	Amazon river	Brazil
Uruguay river	1650	Rio de la Plata	Uruguay
Rio Negro	1550	Amazon river	Brazil
Rio Magdalena	1350	Caribbean Sea	Colombia
Asia			
Yangtze river	5632	East China Sea	China
Yellow river (Hwang Ho)	4150	Bohai Sea	China
Yenisei river	5391	Arctic Sea	Russia
Mekong river	4500	South China Sea	Six Asian countries
Amur river	4354	Sea of Okhotsk	Russia, China
Lena river	4264	Arctic Sea	Russia
Ob river	4016	Arctic Sea	Russia
Brahmaputra river	3848	Gulf of Bengal	India
Indus river	3190	Arabic Sea	Pakistan, India
Ganges river	2700	Gulf of Bengal	India
Euphrates river	2800	Gulf of Persia	Iraq, Syria
Ural river	2534	Caspian Sea	Russia
Pearl river	2400	South China Sea	China
Tigris river	1750	Gulf of Persia	Iraq
Godavari river	1445	Gulf of Bengal	India
Yamuna river	1376	Ganges river	India
Terek river	591	Caspian Sea	Russia
Jordan river	251	Dead Sea	Israel

(continued)

Table 1.1 (continued)

Name	Length in km	Estuary water	Main countries passed
Australia			
Murray river	2652	Gulf of South Australia	Australia
Darling river	1472	Gulf of South Australia, enters Murray river	Australia
Murrumbidgee river	1485	Gulf of South Australia, enters Murray river	Australia
Cooper Creek	1420	Lake Eyre	Australia
Lachlan River	1339	Murrumbidgee river/Murray river	Australia
Flinders River	1004	Gulf of Carpentaria	Australia

Vincente Yanez Pinzon. The first descent on the Amazon from Quito (Ecuador) to the Atlantic Ocean was done by Francisco de Orellana during the years 1541–1542. In the years 1560–1561, the second descent on the Amazon was proceeded by the Spanish conquistador Lope de Aguirre. During 1570–1600, many Jesuit missions were established along the river, but this Christian order became expelled in the year 1759 from Brazil by the Marquis of Pombal. The expedition of C.M. de la Condamine brought the first rubber sample to Europe in the year 1736. In the year 1750, the treaty of Madrid fixed the boundaries between the territory claims of the seafaring powers Spain and Portugal.

The private scientific expedition of the German scientist *Alexander von Humboldt* (1769–1859) during the years 1799–1804 (being accompanied by the French *Aimé Bonpland*) offered first deep insights into the fauna and flora of this region.

In the year 1822, Brazil proclaimed its independence. In 1860 the seeds of the *Cinchona* tree (containing quinine, which was one of the first active compounds against malaria) were exported from Brazil to Europe and from there to Asia.

6. *Big towns along the river*: Only a few bigger towns exist along the Amazon river. The first of them *Pucallpa* (211,611 inhabitants) is situated close to one of the Amazonian spring rivers (Ucayali). The biggest town along the Amazon river is named *Iquitos* (437,376 inhabitants), while the well-known historical towns *Manaus* and *Belem* are not (!) situated directly at the Amazon river but only in the nearness. The second biggest town at the Amazon river is *Parintins* with only 112,000 inhabitants (Fig. 1.1).
7. *Literature*

https://www.britannica.com/place/Amazon-River

Ayres JM, Clutton-Brock TH (1992) River boundaries and species range size in Amazonian primates. Am Nat 140:531–537

Coimbra CEA, Basta PC (2007) The burden of tuberculosis in indigenous people in the Amazonian, Brazil. Trans R Soc Trop Med Hyg 101:635–636

Lopez S, Sierra R (2010) Agricultural change in the Pastaza River Basin: a spatially explicit model of native Amazonian cultivation. Appl Geogr 30:355–369

Fig. 1.1 Basin of the Amazon river and its tributaries. By Kmusser—Own work using Digital Chart of the World and GTOPO data, CC BY-SA 3.0, https://commons.wikimedia.org/w/index.php?curid=4745680

Mehlhorn H (ed) (2016) Encyclopedia of parasitology, 3 vols., 4th edn. Springer, New York

Rosinger A, Tanner S (2015) Water from fruit or the river? Examining hydration strategies and gastrointestinal illness among Tsimane' adults in the Bolivian Amazon. Public Health Nutr 18:1098–1108

Schwartzman S, Zimmerman B (2005) Conservation alliances with indigenous people of the Amazon. Conserv Biol 19:721–727

Silman MR (2011) Plant species diversity in Amazonian forests. In: Bush M, Flenley J, Gosling W (eds) Tropical rainforest responses to climatic change. Springer, Heidelberg, pp 285–314

Vieira CB, de Abreu Corrêa A, de Jesus MS, Luz SLB, Wyn-Jones P, Kay D, Rocha MS, Miagostovich MP (2017) The impact of the extreme Amazonian flood season on the incidence of viral gastroenteritis cases. Food Environ Virol 9:195–207

1.2.2 Amur River

1. *Region*: This river starts its official course when the Shilka and Argun rivers merge. Thus, it has its sources in Siberian Russia passing Mongolia and the Inner Mongolia Autonomous Region of China, forms the border between China and Russian Siberia, and flows finally through Russia to the Strait of Tartary (Sea of Okhotsk).
2. *Name*: Due to its course through several countries respectively territories, this river is differently named: Russian, *Amur*; Chinese, *Hei-lung Chiang*; and Mongol, *Khamuren* = Black river = Black dragon river.
3. *Length*: Starting at the confluence point of the *Shilka* (550 km long) and *Argun* (1600 km long) rivers, the Amur measures 1755 miles (=2824 km) in length, thus reaching in total 4354 miles before entering the Sea of Okhotsk via a bell-shaped, 30-mile (50-km)-long estuary.
4. *Climate*: The climate along this fourth long Asian river is characterized by its rather Northern location between the latitudes 55° and 45° North. Winter temperatures are from the polar type ranging from −4 °F (−20 °C) to −27 °F (−33 °C) in January. Temperatures in the short summer may reach 72 °F (=18 °C). Rainfall depends on the region reaching around 600 mm in mean (50% in the short summertime).
5. *Early discoveries and history*: The Amur basin was exclusively populated by nomads for a long time, who lived from animal breeding and especially from hunting. These groups belonged to groups like Yakuts, Gilyaks, Oroks, and different Mongols and also Manchu groups living in the South of the river, where it was/is warmer/less cold than in the Northern region. Some Manchu tribes entered China and also established in the Amur basin a rather long-lasting empire (=Manchu dynasty, ~1640–1912). Russian influence and settlement started around 1850, and China finally transferred the regions North of the Amur river in 1858 and East of the Ussuri in 1860 to Russia, whose early explorers had been active there during the sixteenth and seventeenth century. Later Russia enlarged its activity especially in Manchuria and came just prior to World War II (1939–1945) into conflicts with the Japanese power becoming more and more active in this region, too. After World War II, China and Russia came into an armed conflict along the Ussuri river in 1969. After the breakdown of the Soviet Union, the governments in Moscow and Beijing established a useful cooperation in these regions.
6. *Big towns along the river*: Compared to the situation in people-rich regions of China and Russia, the towns along the Amur river are rather small: *Blagoveshchensk* 215,000 inhabitants; *Khabarovsk* 580,000 inhabitants; *Komsomolsk* 265,000 inhabitants; and *Tongjiang* 170,000 inhabitants (Fig. 1.2).
7. *Literature*

Fig. 1.2 Basin of the Amur river and its tributaries. By Kmusser—Own work using Digital Chart of the World and GTOPO data, CC BY-SA 3.0, https://commons.wikimedia.org/w/index.php?curid=6077796

Bolgov MV, Korobkina EA, Osipova NV et al (2016) The analysis of long-term variability and estimation of the maximum water levels under conditions of high anthropogenic impact for the Amur River. Russ Meteorol Hydrol 41:577–584

Makhinov AN (2017) The structure of ice and ice cover in lower Amur reaches with branched channel. Water Resour 44:559–567

Polyaeva KV (2017) The first data on parasites of the Amur sleeper *Perccottus glenii* Dybowski, 1877 (Osteichthyes: Odontobutidae) in the Yenisei River basin (Bugach reservoir). Russ J Biol Invasions 8:104–107

Semenov EK, Sokolikhina NN, Tatarinovich EV (2017) Monsoon circulation over the Amur River basin during catastrophic flood and extreme drought in summer. Russ Meteorol Hydrol 42:141

Zubarev VA (2017) Ecological conditions of watercourses in the middle Amur lowland in the areas of drainage reclamation. Water Resour 44:940–951

1.2.3 Brahmaputra River (Jamuna River)

1. *Region*: This river is a transboundary one in Asia flowing through China, India, and Bangladesh. However, its tributary *Yarlung Tsangpo river* has its origin in the Manasarovar Lake (close to the Bhagirath glacier) being situated at a height of 5120 m at the northern side of the Himalaya and flows through Southern Tibet breaking through the Himalaya mountains after having formed great gorges. In the Assam valley, it is called *Brahmaputra* but in Bangladesh *Jamuna*. Then it enters into the Ganges delta merging with the *Padma* (=name of the *Ganges* in Bangladesh) and finally joins the *Meghna* before emptying into the Bay of Bengal.
2. *Name*: The official name is *Brahmaputra* river; however, there are further local names, which always describe only local sections of the river, which have gotten their names in times, when people did not travel to far-off regions. The official name *Brahmaputra* is derived from the Sanskrit terms for *son* (=*putra*) *of Brahma* (one of the main gods in Hinduism). In China it is called *Yarlung*.
3. *Length*: The length of this river as described in literature varies depending on the inclusion of some tributaries or not. Thus there are data reporting lengths from 2900 km to 3848 km. Astonishing is its rather high mean depth of 38 m, which makes it understandable that the yearly occurring floods are very dangerous for people living along the banks of this river.
4. *Climate*: The climate depends on the height, where the river flows, thus reaching from very cold to tropical temperatures. In times of sudden snow melting, the river produces enormous floods, which kill every year large amounts of people living along its banks.
5. *Early discoveries and history*: The region, through which the river and its 11 main tributaries flow, was hidden to European people for a long time, so that historic maps have their earliest data around 1800. Geological investigations have shown that the river changed several times its course by switching back and forth in former times. The regions along this river are apparently the basics of the six cholera epidemics in the past and of the recently ongoing seventh one (see Chap. 3).
6. *Big towns along the river*: Especially the delta is densely populated by at least 130 million of Indian and Bangladeshi citizens, while—of course—the towns in the mountains are much smaller: *Guwahati* (India), 960,000 inhabitants; *Mymensingh* (Bangladesh), 1.4 million inhabitants; and *Shigatse* (Tibet Autonomous Region of China), 700,000 inhabitants (Fig. 1.3).
7. *Literature*

Gupta A (2007) Large rivers: geomorphology and management. Wiley, Hoboken

Mehlhorn H (ed) (2016) Encyclopedia of parasitology, 3 vols., 4th edn. Springer, New York

Rahaman MM, Varis O (2009) Integrated water management of the Brahmaputra Basin: perspectives and hope for regional development. Nat Res Forum 33:60–75

Fig. 1.3 Basin of the Brahmaputra river and its tributaries. By Pfly—Own work, CC BY-SA 3.0, https://commons.wikimedia.org/w/index.php?curid=15068725

Saha AL, Pain GC, Chowdhury K, Mukherjee S, Basu BK (1970) Cholera epidemics (1964–1965) in Brahmaputra Valley of Assam. J Indian Med Assoc 55:299–307

Sarma JN (2005) Fluvial process and morphology of the Brahmaputra river in Assam, India. Geomorphology 70:226–256

Singh VP, Sharma N, Shekar C, Ojha P (2004) The Brahmaputra basin water resources. Springer, Dordrecht

Whitcombe E (2012) Indo-Gangetic river systems, monsoon and malaria. Philos Transact A Math Phys Eng Sci 370:2216–2239

1.2.4 Colorado River

1. *Region*: This river is besides the Rio Grande one of the most important rivers of the Southwestern region of the USA and of Northern Mexico, since it passes from its source at the La Poudre Pass on its way to the Gulf of California in total seven states of the USA and two of Mexico. This river is worldwide famous for its scenic canyons and wild rapids and its passing through 11 US National Parks.
2. *Name*: The river name Colorado is also the name of one of the US states, through which it takes its course passing a mountain region, which is characterized by reddish-colored rocks. The European discoverers liked to use further color names to describe rivers or valleys in this region (e.g., Green river—a tributary of the Colorado—or Black Canyon and Slickrock region). The regions along the river had been settled by several indigenous tribes, which used different names to describe the river: (1) *Havasupai* people, Ha Fay G'am, (2) *Maricopa* people,

Xakxwet; (3) *Mohave* people, Aha Kwahwat; and (4) *Yavapai* people, Hakwata. Even in the early USA, the Colorado river had different names until 1921, when the name Colorado river was fixed.

3. *Length*: The Colorado river has a total length of 2330 km (1450 miles) and thus is the fifth longest river in the USA. The basin size covers 640,000 square kilometers (=247,105 square miles), and the river discharges very large amounts of water into the Gulf of California (ranging from 11.9 cubic meters per second to 10,900 cubic meters per second depending on the season). In former times there had been a much higher water input, but today dams hold back large amounts of water.
4. *Climate*: This river runs through different climate zones reaching from cold temperated ones in the mountains to tropical warm ones.
5. *Early discoveries and history*: First signs of human populations go back to 12,000 years, when *Paleo Indians* of the *Clovis* and *Folsom* cultures had reached the Colorado Plateau. Later followed a so-called Desert-Archaic culture during the years 8000 until 2000 BC. These populations had apparently a nomadic lifestyle collecting plants and hunting animals during their wandering. Later (~2000–700 BC), a so-called *Freemont culture* is documented being followed by the *Puebloan people*, who already had been very active to produce drinking and irrigation water. They were mainly found in the San Juan river region of present times. Especially in the years 600–700 AD, irrigation was used to produce field fruits. In total many cultures had been developed besides each other, and the number of people in the different tribes increased enormously. Around 1025 AD the *Navajo* people dominated the Colorado river basin and stretched their territory to many regions of the today regions of Arizona, New Mexico, Utah, and Colorado. They had apparently learned to use agricultural methods from the Pueblo people, the "state" of which has collapsed in the fourteenth century AD. Another large tribe—the *Mohave*—lived from 1200 AD in the Black Canyon region. In total many different tribes had developed their lifestyle in these regions until in the seventeenth century when the Europeans came—first as explorers and later as invaders—expelling the local population from their grounds and moving them into reservation regions with often bad survival chances. For example, large numbers of the members of the *Mohave* and *Navajo* died either in camps or in battles with American or Mexican invaders. Early Spanish explorers detected in the year 1530 the region under the leadership of *Francisco de Ulloa*, when he and his crew sailed into the Gulf of California. Francisco Vasquez de Coronado's expedition (1540—1542) searched for the fabled cities of gold. Also, in the year 1540, *Hernando de Alarcon* and his fleet found the mouth of the river Colorado. Later in the years 1850–1854, the US Army explored the lower reach of the Colorado river—always looking for the *fabulous gold* "reported" in historical tales.

Indeed, gold was found when adventurers from Georgia discovered it in the year 1859 along the Blue river—a tributary of the Colorado. This finding led to the growing of a mining "boomtown" named Breckenridge being followed by further findings at other places (Ouray, Telluride), which also became important

Fig. 1.4 Basin of the Colorado river and its tributaries. By Shannon1—Own work, CC BY-SA 4.0, https://commons.wikimedia.org/w/index.php?curid=65868008

in their times and thus attracted numerous settlers to this region at the upper Colorado river basin.

6. *Big towns along the river: Grand Junction* (Colorado), 62,000 inhabitants; *Moab* (Utah), 5100 inhabitants; *Page* (Arizona), 6800 inhabitants; *Blythe* (California), 21,000 inhabitants; *Las Vegas* (Nevada), 633,000 inhabitants; and *San Louis Rio Colorado* (Mexico, Sonora), 158,000 inhabitants (Fig. 1.4).
7. *Literature*

Benke AC, Cushing CE (2005) Rivers of North America. Academic, Cambridge, MA
Fowler DD (ed) (2012) Cleaving an unknown world: the Powell expeditions and the scientific exploration of the Colorado Plateau. University of Utah Press, Salt Lake City, UT
Gupta A (2007) Large rivers: geomorphology and management. Wiley, Hoboken
Pritzker B (1998) Native Americans: an encyclopedia of history, culture, and peoples. ABC-CLIO, Santa Barbara, CA
Young RA, Spamer EE (2001) The Colorado River: origin and evolution. Grand Canyon Association

1.2.5 Congo (syn. Zaire) River

1. *Region*: The very old river (1.5–2 million years) has its origin in Central Africa in the mountains and highlands of the East African Rift as well as in the lakes Tanganyika and Mweru. It runs through the Congo rainforest, which is smaller than that of the Amazonas region, but nevertheless the second largest on earth. Its main tributaries are the rivers *Ubangi*, *Sangha*, and *Kasai*. The *Congo* river crosses the equator twice and enters an average of 41,000 cubic meters of water per second into the Atlantic Ocean close to the town Matadi. This river is the main transportation source in Central Africa providing more than 14,500 km of navigable shipping routes. Furthermore, its waterpower is used for more than 40 hydropower plants. Famous are the *Stanley falls* and the *Livingstone falls* being named honoring their British discoverers.
2. *Name*: The Congo river has gotten its name from the ancient, since long disappeared Congo kingdom, which had ruled the region around the "mouth" of the river close to the town of *Boma*. In 1482 the Portuguese explorer and navigator *Diego Cao* was the first European to sight and enter the Congo river region, which today is divided into two African countries: (a) the Democratic Republic of the Congo (Zaire; capital *Kinshasa*) and (b) the Republic of the Congo (capital: *Brazzaville*). The whole region was governed by France since *Pierre Savorgnan de Brazza* claimed this region in 1880 as French colony. However, during the *Congo Conference* (1885 in Berlin), this region became a Belgian colony ruled by Leopold II and was suppressed and exploited by a strict colonialism.
3. *Length*: The Congo river originating by confluence of its source rivers (*Lualaba* and *Luapula*) is the second longest river in Africa after the Nile river measuring about 4320 km in length. Furthermore, this river is one of the deepest ones on earth reaching depths of up to 230 m. There exist more than 4000 islands in the river, wherefrom about 50 have a length of 16 km. The Congo river discharges such large amounts of water into the Atlantic Ocean which are only surpassed by those of the Amazon river. Important tributaries are Lukunga, Aruwimi, Ubangi, Sanga, Lomani, and Kasai rivers.
4. *Climate*: Tropical throughout the year.

5. *Early discoveries and history*: The starting point of Christianization of the people in this region was around 1490, but cultural contacts to Europe remained scarce until European sea powers and Arabian forces started slave trades transporting these poor people until 1865 to South and North American states. Due to the discovery of important natural resources, the whole region was attacked many times by troops of several European countries and later until today by local ethnic groups.
6. *Big towns along the river*: The biggest town situated at the banks of the Congo river is *Kinshasa*, the capital of the Democratic Republic of the Congo with about 11.6 million inhabitants. Starting from *Kisangani*, a town with about 600,000 inhabitants, the tributary river Lualaba is named the Congo river. *Boma* with about 350,000 inhabitants is located about 100 km upstream from the Atlantic Ocean. Its harbor is accessible for seagoing ships, and therefore it is of high importance for the transportation of goods to the Democratic Republic of the Congo. Especially in the small villages along the river the population is hit by numerous parasitic diseases (Table 1.2).
7. *Literature*: The Polish writer Józef Teodor Konrad Korzeniowski (1857–1924) published in 1902 under the pseudonym *Joseph Conrad* the novel *Heart of Darkness*, which describes the suppressive handling of European imperialists, who considered African people as inferiors who can be exploited. The background of this novel is based on his own experiences in Central Africa (Fig. 1.5).

Further Information

Butcher T (2008) Blood river: a journey to Africa's broken heart. Vintage, Penguin Random House, London

Conrad J (1902) Heart of darkness and the secret sharer. Penguin Putnam, Signed Classics, New York

Kelly-Hope LA, Bockarie MJ, Molyneux DH (2012) *Loa loa* ecology in central Africa: role of the Congo River system. PLoS Negl Trop Dis 6(6):e1605

Mehlhorn H (ed) (2016) Encyclopedia of parasitology, 3 vols., 4th edn. Springer, New York

Rimoin A, Hotez P (2013) NTDs in the heart of darkness: the Democratic Republic of Congo's unknown burden of neglected tropical diseases. PLoS Negl Trop Dis 7(7):e2118

1.2.6 Danube River (German: Donau)

1. *Region*: This river has two sources (Breg, Brigach), which are situated in the so-called Black Forest in Central Europe (Germany). From there the Danube takes its course west to eastward and thereby passes or touches in total ten countries before draining into the Turkish Black Sea, which is connected via Bosporus with the Mediterranean Sea complex. Therefore, this river, which is

Fig. 1.5 Basin of the Congo river and its tributaries. By Kmusser—Own work, Elevation data from SRTM, drainage basin from GTOPO, all other features from Natural Earth, CC BY-SA 4.0, https://commons.wikimedia.org/w/index.php?curid=75858826

Table 1.2 Estimated number of parasitic infections among the population of the Democratic Republic of the Congo

1. Ascariasis	23 million
2. Hookworms	31 million
3. Human trypanosomiasis	15,000
4. Lymphatic filariasis	50 million at risk
5. Schistosomiasis	15 million
6. Trichuriasis	26 million

connected by channels with other rivers, is very important for inter-European transportation of mass products.

2. *Name*: The German name *Donau* and the English term *Danube* have their origin in the Indo-European word *danu*, which means fluid, drop, etc., but in the different local languages along the river, there had been developed many different

names being used for centuries depending on their cultures, e.g., *Greek*, istros; *Dacian*, Donaris; *Thraco-Phrygian*, Matoas; and *Latin*, Danuvius.

3. *Length*: The Danube has a total length of 1680 miles (=1770 km) and thus is the second longest European river behind the Volga river (3530 km). Its sources are situated in the Black forest in Southern Germany and are named *Breg* and *Brigach*. These small rivers flow together close to the German town *Donaueschingen*, thus giving rise to the river Danube, which flows from there in southeastern direction through Germany and passes Austria, Slovakia, Hungary, Croatia, Serbia, Bulgaria, Romania, Moldova, and Ukraine, whereby it forms the border between Germany and Austria, Austria and Slovakia, Slovakia and Hungary, Hungary and Serbia, Serbia and Romania, and Moldova and Ukraine before entering the Black Sea.

   In total 31 rivers enter the Danube after its official beginning in Donaueschingen (Germany). In addition to these 31 rivers, the drainage basins of the Danube river (total size 801,463 km^2) receive water from 9 further countries (Bosnia-Herzegovina 4.6%, the Czech Republic 2.9%, Slovenia 2%, Montenegro 0.9%, Switzerland 0.2%, Italy 0.1%, Poland 0.1%, the Republic of Macedonia 0.1%, and Albania 0.1%).

   The Danube river is navigable from the Black Sea (**a**) by smaller ships to Ulm (Germany), (**b**) by larger river boats to Kelheim (Germany), and (**c**) by ocean ships to Braila (Romania). There exist in addition many canal systems, which help to connect also the Atlantic Ocean and Western Europe with the Black Sea:

   - Rhine-Main-Danube channel
   - Rhine-Rhône channel
   - Rhine-Mosel-Loire canal
   - Rhine-Mosel-Seine canal

   Thus, ports like Rotterdam (Netherlands) at the North Sea, Istanbul at the Sea of Marmara, Piraeus in Greece, Genoa in Italy, Marseille in France, and Tripoli in Libya at the Mediterranean Sea can be reached by this Danube river system. The Danube delta is the largest in the European Union covering a surface of 5165 km^2 (1994 square miles).

4. *Climate*: The yearly temperatures along the river belong to the continental type. This includes warm summers and (locally) rather cold winters even leading to icy surfaces on the rivers, which may block or impede shipping activities.

5. *Early discoveries and history*: The valley of the Danube river belongs to the earliest human cultures, since there had been found traces going back to the 6000 BC (e.g., people forming ceramics lived there). Later in the sixth century BC, the king *Darius of Persia* crossed the Danube on his campaign to subdue the *Scythian empire*. Later (around 336 BC), the Greek-Macedonian king *Alexander the Great* (356–326 BC) advanced at the Danube in his fights against the *Thracian Illyrian tribes* before he started his military campaign into the direction to the Near East. Later the *Romans* used the Danube as natural border against Eastern tribes. From 37 AD to ~360 AD, the Danubian Limes (a wall of wood and stones) was built, constantly restored, and thus used as defense wall against penetrations of people from the East.

Since the Danube river was also important as a natural border against penetrating tribes in the following centuries and as a connection line for people and goods, many towns along this river were enormously enlarged and became important and rich due to trade. Thus, it is not astonishing that big kingdoms like those of the Ottomans and Hungary/Austria competed during the fourteenth and nineteenth centuries leading to the Ottoman-Habsburg wars (1526–1791). Especially the Sieges of Vienna by Osman troops in the years 1529 and 1699 had been turning points in the European culture and history of the following centuries until today. In all these cases, the river Danube was used for the transportation of troops, food, and ammunition, and later—in times of peace—the river boats transported trade goods and workers. Besides these economical activities during many centuries, the Danube river was important as food source, since as many as 60 fish species lived in its waters in considerable numbers, e.g., sturgeons, sterlets, salmons, etc., which were sold at high prices.

6. Big towns along the river:

- *Germany*: Ulm (125,000 inhabitants); Regensburg (150,000 inhabitants)
- *Austria*: Linz (205,000 inhabitants); Krems (25,000 inhabitants); Vienna (capital, 1.89 million inhabitants)
- *Slovakia*: Bratislava (430,000 inhabitants)
- *Hungary*: Györ (132,000 inhabitants); Visegrad (1900 inhabitants); Budapest (capital, 1.75 million inhabitants)
- *Croatia*: Vukovar (27,600 inhabitants); Ilok (6700 inhabitants)
- *Serbia*: Novi Sad (232,000 inhabitants); Belgrade (capital, 1.34 million inhabitants)
- *Romania*: Drobeta-Turnu Severin (92,000 inhabitants); Cernavoda (17,000 inhabitants)
- *Bulgaria*: Ruse (145,000 inhabitants); Lom (20,000 inhabitants)
- *Moldova*: No bigger settlements along the river/delta; last towns are Nona and Sulina
- *Ukraine*: No bigger settlements along the river/delta (Fig. 1.6)

7. *Literature*

Bănăduc D, Rey S, Trichkova T, Lenhardt M, Curtean-Bănăduc A (2016) The Lower Danube River-Danube Delta-North West Black Sea: a pivotal area of major interest for the past, present and future of its fish fauna - a short review. Sci Total Environ 545–546:137–151

Brătescu G (1979) Plague prevention and political interests in the Danube region (1750–1850). Sudhoffs Arch 63:25–44

Görlich EJ, Romanik F (1995) Geschichte Österreichs (history of Austria). Tosa, Wien, p 234

Matschke KP (2004) Das Kreuz und der Halbmond. Die Geschichte der Türkenkriege. (The cross and the halfmoon: the history of the Turkish wars). Artemis und Winkler, Düsseldorf, p 360

Mehlhorn H (2016a) Animal parasites, 8th edn. Springer, Heidelberg

Fig. 1.6 Map showing the course (red line) of the river Danube in Europe. Public Domain, https://commons.wikimedia.org/w/index.php?curid=1665391

Mehlhorn H (ed) (2016b) Encyclopedia of parasitology, 3 vols., 4th edn. Springer, New York

Mehlhorn H (2016c) Human parasites, 8th edn. Springer, Heidelberg

Mühlegger JM, Jirsa F, Konecny R, Frank C (2010) Parasites of *Apollonia melanostoma* (Pallas 1814) and *Neogobius kessleri* (Guenther 1861) (Osteichthyes, Gobiidae) from the Danube River in Austria. J Helminthol 84:87–92

Smederevac-Lalić MM, Kalauzi AJ, Regner SB, Lenhardt MB, Naunovic ZZ, Hegediš AE (2017) Prediction of fish catch in the Danube River based on long-term variability in environmental parameters and catch statistics. Sci Total Environ 609:664–671

1.2.7 Darling River

1. *Region*: This Australian river has its sources (rivers *Barwon* and *Culgoa*) in the region of northern New South Wales at the east side of Australia. From there it takes its course southward until it joins the big *Murray river* at the town Wentworth.
2. *Name*: The name of the river was given in the year 1829 by two explorers (Charles *Sturt*, Hamilton *Hume*) honoring their supporter (the British Governor *Sir Ralph Darling*).

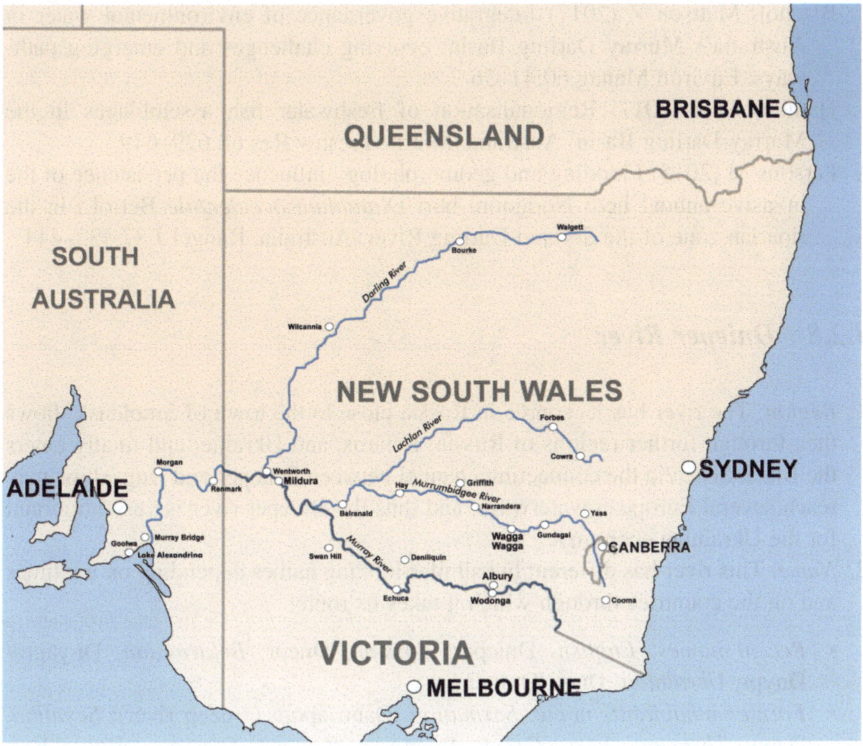

Fig. 1.7 Map showing the course of the Darling river and its three main tributaries in Australia. By Bidgee—Own work, CC BY 3.0, https://commons.wikimedia.org/w/index.php?curid=4182228

3. *Length*: This river has an own length of 1472 km (~915 miles). As tributary of the Murray river, the confluence gives rise to the longest river system in Australia.
4. *Climate*: Continental.
5. *Early discoveries and history*: Following the explorers *Sturt* and *Hume* (in 1828), further expeditions entered this region and got in contact with the local aborigines. For example, those of Major *Thomas Mitchell* in the year 1835 and *Blankowski's* group in the year 1856 collected enormous amounts of fish and other local animals. Astonishing was, and is, that the river dried up 45 times in the period between 1885 and 1960. In such cases of water reduction, several times cyanobacteria "bloomed" and enormously hit the fish life in the water.
6. *Big towns along the river*: *Bourke*, 2000 inhabitants; *Wilcannia*, 550 inhabitants; *Menindee*, 551 inhabitants; and *Wentworth*, 1250 inhabitants (Fig. 1.7).
7. *Literature*

Alston M, Whittenbury K, Western D et al (2016) Water policy, trust and governance in the Murray-Darling Basin. Aust Geogr 47:49–64

Bischoff-Mattson Z (2017) Integrative governance of environmental water in Australia's Murray Darling Basin: evolving challenges and emerging pathways. Environ Manag 60:41–56

Hamilton SH (2017) Regionalisation of freshwater fish assemblages in the Murray-Darling Basin, Australia. Marine Freshw Res 68:629–649

Parsons M (2015) Flooding and geomorphology influence the persistence of the invasive annual herb Noogoora burr (*Xanthium occidentale* Bertol.) in the riparian zone of the dryland Darling River, Australia. Rangel J 37:433–444

1.2.8 Dnieper River

1. *Region*: The river has its source in Russia close to the town of Smolensk; flows then through further regions of Russia, Belarus, and Ukraine; and finally enters the Black Sea. Via the connecting channel between Dnieper and Bug, ships may reach several European waterways, and thus the Dnieper river is very important for the Ukrainian economy.
2. *Name*: This river has different but similar-looking names depending on the times and on the countries through which it takes its route:

   - *Recent names*: *English*, Dnieper; *Russian*, Dnepr; *Belarussian*, Dnyapro/Dnypr; *Ukrainian*, Dnipr/Dniper
   - *Former inhabitants name*: *Sarmatian*, Danu apora (=deep river); *Scyntian*, Danu; *Thracian*, Ister; *Greek*, Danapris; *Romans*, Danaper; *Huns*, Var; *Crimean/Tatar*, Özu; *Bulgars*, Buri-Chai

   This long list shows that this region including the draining basin was often settled by different cultural and powerful populations.
3. *Length*: This river has a total length (depending on citation) of 2201–2285 km (~1400 miles). Its drainage basin comprises 504,000 square kilometers respectively around 195,000 square miles. The mouth of this rivers opens into the *Black Sea* at the Ukrainian border. The river passes 485 km in Russia, 700 km in Belarus, and ~25 km in Ukraine. Its importance as waterway is also shown by the fact that this river has ten big tributaries from the left side and also ten from the right side. However, in total more than 30,000 smaller water sources act as further tributaries showing how important its drainage activity is.
4. *Climate*: The climate of its basin is a typical continental one—hot summers and rather cold winters.
5. *Early discoveries and history*: As it has been cited in Chap. 2 when looking at the different names of people inhabiting this region for centuries, many ethnic different populations had lived in this region due to an apparently agreeable climate that allowed to produce sufficient food for increasing populations. Today remnants of many tribes are still found there. First documents go back to ancient Greek historians, who described this river as *Borysthenes*, which has *Scythian* roots (the Scythians lived there from eighth and seventh century BC).

Fig. 1.8 Map showing the course of the river Dnieper (Dnipro) and its tributaries. By Francis McLloyd, CC BY 2.5, https://commons.wikimedia.org/w/index.php?curid=1729444

6. *Big towns along the river*: *Smolensk* (Russia), 330,000 inhabitants; *Mogilev* (Belarus), 375,000 inhabitants; *Kiev* (Ukraine, capital), 2.93 million inhabitants; and *Dnipro*, 987,000 inhabitants (Fig. 1.8).
7. *Literature*

Cybriwsky RA (2018) Along Ukraine's river: a social and environmental history of the Dnipro. Central European University Press, Budapest

Kalicki T, Sauchyk S, Calderoni G et al (2008) Climatic versus human impact on the Holocene sedimentation in river valleys of different order: examples from the upper Dnieper basin, Belarus. J Int Union Quat Res 189:91–105

Kayukova EP (2017) Water resources of the Crimea. Water Resour 44:886–891

Minkovskaya RY (2014) Balance of water, heat, and solutes in the Dnieper-Bug estuary. Russ Meteorol Hydrol 39:63–68

1.2.9 Don River

1. *Region*: The *Don river* belongs to the major rivers in the Eurasian region. Its basin is situated between the basin of the *Dnieper* in the West and the *Volga* in the East. The *Oka*, which is a tributary of the *Volga*, flows north of the Don. The Don has its source in the town *Novomoskovsk*, which is situated about 120 km south of the Russian capital *Moscow*. At first this river flows southeast to the town on *Voronezh*, and from there it takes its route southwest to its mouth at the Sea of Azov, which has a connection to the Black Sea. The *Volga-Don-canal*, which has a length of about 105 km, connects both rivers and the so-called Tsimlyansk Dam and its reservoir help to raise the water level of the Don and thus enable an easy passage of ships.
2. *Name*: The Russian name of this river is *Don*. In the times of the Scythian population, it had the Greek name *Tanais*, while the Roman poet Plinius the Second (61–137 AD) translated this name as *Silys*.
3. *Length*: In literature the data of the length data vary at a rather low degree. Some sources report 1870 km, while others count 1967 km.
4. *Climate*: In the region along the river occur typical continental temperatures: hot summers—rather cold winters.
5. *Early discoveries and history*: This river was considered by ancient Greek and other geographers as a border between Asia and Europe. The Greek scientist *Plutarch* (45–127 AD) reported that the landscape along the Don river might have been the site of the kingdom of the legendary *Amazons* (=a population of tough and straightforward fighting female warriors), who—according to *Homer's* (eighth century BC) tale on the Trojan war (1194–1184 BC)—should have fought on the side of the Greek heroes (Achilles, Odysseus, etc.). There are also several reports that the Don region suffered later often from epidemics of *plague*, which was called in early times *Black Death*, but is indeed an infection due to the bacterium *Pasteurella* (*Yersinia*) *pestis*. During the fourteenth and fifteenth century, the population was endangered from attacks of wandering troops of the *Crimean Tatars*. In the sixteenth and seventeenth centuries, the so-called Don Cossacks established their government along the Don and defended it for a long time until the eighteenth and nineteenth centuries, when the power of the Russian Tsars was strengthened and they obtained the power. After the overthrow of the Russian Tsar power at the end of World War I after killing the tsar Nikolaus II and his family in the year 1918 in Yekaterinburg, the Russian state and neighboring countries fell totally into the hands of the Russian communist party. This rulership persisted for a long time until the president Mikhail Gorbachev (born 1931) opened Russia for contacts and business with Western and Asian powers in the year 1986.
6. *Big towns along the river*: *Tula*, 500,000 inhabitants; *Voronezh*, 890,000 inhabitants; and *Rostov-on-Don*, 1.1 million inhabitants (Fig. 1.9).
7. *Literature*

Fig. 1.9 Diagram of the route of the river Don and its tributaries in Russia. By No machine-readable author provided. Kmusser assumed (based on copyright claims)—No machine-readable source provided. Own work assumed (based on copyright claims), CC BY-SA 2.5, https://commons.wikimedia.org/w/index.php?curid=1187214

Filippov YG (2015) The impact of the Don river runoff on the water level in the Taganrog Bay. Russ Meteorol Hydrol 40:127

Oparina OS (2013) Current status of the Russian desman habitats in small rivers of the Don River Basin in Saratov oblast and the abundance of this species. Biol Bull Russ Acad Sci 40:854–861

Pleitgen F, Baumeister T (2008) Väterchen Don. Fluss der Kosaken. (Father Don. River of the Cossacks.). Kiepenheuer & Witsch, Cologne

1.2.10 Douro River

1. *Region*: This river has its source in northern-central Spain, flows from there strictly westward, passes the northern region of Portugal, and enters into the Atlantic Ocean in the Porto region.
2. *Name*: This river has its name from the Latin term *Durius*, which is apparently of Celtic origin. This tribe lived there before Roman times and called it apparently *Dubro*. In Roman times this river got the status of a god named *Durius*. In another interpretation it is claimed that this name comes from a Portuguese term today translated as "golden." Today the river is termed *Douro* in Portuguese and *Duero* in Spanish.
3. *Length*: The river has a total length of 897 km (=557 miles) and thus is the third longest river on the Iberian Peninsula just after the *Tagus* (1007 km) and the *Ebro* (930 km).
4. *Climate*: Its source is situated at a height of 2157 m and thus arises in a cold region, before passing moderate to higher temperatures until its mouth.
5. *Early discoveries and history*: The region was for many centuries a pathway of many tribes and for their armies, e.g., wandering Celtic groups or others like Hannibal's Carthage troops coming from North Africa. Later since 1640 the precursor state of Portugal and in the fifteenth century the Spanish kingdom had been established. The populations of both countries speak different languages and exist as separate democratic states now (1910 Portugal became a republic; 1978 Spain; before it was a kingdom). During the times the formation of settlements in South and Central America Portugal and Spain had been strict rivals. For example, the population of Brazil speaks Portuguese, while, e.g., inhabitants of Mexico and people of some South America states speak Spanish.
6. *Important towns along the river*: *Logrono* (Spain), 150,000 inhabitants; *Salamanca* (Spain), 144,000 inhabitants; *Porto* (Portugal), 238,000 inhabitants; and *Vita Nova de Gaia* (Portugal), 302,000 inhabitants (Fig. 1.10).
7. *Literature*

Breuer T (2008) Iberische Halbinsel. In: Geographie, Geschichte, Wirtschaft, Politik. Wissenschaftliche Buchgesellschaft, Darmstadt

Diffie BW, Winius GD (1977) Foundations of the Portuguese empire, 1415–1580. University of Minnesota Press, Minneapolis, MN

Gates D (2001) The Spanish Ulcer: a history of the Peninsular war. Da Capo Press, Cambridge, MA

Newitt MDD (2005) A history of Portuguese overseas expansion, 1400–1668. Routledge, Abingdon, UK

Rodrigues SM, Almeida CMR, Silva D, Cunha J, Antunes C, Freitas V, Ramos S (2019) Microplastic contamination in an urban estuary: abundance and distribution of microplastics and fish larvae in the Douro estuary. Sci Total Environ 659:1071. https://doi.org/10.1016/j.scitotenv.2018.12.273

Wheeler DL (1998) Republican Portugal: a political history, 1910–1926. University of Wisconsin Press, Madison, WI

Fig. 1.10 Diagrammatic representation of the Douro river in Spain and Portugal (darker zone on the left). CC BY-SA 2.5, https://commons.wikimedia.org/w/index.php?curid=436615

1.2.11 Ebro River

1. *Region*: The source of the Spanish river *Ebro* is located in a height of 1980 m close to *Fontibre* in the region of *Cantabria*. By damming the river close to the source, an artificial lake has been produced, which collects amounts of water. From there the water flows—controlled through many gorges—mainly eastward in the *Burgos* province, and its waters become considerably enlarged by many tributaries arriving from the Central Pyrenees. Passing *Zaragoza*, the Ebro has already reached a considerable size. In the final section, the river flows southward, passes considerable large gorges, and thereafter bends somewhat eastward before discharging its water in a delta of the Mediterranean Sea close to *Amposta*. The delta itself has a considerable size covering 340 square kilometers (=130 square miles). The total water is contributed by about 200 tributaries.
2. *Name*: The *Ebro River* is called *Rio Ebro* in Spanish. In Roman times it was named *Iberus*, which sounds similar to the Latin country name *Iberia*. Another cited name was Hiberus, which also contains allusions to *Iberia*—the name of Spain in present times.
3. *Length*: The river has a mean length of 1930 km (~580 miles)—some literature sources vary a bit, but not considerably.

Fig. 1.11 Diagrammatic representation of the basin of the river Ebro in Spain. GPL, https://commons.wikimedia.org/w/index.php?curid=302574

4. *Climate*: The climate varies depending on the height above the sea, where the towns are situated. The mountain regions have a continental climate with dry and rather hot phases, since its mountains block the influx of the humid, moderate, temperated sea air masses. The temperatures in the delta allow splendid agriculture of rice, citrus fruits, etc.
5. *Early discoveries and history*: Early reports go back to the times of the so-called Punic Wars (264–241 BC) between the Roman Republic and the North African town Carthage under the leadership of *Hannibal* (who won the first one, but finally lost the third war (149–146 BC)). The expansion of the Roman territory by establishing a protectorate town named *Saguntum*, which was situated considerably south of the Roman influence zone, was considered to be a treaty violation by Hannibal and led to the Second Punic War (218–201 BC). In medieval times Spain became a Catholic world power always struggling with other sea powers (e.g., Portugal, England, etc.). In modern times (1936 AD), the population of the Ebro region gave rise to the Spanish Civil war, which lasted until 1939.
6. *Big towns along the river*: *Reinosa*, 9300 inhabitants; *Miranda de Ebro*, 35,000 inhabitants; *Calahorra*, 24,000 inhabitants; and *Zaragoza*, 665,000 inhabitants (Fig. 1.11).
7. *Literature*

Bagnall N (1990) The Punic wars. Hutchinson, London
Carr R (2001) Spain: a history. Oxford University Press, Oxford

De Meneses FR (2001) Franco and the Spanish civil war. Routledge, London
O'Callaghan JF (2004) Reconquest and crusade in medieval Spain. University of Pennsylvania Press, Philadelphia
Tusell J, Clark R (2007) Spain: from dictatorship to democracy: 1939 to the present. Wiley, Malden

1.2.12 Elbe River (Labe in Czech)

1. *Region*: This river rises in the Bohemian mountains (formerly called "Riesengebirge" = Czech: Krkonose mountains of the giants), which are situated at the border of the present Czech Republic to Poland. From there its waters flow in a wide arc in the direction to Germany, where the river enters the governmental state of Saxony close to Dresden, which is the capital of the former Kingdom of Saxony (1806–1918) ruled among others by the famous King August the Strong (1670–1733), who had more than 500(!) acknowledged children. From there the Elbe river passes Germany and finally enters the North Sea slightly (50 km north of *Hamburg* at *Cuxhaven*).
2. *Name*: The name Elbe (Czech: Labe) has its source in a medieval tale that mystic (so-called elves = ghostlike, mainly female, nocturnal dwarfs) live in the forests and frighten passing humans, when they suddenly appear and disappear.
3. *Length*: The Elbe river has a total length of 724 miles (=1091 km) and drains an area of 55,620 square miles (=144,060 km^2). Thus, it is the third largest river in Europe. It originates from a region, where several small rivers (Vltava, Melnik, Ohre) merge. Later the Elbe river is enlarged by tributaries such as Vltava (Moldau river), Ohre (Eger river), Havel, and Alster. In the Czech Republic, the Elbe river passes its first 364 km; in Germany the Elbe river passes then 727 km until its mouth at Cuxhaven—a bit north of Hamburg. Several channels connect the Elbe river with the rivers Rhine, Weser, and Oder as well as with the Baltic Sea, and thus the Elbe is one of the most important waterways in Europe being itself navigable for 947 km.
4. *Climate*: The Elbe river runs from South to North through Germany, which is characterized by a typical moderate Central European rather mild climate passing over the year without extreme temperatures during the four clearly defined seasons (spring, summer, autumn, winter).
5. *Early discoveries and history*: Bohemia (now Czech Republic), Saxony, and the regions further downstream have all been highly developed since the early medieval times and contributed to the high cultures in their past and still do it in present times. The river Elbe allowed since many centuries the transport of goods in any direction and is—still today—one of the most important waterways of Central Europe, since it is connected to other German rivers by channels (e.g., the Mittelland Canal connects the Elbe to the Rhine river), so that goods can be transported by waterways through Europe as well as to the rest of the world via the large Seaport of Hamburg.

Fig. 1.12 Map showing (in the center) the course of the European river Elbe (Czech: Labe). By NordNordWest, revision/adaption by Ulamm (talk) 01:00, 27 March 2016 (UTC)—Own work, CC BY-SA 3.0, https://commons.wikimedia.org/w/index.php?curid=47788610

6. *Big towns along the river: Dresden* (capital of the Germany governmental state of Saxony has today 557,000 inhabitants), *Magdeburg* (241,000 inhabitants), and *Wittenberg* (town where Martin Luther (1483–1546) started the reformation of the Catholic Church in the year 1517 by presenting 95 claims for changes being hailed at a church entrance door. This was finally the starting point of the foundation of the Lutheran Church of the present times). Today *Wittenberg* has about 50,000 inhabitants. *Hamburg* has 1.8 million inhabitants, has constructed one of the biggest ports in Europe, and is internationally also known for the world famous Bernhard Nocht Institute for Tropical Medicine, which was already founded around 1900. Close to *Cuxhaven* (50,000 inhabitants), the Elbe river joins the North Sea (Fig. 1.12).

7. *Literature*

Adams MS, Ballin U, Gaumert T, Hale BW, Kausch H, Kruse R (2001) Monitoring selected indicators of ecological change in the River Elbe since the fall of the Iron Curtain (i.e. reunion of both parts of Germany in the year 1989). Environ Conserv 28:333–344

Alexander M (2008) History of the Bohemian Countries. Reclam, Ditzingen

Jochner M, Schwander M, Brönnimann S (2013) Reanalysis of the Hamburg storm surge of 1962. In: Brönnimann S, Martius O (eds) Weather extremes during the past 140 years, Geographica Bernensia G89, pp 19–26

Lüttig G, Meyer KD (1974) Geological history of the River Elbe, mainly of its lower course. In: Centenaire de la Société Géologique de Belgique l'évolution quarternaire des bassins fluviaux de la mer du nord méridionale, Liège, pp 1–19

Mehlhorn H (2016a) Encyclopedia of Parasitology, 3 vols., 4th ed. Springer New York

Mehlhorn H (2016b) Human parasites, 8th edn. Springer, Heidelberg

Milger P (2001) Der Dreißigjährige Krieg. Gegen Land und Leute, Orbis, Munich

Nagy B, Schaer F (2001) Charles IV: Autobiography of Emperor Charles IV; and, his legend of St. Wenceslas: Karoli IV Imperatoris Romanorum Vita Ab Eo Ipso Conscripta; Et. In: Hystoria Nova de Sancto Wenceslao Martyre. Central European University Press, Budapest

Netzband A, Reincke H, Borgemann M (2002) The river Elbe: a case study for the ecological and economical chain of sediments. J Soils Sediments 2:112–116

Preuß O (2016) Hafen Hamburg: Geschichte – Zahlen – Menschen. Wachholtz, Neumünster

Schuller A (2012) Sturmflut über Hamburg. Die Nacht, in der eine Stadt ertrank. Zeitungsgruppe Hamburg (Flood destroying the Hamburg city line).

von Storch H, Gönnert G, Meine M (2008) Storm surges—an option for Hamburg, Germany, to mitigate expected future aggravation of risk. Environ Sci Pol 11:735–742

1.2.13 Ganges River ("Indian": Ganga)

1. *Region*: This river rises in the Himalaya mountains from glaciers (e.g., Gangotri), passes the plains of the plains of the Indian subcontinent, and finally empties its water into the Bay of Bengal sharing the delta with the even larger *Brahmaputra* river.
2. *Name*: The English name Ganges is based on the Hindi term "Ganga." It was and is still today the holy river in India.
3. *Length*: This river, which ranges behind the *Brahmaputra* as the second biggest river in India, is rather short (1560 miles = 2510 km) despite its great importance while flowing through one of the regions with the highest human population on

earth. After rising at the Indian side of the border to the so-called Tibet Autonomous Region (belonging to China), this river gets its water from some bigger rivers (Bhagirathi, Alaknanda, Mandakini, Dhauliganga, Pindar), which receive their water from the Himalayan glaciers. The final Ganges river is formed when the Alaknanda and Bhagirathi rivers have merged at Devprayag. Later the Ganges receives its two main right-bank tributaries: the *Yamuna* river which flows through the Delhi capital region and the *Tons*, which joins the Ganges close to Allahabad. The main left-bank tributaries in Uttar Pradesh are the Gomati, the Chaghara, and the Gomate, while in the state of Bihar, the Gandak, the Chugri, the Kosi, and the Son rivers join the Ganges before it enters the Indian state West Bengal. In this region of West Bengal in India and also in Bangladesh, the Ganges has a local name (*Padna*). Finally, in Bangladesh the Ganges is joined by the bigger *Brahmaputra* (called *Jamuna* in this region). This merged river portion is then called *Padma*, which finally joins with the *Meghna river* close to Chandpur. Finally, the waters flow into the Bay of Bengal via the biggest delta on earth (80,000 km^2) that consists of a giant number of channels (e.g., the largest one is called Meghna estuary).

4. *Climate*: The Ganges river gets its water from melting snow in different Himalaya regions and from often very heavy rainfalls, which may be accompanied by tropical cyclones during the so-called monsoon months (May–June). Thus, the climate along the river from its heights in the Himalaya to the seaside (Gulf of Bengal) ranges between very cold and tropical in the delta region.
5. *Early discoveries and history*: The people living along the river Ganges had been and are still today of mixed origin. Before the Europeans appeared in the geographic region of the present India, many languages had been entered when Turks, Mongols, Burmese, Afghans, Persians, Arabs, etc. arrived. During the seventh century AD, there ruled the feudal empire of Harsha in Northern India. Around the twelfth century AD, the Muslim era started. The people of the European powers came later and mainly did not mix with the different indigenous populations. For centuries England ruled drastically over many local authorities, until in the years 1947 the recent independent state of India and in 1971 Bangladesh (from Pakistan) had been established.
6. *Big towns along the river*: *Kanpur*, 2.8 million inhabitants; *Allahabad*, 1.2 million inhabitants; *Varanasi*, 1.2 million inhabitants; and *Patna*, 2 million inhabitants. The big size and millions of inhabitants of Indian towns along the Ganges river system exposed big problems to the British occupying power from the early beginning, when in the year 1600 Queen Elizabeth I gave the exclusive trade right with India to the British East Indian Company. Although other countries also tried to get power over Indian regions (e.g., France was defeated in the English-French war during the years 1746–1763), England kept for long the power (e.g., Queen Victoria became Empress of India in 1877), until India became an independent state in 1950 (being fixed in the British Commonwealth). Besides many giant ethnic and religious problems, infectious diseases reached giant importance. For example, malaria was in British India the most deadly and debilitating disease in the eighteenth century killing at least 1 million people per

Fig. 1.13 Diagrammatic representation of the basin of the "holy river" Ganges and its tributaries. By Pfly—Own work, CC BY-SA 3.0, https://commons.wikimedia.org/w/index.php?curid=15068725

year out of more than 100 million infected persons, who suffered intensively from this disease. Since the beginning of the registration of malaria cases in the year 1865 in India, it became more and more clear that the monsoon period and malaria had close relations to the transmission of this disease. However, it took until 1897 when the English officer Ross showed in Calcutta that *Anopheles* mosquitoes are able to transmit the agents of disease (=sporozoites of *Plasmodium* species; Ross 1897; Whitcombe 2012; Mehlhorn 2016a, b). However, malaria is not the only problem of the population along the Ganges. There had been and still today (from time to time) occurring epidemics due to water-borne enteritic diseases such as cholera, dysentery, hepatitis, and typhoid fever, which are due to insufficient clearing of sewage entering the Ganges river (Hammner et al. 2006; Batabyal et al. 2014) (Fig. 1.13).

7. *Literature*

Awange JL, Anyah R, Kuhn M, Fukuda Y (2017) Assessing regional climate simulations of the last 30 years (1982–2012) over Ganges–Brahmaputra–Meghna River Basin. In: Climate dynamics: observational, theoretical and computational research on the climate system, vol 49, pp 2329–2350

Batabyal P, Einsporn MH, Mookherjee S, Palit A et al (2014) Influence of hydrologic and anthropogenic factors on the abundance variability of enteropathogens in the Ganges estuary, a cholera endemic region. Sci Total Environ 472:154–161

Burki SJ (1999) Pakistan: fifty years of nationhood. Westview Press, Boulder

Chandra B (2009) History of modern India. Orient Blackswan, Hyderabad

Gain AK (2017) Tidal river management in the south West Ganges-Brahmaputra delta in Bangladesh: moving towards a transdisciplinary approach? Environ Sci Policy 75:111–120

Gosling SN (2017) A comparison of changes in river runoff from multiple global and catchment-scale hydrological models under global warming scenar. Clim Chang 141:577–595

Hammner S, Tripathi A, Kumar Misrah R, Bouskill N, Broadaway SC et al (2006) The role of water use patterns: a sewage pollution in incidence of water-borne/enteritic diseases along the Ganges river in Varanasi, India. Int J Environ Health Res 16:113–132

Khan A (2017) Impact of land-use on groundwater quality: GIS-based study from an alluvial aquifer in the western Ganges basin. Appl Water Sci 7:4593–4603

Kumar D (2017) River Ganges: historical, cultural and socioeconomical attributes. Aquat Ecosyst Health Manag 20:8–20

Mehlhorn H (ed) (2016a) Encyclopedia of parasitology, 3 vols., 4th edn. Springer, New York

Mehlhorn H (2016b) Human parasites, 8th edn. Springer, New York

Ross R (1897) On some peculiar pigmented cells found in two mosquitoes fed on malarial blood. Br Med J 2:1786–1788

Whitcombe E (2012) Indo-Gangetic river systems, monsoon and malaria. Philos Transact A Math Phys Eng Sci 370:2216–2239

1.2.14 Indus River (Sindhu, Darya-e-Sindh)

1. *Region*: This river with its length of 3610 km belongs to the longest ones on earth. It is the longest of Pakistan and has two sources: the rivers *Sengge Zangbo* on the Tibetan Plateau and the *Gar Tsangpo*. The Indus river delta opens its waters into the Arabian Sea and in the so-called Rann of Kutch. This region belongs historically to the oldest human culture societies on earth, since already in the Bronze Age (in the third millennium BC), a major urban civilization occurred there.
2. *Name*: The ancient Indian population named this "border river" *Sindhu*, while the Persians on the other side called it *Hindu*. The name of the present *State of India* is based on the Greek name "*Indos*," the source of which was mentioned in the report of the Greek author *Nearchos* (360–314 BC), who described when and how *Alexander the Great* (356–323 BC) crossed this river in the year 326 BC. Later the Romans adopted this name as *Indus* in various scripts.
3. *Length*: The river has a length of 3610 km (2240 miles) and flows through regions of China, India, and Pakistan. Its source tributaries are situated on the Tibetan Plateau. After their confluence, the "new" river passes regions in Kashmir and Hindukush and flows along the full length of Pakistan before entering the Arabian Sea (=at the Western side of the Indian subcontinent). The delta is situated close to the port of the town of *Karachi* in the Sind province. Thus, the Indus is the longest and most important river of the state of Pakistan.

4. *Climate*: Due to its origin in the high mountains on the Tibet high plateau and its flowing down to tropic regions before entering the Arabian Sea, the climate zones range from very cold to tropical high temperatures, so that the population along the river lives under different climatic conditions. Since regions in the mountains deliver large amounts of water, additional heavy rainfall in other regions may lead to heavy floodings in posterior regions as had occurred in the years 2010 and 2011, when severe floods killed large numbers of humans and animals and destroyed enormous numbers of houses and bridges in many regions.
5. *Early discoveries and history*: The civilization of the Indus valley is very old and goes back to the years around 3300 BC, when already very large habitations occurred in the regions ranging from northeastern Afghanistan to Pakistan and northwest India of our days. Of course, also the coast regions had been densely populated during these early times, and many kingdoms grew up and passed. The Indus river became known to Greek rulers and later to the Romans by reports of the Greek person *Scylax of Caryanda* (late sixth century BC), who was sent to "India" by the Persian King *Darius* during the years 515–516 BC to explore the river being named "Indos" in Greek and "Indus" in Latin.
6. *Big towns along the river*: *Leh*, 31,000 inhabitants; *Skardu*, 30,000 inhabitants; *Sukkur*, 500,000 inhabitants; *Hyderabad*, 6.9 million inhabitants; and *Karachi*, 14.91 million inhabitants (Fig. 1.14).
7. *Literature*

Albinia A (2008) Empire of the Indus: the story of a river. Norton Comp, New York

Gilmartin D (2015) Blood and water: the Indus river basin in modern history. University of California Press, Oakland

Hartmann H (2013) Flooding in the Indus river basin: a spatiotemporal analysis of precipitation records. Global Planet Chang 107:25–35

Hussain Z (2011) Application of the regional flood frequency analysis to the upper and lower basins of the Indus river, Pakistan. Water Resour Manag 25:2797–2822

Mehlhorn H (ed) (2016) Encyclopedia of parasitology, 3 vols., 4th edn. Springer, New York

Mountjoy S (2004) The Indus river. Infobase, New York

Mukhopadhyay B, Khan A (2015) A reevaluation of the snowmelt and glacial melt in river flows within Upper Indus basin and its significance in a changing climate. J Hydrol 527:119–132

Nepal S (2015) Impact of climate change on the hydrological regime of the Indus, Ganges and Brahmaputra river basins: a review of the literature. Int J Water Resour Dev 31:201–218

Possehl GL (1999) Indus age: the beginnings. University of Pennsylvania Press, Philadelphia

Fig. 1.14 Map of the basin of the "holy" Indus river and its tributaries. By Keenan Pepper—Own work, CC BY-SA 4.0, https://commons.wikimedia.org/w/index.php?curid=66085475

1.2.15 *Jordan River*

1. *Region*: Although nowadays this river is rather short, it had and has an enormous importance for the whole world in ancient and present times, since three world religions (Judaism, Christianism, and Islam) have their resources there and are still today the basis of ongoing military and ethnic conflicts. On the other hand, pilgrims of the three worldwide spread religions are able to "import" agents of diseases, which then may spread in this country as has been shown by Elad (2014) when looking at human cases of *anthrax* due to the bacterium *Bacillus anthracis*, which, however, hits animals. This river flows through the following countries or along their borders: the tributaries start in Lebanon (e.g., Mount Lebanon, Mount Hermon) and thus give rise to the river Jordan. From the Southern border of Lebanon, it takes its course southward along the borders of Syria, Israel, state of Palestine, and Jordan finally ending in the sand of the Dead Sea (Lake of Galilee), which is situated about 422 m below sea level in the desert. This is unique, since most rivers on earth enter either one of the oceans or at least in permanently existing lakes.
2. *Name*: The English name *Jordan* has its origin in the Hebrew term *Nahar ha-yardén* = *descender*. This river is especially important for the historically

established religions of Judaism and Christianism, since it is transmitted in historical tales that the people of Israelites entered this God-promised land by crossing the Jordan river, while the New Testament of the Christ reports that Jesus of Nazareth was baptized by Johannes the Baptist using the water of the Jordan river.

3. *Length*: The Jordan river has in contrast to prehistorical times a length of "only" 156 miles (251 km). Thus this river has become rather short especially in recent times = at the end of a "warm time," which will have its peak in about 2000–3000 years.
4. *Climate*: Along the coast of the Mediterranean Sea, the climate is moderately hot, while inside the country = along the river, the climate is characterized by extreme heat and lack of rain, so that desert regions grow constantly.
5. *Early discoveries and history*: This region has a rich history due to the fact that there are the bases of three world religions (Judaism, Christianism, Islam). Jerusalem was the capital of the Jewish kingdom of *David* and *Salomon* around the year 1000 BC. After the destruction of Jerusalem by Babylonian king *Nebuchadnezzar II* (605–562 BC) in the year 587 BC, the Persian King *Cyrus II* reconstructed the temple and the town (520–515 BC). In the year 63 BC, the Romans captured Jerusalem, but the Jewish Herod started 20 BC to reconstruct Jerusalem. In the year 637 AD, Jerusalem was conquered by Arabian tribes introducing there the Muslim religion created by *Abul Kasim Muhammad Ibn Abdallah* (=*Mohammed*, 571–632 AD). Nevertheless, the exercise of the three religions in and around Jerusalem was more or less undisturbed for centuries. In the year 1099 AD, Jerusalem became Christian after European Catholic forces under the guidance of their Christian leader *Gottfried von Bouillon* (first crusade initiated by the Roman Catholic Pope Urban II) entered this "holy town." In 1244 the Arabic Muslim powers conquered Jerusalem again, but could keep it only until 1571 AD, when Turkish rulers established their powership there. In the years 1917–1948, England kept this region as so-called protectorate, since the British and French governments had distributed the whole Middle East region among them by creating new, ethnically mixed countries, which in present times fight against each other in order to become rulers over the local oil fields. The importance of the water of the Jordan river can be recognized that it was one of the reasons for the so-called 6-day war in 1965. Syria, Lebanon, and Jordan attempted to get more water, whereby Israel's water supply would have been decreased by about 11%. The population of the whole region is endangered by both types of leishmaniasis (skin and visceral) due to the transmission activities of sandflies (*Phlebotomus* species).
6. *Big towns along the river*: There are no really big towns along the river, since it mainly passes desert regions (Fig. 1.15).
7. *Literature*

Azmi K, Krayter L, Nasereddin A, Ereqat S, Schnur LF, Al-Jawabreh A, Abdeen Z, Schönian G (2017) Increased prevalence of human cutaneous leishmaniasis in Israel and the Palestinian authority caused by the recent

Fig. 1.15 Diagrammatic representation of the holy river Jordan, which ends in the desert. By Own work—http://www.un.org/Depts/Cartographic/map/profile/israel.pdf, Public Domain, https://commons.wikimedia.org/w/index.php?curid=1901987

emergence of a population of genetically similar strains of *Leishmania tropica*. Infect Genet Evol 50:102–109

Ben-Shimol S, Sagi O, Codish S, Novack V, Barrett C, Fruchtman Y, Berkowitz A, Shemer-Avni Y, Greenberg D (2015) Dramatic increase in laboratory-diagnosed human cutaneous leishmaniasis cases in southern Israel, 2007–2013. Infect Dis 47:161–167

Ben-Shimol S, Sagi O, Horev A, Avni YS, Ziv M, Riesenberg K (2016) Cutaneous leishmaniasis caused by *Leishmania infantum* in Southern Israel. Acta Parasitol 61:855–858

Chaara D, Ravel C, Bañuls A, Haouas N, Lami P, Talignani L, El Baidouri F, Jaouadi K, Harrat Z, Dedet JP, Babba H, Pratlong F (2015) Evolutionary history of *Leishmania killicki* (synonymous *Leishmania tropica*) and taxonomic implications. Parasit Vectors 8:198

Elad D (2014) An unholy disease in the Holy Land: the history of anthrax between the Jordan River and the Mediterranean Sea (1909–2012). Vet J 199:319–323

Gandacu D, Glazer Y, Anis E, Karakis I, Warshavsky B, Slater P, Grotto I (2014) Resurgence of cutaneous leishmaniasis in Israel, 2001–2012. Emerg Infect Dis 20:1605–1611

Mehlhorn H (2016a) Encyclopedia of parasitology, 3 vols., 4th edn. Springer, New York

Mehlhorn H (2016b) Human parasites, 8th edn. Springer, New York

Orshan L, Elbaz S, Ben-Ari Y, Akad F, Afik O, Ben-Avi I, Dias D, Ish-Shalom D, Studentsky L, Zonstein I (2016) Distribution and dispersal of *Phlebotomus papatasi* (Diptera: Psychodidae) in a zoonotic cutaneous leishmaniasis focus, the Northern Negev, Israel. PLoS Negl Trop Dis 10(7):e0004819

Orshan L, Szekely D, Khalfa Z, Bitton S (2010) Distribution and seasonality of *Phlebotomus* sand flies in cutaneous leishmaniasis foci, Judean Desert, Israel. J Med Entomol 47:319–328

Pace D (2014) Leishmaniasis. BIAM 69:510–518

Richter J et al (2011) Mucosal *Leishmania infantum*. Parasitol Res 109:959–962

Schönian G, Kuhls K, Mauricio IL (2011) Molecular approaches for a better understanding of the epidemiology and population genetics of *Leishmania*. Parasitology 138:405–425

1.2.16 Lena River

1. *Region*: This river is besides the *Ob* and Yenisei, the third longest of the great Siberian rivers, which all flow into the *Arctic Ocean*. With respect to the water sources, the Lena is the only one of these three rivers, which gets its water entirely from Russian territories. This is due to the fact that 77% of the surrounding region are covered by continuous permafrost.
2. *Name*: The European name *Lena* has its origin in *Elyu-Ene*, which means “large river.” Other names are *Russian Buryat*, *Evenki*, and *Yakut*.
3. *Length*: The total length of this river, which has its source in the Baikal mountains, is recorded as 4264 km, thus being the second largest arctic river entering the Arctic Sea and largest one, of those which catch all of their water in the Russian territory.
4. *Climate*: Arctic; the 400-km-wide delta is a typical tundra being frozen for about 7 months of the year. Ships may use this river from May to the beginning of October starting at the town *Ust-Kut*.
5. *Early discoveries and history*: The *Lena* was apparently discovered during boat riding in the years 1620–1623 by Russian *fur hunters* led by a guide called *Demia Pyanda*, who explored in the year 1623 a portion of the river of about 2000 km. Another group of ten men guided by *Vasily Bugor* reached in the year 1628 the

Fig. 1.16 Diagrammatic representation of the basin of the river Lena in Europe, which flows into the Arctic Ocean. By Shannon1—Own work, CC BY-SA 4.0, https://commons.wikimedia.org/w/index.php?curid=61596790

Lena river, extorted tributes from the native people, and founded the settlement of *Kirensk* in the year 1632. In the same year, the town *Yakutsk* was founded by *Pyotr Beketov* and 20 men. This settlement became the starting point of further expeditions (e.g., to discover the Lena delta in the year 1632) in later years. Again later (1865–1886) several expeditions were organized by the Russian Imperial Academy of Science. As soon as it was discovered that this region is rich in valuable minerals, the former governments of the Russian tsar established towns and prison camps in dense forest regions. Today these settlements have become economically important towns delivering worldwide important amounts of mineral resources.

6. *Big towns along the river*: *Kirensk*, 12,000 inhabitants; *Olyokminsk*, 9000 inhabitants; *Ust-Kut*, 45,000 inhabitants; and *Yakutsk*, 300,000 inhabitants (Fig. 1.16).

7. *Literature*

Fofonova V, Danilov S, Androsov A et al (2015) Impact of wind and tides on the Lena River freshwater plume dynamics in the summer season. Ocean Dyn 65:951–968

Kirillov AF (2014) Ichthyofauna of the Lena River (Laptev Sea Basin): modern composition and historical formation. J Ichthyol 54:433–445

Kusatov KI, Ammosov AP, Kornilova ZG et al (2012) Anthropogenic factor of ice jamming and spring breakup flooding on the Lena River. Russ Meteorol Hydrol 37:392–397

Tananaev NI, Makarieva OM, Lebedeva LS (2016) Trends in annual and extreme flows in the Lena River basin, Northern Eurasia. Geophys Res Lett 43:10

1.2.17 Limpopo River

1. *Region*: States of South Africa, Botswana, Zimbabwe, and Mozambique. The spring of this river is situated in Krugersdorp close to Johannesburg (South Africa) at the so-called Witwatersrand. Its upper portion is also named *Crocodile river*. As soon as this part of the river has merged with the *Marico river* close to Oliphant's Drift, the river is called *Limpopo*, which finally enters the Indian Ocean close to the town Xai-Xai in the state of Mozambique. This river forms the Northern border to the states of Botswana and Zimbabwe (formerly Rhodesia) and crosses Mozambique on a length of 400 km.
2. *Name*: The name "Limpopo" has its etymological origin from the Sepedi word *diphororo tša meetse*—meaning strong flushing waterfalls. Sepedi language is spoken by a third of the people in the province of Limpopo.
3. *Length*: This is the second longest river of Africa behind the *Nile*, since it reaches a length of 1087 miles (=1750 km).
4. *Climate*: The climate in the four countries through which (or close by) the Limpopo flows is listed below:

   - *Republic of Botswana* (formerly Bechuanaland): semiarid; most of the time, the climate is hot and dry with a rainy season during the summer months from September until April.
   - *Republic of Mozambique*: tropical during rainy season from November until April (80% of yearly rainfall); cool nights during dry season. Average temperatures during the entire year between 25 and 30 °C during the day and 15–25 °C at night.
   - *Republic of South Africa*: temperate climate, due to the varying topography and influence of the oceans the climatic zones range from the desert (Namib) to subtropical climate in the east along the border with Mozambique and the Indian Ocean.
   - *Zimbabwe* (formerly *Rhodesia*): tropical climate with local variations with heat and aridity and also areas with even cool and cold phases in winter. The

rainy season lasts from late October to March, and the hot climate is moderate in increasing altitude.

5. *Early discoveries and history*: In the fifteenth century, the region of South Africa of our days (=former Union of South Africa, 1910–1961) had been settled by black African tribes. Their first contacts to Europeans occurred, when the Dutch *Jan van Riebeeck* built the first European settlement at the *Cape of Good Hope* (close to the present Cape Town). In the years from 1652, the so-called Bures (=farmers from the Netherlands and the German Rhineland) settled in this region after having left their countries due to the fact that they had been protestants and their home country switched to the Catholic religion due to the fact that the new king was a Catholic. However, in the year 1806, this Buren country felt after troop invasion under English government, which suppressed the Bures severely. Thus, they left this region at the Cape in the year 1835 and created in the North the Republics of Natal, Oranje, and Transvaal. Natal became English already in the years 1842–1845, while the others were later overtaken by the English after the "War of Bures" (1899–1902), which had apparently its reason in the finding of masses of gold in their region. The country ranged from 1910 until 1961 as member of the British Empire, then left it, and is now called the Republic of South Africa.
6. *Big towns along or close to the river*: *Johannesburg* (South Africa), 2 million inhabitants; *Gaborone* (Botswana), 240,000 inhabitants; and *Xai-Xai* (Mozambique), 145,000 inhabitants (Fig. 1.17).
7. *Literature*

Balek J (2011) Hydrology and water resources in tropical Africa. Elsevier, Amsterdam, p 91

Erskine VW (1869) Journey of exploration to the mouth of the River Limpopo. J Roy Geogr Soc 39:233–276

Görgens AHM, Boroto RA (1997) Limpopo River: flow balance anomalies, surprises and implications for integrated water resources management. In: Proceedings of the 8th South African national hydrology symposium, Pretoria

Malherbe J, Engelbrecht FA, Landman WA et al (2012) Tropical systems from the southwest Indian Ocean making landfall over the Limpopo River Basin, southern Africa: a historical perspective. Int J Climatol 32:1018–1032

Malherbe J, Landman WA, Engelbrecht FA (2014) The bi-decadal rainfall cycle, southern annular mode and tropical cyclones over the Limpopo River Basin, Southern Africa. Clim Dyn 42:3121–3138

Sara JR (2014) Ecological status of Hout River Dam, Limpopo province, South Africa, using fish condition and health assessment index protocols: a preliminary investigation. Afr J Aquat Sci 39:35–43

1.2.18 Mackenzie River (Nagwichoonjik)

1. *Region*: This North American river flows mainly in northwestern direction covering parts of five Canadian provinces, where important sources are located.

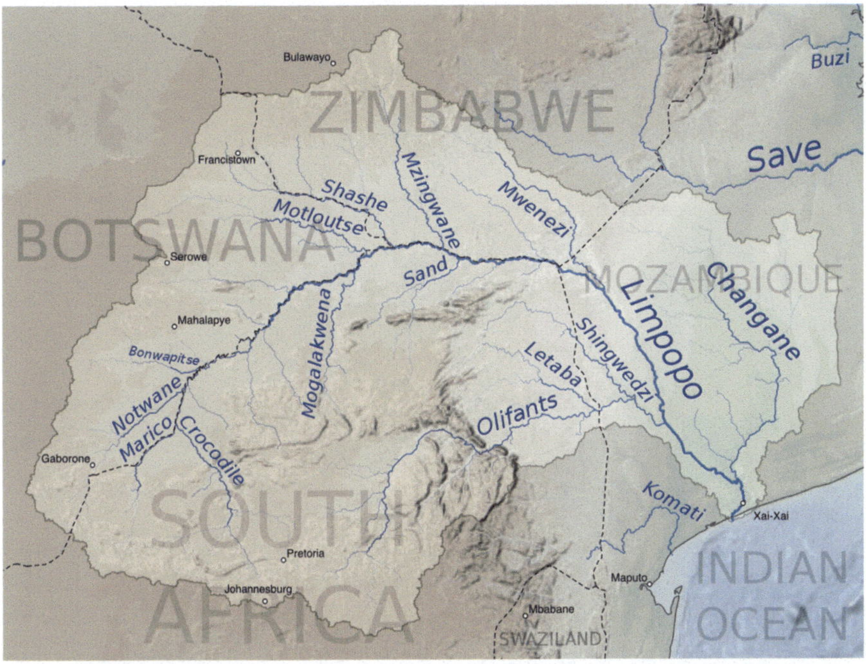

Fig. 1.17 Diagrammatic representation of the river Limpopo in Africa. By Keenan Pepper—Own work, CC BY-SA 4.0, https://commons.wikimedia.org/w/index.php?curid=66115393. Water management of the river Limpopo http://www.fao.org/docrep/008/y5744e/y5744e07.htm

Its main (=final) stem starts from Great Slave Lake and enters its water via a delta into the Arctic Ocean, which was first reached by the English *Sir John Franklin* on August 16, 1825.

2. *Name*: The *Mackenzie river* bears its name honoring the Scottish explorer *Alexander Mackenzie* (1764–1820), who led an expedition in the year 1789. The local population at that time called it *Dehcho* (slavery tribe), which means "big river." Another tribe (*Inuvialuktun* people) called it *Kunkpak*, which can be translated as "great river." French explorers called the river "*fleuve de Mackenzie*" = Mackenzie's river. It is a coincidence that Mackenzie reached the mouth of the river, which he called "Disappointment River" on the starting day of the French revolution 1789 (May 5th), when he noted that he had not found a pathway to the Pacific Ocean.
3. *Length*: If the total length (=counted from its farthest source up to the Arctic Ocean) is considered, the Mackenzie river has a total length of 4241 km (=2635 miles). Thus, it is the second longest river in North America.
4. *Climate*: Northern coldness is common, and thus this region with its enormous size harbors only very few people as permanent population.
5. *Early discoveries and history*: There are archaeological signs that the Mackenzie valley had been one of the important pathways from Asia to America during

Fig. 1.18 Map of the region of the river Mackenzie in North America. By Shannon1—Own work, CC BY-SA 4.0, https://commons.wikimedia.org/w/index.php?curid=63800219

warmer periods. However, the findings of remnants are scarce, since many changes between icy and warm times had happened several times during the thousands of years destroying the surface. The fact is, however, that in more recent times, indigenous people have lived along the river (e.g., remnants are found of the Gwich'in people, who lived at a place called Tsiigehtchic around 400–500 years AD).

6. *Big towns along the river*: This part of Canada contains rather few towns, and thus most of the towns are rather small compared to other regions in Eastern Canada. *Fort Providence*, 727 inhabitants; *Fort Simpson*, 1216 inhabitants; *Norman Wells*, 778 inhabitants; and *Fort Good Hope*, 567 inhabitants (Fig. 1.18).
7. *Literature*

Hodgins BW, Hoyle G (1994) Canoeing north into the unknown: a record of river travel, 1874 to 1974. Dundurn Press, Toronto

Lantz TC (2015) Ecological recovery in an Arctic delta following widespread saline incursion. Ecol Appl 25:172–185

Wolh E (2010) The Mackenzie: river on the brink. In: A world of rivers, environmental change on ten of the world's great rivers. University of Chicago Press, Chicago

Woo M, Thorne RJ (2016) Summer low flow events in the Mackenzie River system. Arctic 69:286

Yang D, Peterson A (2017) River water temperature in relation to local air temperature in the Mackenzie and Yukon Basins. J Arctic 70:47–57

1.2.19 Madeira River

1. *Region*: This river, which has several important tributaries (e.g., Madre de Dios, Mamoré), starts its course in the mountains of Peru and Bolivia, crosses the northern regions of South America, and fuses as biggest tributary with the Amazon river contributing about 15% of the final Amazon water.
2. *Name*: The name *Madeira* has a Portuguese origin given also for a European (Portuguese) island being situated in the Atlantic Ocean close to the Moroccan coast. Inhabitants of the source region in South America call this river *Cuyari* (which means "wood river"). It is reported that the tribe of the *Tupinamba* "wandered" along and/or on this river from the mountain region to that of the Amazon river.
3. *Length*: This river has a length of 3250 km (~2020 miles), and it is documented by a map designed by Emanuel Bowen already in the year 1747.
4. *Climate*: *Cold* to moderate in the region of the sources of the tributaries and *tropical* in the basin of South America, where the rainfall per year often ranges between 49 cm in higher regions and 700 cm in the basin. During rainfall periods the water level of this river may rise for 15 m, so that even ocean vessels may ascend the 1007 km from the coast to the San Antonio Falls near the town *Porto Velho* (Brazil).
5. *Early discoveries and history*: Even though the exploration of the Madeira valley started already in the sixteenth century, the region was not mapped until the 1970s. The region was inhabited by Indians and Mestizos who used the products of the rainforest for their living. In the second half of the twentieth century, farmers and ranchers settled in this region.
6. *Big towns along the river*: *Porto Velho*, 430,000 inhabitants; *Santarem*, 270,000 inhabitants; and *Humaita*, 52,000 inhabitants (Fig. 1.19).
7. *Literature*

Balzino M (2015) Gold losses and mercury recovery in artisanal gold mining on the Madeira River, Brazil. J Clean Prod 102:370–377

Beach CB (ed) (1914) Madeira – a river of South America. FE Compton

Fraser J, Cardoso T, Junqueira A, Falcão N, Clement C (2009) Historical ecology and dark earths in whitewater and Blackwater landscapes: comparing the middle Madeira and lower Negro Rivers. In: Woods WI, Teixeira WG,

Fig. 1.19 Diagram of the basin of the Madeira river in South America. By Shannon—Background and river course data from http://www2.demis.nl/mapserver/mapper.asp, CC BY-SA 4.0, https://commons.wikimedia.org/w/index.php?curid=9617419

Lehmann J, Steiner C, Winkler Prins A, Rebellato L (eds) Amazonian dark earths: Wim Sombroek's vision. Springer, Dordrecht

Herraiz AD, Lima de Alencastro Graça PM, Philip Martin Fearnside PM (2017) Amazonian flood impacts on managed Brazilnut stands along Brazil's Madeira River: a sustainable forest management system threatened by climate change. For Ecol Manag 406:46–52

Molina-Carpio J, Espinoza JC, Vauchel P et al (2017) Hydroclimatology of the upper Madeira River Basin: spatio-temporal variability and trends. Hydrol Sci J 62:911–927

Ziesler R, Ardizzone GD (1979) Amazon river system. In: The inland waters of Latin America. FAO Org United Nations

1.2.20 Main River

1. *Region*: This river has its main extension in Central Germany, while its sources being named as the *White* and *Red Main* rivers are located in the "Fichtelgebirge" and Franconian Mountains. The Main flows from German East regions to the West, where it joins the Rhine river at Mainz close to Frankfurt. It is part of the Main-Danube channel system, which links the Danube and Rhine river for cargo ships that wish to reach the Black Sea after passing the Balcony region respectively the North Sea.
2. *Name*: The Main river has obtained its name from Romans, who called it *Moenia* = "protection wall" being used to stay safe from attacks of German tribes living on the other side of the river. Another historical source reports that in the early Medieval times, the river was called *Moyn*, which sounds like the German name *Main* of the present times.
3. *Length*: The river Main is rather short and measures only 326 miles (=526 km). Ships can be used from Bamberg to Mainz over a distance of 388 km. From there the ships may use the Rhine river to cruise to the Netherlands or to the border of Switzerland. The river Danube, which can be reached via the Main-Danube channel, starts its 2850-km-long course in the German Black Forest passes Austria, Slovakia, Hungary, Croatia, Serbia, Bulgaria, Romania, Moldova, and Ukraine and finally enters the Black Sea.
4. *Climate*: The river valley is favored by the typical Central European climate with clearly defined winter, spring, summer, and autumn seasons. However, occasionally during winter times, the water surface of the Main river may become frozen. These periods lasted much longer in the past centuries than in the present time. In the past the river was often completely frozen up for up to 3 months (e.g., 1768, 1845, 1984, and 1995).
5. *Early discoveries and history*: The region along the Main river was inhabited by *Celtic tribes* during the eighth to fourth century BC being followed by German tribes like the *Cimbri*, who, however, became suppressed by the *Romans* around the times of Jesus Christ. Later the culturally different inhabitants of this region became mixed with people coming from the cold East after crop failure in their habitats.
6. *Big towns along the river*: *Bayreuth*, 72,000 inhabitants, famous for its opera house and *Richard Wagner's* festivals attracting thousands of international people until today; *Würzburg*, 126,000 inhabitants, famous for its historical and cultural buildings, its university, and its wines; *Offenbach*, 124,000 inhabitants, famous for its leather industry; *Frankfurt/Main*, 736,000 inhabitants, famous as

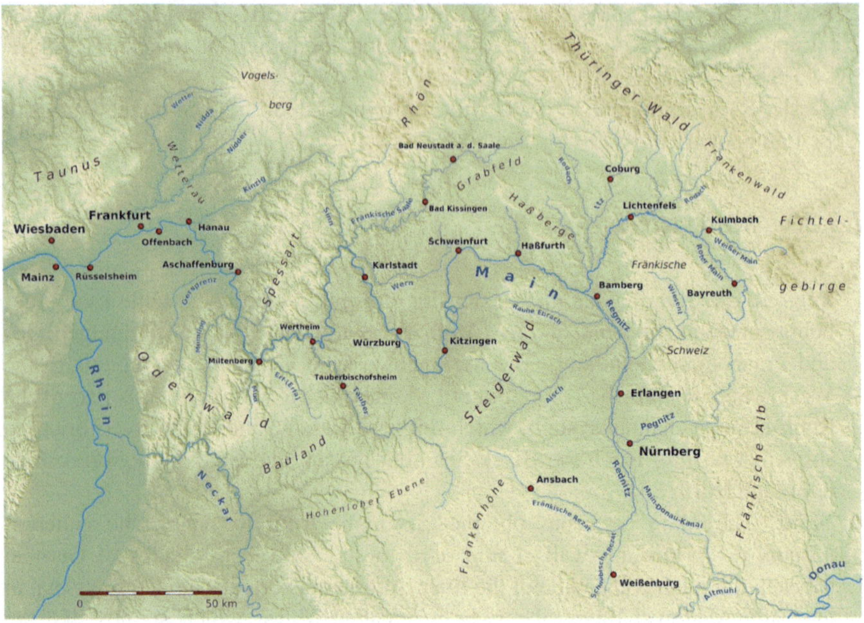

Fig. 1.20 Overview of the Main river flowing in the center of Germany. By BerndH—Own work, CC BY-SA 3.0, https://commons.wikimedia.org/w/index.php?curid=10903145

one of the biggest international bank centers and *Paul's Church*, which had been the meeting room of the first German National Assembly in the year 1848 founding the basis of the German union; *Wiesbaden*, 277,000 inhabitants, seat of the Hessian parliament; and *Mainz*, 213,000 inhabitants. This town is situated on the shore of the river Rhine at the opposite side of the estuary of the Main river. It is the seat of the government of the region Rhineland-Palatinate and hometown of *Johannes Gutenberg* (1394–1445), who was the inventor of the first *printing apparatus* in the year 1445 thus opening a worldwide information system (Fig. 1.20).

7. *Literature*

Arora R (2016) Changing river temperatures in northern Germany: trends and drivers of change. Hydrol Process 30:3084–3096

Sui J (2010) Characteristics of rainfall, snowmelt and runoff in the headwater region of the Main river watershed in Germany. Water Resour Manag 24:2167–2186

1.2.21 Mekong River

1. *Region*: The source of this river is located in the *Tibet Autonomous Region* (being governed by China). It passes several regions in China, Burma, Laos, Cambodia,

Vietnam, and Thailand. Its true source had only been discovered in the year 1994 after some other springs had been suspected before.

2. *Name*: The Mekong river (English name) has different names depending on the country, through which its water flows. In Cambodia it is named *Mékôngk* and *Thonle Than*; in Thai, Myanman, and Lao language *Mae Nam Khing*; in Vietnam *Sông Mê Kông*; and in China *Langcang Jiang* (=turbulent river) or *Lan-ts'ang Chiang*. Other more trivial names are Big water, Dragon running water, Mother river Khong, and River of stones.
3. *Length*: This river is the 7th longest river in Asia and the 12th longest in the world reaching a total length of 2700 miles (4350 km). Its source is located in the Qinghai province of China. It passes the Eastern region of the Tibet Autonomous Region followed by the Yunnan province. Then it becomes the borderline between Burma (Myanmar) and Laos as well as between Laos and Thailand. Finally, the river takes its further course through Laos, Cambodia, and Vietnam before entering the South China Sea a bit south of Chi Minh City, which formerly was named Saigon. Behind Phnom Penh the river is divided into two branches, which are called "Mekong proper" and "Bassac." They both enter via a 70,000 km^2 large, marshy delta into the South China Sea.

   This river drains about 313,000 square miles (810,000 km^2). In Asia these data are only surpassed by the *river Ganges* and the *Yangtze river*. The main amount of water being entered into the sea is mainly based on rainfall in its lower basin; however, the amounts depend on the seasonally changing monsoon winds. In total this river enters each year around 475 billion cubic meters of water into the South China Sea. The Mekong river shows also a natural spectacle since it is connected with the largest freshwater lake in Southeast Asia (=*Tonle Sap*). This lake collects water in the wet season and enters it back into the Mekong during the dry season (=the water runs in different directions)—depending on the dry or rainy season.
4. *Climate*: The temperatures in the lower Mekong basin are rather high throughout the year reaching average daily highs of 32 °C (=89 °F) and lows of 23 °C (74 °F). During the *dry period* in the year, which may be rather long, lack of rain inhibits rice production without irrigation, so that water has to be obtained from the rivers.
5. *Early discoveries and history*: Historical books claim that already *Marco Polo* passed the Mekong in the thirteenth century AD, since he reported details in his memories, but this is not really confirmed. The first documented European was the Dominican missionary *Gaspar da Cruz*. He described that he travelled in the years 1555–1557 by boat on this river in a region, which today belongs to Cambodia. Another documented European expedition was that of the Dutch explorer *Gerrit van Wugshoof* during the year 1641. French explorers had the hope to travel by boat via the Mekong valley to China. But this turned out to be as wrong, since *Francis Garnier* detected as reason in the year 1860 that the Mekong river drops down at the so-called Khone falls (syn. Kônephaeng falls = a 6 mile chain of cataracts) in Southern Laos. These falls were later surrounded by a 14-km-long railway, so that the Mekong river can be used as waterway

connecting the countries, when goods are transported by train, thus avoiding the falls. During spring melt and the season of monsoons from May to October, the Mekong is very dangerous reaching sudden flood waves of 46 feet (14 m). The governments of the countries through which the river takes its course established formally in the year 2001 the so-called Mekong Basin Disease Surveillance Network (MBDS) in order to avoid the spreading of agents of diseases among their large populations (such as cholera, malaria, Japanese encephalitis, hepatitis, dengue fever, visceral leishmaniasis, SARS influenza, etc.).

6. *Important towns along the river*: The fact that the capitals of Laos (*Vientiane* = Viangchan) and Cambodia (*Phnom Penh*) with 800,000 and 2 million inhabitants, respectively, are situated at the banks of the river and that *Ho Chi Minh City* (formerly *Saigon*) with its 8.5 million inhabitants is also not too far from the delta of the Mekong river underlines the economic and cultural importance of this waterway for the whole region (Fig. 1.21).
7. *Literature*

Jaobs JW (1995) Mekong committee history and lessons for river basin development. Geogr J 161:135–148

Le TN (2018) Interplay between land-use dynamics and changes in hydrological regime in the Vietnamese Mekong delta. Land Use Policy 73:269–280

Mehlhorn H (ed) (2016a) Encyclopedia of parasitology, 3 vols., 4th edn. Springer, New York

Mehlhorn H (2016b) Human parasites, 8th edn. Springer, New York

Ngor PB (2018) Flow alterations by dams shaped fish assemblage dynamics in the complex Mekong river system. Ecol Indic 88:103–114

Nguyen-Tien V (2018) Hydropower generation, flood control and dam cascades: a national assessment for Vietnam. J Hydrol 560:109–126

Phommasack B, Jiraphonosa C, Oo MK, Bond CK et al (2013) Mekong Basin disease surveillance (MBDS): a trust-based network. Emerg Health Threat J 6:19944

Sabo JL, Ruhl A, Holtgrieve GW, Elliott V et al (2017) Designing rivers to improve food security in the lower Mekong Basin. Science 358:1270

Ziegler AD, Petney TN, Grundy-Warr C, Andrews RM et al (2013) Dams and disease triggers on the lower Mekong river. PLoS Negl Trop Dis 7:e2166

1.2.22 Meuse River

1. *Region*: This river has its source in France and flows through eastern regions of Belgium, and after passing regions of the Netherlands, it enters into the North Sea via the Rhine-Meuse delta.
2. *Name*: Since this European river passes three countries, it has different historic names: *French*, La Meuse; *Belgian/Walloon*, Mouze; and *German*, Maas. All these names have their origin in the old Celtic name *Mosa*.
3. *Length*: This river measures 925 km (=575 miles) in length on its way through three countries.

Fig. 1.21 Diagrammatic representation of the course of the Mekong river in Asia. By Shannon1—Own work, CC BY-SA 4.0, https://commons.wikimedia.org/w/index.php?curid=65845951

4. *Climate*: Continental on its way, moderate in the delta region.
5. *Early discoveries and history*: Until the year 1301, the region along this river had many rulers depending on marriages of people belonging to the different European Noble families. Beginning from 1301 this river formed the border between the so-called Holy Roman Empire and the Kingdom of France, although—depending on marriages—several regions were ruled by different noble families until finally the democratic countries of our days had been founded

after the wars in the years 1870–1871 and 1914–1918. This river is mentioned in the first stanza of Germany's anthem being written in the year 1841 by W. von Fallersleben on the North Sea island Helgoland saying Germany reaches from the Meuse to the Memel. However, today only the third stanza is the official anthem of Germany, since both rivers now flow in other political countries.

6. *Big towns along the river*: *Sedan* (France), 17,000 inhabitants; *Namur* (Belgium), 111,000 inhabitants; *Liège* (Belgium), 197,000 inhabitants; *Maastricht* (the Netherlands), 121,000 inhabitants; and *Venlo* (the Netherlands), 101,000 inhabitants (Fig. 1.22).
7. *Literature*

Klijn F (2015) Flood risk management for an uncertain future: economic efficiency and system robustness perspectives compared for the Meuse River (Netherlands). Mitig Adapt Strateg Glob Chang 20:1011–1026

Lambert T (2017) Effects of human land use on the terrestrial and aquatic sources of fluvial organic matter in a temperate river basin (The Meuse River, Belgium). Biogeochemistry 136:191–211

Marescaux J (2016) Benthic macro-invertebrate fauna associated with *Dreissena* mussels in the Meuse River: from incapacitating relationships to facilitation. Aquat Ecol 50:15–28

Pierik HJ (2017) Natural levee evolution in the Rhine-Meuse delta, the Netherlands, during the first millennium CE. Geomorphology 5:215–234

Ward PJ, Renssen H, Aerts JCJH, van Balen RT, Vandenberghe J (2008) Strong increases in flood frequency and discharge of the River Meuse over the Late Holocene: impacts of long-term anthropogenic land use change and climate variability. Hydrol Earth Syst Sci 12:159–175

1.2.23 Mississippi River

1. *Region*: This river has its source in the North of the USA and passes to the South exclusively in this country before entering the Gulf of Mexico.
2. *Name*: The river name Mississippi was given by local tribes and later kept by the Spanish explorers like *Hernando de Soto*, who reached this river with his crew in the year 1541 at a place, which is situated close to the present town of Memphis (Tennessee). The indigenous population was named *Indians* by different explorer groups due to the error that they had landed their boats in India.
3. *Length*: The Mississippi has its source in the Lake Itasca in the state of Minnesota. From the source to its delta at the Gulf of Mexico, it passes 2340 miles (=3766 km). Nineteen other larger rivers (e.g., Missouri, Ohio river) contribute their water to the Mississippi, so that this river may finally discharge about 17,000 cubic meters of water per second into the delta. Thus, the Mississippi is the largest river in North America and the eighth biggest in the world. If the length of the Missouri would be added, it would be one of the largest in the world.

Fig. 1.22 Diagrammatic representation of the former "border river of old times" Meuse now flowing through the united countries of Europe. By Bassin_de_la_Meuse.svg: Own Work, svg and inspired by Image:Meuse basin.jpgderivative work: Hans Erren (talk)—Bassin_de_la_Meuse. svg, CC BY-SA 3.0, https://commons.wikimedia.org/w/index.php?curid=11432402

4. *Climate*: Due to the considerable length of this river, it cannot be avoided to flow through completely different temperature zones. Thus during winter the subarctic region in Minnesota reaches mean temperatures of 10° Fahrenheit (=minus 12° Celsius), while the mean temperatures in the subtropical Louisiana range around 55° Fahrenheit (=13° Celsius). The floods of the Mississippi are not controlled yet. There are repeated large floods like that in the year 1925, when 2,722,000 acres had been flooded leaving thousands of people homeless. The reasons for such floods are hurricanes like that called Katrina killing 238 persons in 2005 or several so-called tornados in 2010. The Mississippi helped to spread the early *cholera epidemics* in the years 1832, 1849, and 1866 (see Chap. 4).
5. *Early discoveries and history*: The first European discoverer was apparently the Spanish explorer *Hernando de Soto*, who reached this river in May 1541 and named it "Rio Grande." The purpose was to plunder the local tribes, which, however, defeated the Spanish intruders and killed him and several members of his crew. It took until 1673 when the next European explorers tried their "luck" under the guidance of the French *Louis Jolliet* and a Jesuit priest named *Jacques Marquette*. They started from French Canada with further voyagers in canoe boats, passed the *Fox River*, and paddled on the Mississippi until they came to the "mouth" of the *Arkansas river*. They were later followed by another French explorer (*René-Robert Cavelier*) and his group, who reached as first Europeans the Mississippi delta by boat. In his European arrogance, he declared and claimed the entire Mississippi basin as French new territory without considering the—of course—not amused native inhabitants. However, the lower river region was later taken by the Spanish in 1769, and finally in the *Peace of Paris* (1783), the river was declared as the western border to the newly created USA. After many years of war (1812–1858) and confusion as well as hurly burly among several European powers, the USA of our days has been finally settled as the owner. Today the Mississippi is one of the busiest waterways in the world. Although it produces enormous environmental problems, the USA cannot stop the transports on this waterway, since it is a "blood vein" of business.

   The Mississippi floods are still today not or insufficiently controlled. There are repeatedly large floods like that of the year 1925, when 2,722,000 acres had been flooded leaving hundred thousands of people homeless. The reasons are hurricanes (wind movements with a speed higher than 118 km/h) occurring in the summer months. For example, the hurricane Katrina killed even recently (2005) 238 persons as did a so-called tornado in the year 2010.

   Furthermore, the worldwide occurring cholera epidemics of the years 1832, 1849, and 1866 (see Chap. 4) were distributed by people arriving at the mouth of the river and travelling northward in the USA.
6. *Big towns along the river*: *Minneapolis* (Minnesota), 422,000 inhabitants; *St. Louis* (Missouri), 308,000 inhabitants; *Memphis* (Tennessee), 652,000 inhabitants; *Baton Rouge* (Louisiana), 225,000 inhabitants; and *New Orleans* (Louisiana), 393,000 inhabitants (Fig. 1.23).
7. *Literature*

Fig. 1.23 Map showing the basin of the famous "Old Mississippi river," which was an important "entrance route" in former times. By Shannon1—Own work, CC BY-SA 4.0, https://commons.wikimedia.org/w/index.php?curid=47308146

Allison MA (2012) A water and sediment budget for the lower Mississippi-Atchafalaya river in flood years 2008–2010: implications for sediment discharge to the oceans and coastal restoration in Louisiana. J Hydrol 432–433:84–97

De Jager NR (2016) The upper Mississippi River floodscape: spatial patterns of flood inundation and associated plant community distributions. Appl Veg Sci 19:164–172

Fremling CR (2005) Immortal river: the upper Mississippi in ancient and modern times. University of Wisconsin Press, Madison

Hudson P, Kesel R (2006) Spatial and temporal adjustment of the lower Mississippi: river channel to major human impacts. Ztschr Geomorphol 143:S17–S33

Mehlhorn H (ed) (2016a) Encyclopedia of parasitology, 3 vols., 4th edn. Springer, New York

Mehlhorn H (2016b) Human parasites, 8th edn. Springer, Heidelberg

Morris C (2012) The big muddy: an environmental history of the Mississippi and its peoples from Hernando de Soto to hurricane Katrina. Oxford University Press, Oxford

Pinter N (2010) Historical discharge measurements on the middle Mississippi River, USA: no basis for 'changing history'. Hydrol Process 24:1088–1094

Savage HM, Kothara L (2012) The *Culex pipiens* complex in the Mississippi river basin: identification, distribution and bloodmeal hosts. J Am Mosq Control Assoc 28:93–99

Trimble SW (2013) Historical agriculture and soil erosion in the upper Mississippi Valley hill country. CRC Press, Boca Raton

Yodis G, Kesel R (1993) The effects and implications of base - level changes to Mississippi River Tributaries. Ztschr Geomorphol 37:385–402

1.2.24 Murray River

1. *Region*: Southeastern Australia. The river has its source at a height of 902 m in the so-called Snowy Mountains of the *Great Dividing Range* in the Southern region of Australia in the state of Victoria. From there it flows at the beginning northward forming the border between the inner Australian states Victoria and New South Wales. Close to the town of Wentworth, the *Murray river* is joined by the *Darling river*, which is considerably shorter (1472 km) than the Murray river. Finally, the common water of both rivers enters the South Australian Sea just after having passed the Lake Alexandria. Another important tributary is the *Murrumbidgee river*. This river passes through the Australian states of New South Wales and Victoria.
2. *Name*: The English river name was given honoring the English Colonial Secretary of State and War *Sir George Murray* (1772–1846).
3. *Length*: The Murray river has a length of 2652 km respectively 2530 km (depending on different types of measurements) corresponding to about 1570 miles. Ships may use the river for 1986 km. In order to have a sufficient deepness for ships, the river is artificially dammed up at four places. One of these is *Lake Hume*, which got its name honoring the English discoverer (Lord Hume) of the river. This lake is situated at a height of about 180 m above sea level and is followed by Lake Mulwala at 124 m and Lake Euston at 56 m in height. These lakes help to guarantee the water supply of the big town Adelaide.
4. *Climate*: The Murray river flows in zones with moderate temperatures. Up to 30 °C are common on mostly rainless days. However, water temperatures in the Southern sea, where it enters, increase rarely beyond 20 °C. The winter temperatures are around 10 °C, and rain is common.
5. *Early discoveries and history*: This river was detected in the year 1824 by the English discoverers *H. H. Hume* and *W. H. Hovell*. About 5 years later, another Englishman (*Charles Stuart*) navigated there by boat and named this river honoring Sir George Murray. Along this river numerous aboriginal strains lived at this time. The Ngarrindjeri survived and are today the largest aboriginal

Fig. 1.24 Map showing the course of the Murray river in South Australia. By historicair's original work here on 13:13, 2 August 2007 (UTC). Translated by TastyCakes (talk)—Murray_river_(-Australia)_map-fr.svg, CC BY-SA 3.0, https://commons.wikimedia.org/w/index.php?curid=6048807

community. Already since the year 1918, the water of the Murray river is used for millions of people as drinking water industrial purposes. Thus, this river is of great importance for the Australian economy—but also for tourism and nature friends, since many rare animal species can be found there (e.g., lizards, koalas, Western gray kangaroos, parrots, black swans, etc.). The great importance of this region is also based on the fact that there are located 40% of all Australian farms producing large amounts of food.

6. *Big towns along the rivers*: Australia has only a few large cities (e.g., Sydney, Melbourne). Other towns are rather small. Thus medium-sized towns along the river are Albury/Wodonga (83,000 inhabitants), Echuca (16,000 inhabitants), Mildura (32,000 inhabitants), Murray Bridge (22,000 inhabitants), or Renmark (less than 10,000) (Fig. 1.24).
7. *Literature*

Australian Government: Bureau of Meteorology (2005) Infosheet El Niño and La Niña and Australia's Climate.
Chiew FHS (2010) Comparison of runoff modelled using rainfall from different downscaling methods for historical and future climates. J Hydrol 387:10–23
Grundell R (2012) Interaction between a river and its wetland: evidence from the Murray River for spatial variability in diatom and radioisotope records. J Paleolimnol 47:205–219

Kirby JM (2014) Climate change and environmental water reallocation in the Murray-Darling Basin: impacts on flows, diversions and economic returns to irrigation. J Hydrol 518:120–129

Longest Rivers – Geoscience Australia. http://www.ga.gov.au/scientific-topics/national-location-information/landforms/longest-rivers

Map of Murray River (902–125 m). Bonzle.com

Mehlhorn H (ed) (2016) Encyclopedia of parasitology, 3 vols., 4th edn. Springer, New York

Overton I, Doody T (2013) The river Murray-Darling Basin: ecosystem response to drought and climate change. In: Schwabe K, Albiac-Murillo J, Connor JD, Hassan R, Meza González L (eds) Drought in arid and semi-arid regions. Springer, Cham, pp 217–236

1.2.25 Murrumbidgee River

1. *Region*: This river is a major tributary of the Murray River and flows through the Australian state of New South Wales and the Australian capital Territory (close to the town of Canberra).
2. *Name*: The name Murrumbidgee is a term of the local Australian aboriginal language and can be translated as "big/large water." In former times (~1820 AD), local tribes called this river *Yealambidgie.*
3. *Length*: From its source in the so-called Snowy Mountains until its fusion with the Murray river, the Murrumbidgee river measures 1485 km (923 miles).
4. *Climate*: Subtropical with dry phases, but snowfall is possible in high mountain regions.
5. *Early discoveries and history*: One of the first discoverers of this river had been British military people (John *Oven*, Mark *Curie*), who reached the upper portion of the river during their expedition in the year 1823. Later in 1829 a crew around Charles *Sturt* rowed down the river exploring this region. Afterward these tough men rowed back upstream to their starting site. Beginning in the year 1830, farmers founded settlements in the surroundings of their starting point, thus creating the increasing farming area of our days. However, the river level fluctuated widely. Thus, very important floods occurred from time to time and caused big problems. After construction of the Burrinjuck Dam in the year 1907, the floods were reduced, but big ones in the years 1925, 1950, 1974, and 2012 occurred nevertheless and killed numerous persons.
6. *Big towns along the river*: Real big towns have not been established along the river, but several small- to medium-sized ones: *Wagga Wagga*, 47,000 inhabitants; *Hay*, 2400 inhabitants; *Narrandera*, 3700 inhabitants; *Cooma*, 6700 inhabitants; and *Gundagai*, 1900 inhabitants (Fig. 1.25).
7. *Literature*

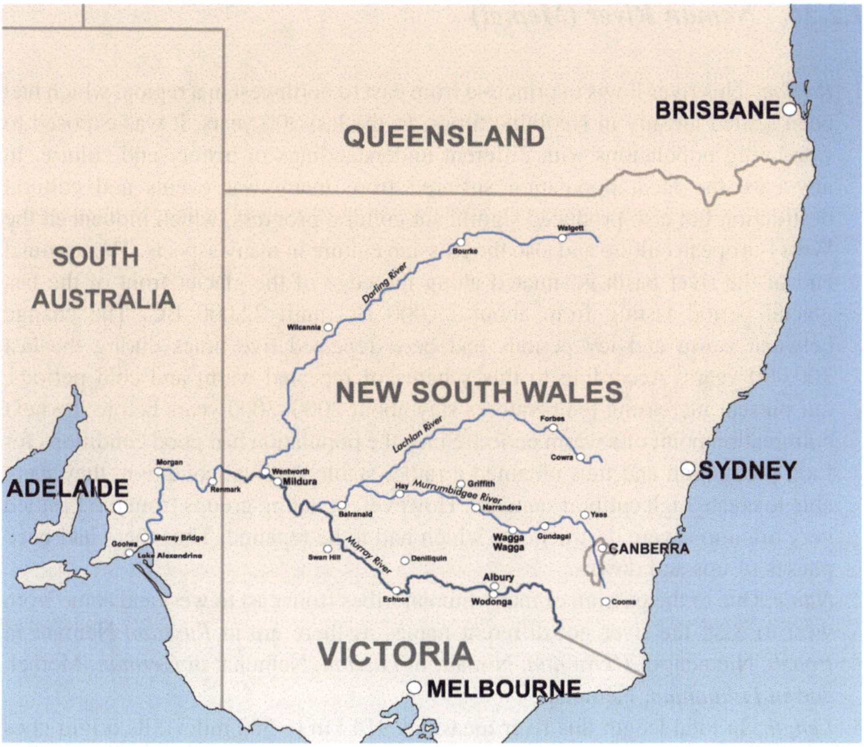

Fig. 1.25 Map showing the course of the Murrumbidgee river in South Australia (right center). By Bidgee—Own work, CC BY 3.0, https://commons.wikimedia.org/w/index.php?curid=4182228

Forbes JP (2015) Recreational fishing effort, catch, and harvest for Murray cod and golden perch in the Murrumbidgee River, Australia. N Am J Fish Manag 35:649–658

Graham S (2009) Irrigators' attitudes towards environmental flows for wetlands in the Murrumbidgee, Australia. Wetl Ecol Manag 17:303–316

Ren S (2014) Modelling impacts of regulation on flows to the Lowbidgee floodplain of the Murrumbidgee River, Australia. J Hydrol 27:1660–1667

Wen L (2011) The influences of climate and hydrology on population dynamics of waterbirds in the lower Murrumbidgee River floodplains in Southeast Australia: implications for environmental water management. Ecol Model 222:154–163

Wen L, Rogers K, Ling J et al (2011) The impacts of river regulation and water diversion on the hydrological drought characteristics in the lower Murrumbidgee River, Australia. J Hydrol 405:382–391

1.2.26 Neman River (Memel)

1. *Region*: This river flows in principle from east to northwest in a region, which had been settled already in Neolithic times. In the last 300 years, it was exposed to wandering populations with different understandings of history and culture. In any case the local population suffered from many war events and cultural destruction but also produced significant cultural progress, which influenced the West European culture and also the Russian culture in many aspects. The terminal end of the river basin is situated along the edge of the glacier front of the last glacial period lasting from about 25,000 BC until 22,000 BC. The change between warm and icy periods had been repeated five times during the last 200,000 years. According to this scheme of repeated warm and cold periods, our present increasing temperatures stay about 2000–3000 years before the next culmination point of a warm period. Since the population had good conditions for food production and thus obtained a rather stable survival condition, they were able to create high cultural standard. However, invading groups from the East led very often to severe destructions, which had to be repaired. Thus there had been phases of ups and downs.
2. *Name*: Due to the passing of many human tribes from east to west and some from west to east, the river got different names as there are in *Russian*, Neman; in *Polish*, Niemen; in *Ukrainian*, Niman; in *Latvian*, Nemuna; in *German*, Memel, and in *Lithuanian*, Nemunas.
3. *Length*: In total length this river measures 914 km (=568 miles). Its origin goes back to the last glacier period, but has never reached higher depths than 5 m at the maximum. Thus, only small river boats can use it. It flows rather slow—mainly not exceeding 1–2 m/s. Nevertheless, under certain circumstances also considerable floods have been reported in the last century.
4. *Climate*: The temperatures along the river are moderate due to its rather close situation at the sea. However, cold east winds may bring also continental winter temperatures.
5. *Early discoveries and history*: Also, in the Neolithic times (seventh to third century BC), people had lived there as was shown by relevant findings of animal bones, the animals of which apparently had been obtained by hunting and/or fishing. To protect Europe from attacks of eastern tribes, the German monk knights (=Deutsch Ritterorden) built castles in this region and at the mouth of the Curonian Lagoon (a portion of the Baltic Sea). Later during the growth of the Russian Empire starting with Peter I. (1672–1725) and his successors, this region became again a protection shield against invading troops from the Far East. However, since this region was passed in eastern and western directions by different people with different cultures, the inhabitants had to suffer from many wars, and thus living standards were low compared to other regions. This river is part of the first stanza of Germany's old national anthem (written 1871 on the island Helgoland by *August Heinrich Hoffmann von Fallersleben*) saying that Germany reaches from "the Maas (Meuse) in the West to the Memel (Neman)" in the Eastern region of Europe.

Fig. 1.26 Diagrammatic representation of the former European "east border river" Neman (Memel), which now is fixed in the European Community. Cropped and labelled version of image:Nemunas.png, Public Domain, https://commons.wikimedia.org/w/index.php?curid=836744

6. *Big towns along the river*: From west to east, the following towns had been founded and enlarged along the river during the centuries: *Sovetsk/Tilsit* (Russia), 42,000 inhabitants; *Grodno* (Belarus), 366,000 inhabitants; and *Kaunas* (Lithuania), 304,000 inhabitants (Fig. 1.26).
7. *Literature*

Moroz MD (2017) Aquatic insects of the Neman River and its tributaries. Entomol Rev 97:30–43

Porter RK, Neele SJ (1813) Retreat of the French Army from Moscow to the banks of the Neimen. Longman, Hurst, Rees, Orme & Browne, London

Vareikis V (2008) Ein zählebiger Mythos oder wer hat das Memelgebiet befreit? (A long living myth – who has liberated the Neman region?). Annaberg Annals Yearbook

1.2.27 Niger River

1. *Region*: The source of the Niger river is situated in the Southeastern highlands of West Africa (in Guinea). From there the water flows through Sudan, Mali, Niger, Benin, and Nigeria before entering the Atlantic Ocean via the Gulf of Guinea (in Nigeria). The Niger floods start yearly from September with peaks in November, but finally end at the beginning of May.

2. *Name*: This river is the third longest one in Africa behind the Nile and Congo (=Zaire) river. Due to varying languages of several ethnical groups in this region, this river is also named *Egerew n'gerewen* (=river of rivers) in *Tuareg*, Orimiri (=great water) in *Igbo*, Joliba (=great river) in *Manding*, Isa Beeri (=great river in *Zama*), Kara in *Hausa*, and Oya in *Yoruba*. Big tributaries are the rivers Sokoto, Kaduna, Benue, Anambra, and Bani. The Niger river is eponymous for the modern states Nigeria and Niger.
3. *Length*: This river has a total length of 4180 km draining a basin of 2,111,700 square kilometers. It discharges giant amounts of water into the Atlantic Ocean reaching from 500 m^3/s up to 27,600 m^3/s depending on the season. The course of this rather long river is rather unique compared to other long rivers, since it takes a long curve in the country although its source is located only 240 km from the Western coast of Africa, where it has its mouth after this long curved country passage. The river's delta covers 25,000 km^2 when reaching the Gulf of Guinea.
4. *Climate*: Being situated between the latitudes 12° and 23° North, it is not astonishing that the climate is a tropical one, so that this river induces yearly constantly considerable floods beginning in September with peaks in November and ending in April/May. This rather long humid period is the reason for the good developmental conditions for bloodsucking insects and the thereby transmitted agents of diseases such as those of malaria, dengue fever, leishmaniasis, Crimean-Congo fever, Rift Valley fever, chikungunya fever, yellow fever, encephalitis, African trypanosomiasis (now reducing), plague, etc. Also, human cases of schistosomiasis and entamoebiasis are common, since the river and its tributaries flood constantly the villages and thus transport infectious stages (=cercariae, cysts, worm eggs, etc.) close to the surroundings or even directly into human dwellings. This tropical climate supports the spreading of numerous agents of diseases, which must be constantly controlled. See Chap. 12.
5. *Early discoveries and history*: The Niger river was presumably known already in the times of the Romans (since it was described by Plinius, AD 23–79). However, it was apparently taken for a part of the Nile river. All this knowledge was merged in the Medieval age, until the so-called African Association was founded in 1788 in London in order to promote exploration of Africa. Thus the Scottish explorer *Mungo Park* discovered the central region of the Niger river and published details in 1799, but it took centuries until deeper insights were obtained (driven by financial expectations; see De Gramonte 1991).
6. *Important towns along or close to the river*: *Bamako* (capital of *Mali*), 2 million inhabitants; *Timbuktu* (founded in the twelfth century), 55,000 inhabitants; *Niamey* (capital of *Niger*), 1.3 million inhabitants; and *Lokaja*, *Onitsha*.
7. *Example of private actions to ameliorate the health situation of the river population*: Konna—Konna—a help for a self-help project at the Niger River in Mali: see Chap. 12 (Fig. 1.27).
8. *Literature*

De Gramonte S (1991) The strong brown god: story of Niger river. Houghton Mifflin, Boston, MA

Fig. 1.27 Map showing the course of the Niger river—a region, which was in former times after World War I (~1918) dominated by France. By Hel-hama—Own work, CC BY-SA 3.0, https://commons.wikimedia.org/w/index.php?curid=27632291

Meek CK (1960) The Niger and the classics: history of a name. J Afr Hist 1:1–17
Mehlhorn H (2016a) Animal parasites, 8th edn. Springer, Heidelberg
Mehlhorn H (2016b) Encyclopedia of parasitology, 3 vols., 4th edn. Springer, New York
Mehlhorn H (2016c) Human parasites, 8th edn. Springer, Heidelberg
Welcomme RL (1986) The Niger river system. In: Davies BR, Walker KF (eds) The ecology of river systems. Springer, New York

1.2.28 Nile River (Al-Nil, Bahr-en-Nil)

1. *Region*: The Nile river is an African river flowing from South to North, thus draining a region of 1,293,000 square miles (3,349,000 km^2). It passes the lakes Victoria and Albert, and finally enters the Mediterranean Sea forming a large delta close to the famous historic town of Alexandria. For a long time, the sources of the Nile had been unknown, although it passed Egypt, which was one of the highest cultured countries in early times. On the search for its sources, the famous English discoverer and missionary *David Livingstone* (1813–1873), who already detected the spring rivers of the Congo river, died during his expedition to the sources of the Nile on May 1, 1873, at the lake Bangweol in the region of the recent state of Zambia.
2. *Name*: The Nile river of the present times has got its recent international name from the Greek term *Neilos* (Latin: *Nilus*), which is probably derived from the Semitic term *Nahael* = river valley. The ancient Egyptians called this river

Ar/*Aur*, while the Coptic population later named it *Laro*, which means "black" referring to the black sediments that are transported by the river Nile during flood times. In Homer's *Odyssey* poem (seventh century BC), this river was called *Aigyptos* like the country, through it flows on its way to the Mediterranean Sea. Present Arabic names are Al-Nil, Al-Bahr, or Bahr Al-Nil.

3. *Length*: The final Nile river of the present days gets its water from three principal sources: the *Blue Nile* (Arabic, *Al-Bahr Al Azrag*; Amharic, *Abay*) and the *Atbara* (Nahr Atbara), both flowing from the highlands of Ethiopia, and the *White Nile* (Arabic, Al-Bahr Al Abyad). It has a long history since it connects different populations and cultures of the recent countries Tanzania, Burundi, Rwanda, the Democratic Republic of the Congo, Kenya, South Sudan, Ethiopia, Sudan, and finally Egypt. Some people believe to have indicators or even proofs that the precursor of the Nile river of the present times existed already 30 million years ago; however, it was much shorter at the beginning and had its sources in African regions between 18° and 20° North.

   Today the Nile river is considered as one of the both longest rivers in the world besides the Amazon river in South America. Both reach about 6,900 km (=4287 miles) in length depending on the counting of their spring sources. One of the important Nile river sources is the *Kagera river*. The Nile river passes the Lake Victoria and Lake Albert as *White Nile* before joining the *Blue Nile* in Khartoum (capital of Sudan), which comes from the Ethiopian Plateau. The waters of both rivers form the final (=united) Nile river, which stretches over a length of 830 miles (=1336 km) from Khartoum (capital of Sudan) to Lake Nasser. This second part of the Nile river includes Lake Nasser, which contains the water that in present times is held back by the Aswan High Dam, which in its present size was built by help of Russian engineers during the presidency of the famous officer Nasser, who reached liberty from England power in the years 1960–1971. This Lake Nasser (also called lake) is the second largest man-made lake in the world covering about 2600 square miles including more than 300 miles of the precedent Nile river course. From there the distance to Cairo is about 600 miles (=965 km). North of Cairo the Nile river flows into the Mediterranean Sea via a large delta, which has a length of about 100 miles from South to North and a width of 155 miles (thus covering 24,000 km^2) between Port Said and Alexandria. There exist several branches of the Nile river which guarantee a constant water supply. Thus, the farmers are able to get several harvests per year, which guarantee sufficient food for the constantly growing Egyptian population, which has reached about 97 million people in recent times.

4. *Climate*: In present days the Nile river is running through different climate zones. While the Nile basins in Sudan and Egypt are almost completely rainless in Northern winter months, the Southern parts and the Ethiopian regions may suffer from heavy rainfalls reaching 1.5 m at the top during Northern summertimes. On the other hand, the Northeastern heights of additional trade winds occurring between October and May often induce severe aridity. Tropical climate temperatures and intense rainfalls occur in the East African lake regions and in

Southwestern Ethiopia. Concluding it can be stated that there are many various climate zones along this long river. Due to this length of the Nile river and due to its passing through several climate zones, several phenomena had not been understood for long, e.g., the rising of floods in summer at different times due to heavy tropical rains in Ethiopia, while in November and December, the level of the river in Cairo decreases rapidly and reaches lowest levels in March to May. Both arms (white and blue) of the Nile river contribute different amounts of the final water level. The *White Nile* river brings a constant supply of water throughout the year (e.g., in April–May, about 80% of the water is brought by the White Nile river). The *Blue Nile* river contributes considerably to the typical Nile floods, which start in June and reach a maximum in Khartoum in September, where water levels, which rise to more than 20 feet (~6 m), are often noted.

5. *Early discoveries and history*: Egypt and its culture belong to the earliest groups of known human high cultures worldwide being followed later by the Greek and the Romans, which, however, also dropped down after several centuries of high standards. In the eighteenth and nineteenth centuries AC, the rediscovery of details of the Egyptian high culture was started by European discoverers (English, German, French) and opened unbelievable insights into the long-forgotten culture. Today the religious groups in Egypt consist mainly of Sunnis besides few Shiites and Christian Copts (10%).
6. *Big towns along the river*: *Khartoum* (Sudan), 2 million inhabitants; *Aswan* (Egypt), 300,000 inhabitants; *Cairo* (Egypt), 9.5 million inhabitants; and *Alexandria* (Egypt), 4.55 million inhabitants (Fig. 1.28).
7. *Literature*

Abdel-Ghaffar F, Morsy K, El-Ganainy S, Ahmed M, Gamal S, Bashtar AR, Al Quraishy S, Mehlhorn H (2015) Twelve myxosporean species of the family Myxobolidae infecting freshwater fishes of the River Nile, Egypt, with the description of four novel species. Parasitol Res 114:2985–2998

Abdel-Ghaffar F, Morsy K, Bashtar AR, El-Ganainy S, Gamal S (2013) *Thelohanellus niloticus* sp. nov. (Myxozoa: Myxosporea), a parasite of the Nile carp *Labeo niloticus* from the River Nile, Egypt. Parasitol Res 112:379–383

Amer S, Zidan S, Adamu H, Ye J, Roellig D, Xiao L, Feng Y (2013) Prevalence and characterization of *Cryptosporidium* spp. in dairy cattle in Nile River delta provinces, Egypt. Exp Parasitol 135:518–523

Elmadhoun WM, Msmar AH, Elnoby OA, Noor SK, Suliman AA, Bushara SO (2013) Situation analysis of schistosomiasis and soil-transmitted helminthes in River Nile State, Sudan. Trans R Soc Trop Med Hyg 107:195–199

Grimal N (1992) A history of ancient Egypt. Blackwell Books, London

Mehlhorn H (2016a) Animal parasites, 8th edn. Springer, Heidelberg

Mehlhorn H (2016b) Encyclopedia of parasitology, 3 vols., 4th edn. Springer, New York

Mehlhorn H (2016c) Human parasites, 8th edn. Springer, Heidelberg

Fig. 1.28 Showing the river Nile—its borders had been home of one of the highest educated human societies. By Hel-hama—Own work, CC BY-SA 3.0, https://commons.wikimedia.org/w/index.php?curid=27624659

Redford DB (ed) (2001) The Oxford encyclopedia of ancient Egypt. Oxford University Press, Oxford

Wilkinson RH (2010) The rise and fall of ancient Egypt: the history of a civilization from 3000BC to Cleopatra. Bloomsbury, London

Wilkinson RH (2003) The complete gods and goddesses of ancient Egypt. Thames and Hudson, London

Zarroug IM, Elaagip AH, Abuelmaali SA, Mohamed HA, ElMubarak WA, Hashim K, Deran TC, Aziz N, Higazi TB (2014) The impact of Merowe Dam on *Simulium hamedense* vector of onchocerciasis in Abu Hamed focus - Northern Sudan. Parasit Vectors 7:168

1.2.29 Oder River

1. *Region*: This European river rises in the Czech Republic close to the town of *Olomouc* (formerly *Olmütz*), flows then for 725 km (=451 miles) through Western Poland, and forms later for 187 km the border between Poland and Germany. Finally, it flows divided into three branches (Dziwna, Swina, Peene) into the Bay of Pomerania belonging to the Baltic Sea. Another river named *Oder* also exists in the center of Germany. Its spring is situated at the mountain Brocken in the Harz forest reaching a length of only 60 km.
2. *Name*: During history this river changed often its name, *Oder* (German), *Odra* (Polish), and *Wodra* or *Odra* (Sorbian). Around 1534 it had been called in Latin *Viadrus*.
3. *Length*: The Oder has a total length of 840 km (=451 miles), and thus it is the third longest river in Poland behind the *Wisla* = *Vistula* (in German: Weichsel) with 1090 km and *Warta* (=Warth) with 762 km.
4. *Climate*: Continental (warm summers, cold winters).
5. *Early discoveries and history*: This river was already known to the Romans as *Kadrus* or *Vidua*, since at these times there was already intense trade activity dealing with *ambergris* (=intestinal excretions of whales being used as ingredient for perfumes). Before Slavic tribes settled along the river, the Oder and the neighboring region had been under German influence, which lasted for long (e.g., under the Prussian King *Frederick the Great* (1712–1786), who ordered to build a straight channel in the Oderbruch region close to the town *Küstrin* in the years 1746–1753). Later a drainage system was constructed in the town of *Breslau*, and in 1887–1891 the *Oder-Spree-Canal* was built. After World War I (1914–1918=, lost by the Germans and Austrians, the *Treaty of Versailles* ruled out that the Oder region became subject to an International Commission. However, in the midtime of *World War II* (1939–1945), the allies (France, England, the USA) decided at the Tehran Conference that the rivers Oder and Neisse should be the new Eastern border of Germany to Poland. This was finally confirmed at the Potsdam conference in the year 1945 after *World War II*. This led to the fact that a big part of the German population fled from there at the end of World War II. In the year 1990, the reunited Germany and the Republic of Poland

Fig. 1.29 Basin of the river Oder, which had been border between Central Europe and Russian influence for centuries. It runs now in the East region of the European Community. By NordNordWest—Own work, CC BY-SA 3.0, https://commons.wikimedia.org/w/index.php?curid=3588973

signed a treaty recognizing the river Oder as border. However, both countries are now members of the European Union, where inner borders exist no longer.

6. *Big towns along the river*: *Ostrava* (Ostrau), 290,000 inhabitants; *Opole* (Oppeln), 188,000 inhabitants; *Wroclaw* (Breslau), 640,000 inhabitants; *Frankfurt/Oder*, 58,000 inhabitants; and *Szczecin* (Stettin), 405,000 inhabitants (Fig. 1.29).
7. *Literature*

Clarke R (2006) The origin of ambergris. Lat Am J Aquat Mamm 5:7–21

Pastuszak M, Sitek S, Grelowski A (1998a) The exceptional Oder flood in summer 1997. Nutrient concentrations in the Swina Strait during the years

1996 and 1997 - with emphasis on the flood event. Dtsch Hydrog Z 50:183–202

Pastuszak M, Siegel H, Sitek S, Gerth M, Tschersich G, Nagel K, Grelowski A (1998b) Impact of water temperature on nutrient concentrations in the Oder estuary in 1996–1998. Dtsch Hydrog Z 51:423–439

Schlögel K, Halicka B (eds) (2007) Oder – Odra. Blicke auf einen europäischen Strom (Oder – Odra. Views of an European stream). Peter Lang, Frankfurt/Main

Witt G, Gründler P (2005) The consequences of the Oder flood in 1997 on the distribution of polycyclic aromatic hydrocarbons in the Oder River. Clean Soil Air Water 33:301–314

1.2.30 Okavango River

1. *Region*: This river passes through several regions in southern Africa. It has its source in Angola and flows at first southeastward and further in the south, where it represents a part of the border between the two countries Angola and Namibia. Afterward it enters Botswana and drains finally into the *Moremi Game Resort*. Thus, it is a river that does not enter the sea but finishes in a swamp in the Kalahari desert being either named *Okavango delta* or *Okavango alluvial fan* (similar to the *Jordan river* that ends in a desert region). The basin covers about 530,000 square kilometers. In the rainy season, however, an outflow is often established to the *Boteti river*. In former colder periods of the earth, the Kalahari desert was a giant lake (Makgadikgadi) being filled by water of the Okavango.
2. *Name*: In former times the present name was spelled either *Okovango* or *Okovanggo*. The Portuguese name is *Rio Cubango*.
3. *Length*: The *Okavango river* has in some reports a length of 1600 km (~990 miles) and in others 1700 km (~1100 miles).
4. *Climate*: The region of the Okavango is tropical warm with rainfall in January in Angola. This water, however, needs several months to reach Botswana between June and August.
5. *Early discoveries and history*: Archaeological records show that northern Botswana has been occupied by different human populations at least for the past 100,000 years—most probably for much longer.

   The Delta's first inhabitants apparently had been hunter-gatherers as it can be concluded from findings at historic fireplaces. Perhaps these people had been the ancestors of some of today's *Bushmen, San*, or *Basarwa* tribes. With the introduction of livestock and crop farming by early *Bantu people*, permanent settlements had been established long before present times.

   The *Tawana people*, for example, originally entered Ngamiland on hunting forays. They established their first settlements around 1800 AD. Over the next 100 years, they moved their capital not less than seven times before their last move to Maun in the year 1915. Livestock diseases apparently have played an important role in the population dynamics of Ngamiland. For example, in 1896 the cattle plague epidemic due to a paramyxovirus decimated the cattle population and sleeping sickness (due to the protozoans of the genus *Trypanosoma*) in the 1940s and 1950s and led to the abandonment of several villages in the Delta. Historically conflicts have also played a role in the region's population dynamics with raids by the Matabele people in the second half of the 1800s forcing the residents of many settlements to flee. The German-Herero conflicts from 1904 to 1906 lead to the result that many *Herero* people escaped to Ngamiland. More recently in 1969 and 1970, the *Hambukushu* people have been relocated to the Etsha settlements of the panhandle to evade conflicts in Angola.

6. *Big towns along the river*: There exist no big towns along the river, which passes only along small villages with few inhabitants. However, the *Okavango delta* region is today very attractive to tourists for safaris studying game animals (Fig. 1.30).
7. *Literature*

Basedau M (2005) Resource politics in sub-Saharan Africa. GIGA, Hamburg
Mehlhorn H (2016a) Animal parasites, 8th edn. Springer, New York
Mehlhorn H (ed) (2016b) Encyclopedia of parasitology, 3 vols., 4th edn. Springer, New York
Mehlhorn H (2016c) Human parasites, 8th edn. Springer, New York
Milzow C (2009) Regional review: the hydrology of the Okavango Delta, Botswana – processes, data and modelling. Hydrogeol J 17:1297–1328
Varis O, Tortajada C, Biswas AK (2008) Management of transboundary rivers and lakes. Springer, New York

1.2.31 Oranje River

1. *Region*: This river flows from the East to the West of Africa. It is behind the Sambesi, the second longest river in the Southern region of Africa.
2. *Name*: The river name *Oranje* has its origin in a Dutch term for orange (given by the settlers coming from the Netherlands), which apparently has its origin due to the fact that its water transports massive amounts of reddish/yellowish sand. The name Oranje is mainly used in the state of Namibia, while it is named *Senqu* in Lesotho and *Gariep* in South Africa.
3. *Length*: This river has two rivers (*Vaal* and *Upper Oranje*) as main sources. In total it reaches a length of 2160 km (=1367 miles). After leaving the *Lesotho* regions, it takes its course westward, where it forms the border between *Namibia* and *South Africa* for 500 km. Finally, this river enters close to a town, which has the Dutch name *Oranjemund* (engl.: mouth of the Oranje river) the Southern region of the Atlantic Ocean. This river transports big amounts of sand from inner regions to the mouth of the river as does a south-north seawater streaming (=*Benguela stream*) due to constant southwest winds. These two facts enable the formation of giant sand dunes occurring along the landscape of the Namib desert.

Fig. 1.30 Portion of the map of Central Africa with the Okavango river in the center. By the United Nations—http://www.un.org/Depts/Cartographic/map/profile/okavango.pdf, Public Domain, https://commons.wikimedia.org/w/index.php?curid=3892084

4. *Climate*: Since the river has its source at a height of 3150 m above sea level, it passes during its 2160-km-long down-run several climate zones from rather cold to tropical desert heat.
5. *Early discoveries and history*: The first European farmers started their settling during the 1690s in this region after leaving their European countries (the Netherlands, Germany, Switzerland) as consequences of religious wars in their homelands (Catholicism versus Calvinism). The river was named *Oranje River* in the year 1779 by Colonel Robert Gordon, who was commander of the Dutch East India Company garrison at Cape Town, honoring the Dutch Ruler House.
6. *Big towns along the river*: Since the climate of this region is not very suitable neither for farming nor for industrial activities, the number and size of towns

Fig. 1.31 Map of the basin of the Oranje river in Africa. By Keenan Pepper—Own work, CC BY-SA 4.0, https://commons.wikimedia.org/w/index.php?curid=66101365

along the river are rather small (e.g., *Upington* with 74,000 inhabitants is one of the biggest) (Fig. 1.31).

7. *Literature*

Cambray JA, Davies BR, Ashton PJ, Agnew JD, De Moor FC, Skelton PH (1986) The Orange-Vaal river system. Springer, Dordrecht

Kruchen T (2012) Orange-Sengu: artery of life. Brandes and Apel, Amsterdam

Mehlhorn H (2016) Human parasites, 8th edn. Springer, Heidelberg

Ramollo PP (2011) Freshwater fish abundance and distribution in the Orange River, South Africa. J Fish Int 6:13–14

Swanevelder CJ (1981) Utilising South Africa's largest river: the physiographic background to the Orange River scheme. GeoJournal 2(S2):29–40

1.2.32 Orinoco River

1. *Region*: The river was first discovered by *Christopher Columbus* and his crew on August 1, 1498, on his third voyage to the "Americas." The source is located in the region of the Cerro-Delgado-Chalbaud mountains (being discovered in the year 1951). Another = second source is located close to the border between Venezuela and Brazil being lately discovered just in the year 1951 by cooperation

of Venezuelan and French explorers. Its draining basin opens into the Atlantic Ocean via the *Delta Amacuro* in Venezuela, wherein many other rivers release their water.

2. *Name*: The Spanish name of the river *Orinoco* is transferred in most languages, but also the writing *Orinoquia* can be found.
3. *Length*: The length of this river is documented in several publications to reach 2140 km (~1330 miles) (see Table 1.1 at the top of this chapter). Its basin size is ~880,000 square kilometers.
4. *Climate*: The whole region of the river belongs to the tropical zone. Thus, it is not astonishing that this river is the fourth largest in the world with respect to the amounts of water discharged into the sea. In average it discharges ~37,000 cubic meters of water per second with maximum volumes of up to 54,000 cubic meters per second in the rainy season.
5. *Early discoveries and history*: As already shown above, *Christopher Columbus* (1443–1506) discovered this region in 1498. More details of the delta were explored later in the early sixteenth century by German expeditions guided by *Ambrosius Ehinger* (syn. *Dalfinger*) (born before 1500). He died in 1533, but before he was the first European ruler in Venezuela and worked in the name of the famous German business family *Welser* (who had their center in the town of Augsburg, Germany). Somewhat later Spanish expeditions followed, when people like *Ambrosius de Berrio* sailed down the river in hope to find the "wonderland of gold," which was also searched by the English (queen lover) Sir *Walter Raleigh* (1552–1618), but never found. Starting in the year 1800, the famous German scientist *Alexander von Humboldt* (1769–1859) and his French companion *Aimé Bonpland* stayed in this region and documented enormous amounts of new plant and animal species. Until today this region of the Orinoco river is one of the regions with the worldwide highest density of animals and thus belongs to the world natural heritage.
6. *Big towns along the river*: *Puerto Ayacucho*, 105,000 inhabitants; *Ciudad Bolivar*, 407,000 inhabitants; and *Ciudad Guyana*, 878,000 inhabitants (Fig. 1.32).
7. *Literature*

Helferich G (2004) Humboldt's cosmos: Alexander von Humboldt and the Latin American journey that changed the way we see the world. Gotham Books, New York

Humboldt A von (1853) Personal narrative of travels to the equinoctial regions of America during the years 1799–1804. Henry G. Bohn, London

Humboldt A von (1807) Ansichten der Natur (Aspects of nature). Verlag der Cottaschen Buchhandlung, Stuttgart

Mehlhorn H (ed) (2016) Encyclopedia of parasitology, 3 vols., 4th edn. Springer, New York

Fig. 1.32 Map showing the basin of the Orinoco river in northern South America. By Guiana_shield_map-fr.svg: Sémhur (talk)Cuenca_del_Orinoco.png: Fev (talk)derivative work: Shadowxfox—This file was derived from:Guiana shield map-fr.svgCuenca del Orinoco.png, CC BY-SA 4.0, https://commons.wikimedia.org/w/index.php?curid=43496066

1.2.33 Paraguay River

1. *Region*: This river belongs to a river system at the East coast of South America discharging water from regions of South Brazil, Bolivia, Paraguay, and Argentina via the river *Parana* into the *Rio de la Plata*. Its source is situated in the Brazilian state *Mato Grosso* (see map of *Parana river* Fig. 1.33).
2. *Name*: The English name of this river derives from *Guarani*, an indigenous language in South America. In Spanish it is called *Rio Paraguay* and in Guarani it is named *Yspry Paraguai*. This name was also used for the present official name of the country: *Republica del Paraguay*. This region has already been discovered by European sailors in the years 1524–1525 and became a Spanish territory in the year 1537, which was added to the Spanish viceroyalty *Peru* in the year 1542 and later to the viceroyalty *Buenos Aires* respectively La Plata. Finally, in the year 1811, it became independent. After several wars against Argentina, Brazil, and Uruguay, Paraguay lost considerable amounts of its territory, so that it harbors today only 6.95 million inhabitants.

3. *Length*: The *Paraguay river* is the largest tributary river of the *Parana river* with a length of 2621 km (1629 miles). Its source is situated in the Brazilian state of Mato Grosso. From there it runs in southwestern direction, passes the Brazilian city *Caceres*, then flows southward, and passes the giant Pantanal wetlands and along the city of *Corumba*. Further on, it runs along the border between the states of Brazil and Bolivia crossing the Brazilian states of *Mato Grosso* and *Mato Grosso del Sul* for a short distance. Having reached *Puerto Bahia Negra*, the river represents the border between Paraguay and Brazil. From there it flows southward and joins the Apa river. Afterward, the Paraguay river runs Southeast and Southwest in curves, thereby dividing Paraguay practically into two halves. While the Western part is semiarid and thus largely inhabited, the Eastern part offers space for 96–98% of the entire population. After flowing down for further 400 km and after confluencing with the *Pilcomayo River*, it passes the town *Asuncion*, which is the capital of Paraguay (525,000 inhabitants). From there the river takes another 271 km (~170 miles) before joining the *Parana river*, which transports all water into the Rio de la Plata.
4. *Climate*: The river flows through a broad spectrum of biological habitats reaching from tropical (Pantanal wetlands) to dry, semiarid regions (Gran Chaco).
5. *Early discoveries and history*: The river was already discovered in the very early times of the colonization of the South American continent.
6. *Big towns along the river*: *Cáceres* (Brazil), 88,000 inhabitants; *Corumba* (Brazil), 92,000 inhabitants; *Puerto Bahia Negra* (Paraguay), 3900 inhabitants; and *Asuncion* (Paraguay, capital), 525,000 inhabitants.
7. *Literature*

Fraser LH, Keddy PA (2005) The world's largest wetlands. Ecoscience 13:559–560

Gleick PH (ed) (1993) Water in crisis: a guide to the world's freshwater resources. Oxford University Press, Oxford

Ioris AAR, Irigaray CT, Girard P (2014) Institutional responses to climate change: opportunities and barriers for adaptation in the Pantanal and the upper Paraguay River Basin. Clim Chang 127:139–151

Krepper CM, Garcia NO, Jones PD (2006) Paraguay river basin response to seasonal rainfall. Int J Clim R Meteorol Soc J Clim Sci 26:1267–1278

UN (1978) United Nations register of international rivers. Pergamon Press, New York

Varis O, Tortajada C, Biswas AK (2008) Management of transboundary rivers and lakes. Springer, New York

1.2.34 Parana River

1. *Region*: This South American river, which has its origin, when the tributaries *Paranaiba* and the *South American Rio Grande* fuse in southern Brazil. From

there it takes its course at first in southwestern direction for about 620 km (~385 miles), and then it enters the country Paraguay. Then for about 200 km, the Parana forms a natural border between the countries *Paraguay* and *Brazil* until its confluence with the Iguazu river. Shortly behind this fusion site, the Itaipu dam stops the river's flood. By the way this dam is the second largest in the world just behind the so-called Three Gorges Dam in the Republic of China. As soon as the Parana has fused with the *Iguazu river*, the Parana represents the border between the two countries Paraguay and Argentina. After flowing in principle southward, it fuses with the *Paraguay river*, which is considered as tributary of the *Parana*. After this confluence the water of both rivers flows further southward for 820 km through the country Argentina, turns somewhat back to the east, and joins finally close to the town Rosario the *Uruguay river*. The water of these three big rivers (and several smaller ones) finally enters the *Rio de la Plata*, which is considered by some scientists as own river (entering the Atlantic Ocean), but by others just as a giant bay (see Fig. 1.33).

2. *Name*: The *Parana river* is known under different local names, *Spanish* and *Portuguese*, Rio Parana, and *Local Guarani*, Ysyry Parana. The term Parana has its origin in the language of the local tribe *Tupi* and means "sized as the sea."
3. *Length*: This river with its two sources (Paranaiba river and Rio Grande) and two very big tributaries (Paraguay and Uruguay rivers) has a length of about 4880 km (~3030 miles). Thus, it is behind the Amazon river, the second longest river in South America.
4. *Climate*: Due to the enormous length, this river and its tributaries flow through moderate, subtropical, and tropical zones.
5. *Early discoveries and history*: The first European to travel the "Father of the Waters" (from the Guaraní language) was Sebastian Cabot in 1526. In 1525 he assumed charge of a three-ship Spanish expedition intended to develop trade with the Orient. However, he diverted the expedition to the South American continent due to reports of fabulous wealth in the Río de la Plata region. After 3 years of fruitless exploration, he returned to Spain where he was judged responsible for the failure of the expedition and was banished to Africa. The Paraná soon after became the principle explorers' route into the interior of the South American continent.
6. *Big towns along the river*: *Buenos Aires*, (Argentina), 2.89 million inhabitants; *Rosario*, (Argentina), 948,000 inhabitants; and *Corrientes*, (Argentina) 322,000 inhabitants.
7. *Literature*

Alonso X (2018) Macrophytes as potential biomonitors in peri-urban wetlands of the middle Parana River (Argentina). Environ Sci Pollut Res Int 25:312–323

Barros IT (2017) Environmental risk assessment in five rivers of Parana River basin, Southern Brazil, through biomarkers in *Astyanax* spp. Environ Sci Pollut Res Int 24:16228–16240

Iriondo MM, Paggi JC, Parma MJ (2007) The middle Parana River: limnology of a subtropical wetland. Springer, New York

Fig. 1.33 Map showing the common giant basin of the rivers Parana, Uruguay, and Paraguay and the Rio de la Plata delta and several other tributaries. By Kmusser—Own work, Elevation data from SRTM, drainage basin from GTOPO [1], all other features from Vector Map, CC BY-SA 3.0, https://commons.wikimedia.org/w/index.php?curid=11713483

Rèvy JJ (2018) Hydraulics of great rivers: the Paraná, the Uruguay and the La Plata estuary. Forgotten Books, London

1.2.35 *Pearl River (Zhu Jian, Canton River)*

1. *Region*: Vietnam, Southern China.
2. *Name*: In China the *Pearl River* is also known as *Zhu Jian*, *Pearl river*, or *Canton river*. The name Canton corresponds to the former name (until World War II) of this big town (13 million inhabitants), which today is called *Guangzhou* being situated close to the river's delta, where three rivers (Xi = West, Dong = East,

Bei = North) enter their water into the South China Sea. The old name Pearl River was apparently given due to the fact that lots of shiny pearls had been found in former times in larger numbers at the river's shore in Canton = Guangzhou.

3. *Length*: The *Pearl River* has a length of about 1500 miles (=2400 km) and is China's third longest river after *Yangtze river* and *Yellow river*. It drains the Guangdong and Guangxi provinces; parts of the regions of Yunnan, Guizhou, Hunan, and Jiangxi in China; and also parts of Vietnam's Northeast Cao Bang and the Lang Song region.
4. *Climate*: The river delta, which is now lined by an urban agglomeration of cities with millions of inhabitants, has a humid subtropical climate. However, only during the monsoon months between April and September heavy rainfalls occur.
5. *Early discoveries and history*: The Pearl River and the town (region) of Guangzhou have a long-lasting history and had and have great importance for the old and present China. The first city at this place was built in the year 214 BC at the gorge of the Pearl River being named *Panyu* and became the capital of the so-called Nanyue kingdom in the year 206 AD. After growing it became the capital of the Guang prefecture in the year 226 AD and the official name became Guangzhou.

   Due to its position at the China South Sea, this region and town has often been visited by traders but was also often attacked by pirates of various nations originating from regions in India, Arabia, and Persia. The Portuguese were the first Europeans that arrived in the early sixteenth century and named the river *Rio das Perolas*. They gained after many fights with European nations the trade monopoly in the year 1511, which was broken by English forces in the seventeenth century. After 1760 all foreign trade in China was restricted to Guangzhou, which in the 1770s became the exclusive place for *British guided opium imports*. The Qing government, however, banned the opium trade and destroyed thousands of tons of the drug in Guangzhou due to huge social problems as consequence of the opium abuse. In revenge the English sent a fleet and thus started the first of three *opium wars*. The thus derived struggles lasted until 1842, when the English dictated the *Treaty of Nanjing* forced the Chinese government to cede Hong Kong to the British (which lasted until 1997, when Hong Kong reached a special status as new part of China). Then Guangzhou became one of the five treaty ports for unrestricted trade (with all problems). During this period Guangzhou, however, became famous for its strict resistance and rebellion to the British troops. During the years 1917–1919, Guangzhou was seat of the Kuomintang Government being based on the National Republican Party of China after having overthrown the Manchu Dynasty. The program of this *Kuomintang Government* was developed by the physician *Sun Yat-sen* establishing the three basics *nationalism, democracy*, and *socialism*. One of the recent, most active universities of Guangzhou is named Sun Yat-sen University being founded by Peter Palmer, the American preacher, and his followers, who attracted among others also many German physicians, so that until World War II teaching of students was often held in German.

Fig. 1.34 Diagrammatic representation of the basin of the famous Pearl river. By Kmusser—own work, CC BY-SA 3.0, https://commons.wikimedia.org/w/index.php?curid=5235047

6. *Big towns along the river*: *Guangzhou*, 13 million inhabitants; *Dongguan*, 8.5 million inhabitants; *Shenzhen*, 12.5 million inhabitants; and *Fuzhou* being bordered by Pearl and Yangtze rivers, 4 million inhabitants (Fig. 1.34).
7. *Literature*

Gu X, Zhang Q, Singh VP et al (2017) Hydrological response to large-scale climate variability across the Pearl River basin, China: spatiotemporal patterns and sensitivity. Glob Planet Chang 149:1–13

Lin Q (2017) Correlation between hydrological drought, climatic factors, reservoir operation, and vegetation cover in the Xijiang Basin, South China. J Hydrol 549:512–524

Liu Y, Mehlhorn H, Wu Z (2019) Schistosomiasis control: successful experiences from China: Parasitology Research Monographs, vol 11. Springer, New York, Heidelberg

Huang C, Zhang Q, Singh VP, Gu X, Shi P (2017) Spatio-temporal variation of dryness/wetness across the Pearl River basin, China, and relation to climate indices. Int J Climatol 37:318–332

Mehlhorn H, Wu Z, Ye B (eds) (2014) Treatment of parasitosis in traditional Chinese medicine: Parasitology Research Monographs, vol 6. Springer, New York, Heidelberg

Mehlhorn H (2016a) Encyclopedia of parasitology, 3 vols., 4th edn. Springer, New York

Mehlhorn H (2016b) Human parasites, 8th edn. Springer, Heidelberg

Zhang Q (2017) Response of vegetation to different time-scales drought across China: spatiotemporal patterns, causes and implications. Glob Planet Chang 152:1–11

Zong Y, Huang G, Switzer AD, Yu F, Yim WWS (2009) An evolutionary model for the Holocene formation of the Pearl River delta, China. The Holocene 19:129–142

1.2.36 Po River

1. *Region*: This river flows eastward across Northern Italy after leaving its source in the *Cottian Alps* and joins the Adriatic Sea (a part of the Mediterranean Sea) close to the water town *Venice*. The Po basin receives water from tributaries coming from Italy, France, and Switzerland.
2. *Name*: The name of the *Po river* has its origin in the Latin term *Padus* or in the ancient Greek adaption *Pados*. The local (Ligurian) name is/was *Bodincus* respectively *Bodencus*, which means "bottomless."
3. *Length*: The Po River is the longest river in Italy. However, there are published different data of the length (652 km = 405 miles or 682 km = 422 miles) depending on the addition of the tributary river *Maira*. The maximum width is 503 m at very rare sites. In total the Po drains an area of 74,000 square kilometers by help of a very large number of often very small tributary rivers or even brooks.
4. *Climate*: There is a typical Mediterranean climate, since the river flows in its larger parts parallel to the Alpine mountains, from where many tributaries arrive. Due to the fact that the Alpine mountains are high and covered with snow, the Po discharges in spring time large amounts of water leading often to heavy floods. This led to the fact that many high levees are needed along its banks.
5. *Early discoveries and history*: This region had already a long history, before the Roman Empire was erected and had reached tops of culture at these times. The historical people before the Roman Empire had been the tribes of *Celts* (sixth century BC), *Ligures* (fifth century BC), *Etruscans* (first century BC), *Veneti* (third century BC), *Umbri* (fifth century BC), etc.
6. *Big towns along the river*: *Torino*, 886,000 inhabitants; *Ferrara*, 132,000 inhabitants; *Cremona*, 72,000 inhabitants; *Piacenza*, 102,000 inhabitants; and *Milan* (connected by a channel), 1.35 million inhabitants (Fig. 1.35).
7. *Literature*

Bloch R (1969) The ancient civilization of the Etruscans. Cowles Book, New York

Goffart WA (2006) Barbarian tides: the migration age and the later Roman Empire. University of Pennsylvania Press, Philadelphia

Grove AT, Rackham O (2003) The nature of Mediterranean Europe: an ecologic history. Yale University Press, New Haven, CT

McKinney FK (2007) Pleistocene and Holocene sediments. In: The northern Adriatic ecosystem. Columbia University Press, New York

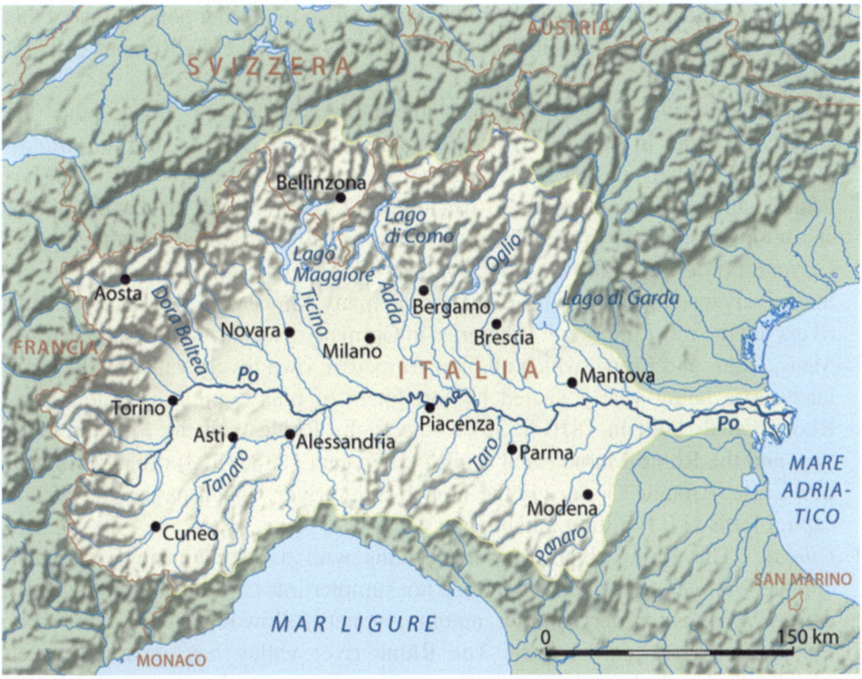

Fig. 1.35 Map of the Po delta in Italy crossing Italy from the west to the east. By NordNordWest—Own work using:GTOPO-30 Elevation Data by USGS, CC BY-SA 3.0, https://commons.wikimedia.org/w/index.php?curid=3806908

Swaddling J, Perkins P (2009) Etruscan by definition: the culture, regional, and personal identity of the Etruscans: papers in honor of Sybille Haynes. British Museum, London

Zwingle E (2002) Italy's Po River punished for centuries by destructive floods, northern Italians stubbornly embrace their nation's longest river, which nurtures rice fields, vineyards, fisheries—and legends. Natl Geogr 201(5):92–115

1.2.37 Rhine River

1. *Region*: This river, which apparently exists since 14–15 million years, has its sources in the Swiss Alps relatively close to the town Chur (e.g., in the lake Toma in a height of 2300 m). From there it flows in Central Europe passing the following countries from South to North, Switzerland, Liechtenstein, and Austria, passes the Lake Constance (539 km^2), then forms the border between France and Germany followed by a long passage through Germany, and finally enters the Netherlands, where it splits up into a number of wide branches called Lek, Waal, Merwede, etc. before finally entering the North Sea (Fig. 1.1).

2. *Name*: The recent name has its origin in terms given by former populations of this region; e.g., Celtics named it *Renos*, and later invaders like the Romans called it *Rhenus*.
3. *Length*: This river is legalized as an international waterway already since the *treaty of Vienna* in the year 1815 (just after Napoleon's defeat in the *Battle of Nations* at Leipzig in 1813). An original length of 820 miles (=1320 km) has been documented, but it shrunk to 765 miles (1230 km) due to the construction of shortening waterways in the delta region. Starting from Rheinfelden at the Swiss-German border, the following 540 miles (=870 km) are navigable and thus are intensively used for shipping of goods of many kinds. The following smaller rivers are the most important tributaries in Germany, Neckar, Main, Lahn, Mosel, Maas, Ruhr, and Lippe, which all have formed an own lovely valley with scenic landscapes being yearly visited by foreigners of many countries. The Rhine-Rhône canal (built in 1810–1833) connects the Rhine with the Mediterranean Sea, and the Rhine-Marne canal (built in the years 1838–1853) offers a connection to the Atlantic Ocean via the French Seine river, which passes the French capital Paris (11.3 million inhabitants).
4. *Climate*: The Rhine river flows in regions with a typical central European climate—moderate winters and not too hot summertimes. In principle four clearly defined seasons (spring, summer, autumn, winter) follow repeatedly each other.
5. *Early discoveries and history*: The Rhine river valley has apparently been populated already for a long time by humans. For example, even skeletal remnants of the human precursor *Homo neanderthalensis* have been detected in 1856 in a grotto close to the recent governmental town Düsseldorf in Western Germany. These human precursors lived—as was later shown—during the early Pleistocene period in wide regions of Europe but also in Asia and Africa. Fishing in rivers was apparently an important way to secure the supply of their food all over the year. But the so-called Neanderthal population disappeared, when the precursors of the modern human population invaded these regions attracted by the all year long acceptable, rather mild climate. In the times around 50 years BC, the Rhine river was used as a border by the Romans to keep away the German tribes from entering the growing Roman empire, which had reached its widest eastern extent at this time. In present times the Rhine river is culturally, economically, and historically one of the most important rivers in the world, although riverine fish have seriously declined due to pollution and straightening of the banks.
6. *Big towns along the river*: *Basel* (Switzerland, pharmaceutical industry), 171,000 inhabitants; *Mannheim* (Germany, industry), 338,000 inhabitants; *Ludwigshafen* (chemical industry), 169,000 inhabitants; *Mainz* (local government), 215,000 inhabitants; *Koblenz* (there the Mosel river enters the Rhine river and thus offers ship connections to France), 114,000 inhabitants; *Cologne* 1.1 million inhabitants; *Leverkusen*, 164,000 inhabitants (government, chemical industry); and *Düsseldorf* (center of banking and industry), 617,000 inhabitants. The river flows south-northward through the Middle-Rhine mountains, thus forming during millions of years a scenic valley lined by large and famous vineyards and numerous medieval fortresses that go back to the eleventh and twelfth century,

during which the noble fighter cast (=so-called knights) had its greatest importance. Today masses of international visitors come and "dive" into the medieval culture and the present vines. This region has in former times always seen large wandering populations coming from the North or South (especially from the Balcony region). Thus, it is not astonishing that the so-called Nibelungenlied (=tale of the Nibelungs) had its origin there around the year 1100 AD. This poem is based on old tales, but was written down by Christian monks for the first time and then was presented underlined by guitar music to the nobles of their times living in their castles along the river Rhine. The contents of the "Nibelungenlied" draw a picture of love, hate, loyalty, revenge, and finally downfall of a whole tribe in early times, when wandering warrior populations like the Huns tried to enter Western Europe (Germany, France, Italy) around 375–451 AD, but were finally defeated in the year 451 AD on the Katalaunic fields (being situated in France close to Châlons-sur-Marne (Fig. 1.36)).

7. *Literature*

Breeze DJ (2011) The frontiers of imperial Rome. Pen & Sword, Barnsley

Cohen KM, Stouthamer E, Berendsen HJA (2002) Fluvial deposits as a record for later quaternary neotectonics activity in the Rhine-Meuse delta, the Netherlands. Neth J Geosci 81:389–405

Kooistra LI, van Dinter M, Dütting MK, van Rijn P, Cavallo C (2013) Could the local population of the lower Rhine delta supply the Roman army? Part 1: the archaeological and historical framework. J Archaeol 4:5–23

Lenders HJR (2017) Fish and fisheries in the lower Rhine 15501950: a historical-ecological perspective. J Environ Manag 202:403–411

Mehlhorn H (2016) Human parasites, 8th edn. Springer, Heidelberg

Peacock DPS (1978) The Rhine and the problem of Gaulish wine in Roman Britain. In: Du Plat Taylor J, Cleere H (eds) Roman shipping and trade: Britain and the Rhine provinces. The Council for British Archaeology, London, pp 49–51

Schmitz MG (ed.) (2006) Sensation im Neandertal. Johann Carl Fuhlrott und die 150 Jahre alten Knochenfunde aus der "Feldhofer Grotte". Reprint der Erstveröffentlichung von 1859. Selbstverlag M.-G. Schmitz, Kelkheim/Taunus (Finding of the bones of the Neandertal people)

Sures B, Knopf K, Würtz J, Hirt J (1999) Richness and diversity of parasite communities in European eels *Anguilla anguilla* of the river Rhine, Germany, with special reference to helminth parasites. Parasitology 119:323–330

Thielen F, Münderle M, Taraschewski H, Sures B (2007) Do eel parasites reflect the local crustacean community? A case study from the Rhine River system. J Helminthol 81:179–189

Wolters R (2006) The Romans in Germany, 5th edn. C.H. Beck, Munich

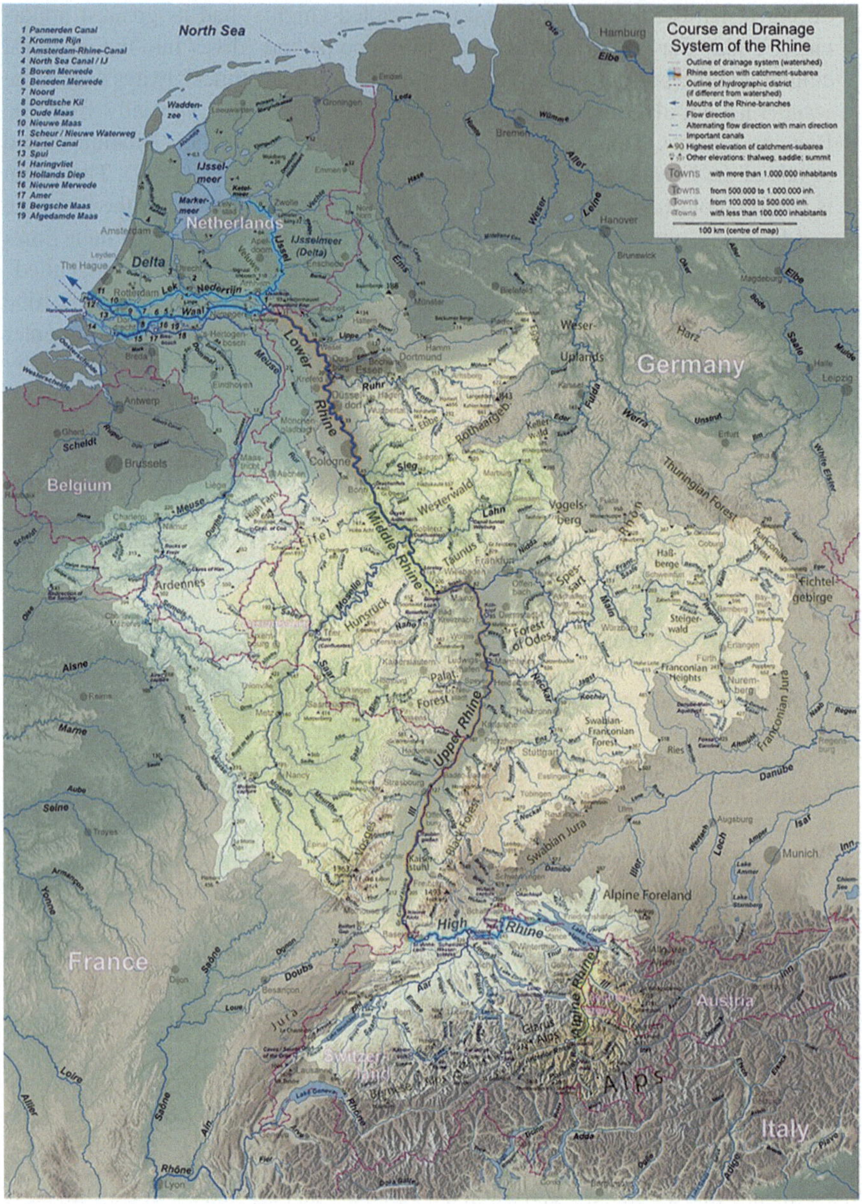

Fig. 1.36 Map of the countries (Switzerland, Germany, the Netherlands) passed by the river Rhine, which enters finally the North Sea. By WWasser—Own work, CC BY-SA 3.0, https://commons.wikimedia.org/w/index.php?curid=24474561

1.2.38 Rhône River

1. *Region*: This French river has its source at a glacier in the Swiss Alps. Before reaching the town Brig, further water is entered by the *Massa river* flowing down from the *Aletsch glacier*. Further on, it passes the *Lake Geneva* and finally leaves Switzerland after a total of 290 km and enters France and passes *Lyon*, where the river *Saône* joins it, thus reaching a considerable size. Southward from Lyon, the Rhône flows through the region between the Alps and the *Central Massif*. Finally, at the town of Arles, the Rhône is divided into two major branches, thus giving rise to the so-called Camargue delta (French: grand et petite Rhône). These branches enter their water into the Mediterranean Sea. During snow melting times, the Rhône has extreme current speeds of up to 10 km (=6 miles) per hour. It discharges at the maximum up to 13,000 m^3/s (mean 1710 m^3/s) into the Mediterranean Sea.
2. *Name*: The French name *Rhône* has its origin in Roman times and is based on the Gaulish name *Rodonas*, which contains a term, which describes "a type of rolling." The old German adaption of the Latin term was *Rotten*, in Italian it is *Rodane*, and at local sites it is named Occitan *Rose*.
3. *Length*: The Rhône has a total length of 812 km (505 miles).
4. *Climate*: In the Alpine mountains, which have a considerable height, the region has a clear continental climate, but along the Mediterranean Sea, smooth and mild temperatures are dominant.
5. *Early discoveries and history*: The lower Rhône region was dominated by the Romans for many centuries, who later expanded their dominance over the entire region of the present France and parts of Northern Spain. To attain this dominance, they had to overcome *Celtic tribes* such as *Seduni, Seobriges, Helvetii*, etc., which lived near the Rhône. Furthermore, the *Carthaginians* under *Hannibal* (247–183 BC) used this region as starting point to enter the Roman territory. However, the Romans succeeded for many centuries, and their cultural impact is still today present in the states of Italy, Spain, Germany, and Great Britain.
6. *Big towns along the river*: *Lyon*, 515,000 inhabitants; *Valence*, 62,000 inhabitants; *Avignon*, 92,000; and *Arles*, 53,000 inhabitants (Fig. 1.37).
7. *Literature*

Bravard JP (2012) Metamorphosing rivers across historical cities of Europe: the Rhône River, Lyon and Vienna, France. Quat Int 279:65–66

Brown AG (2018) Natural vs anthropogenic streams in Europe: history, ecology and implications for restoration, river-rewilding and riverine ecosystem services. Earth Sci Rev 180:185–205

Prichard SB (2011) Confluence: the nature of technology and the remaking of the Rhône. Harvard University Press, Harvard

Sastre V et al (2010) Morphology and recent history of the Rhône River Delta in Lake Geneva (Switzerland). Swiss J Geosci 103:33–42

Vella MA, Ghilardi M, Diouf O, Parisot JC, Hermitte D et al (2013) Géoarchéologie du Rhône dans le secteur du pont Saint-Bénézet (Avignon,

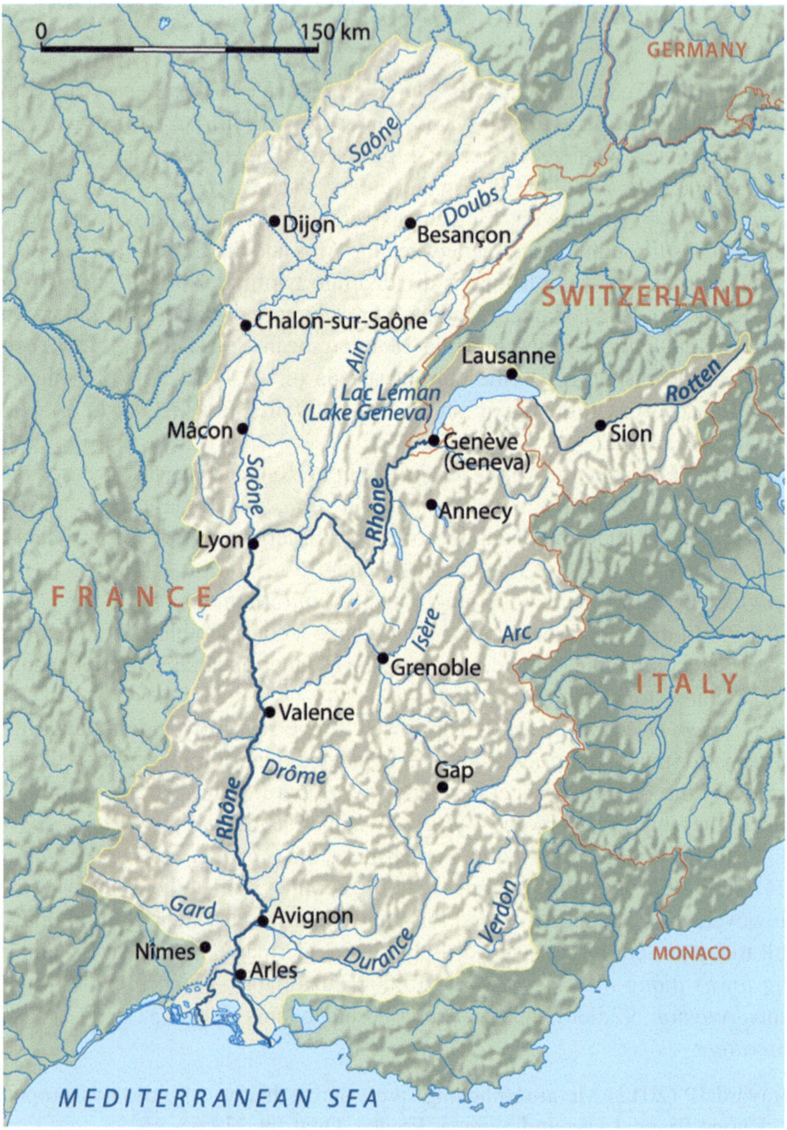

Fig. 1.37 Map showing the basin of the Rhône river in France. By NordNordWest—Own work using:GTOPO-30 Elevation Data by USGS, CC BY-SA 3.0, https://commons.wikimedia.org/w/index.php?curid=3769538

Provence, France) au cours de la seconde moitié du deuxième millénaire apr. J.-C.: Etude croisée de géographie historique et des paléoenvironnements. Géomorphologie: relief, processus, environnement, Groupe français de géomorphologie (GFG) 19:287–309

1.2.39 Rio de la Plata (engl. Plate River)

1. *Region*: This "river" and its tributaries drains about 1.2 million square miles (=3.2 million km^2) in South America (see Fig. 1.33). It enters finally into the Southern Atlantic Ocean. In reality the Rio de la Plata is just the mouth delta of at least three big rivers, since it receives its water from the *Parana river* which has a length of 3560 km, from the *Uruguay river* and from the *Paraguay river* (see Sects. 1.2.33 and 1.2.34). Some geographers consider this opening not as a separate river, but interpret it as a portion of a large ocean bay, into which the water of the above cited rivers enter their water freight. In any way this water system (bay or river) is lined by the countries of Uruguay in the North and by Argentina in the South. Along this "river" (respectively "bay"), the important ports of *Montevideo* (Uruguay), *Buenos Aires*, and *La Plata* (both Argentina) connect the towns of the inner waterfront with the world.
2. *Name*: Its name has its origin in the Spanish term plata for a flat region.
3. *Length*: Independently from the consideration whether the Rio de la Plata is a separate river (as had been considered by the first explorers) or just the "mouth" of a river system, the Rio de la Plata has a length of 180 miles (290 km) and measures 136 miles (219 km) from shore to shore at the opening to the South Atlantic Ocean.
4. *Climate*: The basins of the supplying rivers Uruguay (=Alto Parana, Paraguay) and Parana are characterized by a mainly hot and humid climate in some phases of the year, but the weather periods in April and May are rather arid, while the summer period from October to March is rather rainy. Summer temperatures may be high (up to 40–42 °C), while winters are occasionally rather cold (around 3 °C).
5. *Early discoveries and history*: This region was discovered by *Spanish sailors* under the guidance of *Juan Diaz de Solis* in the sixteenth century (1516) followed by Portuguese adventurers (*Ferdinand de Magellan*) in 1520 as well as by Italians under the commander *Sebastian Cabot* in the years 1526–1529. Remaining sailors of these nations interbred with the different indigenous populations. The present riverine population belongs therefore largely to the type of mestizo or is mixed with people of Spanish and Guarani origin. The Europe-derived inhabitants of both countries (Uruguay, Argentina) live since long mainly along the shores of the Rio de la Plata. Thus, Buenos Aires was founded already in the year 1580. Starting in the year 1610, Jesuit priests established more than 30 missions but were expulsed in the year 1767.
6. *Big towns along the river bay*: *Buenos Aires* (14.73 million inhabitants), *Montevideo* (1.3 million inhabitants), and *La Plata* (830,000 inhabitants).
7. *Literature*

Cabreira AG, Madirolas A, Alvarez Colombo G, Acha EM, Mianzan HW (2006) Acoustic study of the Río de la Plata estuarine front. ICES J Mar Sci 63:1718–1725

Garavaglia JC, Gelman JD (1995) Rural history of the Rio de la Plata, 1600–1850: results of a historiographical renaissance. Lat Am Res Rev 30:75–105

Guerrero RA et al (1997) Physical oceanography of the Río de la Plata Estuary, Argentina. Cont Shelf Res 17:727–742

Hurtado AM (2004) The Guaraní under Spanish rule in the Rio de la Plata. Barbara Ganson. J Anthropol Res 60:147–149

https://www.britannica.com/place/Rio-de-la-Plata

Kucklick C, Pillitz C, Gaede PM (eds) (2003) Das Schweigen am Silberfluss – Südamerika. Abenteuer zwischen Anden und Atlantik. Gruner + Jahr, Hamburg

Trevin JO, Day JC (1990) Risk perception in international river basin management: the Plata Basin example. Nat Resour J 30:87–105

1.2.40 Rio Grande River

1. *Region*: This river has its source at a height of 3700 m in the continental dividing range in the South of the US state of *Colorado*, crosses north-southward the US state of *New Mexico*, and represents for a considerable number of miles the border between the US states of *Texas* and *Mexico*. Leaving the mountains it flows in Texas at first into the east direction and later strictly southward before discharging its water into the *Gulf of Mexico*.
2. *Name*: Besides the American name *Rio Grande*, this river had/has several other names depending on the different ethnic languages of the populations that had/have their homes close to the river—*Spanish*, Rio Bravo del Norte or Rio Bravo; *Navajo Indian*, Tooh Ba'aadii; and *Kòtsoi Indian*, Jicarilla Apache.
3. *Length*: This river has a length of 1431 or up to 2008 km (respectively, 889 or 1248 miles) depending on the adding of the length of tributaries or not. Since the river's bed is rather flat (with a maximum depth of 18 m at few places), ocean ships cannot enter its mouth—only small boats can be used at some sites.
4. *Climate*: This river, which flows mainly from north to south, passes thereby several rather different climate zones. For example, in the north there are rather dry and cold mountain zones, while the southern zones are rather warm and humid.
5. *Early discoveries and history*: The region along this river was settled for many centuries by so-called Indian tribes, before the Americas had been discovered. At first different European nations (England, France, Spain, Portugal) had sent their discoverers, which were later replaced by troops officially protecting the settlers arriving from different countries. These settlers mainly came from Spain and Portugal on one side and from England and Ireland on the other. Later people from many other countries arrived *willingly* from Europe or as *slaves* "imported" by Christian European seafarer nations, which fought each other for many centuries besides depressing and diminishing the original inhabitants, who had

been erroneously called Indians. After the American Civil War (Southern against Northern states) during the years 1861–1865 and after the war between US troops against those of Mexico in the years 1846–1848, Texas was admitted as a state of the USA. By the way Mexico had abolished slavery already in 1828, while it took much longer in the USA. In 1944 the USA and Mexico signed a treaty regarding the river, and according to this contract, several regions along the border became National Parks.

6. *Big towns along the river*: from North to South, the following towns had been settled at the river—*Albuquerque*, 560,000 inhabitants; *Las Cruces*, 102,000 inhabitants; *El Paso*, 683,000 inhabitants; *Ciudad Juarez*, 1.4 million inhabitants; *Laredo*, 257,000 inhabitants; and *Brownsville*, 184,000 inhabitants (Fig. 1.38).
7. *Literature*

Benke AC, Cushing CE (2005) Rivers of North America. Academic, Cambridge, MA

Diaz GT (2015) Border contraband: a history of smuggling across the Rio Grande. University of Texas Press, Austin

Horgan P (1991) Great River: the Rio Grande in North American history, 4th edn. Wesleyan University Press, Hanover

Kelley P (1986) River of lost dreams: navigation on the Rio Grande. University of Nebraska Press, Lincoln

1.2.41 Saint Lawrence River

1. *Region*: This river, which has its source in the North river in Minnesota, is one of the biggest rivers in North America and connects the so-called Great Lakes with the Atlantic Ocean. In principle it flows in the middle latitudes mainly in northeastern direction, thus passing the USA and Canada.
2. *Name*: This river has different names depending on the local tribes and on the naming by European adventurers or international military powers. Indians of the Mohawk tribe called it in their language "big waterway." In French it is called *Fleuve Saint Laurent*, since the discoverers arrived in the estuary on the day of the holy *Saint Lawrence of Rome* (died 258 AD) being named by the early European Catholic society.
3. *Length*: There are different ways of counting the length. If the total length is counted from the farthest head water to the mouth at the Atlantic Ocean, the water passes in total 3058 km (~1900 miles), but if its length measurement starts only at the outflow of the Lake Ontario, the length is just 1197 km (~744 miles).
4. *Climate*: The climate in this region is typical for northern continental regions—warm in summertime, but very cold in winter.
5. *Early discoveries and history*: First reports go back to the eleventh century, when *Norse people* (Viking skippers) explored with their boats the Gulf of the St. Lawrence river. Then it took around 400 years until European explorers such as the English *John Cabot* (~1450–~1500) or the Spanish brothers *Caspar* and *Miguel Corte-Real* (~1450–~1500) sailed into the same mouth of the river.

Fig. 1.38 Map of the Rio Grande and its tributaries. By Kmusser—Own work, Elevation data from SRTM, drainage basin from GTOPO [1], U.S. stream from the National Atlas [2], all other features from Vector Map, CC BY-SA 3.0, https://commons.wikimedia.org/w/index.php?curid=11218868

The first European, who sailed upstream the St. Lawrence river, was the French *Jacques Cartier* in the year 1535. Later followed Basque whalers, but their fleet was destroyed by King Philipp II of Spain. Until the early seventeenth century, the French introduced the name *Rivière du Canada*, which was also used by the explorer team of the French *Samuel de Champlain* (1567–1635). Despite all these different fighting actions between French and English troops, the river was used to explore the North American interior country. At this time the river itself could

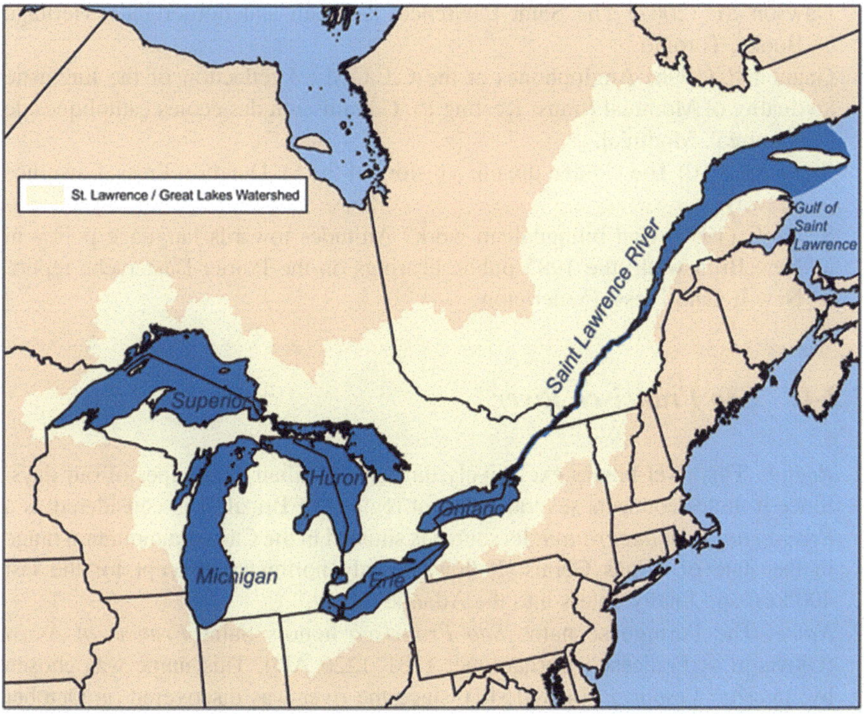

Fig. 1.39 Diagrammatic representation of the region through which the St. Lawrence river takes its course. By No machine-readable author provided. Kmusser assumed (based on copyright claims)—No machine-readable source provided. Own work assumed (based on copyright claims), CC BY-SA 2.5, https://commons.wikimedia.org/w/index.php?curid=276502

only be used by ships until Montreal due to the virtually impassable *Lachine Rapids*. However, in the year 1825, the Lachine Canal was opened, and thus ships could pass the region of the rapids. From this time on, many further canals and locks were built to enlarge the range of navigability, but only starting in the year 1959, the now called *Saint Lawrence Seaway* was ready to be used by ocean-going vessels to reach the Lake Superior. The opening ceremony was held by US president Dwight Eisenhower (1890–1969) and by the English Queen Elizabeth II (born 1926). This shows that the French influence is rather low, although lots of Canadian inhabitants have French roots and about 21.3% of the Canadian population speak French (e.g., in the Quebec region).

6. *Big towns along the river*: *Montreal*, 1.7 million inhabitants; *Trois-Rivieres*, 131,000 inhabitants; and *Quebec*, 532,000 inhabitants (Fig. 1.39).
7. *Literature*

Benke AC, Cushing CE (2005) Rivers of North America. Academic, Cambridge, MA

Dawson SE (2007) The Saint Lawrence: its basin and borderlands. Heritage Books, Toronto

Gagnon R (1996) Anglophones at the C.E.C.M.: a reflection of the linguistic duality of Montréal (trans: Keating P). Commission des écoles catholiques de Montréal, Montréal.

Stagg R (2010) The Golden dream: a history of the St. Dundurn Press, Lawrence Seaway

Steele C (1990) Can bilingualism work? Attitudes towards language policy in New Brunswick: the 1985 public hearings on the Poirier-Bastarache report. New Ireland Press, Fredericton

1.2.42 São Francisco River

1. *Region*: This river passes exclusively through Brazilian landscapes of our days. Since it unites/connects several different regions of Brazil, it is considered as a *River of national integration*. Its source is situated in the Canastra mountain range in the state of Minas Gerais. It flows mainly northward (except for the last 400 km) and finally enters into the Atlantic Ocean.
2. *Name*: The Portuguese name *Sao Francisco* honors Saint *Francis of Assisi* (Giovanni = Francesco Bernardone; 1181–1226 AD). This name was chosen by *Amerigo Vespucci* (1451–1512), since the river was discovered on October 4, 1501, the day of feast. Indigenous people of this region had named this river *Opara* before the European colonization started. Later on the river got the Portuguese name *Velho Chico* meaning *Old Frank*.
3. *Length*: This river has a total length of 2914 km (~1811 miles). It runs south-northward and finally south-eastward. This river receives water from 168 tributary rivers showing that it drains an enormous region. It is the fourth longest river in South America. The river is navigable all year long between Pirapora and the twin cities Pernambuco and Bahia.
4. *Climate*: Tropical.
5. *Early discoveries and history*: The Italian *Amerigo Vespucci* (1451–1512)—his first name became eponymous for both "Americas," landed on October 4, 1501, at the river's mouth. Since this day was dedicated to the holy martyr Francisco, the river was baptized using this first name. The discovery of the full length of this river took many years until the British Explorer Sir *Francis Burton* (1821–1890) was sent to the town Santo in Brazil, from where he started expeditions into the highlands. For example, he and contemporaries undertook canoeing tours from the source to the insurmountable waterfalls named Paulo Alfonso.
6. *Big towns along the river*: *Petrolina* (Pernambuco), 217,000 inhabitants; *Juazeiro* (Bahia), 150,000 inhabitants; and *Paulo Alfonso*, 93,000 inhabitants (Fig. 1.40).
7. *Literature*

Fig. 1.40 South-North course of the Sao Francisco river in Brazil. By Shannon1—Own work, CC BY-SA 4.0, https://commons.wikimedia.org/w/index.php?curid=65924215

da Costa EP, Pinto CC, Soares ALC, Melo LDV, Oliveira SMAC (2017) Evaluation of violations in water quality standards in the monitoring network of São Francisco River basin, the third largest in Brazil. Environ Monit Assess 189:590

Dos Santos Silva J, Caranha L, Moura Santos FK, Dos Santos AP, Rodrigues da Silva LO, Ferreira Rangel E (2017) Sand fly (Diptera, Psychodidae, Phlebotominae) abundance and diversity in areas affected by the São Francisco River transposition project in Ceará State, Brazil. Parasit Vectors 10:403

Favre TC, Fernandez MA, Beck LC, Guimarães RJ, Pieri OS, Thiengo SA (2016) Assessment of schistosomiasis in the semi-arid Northeast region of Brazil: the São Francisco River large-scale water transposition project. Rev Soc Bras Med Trop 49:252–257

Hacket J (2005) Investigation of fish mortality on the San Francisco river, Minas Gerais, Brazil. Technical Report, World Fisheries Trust

Reis RE, Albert JS, Di Dario F, Mincarone MM et al (2016) Fish biodiversity and conservation in South America. J Fish Biol 89:12–47

Silva JDDF, Pinheiro MCC, Sousa MS, Gomes VDS, Castro IMN, Ramos Junior AN, Bezerra FSM (2017) Detection of schistosomiasis in an area directly affected by the São Francisco River large-scale water transposition project in the Northeast of Brazil. Rev Soc Bras Med Trop 50:658–665

1.2.43 Seine River

1. *Region*: This French river rises on the plateau of Langres in the northeastern region of France in the neighborhood of the town Dijon. From there it flows in the western region of France, passes the French capital *Paris*, and finally enters the *English-French Sea Channel* at the towns of *Le Havre* (on the right bank) and *Honfleur* (on the left bank). Ocean ships may sail until *Rouen*, which is situated around 120 km (=75 miles) from the sea.
2. *Name*: The name of this river has its base in the name of a Gallo-Roman goddess called in Latin *Sequana*, who was apparently also adored by a Celtic tribe called Sequannians, which lived in the region of the French town Besançon of our days.
3. *Length*: The Seine river has a total length of 777 km (=483 miles). There are eight tributaries (e.g., Yonne, Risle, Marne, Oise).
4. *Climate*: Central European climate.
5. *Early discoveries and history*: Since about 570 km of the Seine can be reached by upstream sailing ships, this river was one of the most important pathways for many centuries of penetrating tribes but also used for trade in times of peace. The latter times were scarce during the last 2000 years. Thus, troops of the Romans, Huns, Vikings, English Kings, Germans, Spanish, and Italians entered and left this region. In all cases the river Seine was the pathway to conquer at different places. On the other hand, the river was and is also used for transportation of goods into the world and in regions, where French troops entered foreign countries.
6. *Big towns along the river*: *Paris* (capital of France), 2.2 million inhabitants; *Le Havre*, 170,000 inhabitants; *Honfleur*, 7700 inhabitants; and *Rouen*, 110,000 inhabitants (Fig. 1.41).
7. *Literature*

Ayrault S, Le Pape P, Evrard O et al (2014) Remanence of lead pollution in an urban river system: a multi-scale temporal and spatial study in the Seine River basin, France. Environ Sci Pollut Res 21:4134–4148

Azimi S (2016) Influence of the water quality improvement on fish population in the Seine River (Paris, France) over the 1990–2013 period. Sci Total Environ 542:955–964

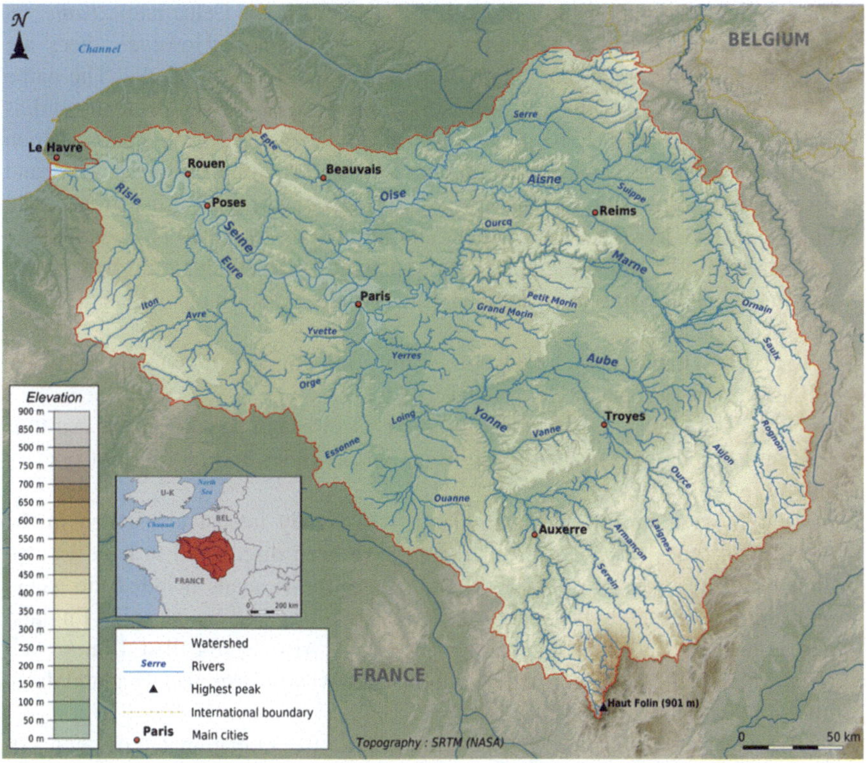

Fig. 1.41 Diagrammatic representation of the basin of the Seine river. By Paul Passy (Chamois rouge)—Own work; Topographic data: NASA Shuttle Radar Topography Mission edited with GRASS GIS, Quantum GIS and Inkscape; Hydrology: PIREN-Seine; International borders: Natural Earth, CC BY-SA 3.0, https://commons.wikimedia.org/w/index.php?curid=20225927

Bonnet E, Amalric M, Chevé M (2014) Mapping mental representations of industrial risk illustrated with the populations of the estuary of the Seine River, France. Cartographica 49:102–113

Edwards-May D (2010) Inland waterways of France. St. Ives, Cambs

1.2.44 Thames River

1. *Region*: This river has its source at *Thames head* in the county Gloucestershire, flows through the southern parts of England, passes *London* (the capital of England), and enters finally the North Sea at *Southend-on-Sea* in the county Essex. This river is very old and is already documented to have existed 58 million years ago and surpassed several ice ages from 500,000 BC years to the one at 20,000 BC.

2. *Name*: The traditional name *Thames* has its sources in the Celtic name *Tamessa*, which was transformed in middle English into *Temese*. However, there are several other interpretations. In any way this is said to mean "dark." The name is very old and well known, e.g., it is mentioned as *Tamesis* by Romans and as *Tamisiam* in the year 1210 in the text of the *Magna Carta* document, which regulated the peace between the English king and a group of rebel barons and promised the protection of church rights, protection for the barons from illegal imprisonment, access to swift justice, and limitations on feudal payments to the crown. However, not the whole river is named Thames, but a portion (when passing the famous university town Oxford) is called *Isis*, which refers to a former Egyptian goddess.
3. *Length*: The Thames has a total length of only 346 km (215 miles) and thus is one of the shortest European rivers (see Table 1.1).
4. *Climate*: The climate is influenced by the typical island position in the sea thus leading to moderate cold temperatures in winter times and not too hot ones in summer.
5. *Early discoveries and history*: England was already intensively settled in the pre-Roman times (e.g., early settlements have been discovered to have existed already in the times of 3300–2700 BC). Later the Romans invaded the island (e.g., Julius Caesar). In the year 54 BC ,the emperor *Claudius* occupied England and constructed fortifications along the river. Also a so-called Limes wall (*Hadrian's wall*) had been constructed in the years to protect England from invasions of Scottish tribes.

   The water of the Thames was in former times often contaminated by agents of diseases. Especially disastrous had been four cholera epidemics in the years between 1832 and 1865. These and several typhoid diseases killed hundred thousands of inhabitants. One of them was Prince Albert, who died in 1861 due to typhoid, which had been spread by dirty Thames water close to Windsor Castle. Starting in the Medieval times England became a dominating naval power creating an Empire (British Commonwealth) around the world, which finally broke down when India left it after the Second World War and South Africa in the year 1961).
6. *Big towns along the river*: *London* (capital of Great Britain), 8.8 million inhabitants; *Oxford*, 161,000 inhabitants; and *Reading*, 156,000 inhabitants (Fig. 1.42).
7. *Literature*

Ackroyd P (2007) Thames: sacred river. Chatto & Windus, London

Graham F, Embleton R (2003) Hadrian's wall in the days of the Romans. Wrens Park, Staffordshire

Kitson PR (1996) British and European river names. Trans R Philos Soc 94:73–118

Milne G, Bates M, Webber MD (1997) Problems, potential and partial solutions: an archaeological study of the tidal Thames, England. World Archaeol 29 (1):130–146

Rousseau GS, Haycock DB (2003) Coleridge's choleras: cholera morbus, asiatic cholera, and dysentery in early nineteenth-century England. Bull Hist Med 77:298–331

Fig. 1.42 Location of the Thames river in South England, CC BY-SA 3.0, https://commons.wikimedia.org/w/index.php?curid=134762

Sinclair M (2007) The Thames: a cultural history. Oxford University Press, Oxford, NY
Thomas AJ (2015) Cholera – the Victorian plague. Pen & Sword, Barnsley

1.2.45 Tiber River

1. *Region*: This river, which has its source in the so-called Apennine mountains at the Mount Fumaiolo in the center of Italy, flows through the regions of Tuscany, Umbria, and Lazio and finally enters into the Tyrrhenian Sea (a part of the Mediterranean Sea) draining a region of 6709 square kilometers.
2. *Name*: The name of the Tiber river derives from the Latin term *Tiberis*, which was the name of a legendary king of an Indo-European population being translated into the Italian language as *Tevere*. Popularly this river is also known as *flavus* due to its yellowish water color.
3. *Length*: The river has a length of 406 km (~252 miles).
4. *Climate*: Mediterranean mild to warm.
5. *Early discoveries and history*: Italy of our days was for several centuries in the times of the Roman Empire (~60 BC–394 AD) the center of European power and culture. Thus, even today it is possible to find considerable monuments of the glorious times everywhere in Italy (and also everywhere in Europe). The Tiber is in history (fairy tales) claimed to be the river into which the little twin boys *Romulus* and *Remus* had been thrown by enemies. Later after being saved by a female wolf and grown up, they founded the town of Rome. However, *Romulus*, who became the first Roman king, should have killed his brother during

Fig. 1.43 Diagrammatic representation of the short, but famous Tiber river in Italy. Public Domain, https://commons.wikimedia.org/w/index.php?curid=101371

foundation time of Rome. In the times of Roman emperors, executed criminals were thrown into the river especially during the period of the often occurring floods of up to 2 m in height (=6–7 feet).

6. *Big towns along the river*: *Ostia*, 200,000 inhabitants; *Rome* (capital of Italy), 2.9 million inhabitants; and *Citta di Castello*, 40,000 inhabitants (Fig. 1.43).
7. *Literature*

Boatwright MT (2000) Hadrian and the cities of the Roman empire. Princeton University Press, Princeton

Everett-Heath J (2005) Tiber. Oxford University Press, Oxford

Garnsey P, Saller R (1987) The Roman empire: economy, society and culture. University of California Press, Berkeley

Kelly C (2006) The Roman empire: a very short introduction. Oxford University Press, Oxford

Platner SB (1929) A topographical dictionary of ancient Rome. Thomas Ashby, Oxford, pp 536–538

1.2.46 Tigris-Euphrates River System

1. *Region*: The Middle East in the so-called two-stream region. Both rivers start on top of 1800–3000-m-high mountains and follow roughly parallel courses through

the heart of the Middle East. The lower region that they pass just before entering the Gulf of Persia is/was called *Mesopotamia* (Greek: land between the rivers). Before entering the sea, the rivers merge and form the so-called Shat-el-Arab in the Southeastern corner of Iraq.

2. *Names*: (a) *Euphrates*: *Arabic*, Al-Furat; *Turkish*, First; *Sumerian*, Buranun; *Bible*, Perath; (b) *Tigris*, *Arabic*, Dijah; *Turkish*, Dicle; *Sumerian*, Idigna; and *Bible*, Hiddekel.
3. *Length*: (a) The *Euphrates* measures 1750 miles (=~2800 km) in length. Its sources are the small rivers Krishna and Murat, which start in Eastern Turkey close to the source of the Tigris. From there it passes Northern Syria and Iraq and flows finally from there to the Persian Gulf. (b) The *Tigris* measures 1180 miles (=~1750 km) in length and has its source in Lake Mazor in Eastern Turkey and passes the same countries as the Euphrates and enters together with the Euphrates via the Shat-el-Arab into the Gulf of Persia (Iran).
4. *Climate*: In the high Taurus and Zao mountains, the temperatures are below 0 °C, so that in spring, the melting snow contributes considerable amounts of additional water. When reaching the alluvial plain close to Samarra, the temperatures increase and induce a continental subtropical climate reaching average temperatures in summer higher than 90 °F (32 °C), but lower than 50 °F (10 °C) in winter. On the Mesopotamian plain (terminal region before the rivers enter the Persian Gulf), the summer temperatures are extremely high (120 °F = 49 °C) during the day, but only 22 °C at night.
5. *Early discoveries and history*: The region between both rivers is one of the oldest human high cultural regions, if not the very oldest one. Around 3000 years BC, the towns of the Sumerians (Ur Uruk, Kish, Lagash) became important and were followed by the empires of Umma (2365–2350 BC), Uruk (2070 BC), and Babylonia (~1700 BC). The latter was destroyed several times, but often reconstructed (e.g., destroyed by Nebuchadnezzar II 605–562 BC and reconstructed in 539 BC by the Persian King Kyros II). In the year 331 BC, this region became included into the short-lasting Empire of the Greek *Alexander the Great* (356–323 BC). Then this region became prey of many other invading tribes during the following centuries (e.g., 321 BC Seleukids, 140 BC Parthery, 228 BC Sassanids, 635 BC Arabs).

   The Chaldean King Nebuchadnezzar II (605–562 BC) became historically one of the most famous kings due to the fact that he destroyed Jerusalem and the state of Judea in 597 BC and took the nobles of Judea in *Babylonic captivity* until 587 BC. Babylon is known also by the famous historical tale of the so-called Babylonian language confusion, which—according to the bible text (Moses 11)—should have been the punishment of God (Jahve) for the attempt to erect the *Babylonian tower* that should reach from earth to heaven, which was considered by local contemporary "priests" as blasphemy.
6. *Big towns along the rivers*: *Mosul*, 665,000 inhabitants, city at the banks of Tigris river close to the ruins of Ninive, Tarbis, and Nimrud; *Baghdad*, 7.7 million inhabitants, capital of Iraq since 1920 located at the banks of Tigris river; *Halep* (formerly *Aleppo*), 2.1 million inhabitants; Aleppo is known due to the so-called

Fig. 1.44 Diagrammatic representation of the region between the two famous rivers of one of the greatest human high cultures. By No machine-readable author provided. Kmusser assumed (based on copyright claims)—No machine-readable source provided. Own work assumed (based on copyright claims), CC BY-SA 2.5, https://commons.wikimedia.org/w/index.php?curid=323268

Aleppo bump due to the infection of humans with *Leishmania major* stages transmitted by bites of sandflies of the genus *Phlebotomus*. *Ramadi* (Iraq), 375,000 inhabitants, located at the banks of Euphrates river; *Fallujah* (Iraq), 275,000 inhabitants (Fig. 1.44).

7. *Literature*

Altinbilek D (2004) Development and management of the Euphrates-Tigris basin. Int J Water Ressour Manag 20:15–33

Cunningham B (1938) The régime of the rivers Euphrates and Tigris. Nature 142:373–374

Isaev VA, Mikhailova MV (2009) The hydrology, evolution and hydrological regime of the mouth area of the Shatt-al-Arab river. Water Resour 36:380–395

Mehlhorn H (ed) (2016) Encyclopedia of parasitology, 3 vols., 4th edn. Springer, New York

Woods C (2005) On the Euphrates. Zschr Assyrol 95:7–45

1.2.47 Uruguay River

1. *Region*: This South American river has its source in the region of the *Serra do Mar* belonging to Brazil. Thus, it flows from northern regions of South America to the South of this continent and forms parts of the boundaries of Brazil, Argentina, and Uruguay. Its first tributaries are the *Canoas River* and the *Pelotas River*, which give together rise to the river's first water. Together with the *Parana River* the Uruguay River forms the *Rio de la Plata* estuary, which is considered by some specialist as own river. One of its largest tributaries is the *Rio Negro*, which enters this river about 100 km before entering the Rio de la Plata (see Fig. 1.33, Parana river).
2. *Name*: The onomatopoeic name *Uruguay* has its origin in the understanding of European settlers when hearing the river name in the language of the local *Guarani* people. However, several interpretations exist. One of them explains that *Uru* is an indigenous bird species, and another interpretation is that *Urugua* describes the local snail *Pomella megastoma* (a snail with a large mouth).
3. *Length*: This north-to-south flowing river has a length of 1838 km (1142 miles), and thus it is not much longer than the Rio Negro, which is its confluent partner reaching a length of 1555 km.
4. *Climate*: Tropical and subtropical.
5. *Early Discoveries and history*: This river was discovered by European settlers coming as well from Spain as from Portugal. Both nations use the same name for the river; however, the pronunciation is different.
6. *Big towns along the river*: *Uruguiana*, 125,000 inhabitants, and *Fray Bentos*, 24,000 inhabitants.
7. *Literature*

Davies BR, Walker KF (eds) (1986) The ecology of river systems. Springer, Dordrecht

Fuentes CM (2016) Downstream passage of fish larvae at the Salto Grande Dam on the Uruguay River. River Res Appl 32:1879–1889

Rivas-Rivera N, Eguren G, Carrasco-Letelier L (2014) Screening of endocrine disruption activity in sediments from the Uruguay River. Ecotoxicology 23:1137–1142

Suarez R, Pineiro G, Barceló F (2018) Living on the river edge: the Tigre site (K-87) new data and implications for the initial colonization of the Uruguay River basin. Quat Int 473:242–260

1.2.48 Volga River

1. *Region*: East Europe—its sources are situated in the Valdai Hills Northwest of Moscow (the capital of Russia). From a rather low height (748 feet = 228 m), it passes slowly to the Caspian Sea (=92 feet below sea level) receiving water from

about 200 tributaries. This waterway is the principal one of Western Russia, and its basin sprawls across two fifths of the European part of Russia. The river basin drains around 533,000 square miles (=1.38 million m^2).

2. *Name*: The Volga river of our days was named *Ra* in ancient Greek or *Itil* respectively *Etil* in Tatar language.
3. *Length*: The main Volga river has a length of 3688 km (~2300 miles) and is dammed up several times before it enters the Caspian Sea via a giant (150-km-wide) delta. This river, which is the longest and water-richest river in Europe, is connected with the river Don by a 101-km-long channel, and thus ships can pass from the Caspian Sea to the Black Sea. There is also another pathway (Lenin channel) of 128 km of length connecting, e.g., lately Moscow with the Volga river. The Volga river is navigable for about 2000 miles (=3220 km). Also 70 tributaries can be used by ships.
4. *Climate*: Due to the rather Northern spring of the Volga river, it passes through several climate zones. From its source to the Kama river, there is a zone characterized by cold and snow-rich winters and warm but humid summers. This zone is followed by another zone with hot and dry summers and cold winters with little snow. In the following Southern and Eastern zones, temperatures are higher, but only low-graded rain occurs.
5. *Early discoveries and history*: Early descriptions of the Volga river regions are known for a long time, e.g., those of Ptolemy (second century AD) are followed later by Slavic and Arabic contributors in the tenth to eleventh century. Also the English John Perry measured the flow speed of this river in the year 1700. In the following century 1780–1894, the knowledge of the river's parameters became more and more clear. Around 650 AC the so-called Volga-Bulgarian empire was founded by a Turkish tribe that became Islamic around 900 AD. It existed until the Mongolian tribes entered this region in the years around 1237 AD. Their successors (Tatars) are still there counting around 5.5 million people, who are also called Turk people. Also, a German population settled along the banks of the Volga river. This republic was terminated by the socialistic Russian government during World War II (1939–1945), and the Germans were detained in working camps in Siberia. Starting with the modern Russia of Gorbachev (born 1931), their descendants got the chance to move to Germany, where their integration was successful in contrast to that of other ethnical groups.
6. *Big towns along the river*: *Volgograd* (until 1925: Zarizyn, 1925–1961: Stalingrad), today about 1 million inhabitants; during World War II, the German Army lost the decisive battle in 1942/43 in wintertime; only about 5000 out of the 90,000 German survivors returned to Germany in the year 1955 due to the activities of the famous Dr. Konrad Adenauer (1876–1967), who was the first chancellor (1948–1963) of the Federal Republic of Germany after World War II. *Gorky* (Nizhny Novgorod from 1932 to 1990) 1.25 million inhabitants; *Ulyanovsk* (formerly Simbirsk) 620,000 inhabitants; and *Astrakhan* 520,000 inhabitants, port in the delta (Fig. 1.45).

Fig. 1.45 Basin of the Volga river in Russia. CC BY-SA 2.5, https://commons.wikimedia.org/w/index.php?curid=1122527

7. *Literature*

Golub VB (2017) Plant communities of the lower reaches of the Volga River on soils with strongly seasonal dynamics of salinization. Arid Ecosyst 7:23–30

Kuzmina ZV (2017) Changes in terrestrial ecosystems of the lower Volga region caused by river flow regulation. Arid Ecosyst 7:234–242

Mehlhorn H (ed) (2016) Encyclopedia of parasitology, 3 vols., 4th edn. Springer, New York

Obukhova OV (2017) Dynamics of potentially pathogenic microflora of the water and pike perch in the Volga Delta. Contemp Probl Ecol 10:563–574

1.2.49 *Yamuna River (syn. Jumna)*

1. *Region*: India—it is the most important subsidiary river of the Ganges in India.
2. *Name*: This river is also named Jumna or Jamuna, which means "twin" thus looking at its final confluence with the river Ganges.
3. *Length*: This river is in total 1376 km long and has its source in a height of 6320 m close to the famous Yamnoty Shrine in the Uttarakhand region of the Himalaya mountains. From there it passes the regional states Haryana and Uttar Pradesh and the region around Delhi. Close to the town Allahabad, it enters the Ganges river and contributes there more water than the Ganges itself (~2900 m^3/s to 1800 m^3/s, respectively).
4. *Climate*: Very cold at the source to tropical at the confluence point with the Ganges river.
5. *Early discoveries and history*: The river was first described by the Macedonian Nicator Seleukos, who was member of the Seleukos clan, who reigned after Alexander the Great (356–323 BC) in Macedonia from 312 BC until 64 BC. Seleukos was member of Alexander's expedition team in this region, and his reports concerning the rivers in this region were noted by Greeks and later by the Romans, who called this river Jomanes. People in the region had further names such as Djemma or Zemna. On the other hand, Yamuna is today in India a rather common first name of girls. This river had and has even today an important religious significance. For example, the goddess Yami is the sister of the god of death (Yama) and the daughter of god of sun Surya. Thus, millions of people pray every year along the Yamuna and Ganges rivers. Also, the Taj Mahal—the mausoleum of Mumtaz Mahal, the wife of the Mughal emperor Shah Jahan—is situated at the banks of the Yamuna river.
6. *Big towns along the river*: Allahabad (2.2 million inhabitants); Agra (1.84 million inhabitants); Delhi (28.5 million inhabitants); and Mathura (500,000 inhabitants) (Fig. 1.46).
7. *Literature*

Dubey M (2014) Exploring bacterial diversity from contaminated soil samples from river Yamuna. Microbiology 83:585–588

Gary C (2008) Troubled waters: religion, ethics, and the global water crisis. Rowman & Littlefield, Lanham

Jaiswal N, Malhotra A, Malhotra SK (2016) Bioinvasion: a paradigm shift from marine to inland ecosystems. J Parasit Dis 40:348–358

Koch E (2006) The complete Taj Mahal: and the riverfront gardens of Agra, 1st edn. Thames & Hudson, London

Mehlhorn H (ed) (2016a) Encyclopedia of parasitology, 3 vols., 4th edn. Springer, New York

Mehlhorn H (2016b) Human parasites, 8th edn. Springer, Heidelberg

Parween M (2017) Waste water management and water quality of river Yamuna in the megacity of Delhi. Int J Environ Sci Technol 14:2109–2124

Rothfarb E (1998) In the land of the Taj Mahal. Henry Holt, New York

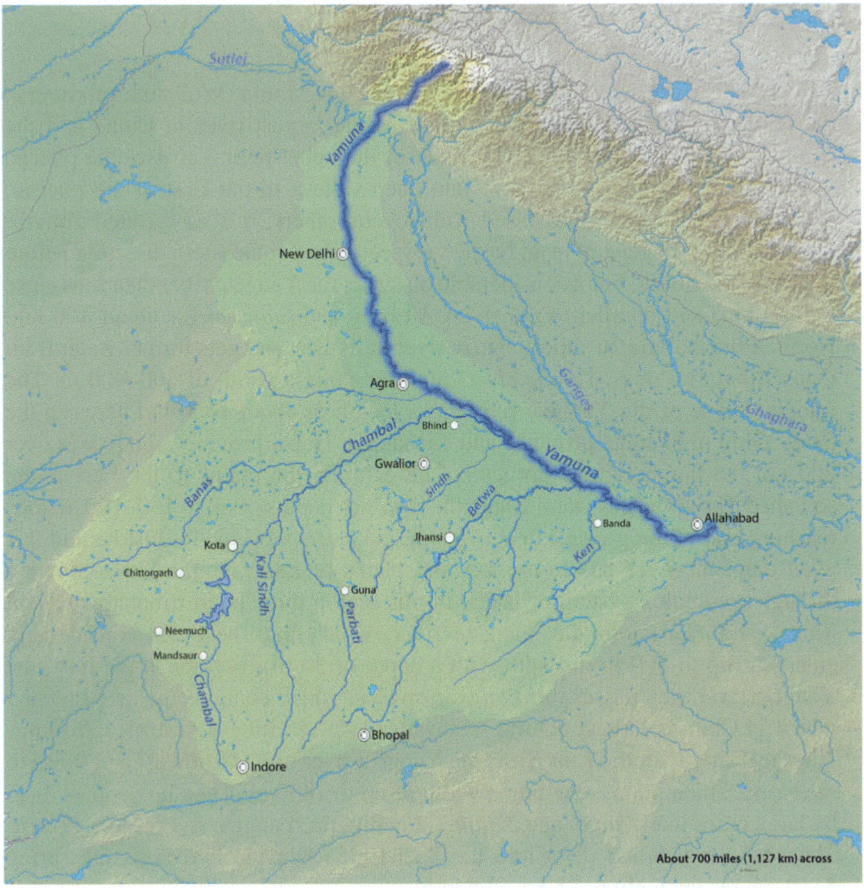

Fig. 1.46 Diagrammatic (thick-lined) representation of the course of the Yamuna river, a tributary of the Ganges entering it close to Allahabad. By Shannon—Background and river course data from http://www2.demis.nl/mapserver/mapper.asp, CC BY-SA 4.0, https://commons.wikimedia.org/w/index.php?curid=9971222

Sharad SK, Agarwal PK, Singh VP (2007) Hydrology and water resources in India. Springer, Cham

Sharma AP (2017) Patterns of fish diversity, community structure and ecological integrity of River Yamuna, India. Aquat Ecosyst Health Manag 20:30–42

Singh AK (2014) Hazard assessment of metals in invasive fish species of the Yamuna River, India in relation to bioaccumulation factor and exposure concentration for human health implications. Environ Monit Assess 186:3823–3836

1.2.50 Yangtze River (Yangtze Kiang)

1. *Region*: This river, which has its source in the Ulan Moron (Wulanmoron) river at the border of Tibet Autonomous region, is the largest river in China and the country's principal waterway. The water of the upper region crosses the plateau of Tibet and descends from there into deep valleys in the East of the plateau. When this basic stream has fused with several others, it is now called *Yangtze* river and flows in general in an Eastern respectively Southeastern direction before it turns to the South. For a considerable distance, the Yangtze river then runs close (25–50 km) and parallel to the rivers *Mekong* and *Salween* for about 400 km. North of the latitude 26°, the Yangtze river turns East and gets further water from many tributaries (e.g., *Yalong river*), thus reaching a width of 300–400 m. The middle course of the Yangtze river has a length of about 1000 km between the cities Yibin in Sichuan province and Yichang in Hubei province and passes over 200 km the famous green gorges before entering the plains to the East at the so-called Three Gorges Dam, where the river has depths of up to 150–180 m, thus reaching one of the greatest depths among all rivers worldwide. Just behind this giant dam, the river passes the lowland plains of East Central China, flowing through a complex system of lakes. In this region three main tributaries (*Yuan*, *Xiang*, and *Han*) join the Yangtze river, which thus becomes considerably enlarged (up to 800 m in width with a depth of 30 m). Further in the East, the Yangtze river then passes the *Jiangxi province*, which contains the lake Poyang, which is China's largest natural freshwater lake reaching a size of 3500 km^2. After merging with the *Kan river*, the Yangtze river turns Northeast and flows to the North China plains reaching a width of up to 1800 m. Then it becomes lined by large cities like *Wuhan* and *Nanjing*. Finally, the Yangtze river enters close to Shanghai via a broad delta into the East China Sea via two arms measuring 5–10 km respectively 6–15 km in width.
2. *Name*: The Yangtze river, which in Chinese is also called *Pinyin*, has got its international name from the ancient kingdom of *Yang* being mainly used by the people in the West. In China the name *Chang Jiang* (long river) is mainly used, but also *Da Jiang* (great river) or just *Jiang* (river).
3. *Climate*: The Yangtze river runs through several different climate zones reaching from very cold regions in the high mountains to subtropical and tropical zones at its mouth (delta region) with monsoon rains introducing extensive floods that led from time to time to national flood disasters with many thousands of human deaths (e.g., in 1870, 1898, 1931, 1949, 1954).
4. *Length*: The Yangtze river with its source on the plateau of Tibet has a length of about 6300 km and thus ranges behind the Amazonas and Nile river as the third longest river in the world. Its basin extends for 3200 km from West to East and for 1000 km from North to South draining an area of 1,808,500 km^2. It passes ten provinces or regions and receives water from eight principal tributaries (Yalung, Min, Jialing, Han, Wu, Yuan, Xiang, and Gan).

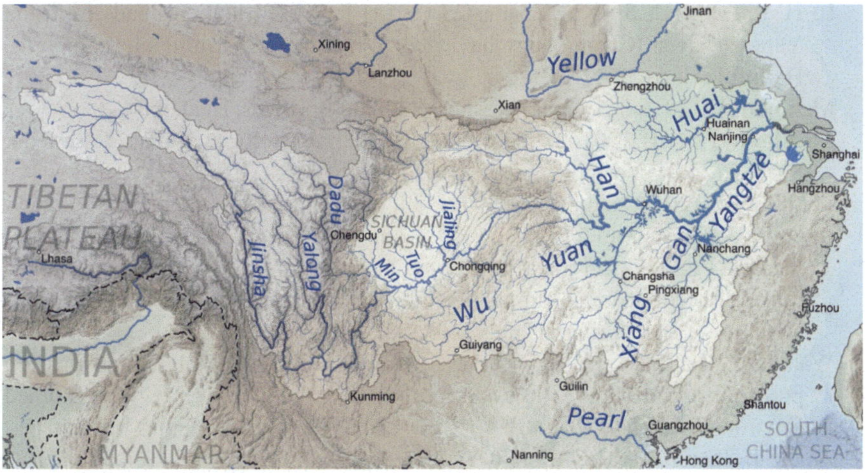

Fig. 1.47 Basin of the Yangtze river in China. By Keenan Pepper—Own work, CC BY-SA 4.0, https://commons.wikimedia.org/w/index.php?curid=66050281

5. *Early discoveries and history*: The region, which covers the China of our days, was nearly unknown in Europe until the Venetian merchant *Marco Polo* (~1271–1295) came back from a longstanding journey in this region (~1271–1295). He reported exotic details of the unknown but highly developed China and reached popularity like the movie stars of our days. The population of this region was based on many different ethnics which were united several times during the centuries and separated again until in the years after World War II the People's Republic of China was established by *Mao Zedong* (1893–1976) and separated from the self-appointed "National China" on Taiwan Island reigned first by the army general Chiang Kai-shek (1887–1975). The People's Republic of China is today the most densely populated country on earth besides India—both containing around 1.5 billion inhabitants.
6. *Big towns along the river*: Due to its high importance, many big and even huge cities are located along the Yangtze river. *Chongqing* is one of the biggest with about 7.6 million inhabitants as well as *Wuhan* counting about 8.6 million inhabitants. *Nanjing* with 8.3 million inhabitants is located further downstream, and *Shanghai* with more than 21 million inhabitants is situated close to the mouth of the river (Fig. 1.47).
7. *Literature*

Chen Z, Zong Y, Wang Z, Wang H, Chen J (2008) Migration patterns of neolithic settlements on the abandoned Yellow and Yangtze River deltas of China. Quat Res 70:301–314

Guo Y, Lai S, Zhang J, Liu Q, Zhang H, Ren Z, Mao D, Luo C, He Y, Wu H, Li G, Ren D, Liu X, Chang Z (2018) Mosquito population dynamics during the construction of Three Gorges Dam in Yangtze river, China. Acta Trop 182:251–256

Li HM, Tang HJ, Sh XY, Zhang CS (2014) Increased nutrient loads from the Yangtze river (Changjiang) have led to increased harmful algae blooms. Harmful Algae 39:92–101

Mehlhorn H (2016a) Animal parasites, 8th edn. Springer, Heidelberg

Mehlhorn H (ed) (2016b) Encyclopedia of parasitology, 3 vols., 4th edn. Springer, New York

Mehlhorn H (2016c) Human parasites, 8th edn. Springer, Heidelberg

Sun LP, Wang W, Zuo YP, Zhang ZQ, Hong QB, Yang GJ, Zhu HR, Liang YS, Yang HT (2017) An integrated environmental improvement of marshlands: impact on control and elimination of schistosomiasis in marshland regions along the Yangtze river, China. Infect Dis Poverty 6(1):72

Wang DQ, Tang LH, Hu ZC, Zheng X, Hang WK (2013) Malaria transmission potential in the Three Gorges reservoir of the Yangtze river, China. Biomed Environ Sci 26:54–62

Zhang J, Zhang ZF, Liu SM, Wu Y, Xiong H, Chen HT (1999) Human impacts on the large world rivers: would the Changjiang (Yangtze River) be an illustration? Glob Biochem Cycles 13:1099–1105

Zhao Y, Lee H, Zhao H, Cai S, Huang X (2017) Relationship between ancient bridges and population dynamics in the lower Yangtze River Basin, China. PLoS One 10:1371

Zhou YB, Liang S, Chen Y, Jiang QW (2016) The Three Gorges Dam: does it accelerate or delay the progress towards eliminating transmission of schistosomiasis in China? Infect Dis Poverty 5(1):63

1.2.51 Yellow River (Hwang Ho River)

1. *Region*: The source of this river is situated in the Bayan Har Mountains in the Qinghai province in Western China, and its water flows to the shore of the Bohai Sea close to the town *Dongying* in the Shandong province. The estuary basin has an east-west diameter of about 1900 km and a north-south length of 1100 km and thus is of enormous size.
2. *Name*: The modern name *Yellow river* describes the fact that the water of the river appears muddy yellow (due to amounts of silt) entering several times during the year. Older Mongolian sources described it as *Black river*, while the translated Tibetan name is *River of the peacock*. The name "Black river" describes mainly the river region, where loess soil (silt) has not yet entered.
3. *Length*: The river has a total length of 5464 km (3395 miles), and its draining area covers about 752,546 square kilometers = 290,560 square miles. The Yellow river passes through seven provinces and two autonomous regions: from West to East follow each other Qinghai, Gansu Autonomous Region, Ningxia, Inner Mongolia Autonomous Region, Shoanxi, Shanxi, Henan, Shandong. In total this river has 17 considerable tributaries, which enter enormous amounts of water.
4. *Climate*: Due to the different heights of the riverbed (=4800 m close to the source and zero meters when entering the sea), it passes through various climate zones,

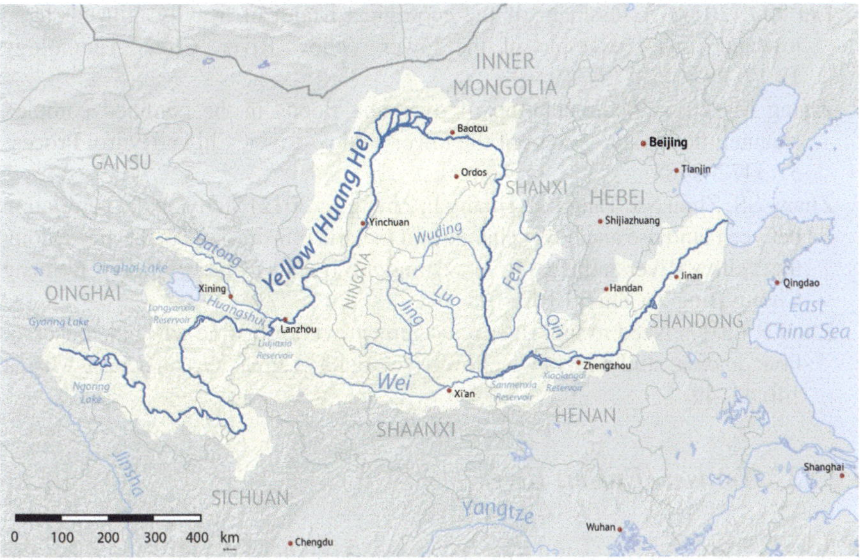

Fig. 1.48 Basin of the Yellow river in China. By Shannon1—Created using Natural Earth and NASA SRTM data, both public domain, CC BY-SA 4.0, https://commons.wikimedia.org/w/index.php?curid=9617583

which are endangered by enormous rainfall and extreme amounts of water running downhill. Thus, in history until today, a huge number of houses have been destroyed, and millions of people have been killed by floods.

5. *Early discoveries and history*: The river has its place since long in Chinese history tales dealing with the giant *Kua Fu*, who allegedly once swallowed all the river's water. The fact is that the river changed his bed several times during the centuries and the history is full of devastating floods killing in total millions of people. In present times many efforts have been taken to strengthen levees and control the floods by different methods in order to avoid disastrous, life-threatening situations in times of extremely high water levels.
6. *Big towns along the river*: *Lanzhou*, 3.7 million inhabitants; *Yinchuan*, 2 million inhabitants; *Hohhot*, 2.9 million inhabitants; *Zhengzhou*, 9.6 million inhabitants; and *Dongying*, 2.1 million inhabitants (Fig. 1.48).
7. *Literature*

Grousset R (1959) The rise and splendor of the Chinese empire. University California Press, Berkeley, CA

Li T (2018) Assessment of ecosystem health of the Yellow River with fish index of biotic integrity. Hydrobiologia 814:31–43

Liang C (2016) The ecological benefit - loss evaluation in a riverine wetland for hydropower projects. A case study of Xiaolangdi reservoir in the Yellow River, China. Ecol Eng 96:34–44

Liu XQ (2017) Assessment of the ecological health of heavily utilized, large lowland rivers: example of the lower Yellow River, China. Limnology 18:17–29

Zhang M (2017) Sedimentation of overbank floods in the confined complex channel: floodplain system of the lower Yellow River, China. Hydrol Process 31:3472–3488

Zhang SS, Zhou SS, Tang LH, Huang F, Zheng X (2012) Study on the correlation between land use and cover change and malaria transmission in the areas along the Yellow River and Huai River. Zhongguo Ji Sheng Chong Xue Yu Ji Sheng Chong Bing Za Zhi 30:102–108

Zhou R (2017) Need to link river management with estuarine wetland conservation: a case study in the Yellow river delta, China. Ocean Coast Manag 146:43–49

1.2.52 Yenisei (Enisei) River

1. *Region*: This river of Central Russia is the world sixth largest one with respect to the amount of water discharge (half of its water comes from snow). It has its origin close to the town *Kazyl* in the Russian republic of Tyva, where the Great Yenisei (*By-Khem*) and the Small Yenisei (*Maly*) merge. From this confluence place, the river flows along the border between Eastern and Western Siberia finally entering the very cold (often frozen) *Kara Sea* (Arctic Ocean). The ultimate source, however, is the *Selenga* river, which comes from Western Mongolia and flows through the Baikal lake and finally into the Angara tributary, which is only one of up to 20,000 (or more) small tributaries at all. All the major tributaries have their rise on the large Siberian plateau. In general, the Yenisei flows from South to North leading to the Yenisei Gulf of the so-called Kara Sea. The river is mostly frozen from October to the end of May.
2. *Name*: The name of the Yenisei river is also spelled as *Yenisey* or *Enisei* respectively *Evenk loanesi*, which means "great river".
3. *Length*: Depending on the counting of the Great Yenisei as early tributary, the total length of this river is 2167 miles (=3487 km). Other counts including the rivers Angara and Selenga reach a length of 5391 km in total.
4. *Climate*: The Yenisei basin bay in its Northern regions is characterized by a subarctic climate with long icy periods (−28 to −32 °C). The central and Southern regions have mainly continental conditions with rather warm periods and temperatures of 18–20 °C in July. In the North there is an annual rainfall (snowfall) of about 400–500 mm and 500–750 mm in the central region and up to 1100 mm in the mountains. The rain falls mainly (80–90%) in the warmer months (late summer/autumn). In the Northern regions there is also permafrost.
5. *Early discoveries and history*: For a long time, the Yenisei region was not settled by humans. First reports go back reported that people joined the valley in the eleventh century. Russians started to settle there in small numbers in the period of 1607–1619. Towns like Krasny Yar (Krasnoyarsk) were founded in 1627 and the

(now much bigger) town Irkutsk in 1652. Trials to explore the region were operated by help of ships (e.g., Northern expedition 1733–1742, Lower Yenisei expedition 1894–1896 and 1907–1912). Today scientific and industrial expeditions as well as founding of towns are ongoing due to the discovery of rich mineral resources in this region.

6. *Big towns along the river*: *Irkutsk* 590,000 inhabitants, center of colonization of East Siberia in the eighteenth and nineteenth century, now universitarian center, and center of metal and textile industries and food production; *Krasnoyarsk* 980,000 inhabitants, town on the Upper Yenisei, station of the Trans-Siberian Railway, region with big reserves of copper, coal, gold, and molybdenum; and *Bratsk* 250,000 inhabitants. Close to Bratsk the Gulag Angara prison camp had been constructed in communistic times for 44,000 prisoners for railway construction, e.g., of the Baikal-Amur Mainline, now being the location of one of the biggest hydroelectric plants and site of oil production (Fig. 1.49).
7. *Literature*

Drits AV (2017) Distribution and grazing of dominant zooplankton species in the Ob estuary: influence of the runoff regime. Estuar Coasts 40:1082–1095

Hrebtov M (2017) Numerical study of winter diurnal convection over the city of Krasnoyarsk: effects of non-freezing river, undulating fog and steam devils. Boundary-Layer Meteorol 163:469–495

Ponomareva YA (2016) Ratio between living and dead cells and the size structure of the Yenisei River phytoplankton downstream of the Krasnoyarsk Hydroelectric Power Station. Contemp Probl Ecol 9:582–589

1.2.53 Yukon River

1. *Region*: This North American river gets its main head water from the Atlin Lake and the Tagish Lake in neighborhood of the border between *British Columbia* and the *Yukon territory*. From there it flows through the central *Yukon territory* of Northwestern Canada and through the central region of the US American state Alaska. Finally, it discharges its water into the *Bering Sea*, which is the most northern portion of the Pacific Ocean. This sea is named honoring the Danish researcher *Vitus Jonassen Bering* (1680–1741), who sailed in the year 1728 through a special region of this sea, which now is called *Bering Strait* and leads to the Northern Polar Sea. He was the head of the so-called Great Nordish Expedition.
2. *Name*: The name *Yukon* has its origin in *chųų gąįį han*, which means "white water" in Gwich'in, the language of the native Alaskan people and refers to the pale color of the glacier runoff in the Yukon river.
3. *Length*: The Yukon river has a total length of 3190 km (1980 miles), and its head water drains an area of about 850,000 square kilometers (~328,000 square miles).
4. *Climate*: This region has a so-called subarctic climate, which is characterized by relatively "warm" but very short summertimes (only around 16 °C = 60 °F) in

Fig. 1.49 Basin of the Russian Yenisei river. By Kmusser—Own work using Digital Chart of the World and GTOPO data, CC BY-SA 3.0, https://commons.wikimedia.org/w/index.php?curid=4792956

mean. The winter period is very long, and temperatures range in January between −21 °C and −31 °C with minus extremes up to −63 °C.

5. *Early discoveries and history*: For long times this very cold region was settled exclusively by North American Indian tribes. During the mid-nineteenth century, North European explorers arrived as well as Russian fur hunters and people on the search for mineral wealth (gold, etc.). Thus, in the follow-up settlements like those of *Fairbanks*, *Whitehorse*, and *Dawson* flourished, but later often lost their importance. The delta mouth was first described in 1831 by Russian fur hunters. Further on, Russians mapped about 970 km of the lower river. Early settlements of Europeans became attractive for Indians, so that Europeans often withdrew from certain regions, but came back later and explored the wealth hidden in the soil.

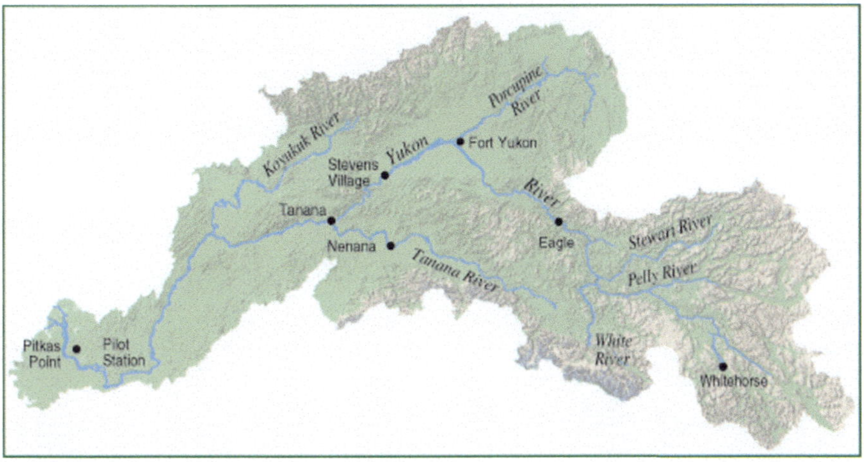

Fig. 1.50 Diagrammatic representation of the basins of the Yukon and Tanana rivers in British Columbia, Canada. By United States Geological Service—http://agdc.usgs.gov/data/usgs/water/yukon.html, Copyrighted free use, https://commons.wikimedia.org/w/index.php?curid=1411638

6. *Big towns along the river*: There are no real big towns—*Dawson*, 1300 inhabitants; *Fairbanks*, 31,500 inhabitants; *Whitehorse*, 27,000 inhabitants, *Keno City*, 24 inhabitants; and *Mayo*, 200 inhabitants (Fig. 1.50).
7. *Literature*

Ken S, Coates KS (1991) Best left as Indians. Native-white relations in the Yukon Territory, 1840–1973, Montreal. McGill-Queen's University Press, Kingston

Golder FA, Stejneger L (2018) Bering's voyages. An account of the efforts of the Russians to determine the relation of Asia and America. Franklin Classics

Greenhous B (ed) (1987) Guarding the goldfields. The story of the Yukon field force. Dundurn, Ottawa

Stejneger L, Golder FA (2018) Bering's voyages: the log books and official reports of the first and second expeditions, 1725–1730 and 1733–1742. Franklin Classics

Wharton D (1972) The Alaska gold rush. Indiana University Press, Bloomington

Chapter 2
Conquering the God of Plague in China: A Tale of Over 60 Years

Wei Wang

Abstract Schistosomiasis, caused by the blood fluke (trematodes) of the genus *Schistosoma*, is a neglected tropical disease that affects over 280 million people in the tropical and subtropical regions, with a further 800 million at risk of infection. *S. mansoni*, *S. japonicum*, and *S. haematobium* are the most significant species that infect humans causing schistosomiasis. China is only endemic for *S. japonicum*. The control campaign for schistosomiasis japonica in China was launched by Chairman Mao Zedong in 1950s after the People's Republic of China was founded in 1949, and the control efforts have achieved great success in controlling the transmission of schistosomiasis japonica in China. Historically, there are three shifts of national schistosomiasis control strategy in China, including snail control-based integrated control strategy, morbidity control-based integrated control strategy, and the currently implemented integrated control strategy with emphasis on control of source of *S. japonicum* infections. Many immunodiagnostics have been developed, which facilitates the identification of infections and plays an important role for accelerating the progress toward the elimination of schistosomiasis japonica in China. Since the introduction of praziquantel in the early 1980s, the agent has replaced other schistosomicides to become the only drug of choice for the treatment of *S. japonicum* infections in both humans and bovines. Although there is concern about emergence of praziquantel resistance in the parasite following long-term, repeated, and extensive application, the efficacy of praziquantel remains satisfactory, and there is no evidence of reduced susceptibility detected in main endemic foci of China. In addition, a national schistosomiasis surveillance network has been established in China, which provides valuable data to develop and adjust the national schistosomiasis control strategy. Novel tools, such as the geographic information

W. Wang (✉)
Key Laboratory of National Health Commission on Parasitic Disease Control and Prevention, Wuxi City, Jiangsu Province, China

Jiangsu Provincial Key Laboratory on Parasites and Vector Control Technology, Wuxi City, Jiangsu Province, China

Jiangsu Institute of Parasitic Diseases, Wuxi City, Jiangsu Province, China
e-mail: wangwei@jipd.com

H. Mehlhorn, S. Klimpel (eds.), *Parasite and Disease Spread by Major Rivers on Earth*, Parasitology Research Monographs 12,
https://doi.org/10.1007/978-3-030-29061-0_2

system/remote sensing (GIS/RS), Google Earth, and intelligent device to detect *S. japonicum*-infested water, have been introduced for schistosomiasis control in China. Since both the prevalence and intensity of *S. japonicum* infection have been reduced to an extremely low level in China, a highly sensitive surveillance-response system is therefore required, which would facilitate the elimination of this neglected tropical disease in China. Currently, China is transferring the successful experiences from schistosomiasis control to foreign countries endemic for schistosomiasis. Let's expect the coming of the miracle "Farewell to God of Plague" as soon as possible in both China and across the world.

Keywords Schistosomiasis japonica · Diagnosis · Treatment · Praziquantel · Surveillance-response system · China · God of plague

2.1 Introduction

Schistosomiasis, caused by blood flukes (trematode) of the genus *Schistosoma*, is a chronic debilitating parasitic disease ranking second only to malaria in terms of its socioeconomic and public health importance around the world (Colley et al. 2014). This neglected tropical disease is estimated to affect over 280 million people living in 78 countries (Steinmann et al. 2006), and over 280,000 deaths occur each year due to schistosomiasis-related disorders (King 2010). In addition, this parasitic disease causes a total number of Disability Adjusted Life Years (DALY) lost at 1.532 million per year (King et al. 2005). Five species of the parasite are found to infect humans, including *Schistosoma haematobium*, *S. mansoni*, *S. japonicum*, *S. megonki,* and *S. intercalatum*, and the major disease-causing species are *S. haematobium*, *S. mansoni*, and *S. japonicum* (McManus et al. 2018).

S. japonicum is the only species of blood fluke that occurs in China (Zhou et al. 2010). The description of schistosomiasis in China dates back more than two millennia, as proved by the identification of *S. japonicum* eggs from two ancient corpses excavated in 1971 and 1975 (Zhou et al. 2005a); however, schistosomiasis was recognized in China until the early twentieth century due to the limitations of scientific development. Before the founding of the People's Republic of China in 1949, there were no specific schistosomiasis control programs due to political systems, socioeconomic levels, and S & T status, and some field investigations were conducted to preliminarily understand the epidemiology of schistosomiasis in the country, and the harm of the disease to human health was gradually recognized. After the founding of the People's Republic of China, the central government prioritized schistosomiasis control, and great success has been achieved following the control efforts for more than 60 years. In this chapter, the discovery, epidemiology, diagnosis and treatment of schistosomiasis, as well as the evolution of Chinese national schistosomiasis control strategy, Chinese schistosomiasis surveillance networks, and novel tools used for schistosomiasis elimination programs were described.

2.2 Historical Background

2.2.1 Discovery of Schistosomiasis in China

In 1905, when the resident medical officer John Catto dissected the body of a Chinese man from Fukien Province that presumably had never away from China before and died of cholera, Dr. Catto found specific schistosome worms and ova in the mesenteric vessels, and therefore characterized this trematodes as a novel species of *Schistosoma*, *S. cattoi* (Catto 1905). This case was presumed to Foochow or Amoy, where schistosomiasis was not endemic (Faust and Meleney 1924). However, this new species *S. cattoi* was considered as *S. japonicum*, discovered by Dr. Katsurada in the Philippines (Sewell 1919). The first clinical case infected with *S. japonicum* in China was diagnosed by an American physician in Changde City, Hunan Province, in 1905 (Logan 1905). Since then, human *S. japonicum* infections were successively reported along the Yangtze River basin (1907 in Anhui, 1908 in Zhejiang, 1910 in Hubei and Shanghai, 1911 in Guangdong, and 1913 in Jiangsu) (Wang 2006). However, schistosomiasis was not paid much attention to at that time.

2.2.2 Epidemiology of Schistosomiasis from 1920s to 1949

In the early 1920s, Faust and colleagues investigated the prevalence of human *S. japonicum* infections in Soochow and Jiaxing, eastern China (Faust 1924), and a monograph entitled *Studies on Schistosomiasis Japonica* was published (Faust and Meleney 1924). This is the first monograph relating to schistosomiasis japonica research in China. Since then, Chinese scholars and foreign experts working in China (mostly doctors) started the schistosomiasis japonica research and achieved plenty of important findings, which provided valuable data for the initiation of the national schistosomiasis control program in China. In 1923, the government started to collect schistosomiasis control data, and launched schistosomiasis control activities since 1929. In 1933, the first official schistosomiasis control team was built in Kaihua, Zhejiang Province, and nationwide investigation of communicable diseases including schistosomiasis was conducted in 1935. However, these government-led schistosomiasis control activities ceased since 1937 because of the World War II.

In early 1930s, a 42.6% (232/544) prevalence of human *S. japonicum* infections in Chih-huai-pan, Kaihua, Chekiang Province (Kan and Yao 1934), and results from another study in the same area showed a 49% egg-positive rate in 1234 local villagers tested and a 77% positive rate of intracutaneous test (Kan and Kung

1936). In 1936, schistosomiasis japonica was found to be endemic in Futsing, Fukien Province, southeastern China, and 29 snail habitats were identified in Futsing (Tang 1936). A further investigation in 1939 showed a 19.6% egg-positive rate in 102 local villagers with spleen discomfort, and a 5.5% egg-positive rate in 566 inpatients from local hospitals; these patients were diagnosed in 26 sites across Futsing, and the number of snail habitats increased from 29 to 76 (Tang 1939). On August 1934, 12 out of 16 British and 2 out of 16 Chinese sailors belonging to a British gunboat got infections with *S. japonicum* by bathing off the ship at Chinkiang in the lower reaches of the Yangtze River (Spiridion 1936). In 1939, *S. japonicum*-infected smooth-shelled *Oncomelania hupensis* was detected in irrigation ditches and *S. japonicum* eggs were found in the fecal samples from the local residents living in Tali-Fu, west of Yunnan Province, in which schistosomiasis was proved to be endemic (Cecil and Robertson 1940). During the period between 1930 and 1936, a total of 39 patients with schistosomiasis were admitted to Nanking central hospital, and these cases were predominantly men (97.4%), aged at 10–30 years (69.2%), and soldiers (20.5%) and labors (20.5%), most of whom (61.5%) were infected due to bathing in the Yangtze River in summer (Chu 1937). Moreover, epidemiological surveys were conducted in various areas of China that were endemic for schistosomiasis, and high prevalence rates of human *S. japonicum* infections were reported across the country (Kuo 1946; Carroll 1946; Yao 1936). Approximately 32.8 million people were estimated to be infected with *S. japonicum* in 1948 (Mao 1948); however, the number of people infected and at risk of infection was assumed to be highly underestimated (Stoll 1947; Wright 1950). A considerable number of residents died of *S. japonicum* infections, and many endemic villages left deserted, with the emergence of many "village without villagers," "widows villages," and "big-belly village" present in China (Xiao-Nong et al. 2018). From 1949 to 1950, a total of 33,891 Chinese soldiers were found to be infected with *S. japonicum* and lost their working capacity due to taking military exercises or swimming in water in endemic areas (Minggang and Zheng 1999).

2.2.3 The Call for Farewell to the GOD of Plague

Available data showed that schistosomiasis was endemic in 400 counties of China, with an estimated 11.8 million people infected and more than 100 million people at risk of infection, and a total area of 14.5 billion m^2 was found to breed *Oncomelania* snails, the intermediate host of *S. japonicum* at early 1950s (Fig. 2.1) (Anon 1977; Chen 1989a; Mao 1986). Being deeply concerned about the sufferings of the working people, since 1949 when the People's Republic of China was founded, leaders of the new country prioritized the control of schistosomiasis japonica. In 1950, the Ministry of Health organized an epidemiological survey of schistosomiasis in rural areas of the country, and in 1951, the *National Schistosomiasis Control Scheme in China* was issued. In 1953, Mao Zedong, Chairman of the People's Republic of China, stated that the State Council must emphasize schistosomiasis

Fig. 2.1 Geographical distribution of schistosomiasis in the People's Republic of China

control because of its severe damages, and in 1955, Chairman Mao visited schistosomiasis-endemic areas and stated that the Ministry of Health should eradicate schistosomiasis as a political task. On February 1956, Chairman Mao called for the integration of all resources for eradication of schistosomiasis in China at the supreme state conference. Since then, strong political will urged the building of leading group for schistosomiasis control at national, provincial, city and county levels, research committees and professional schistosomiasis control institutions, the formulation of directives and policies for schistosomiasis control, financial support, and mass participation in schistosomiasis control activities (Sandbach 1977). In Yujiang County, Jiangxi Province, a hyper-endemic area, a package of schistosomiasis control activities was employed, including snail control, chemotherapy, and health education since the call, and the integrated control efforts resulted in eradication of schistosomiasis in the county in 1958 (Dan-Dan et al. 2018). This county, therefore, became the first endemic area where schistosomiasis was eradicated in China. In view of this great achievement, a report that detailed the experiences of

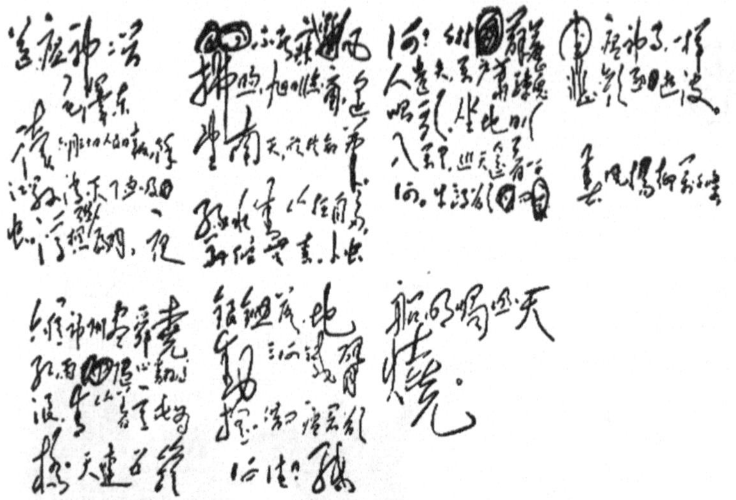

Fig. 2.2 The handwriting draft of the two poems "Farewell to the God of Plague" by Chairman Mao Zedong

schistosomiasis control in Yujiang County was published in *People's Daily*, the official newspaper of the Central Committee of the Communist Party of the People's Republic of China. After reading this report, Chairman Mao was very excited, and wrote the famous two poems *Farewell to the God of Plague* (Fig. 2.2). Since then, schistosomiasis was also called "god of plague" in China.

2.2.4 Schistosomiasis Control from China's Reform and Opening-Up to Present

In 1978, the China launched an economic reform and initiated the "Reform and Opening-up" policy. At the initial stage of economic reform, the commercialization of healthcare weakened the schistosomiasis prevention and control, and less funding was allocated to schistosomiasis control (Collins et al. 2012). During the period from 1992 through 2001 when the World Bank Loan Project for Schistosomiasis Control was implemented, Chinese government obtained adequate financial support from the World Bank, which greatly facilitated the progress toward control of transmission of schistosomiasis japonica in China (Xianyi et al. 2005; Changsong et al. 2002; Zhang et al. 2012). Following the termination of World Bank Loan Project for Schistosomiasis Control, the shrunk funding used for schistosomiasis control, together with changes of natural and social factors, resulted in a resurgence of schistosomiasis in China at the early twenty-first century (Liang et al. 2006). Confronting the challenges of resurging schistosomiasis, Chinese government made schistosomiasis one of the four highest priorities in communicable disease control (schistosomiasis,

HIV/AIDS, hepatitis B, and tuberculosis) in 2004, and issued the *National Mid- and Long-term Plan for Schistosomiasis Prevention and Control in China (2004–2015)*, which proposed a goal that reduced human prevalence to less than 5% in 2008 and to less than 1% in 2015 (Wang et al. 2008). In 2014, schistosomiasis was further included in the top five diseases (schistosomiasis, viral hepatitis, tuberculosis, cancer, and severe mental disorders) by the central government of China (http://www.gov.cn/guowuyuan/2014-09/01/content_2743567.htm). Under the strong political support, the implementation of integrated control interventions resulted in schistosomiasis infection control (<5% prevalence of *S. japonicum* infections in both humans and livestock, and no outbreak of acute schistosomiasis) in 2008 (Hao et al. 2008) and transmission control (<1% prevalence of *S. japonicum* infections in both humans and livestock, no outbreak of acute schistosomiasis, and no *S. japonicum*-infected *Oncomelania hupensis* snails for two successive years) in 2015 in China (Zhang et al. 2016a). In 2014, the central government proposed an ambitious goal, to achieve schistosomiasis transmission interruption (no local *S. japonicum* infections in humans and livestock for successive 5 years, no *S. japonicum*-infected for more than 5 successive years, and establishment of a sensitive and effective surveillance-response system) by 2020 and schistosomiasis elimination (no local *S. japonicum* infections in humans, livestock, and *Oncomelania hupensis* for successive 5 years after transmission interruption) by 2025 in China (Lei and Zhou 2015). By the end of 2017, 5 out of the 12 provinces that are endemic for schistosomiasis achieved elimination of schistosomiasis, Sichuan Province achieved transmission interruption, and the remaining 6 endemic provinces achieved transmission control; among the 450 endemic counties, 50.89% (229/450), 30.89% (139/450), and 18.22% (82/450) achieved schistosomiasis elimination, transmission interruption, and transmission control, respectively (Li-Juan et al. 2018). In addition, 37,601 cases with *S. japonicum* infections were estimated by 2017, including 1 acute case and 29,407 advanced cases, and 1 bovine was egg positive (Li-Juan et al. 2018). The latest data demonstrate a continuous decline in the endemic situation of schistosomiasis in China, and the country is moving to schistosomiasis transmission interruption and elimination.

2.3 Shift of Chinese National Schistosomiasis Control Strategy

Before the founding of the People's Republic of China in 1949, there was no national schistosomiasis control strategy in the country, and the government only led a few epidemiological surveys of schistosomiasis. After the founding of new China, the Communist Party of China and the government highly emphasized schistosomiasis control (Xiao-Nong et al. 2018). From the initiation of national schistosomiasis control programs until now, there are three shifts in the national schistosomiasis

Fig. 2.3 Integrated measures used for *Oncomelania hupensis* snail control in China during the period between middle 1950s and early 1980s. (**a**) Land reclamation; (**b**) soil burying; (**c**), tractor ploughing; (**d**) chemical treatment

control strategy in China, and all strategies were developed with high adaptation to active circumstances (Xu et al. 2016a).

At the first stage (from middle 1950s to early 1980s), a snail control-based integrated strategy was employed in China, which mainly depends on snail control with chemical treatment and environmental improvements, together with individual protection (Collins et al. 2012). During this stage, mass campaigns were launched to eliminate *Oncomelania hupensis* snails by means of agricultural production, hydraulic construction, and chemical treatment (Fig. 2.3). The implementation of the snail control-based integrated strategy resulted in a remarkable decline in human *S. japonicum* infections, number of severe cases and snail habitats, and the snail habitats reduced by 11 billion m^2 until 1981 (Mao 1987; Chen 1989b). In 1985, schistosomiasis was eradicated in Shanghai and Guangdong, where

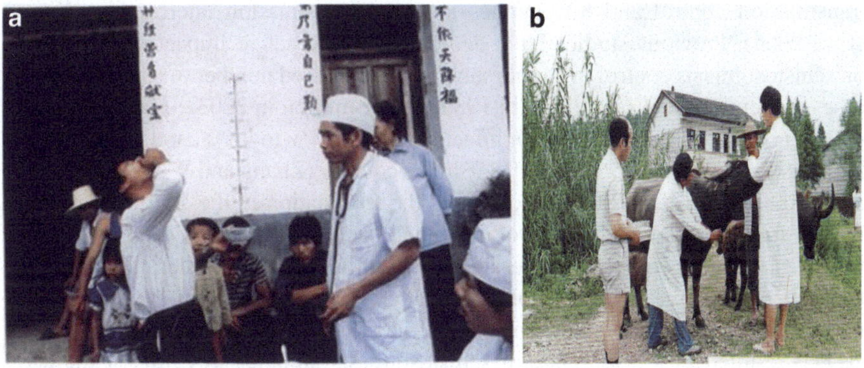

Fig. 2.4 Chemotherapy with praziquantel in China. (**a**) Patients receiving oral administration with praziquantel; (**b**) bovines given praziquantel treatment by gavage

schistosomiasis was highly prevalent historically (Zhou et al. 1994). Experiences from the snail control-based integrated strategy demonstrated that the goal of schistosomiasis eradication might be achieved through snail control in plain regions with waterway networks where water levels are under control, and in regions with relatively high economic development; however, the effectiveness of the snail control-based integrated strategy seemed unsatisfactory in marshland and lake regions and mountainous and hilly regions where the settings are complex and the economic development is not high because of the high cost and complicated procedures of mollusciciding treatment and environmental improvements (Mao 1987; Chen 1989b). Although huge progresses had achieved in schistosomiasis control during this stage, the number of acute schistosomiasis cases remained high at middle and late 1980s and the snail habitats reduced from 2.75 billion m^2 in 1980 to 3.47 billion m^2 in 1988 (Chen 1989b). More importantly, a resurgence of schistosomiasis was found in some endemic foci of China (Chen 1989b).

At the second stage (from middle 1980s to 2003), a morbidity control-based integrated strategy was employed in China, which mainly depends on synchronous chemotherapy of humans and bovine with praziquantel (Fig. 2.4), snail control, and health education (Xu et al. 2016a). With the advent and extensive use of praziquantel, a highly effective and lowly toxic oral schistosomicide praziquantel (Chen 2005, 1984; Chen et al. 1983), the national schistosomiasis control strategy from transmission control to morbidity control (Qing-Wu et al. 2002). Such a strategy has been proved effective to reduce the prevalence and intensity of *S. japonicum* infections, with a reduction in the number of schistosomiasis cases from 1.7 million in 1992 to 0.83 million in 2001 in China, and resulted in eradication of schistosomiasis in Fujian (1987), Guangxi (1989), and Zhejiang (1995) at this stage (Chen 1986; Sleigh et al. 1998; Wu et al. 2005). During the period between 1992 and 2001, the World Bank Loan Project for Schistosomiasis Control implemented in China was conducted in China (Xianyi et al. 2005). During the 10-year period, 47 out of the 219 counties endemic for schistosomiasis that were covered by the project achieved transmission control and 82 counties achieved transmission interruption (Xianyi et al. 2005). Previous studies have demonstrated the active impact of the project on schistosomiasis control in China, such as the estimated number of schistosomiasis cases reducing from 1.64 million in 1989 to 0.87 million in 2003, a reduction in the prevalence of human *S. japonicum* infections from 9.7% to 4.9%, and a reduction in livestock infection from 13.2% to 9.1% in the country (Zhang and Wong 2003; Yuan et al. 2000; Zhou et al. 2005b). Following the termination of the World Bank Loan Project for Schistosomiasis Control in China, inadequate financial support to schistosomiasis control, reform of professional institutions, as well as alteration of natural factors such as large flooding along the Yangtze River basin (Wu et al. 2008) and water resources development (Zheng et al. 2002; Yang et al. 2005a; Wang et al. 2009a) resulted in a resurgence of schistosomiasis japonica in China at the early twenty-first century (Liang et al. 2006). Experiences from the morbidity control-based integrated strategy demonstrate that such a strategy is effective to rapidly reduce the prevalence of *S. japonicum* infections and achieve remarkable effectiveness for schistosomiasis control within a short period of time (Xu et al. 2016a); however, the snail habitats were not completely altered, and reinfections with *S. japonicum* in both humans and livestock still occurred, notably in hyper-endemic areas. Since monochemotherapy with praziquantel fails in the prevention of *S. japonicum* reinfections and interruption of schistosomiasis transmission, the achievements of schistosomiasis control are difficult to be sustainable (Wang and Liang 2015; Wu and Huang 2013; Ross et al. 2015).

From 2004 to present, an integrated strategy with emphasis on control of the source of *S. japonicum* infection was employed, which consists of removal of cattle from snail-infested grasslands, replacement of bovine with mechanized farm equipment, improved sanitation, providing boats with fecal-matter containers, chemotherapy, snail control, and intensive health education (Fig. 2.5) (Wang et al. 2009b). At the early twenty-first century, schistosomiasis became one of the major public health problems again in China due to re-emerging schistosomiasis across the country (Wang et al. 2008). Since 2004, the integrated strategy with emphasis on control of the source of *S. japonicum* infection was implemented in China, and pilot studies showed high effectiveness of this integrated strategy for schistosomiasis control (Wang et al. 2009c). Results from further expanded studies indicate that the integrated strategy with emphasis on control of the source of *S. japonicum* infection is effective to reduce the transmission of *S. japonicum* in all types of endemic areas in China (Liu et al. 2017; Chen et al. 2014; Zhou et al. 2011, 2013a; Hong et al. 2011, 2013; Sun et al. 2011a, 2017), and such a strategy has remarkable long-term effectiveness on the transmission of schistosomiasis japonica in China (Wang et al. 2017). A recent meta-analysis reveals that the integrated strategy with emphasis on control of the source of *S. japonicum* infection is highly effective to control the transmission of *S. japonicum* in China and shows more effective to control *S. japonicum* infections in both human (log*RR* difference = 1.32, 95% *CI*: 0.78–1.86) and *Oncomelania hupensis* snails (log*RR* difference = 1.53, 95% *CI*: 0.76–2.31) relative to the morbidity control-based integrated strategy (Qian et al. 2018). The 12-year implementation of this integrated strategy resulted in a 90.84%

Fig. 2.5 Integrated measured used to control the source of *Schistosoma japonicum* infections. (**a**) Replacement of bovines with machines; (**b**) forbidding bovines to pasture in marshlands; (**c**) raising bovines in fences; (**d**) building public toilets for boatmen and fishermen; (**e**) building household lavatory; (**f**) intensified management of boatmen and fishermen

reduction in the number of human schistosomiasis cases and a 98.78% reduction in the bovine infection with *S. japonicum*, and transmission control was achieved in China (Zhang et al. 2016a; Wang et al. 2017). These data demonstrate that the integrated strategy with emphasis on control of the source of *S. japonicum* infection is the optimal strategy for schistosomiasis control in the current socioeconomic circumstances and facilitates the progress toward elimination of schistosomiasis in China (Xiao-Nong et al. 2018).

2.4 Three Nationwide Epidemiological Surveys on Schistosomiasis Japonica

Periodic national epidemiological surveys are of great importance to understand the endemic status of schistosomiasis and evaluate the effectiveness of the national schistosomiasis control program, which may provide insight into the development and alteration of the national schistosomiasis control strategy. During the period from the initiation of the national schistosomiasis control programs until now, there are three nationwide epidemiological sampling surveys on schistosomiasis japonica conducted in China.

In 1989, the first nationwide epidemiological sampling survey on schistosomiasis was conducted in eight provinces that were endemic for schistosomiasis in China, using a stratified, randomized, cluster sampling method, including Hunan, Hubei, Jiangxi, Anhui, Jiangsu, Zhejiang, Yunnan, and Sichuan, and a total of 383 sampling sites were assigned covering 556,357 subjects (Zhen 1993). All subjects were subjected to Kato-Katz and miracidia hatching test (three slides for one stool sample), and hepatomegaly and splnomegaly were detected using an A-mode ultrasonography, while livestock infection was identified using miracidia hatching test. The overall prevalence of human *S. japonicum* infection was 10.2%, and the prevalence was 14.4%, 13.5%, 5.5%, 8.6%, 0, 0, 4.4%, and 16.8% in Hunan, Hubei, Jiagnxi, Anhui, Jiangsu, Zhejiang, Yunnan, and Sichuan, respectively, and 0.1% in children at ages of 10–14 years, while 14%, 12.4%, 3.1%, 2.6%, and 28.6% infection rates were detected in cattle, buffaloes, pigs, dogs, and sheep. Based on the results of the sampling survey, it was estimated that there were 1.52 million people having schistosomiasis and 57.6 thousand advances cases in China, and the snail habitats were estimated to be 3.675 billion m^2, with 95.15% in Hunan, Hubei, Jiangxi, Anhui, and Jiangsu (Zhen 1993).

During the period between September and November, 1995, the second nationwide epidemiological sampling survey on schistosomiasis was conducted in 195 sites covering 314,177 residents and 14,543 bovines sampled from 8 provinces that were endemic for schistosomiasis in China, using a stratified and clustering random sampling technique, including Hunan, Hubei, Jiangxi, Anhui, Jiangsu, Zhejiang, Yunnan, and Sichuan (Office of Endemic Diseases Control, Ministry of Health, Expert Advisory Committee on Schistosomiasis, Ministry of Health, Institute of Parasitic Diseases, Chinese Academy of Preventive Medicine 1998). The overall prevalence of *S. japonicum* infection was 4.89% in humans and 9.06% in bovines, and the human infection rate was 7.84% in Hunan, 7.19% in Hubei, 6% in Yunnan, 4.54% in Anhui, 3.8% in Jiangxi, 1.92% in Sichuan, and 0.03% in Jiangsu, respectively. Based on the results of this sampling survey, it was estimated that there were 865,084 cases with schistosomiasis, 100,251 bovines with *S. japonicum* infections, and 55,961 advanced cases in China (Office of Endemic Diseases Control, Ministry of Health, Expert Advisory Committee on Schistosomiasis, Ministry of Health, Institute of Parasitic Diseases, Chinese Academy of Preventive Medicine 1998). In addition, the highest prevalence of human infections was found in marshland and lake regions (6.63%), followed by in mountainous and hilly regions (2.75%), and the lowest prevalence was found in plain regions with waterway networks (0.02%) (Office of Endemic Diseases Control, Ministry of Health, Expert Advisory Committee on Schistosomiasis, Ministry of Health, Institute of Parasitic Diseases, Chinese Academy of Preventive Medicine 1998). These data demonstrate a remarkable decline in the prevalence of *S. japonicum* infections in both humans and bovines relative to the first nationwide sampling survey.

In 2004, the third nationwide schistosomiasis sampling survey was conducted covering 239 samples sites from all 7 schistosomiasis-endemic provinces of Anhui, Hubei, Hunan, Jiangsu, Jiangxi, Sichuan, and Yunnan (Zhou et al. 2007a). All participants were screened for serum-specific IgG antibody against *S. japonicum* using a commercial enzyme-linked immunosorbent assay (ELISA), and the sero-positives were tested with Kato-Katz technique (three slides for one fecal sample). Of the 250,987 participants tested with ELISA, the sero-prevalence of *S. japonicum* infection was 12.2%; of these persons, 94.2% submitted a stool specimen, and 9.1% of these specimens were egg positives. Therefore, the estimated prevalence of human *S. japonicum* infection was 2.5%, and the estimated prevalence rates were 4.2%, 3.8%, 3.1%, 2.2%, 1.7%, 0.9%, and 0.3% in the provinces of Hunan, Hubei, Jiangxi, Anhui, Yunnan, Sichuan, and Jiangsu, respectively. The highest prevalence rate was found in the lake and marshland region (3.8%), followed by in the mountainous and hilly regions (1.1%) and the lowest rate was seen in the plain regions with waterway networks (0.06%). Based on the results of the sampling survey, it was estimated that there were 726,112 (95% *CI*: 714,497–737,728) persons infected with *S. japonicum* in China in 2004 (Zhou et al. 2007a).

Results from three nationwide sampling surveys demonstrate that schistosomiasis is mainly concentrated in the marshlands along the Yangtze River, and the relative importance of bovine increases in the transmission of *S. japonicum*, although human prevalence appears a remarkable decline since 1989 (Table 2.1). These three nationwide epidemiological surveys have enabled the dynamic trends of schistosomiasis epidemics and have assisted in analyzing subtle changes in endemicity. The results can be summarized mathematically allowing the variation in efficacy of control

Table 2.1 Comparison of the major endemic indicators of schistosomiasis japonica among the three nationwide sampling surveys in China

Index	The first sampling survey	The second sampling survey	The third sampling survey
Human prevalence of *S. japonicum* infection (%)	10.2	4.89	2.5
Estimated number of schistosomiasis cases in China	1.52 million	865,084	726,112
Estimated number of advanced cases in China	57.6 thousand	55,961	N/A
Bovine infection rate (%)	14	9.06	N/A
Estimated snail habitats in China (m^2)	3.675 billion	3.4 billion	3.52 billion

measures to be readily determined and enabling control strategies to be adjusted and updated (Li et al. 2005a).

2.5 Diagnosis

Currently, the definite diagnosis of *S. japonicum* infections still depends on detection of schistosome eggs, and parasitological techniques are still the diagnosis of choice in national schistosomiasis control programs (Gray et al. 2011). In China, miracidium hatching test and Kato-Katz technique serve as gold standards for definite diagnosis of *S. japonicum* infection; however, parasitological methods are labor-intensive and time-consuming, and have declining compliance rates, resulting in a low coverage (Zhu et al. 2014; Yu et al. 2007; Zhou et al. 2007b). In addition, Kato-Katz technique is reported to underestimate the prevalence of *S. japonicum* due to a high missing diagnosis rate (Lin et al. 2008a).

Immunodiagnostics, which have shown a high sensitivity, and simple and rapid procedures, have been accepted as an excellent epidemiological tool for the screening of target populations in schistosomiasis-endemic areas, and are also useful for the surveillance of cure after chemotherapy, and for periodic control of transmission of the infection after it has been eliminated in an area (Gray et al. 2011). During the past six decades, Chinese scientists make great contributions to immunodiagnosis of schistosomiasis japonica and develop a large number of immunodiagnostic assays (Zhu 2005). Circumoval precipitation test, indirect hemagglutination assay (IHA), enzyme-linked immunosorbent assay (ELISA), dot immunogold filtration assay (DIGFA), and dye dipstick immunoassay have been developed to screen the target populations for chemotherapy (Wen et al. 2005; Xu et al. 2011a, b; Lin et al. 2008b; Zhu et al. 2002; Cai et al. 2012; Wang et al. 2012a). Currently, IHA, ELISA, and DDIA are three common immunodiagnostic assays used for identifying human *S. japonicum* infections (Xu et al. 2011b). These immunological techniques have shown a high sensitivity and specificity in the diagnosis of schistosomiasis japonica and play important roles in the national schistosomiasis control programs of China (Zhang et al. 2016b).

Recently, novel PCR-based molecular biological assays have been developed to detect *S. japonicum* infection (Fung et al. 2012; Lier et al. 2009; Wang et al. 2011; He et al. 2018; Weerakoon et al. 2016, 2017; Xu et al. 2010), and loop-mediated isothermal amplification (LAMP) assay has been proved effective for sensitive and rapid detection of *S. japonicum*, which provides a potential tool for the early diagnosis and therapy evaluation for *S. japonicum* infection in humans (Xu et al. 2015; Xia et al. 2009).

2.6 Treatment

During 60-year control activities, many chemicals have been screened and developed for the treatment of schistosomiasis japonica at various stages in China, including potassium antimony tartrate, furapromidum, nitrofuran, niridazole, amoscanate, praziquantel, and artemisinins (Xiao et al. 2010). Currently, praziquantel is virtually the only choice of drug for the treatment of human *S. japonicum* infections in China (Chen 2005). In the national schistosomiasis control program of China, it is defined that mass chemotherapy with praziquantel is performed in highly endemic regions (human prevalence > 15%), selective chemotherapy given in moderately endemic region and expanded chemotherapy given lowly endemic regions (Wu and Huang 2013). Although there is a concern about development of praziquante-resistant isolates of *S. japonicum* following long-term extensive application (Wang et al. 2012b), evaluation of the susceptibility of *S. japonicum* to praziquantel in highly, moderately, and lowly endemic regions of China demonstrate that the current efficacy of praziquantel against *S. japonicum* is still high and has not changed after more than three decades of repeated, expanded chemotherapy in the main endemic areas of China, suggesting that no evidence of tolerance or resistance to praziquantel in *S. japonicum* in China (Wang et al. 2010, 2012c; Seto et al. 2011).

Artemisinin (qinghaosu), extracted from the Chinese herb *Artemisia annua* L., and its two major derivatives—artemether and artesunate—have been shown active against *S. japonicum* (Liu et al. 2012, 2014; Xiao 2005). Administration of artemether or artesunate at multiple doses over 1- or 2-week intervals was found to result in protection rates of 65–97% for preventing human *S. japonicum* infections, while combined treatment with praziquantel and artemisinin derivatives (artemether or artesunate) resulted in a higher protection rate of 84% (95% *CI*: 64–91%) than praziquantel monotherapy for treatment, and praziquantel and artesunate in combination had a great protection rate of 96% (95% *CI*: 78–99%) for preventing schistosomes infection (Xiao et al. 2005; Hua et al. 2010; Li et al. 2005b; Hou et al. 2008; Liu et al. 2011). In addition, two antimalarial agents mefloquine and dihydroartemisinin also show activity against *S. japonicum* infections in animal models and in vitro assays (Keiser et al. 2009; Xiao et al. 2009a, b; Xiao 2013; Li et al. 2011a, b, c), which deserve further clinical test. More interestingly, the sensitivity of artemisinin derivatives dihydroartemisinin, artemether, and artesunate does not reduce in praziquantel-resistant *S. japonicum* (Wang et al. 2014a, b).

In addition to artemisinin (Zhang et al. 2014), experimental and clinical studies have shown the potential of traditional Chinese herbs for the treatment of *S. japonicum* infections, notably hepatic fibrosis caused by *S. japonicum* infections (Mehlhorn et al. 2014). However, randomized, controlled clinical trials are needed to assess the efficacy and side effects of these traditional Chinese herbs for human *S. japonicum* infections.

2.7 Chinese Schistosomiasis Surveillance Network

In China, surveillance of schistosomiasis involves detection of infections in both humans and animals, snail survey, and investigation of natural and social factors associated with schistosomiasis transmission (Zhou et al. 2005c). Currently, Chinese schistosomiasis surveillance network includes three categories (Zhu et al. 2011).

1. Nationwide cross-sectional sampling survey. From the initiation of the national schistosomiasis control program until now, three nationwide sampling surveys on schistosomiasis had been conducted (Zhen 1993; Office of Endemic Diseases Control, Ministry of Health, Expert Advisory Committee on Schistosomiasis, Ministry of Health, Institute of Parasitic Diseases, Chinese Academy of Preventive Medicine 1998; Zhou et al. 2007a), and the results were of great significance to understand the epidemiology of schistosomiasis across China and provided important decision-making evidence to formulate national schistosomiasis control strategy at various stages and evaluate the effectiveness of the strategy.
2. Regional surveillance, including surveillance of schistosomiasis in five provinces of China which have reached the national criteria for elimination of the disease (Wu et al. 2005), and monitoring of the transmission of *S. japonicum* in potential risk regions of China (Qian et al. 2015). In the five provinces with interruption of schistosomiasis transmission, snails, *S. japonicum* infections in local residents, livestock and wild animals, and imported sources of *S. japonicum* infections were monitored, and imported cases with *S. japonicum* infections and residual snail habitats were detected in these five provinces, which may lead to re-emerging schistosomiasis (Wen et al. 2011). The surveillance results indicated the consolidation of schistosomiasis control achievements; however, factors leading to resurgence of schistosomiasis still exist in these transmission-interrupted areas. Further intensified surveillance is needed to eliminate the potential risk factors, so as to prevent re-emerging schistosomiasis. To assess the impact of climate warming and water resource development on the transmission of schistosomiasis, ten schistosomiasis potential foci were assigned in ten non-endemic areas from five provinces of Jiangsu, Shandong, Anhui, Hubei, and Chongqing since 2008, to investigate suspected settings and floating humans and livestock and monitor imported snails and sources of *S. japonicum* infections (Qian et al. 2015). Results from the 5-year evaluation between 2008 and 2012 showed that there were sources of *S. japonicum* infections imported into the potential schistosomiasis foci and there was a high possibility of importation of snails into the potential schistosomiasis foci (Dang et al. 2014). These findings provided the target for schistosomiasis surveillance in potential foci, and provided insights into the development of interventions to effectively reduce the risk of schistosomiasis transmission.
3. Sentinel surveillance sites of China (Zhu et al. 2011). During the period between 1990 and 1998, 14 sentinel surveillance sites were assigned by the Ministry of Health in 6 provinces of Hubei, Hunan, Jiangxi, Anhui, Sichuan, and Yunnan where schistosomiasis was highly endemic, and from 2000 to 2004, 21 sentinel

sites were set in 7 endemic provinces of Hubei, Hunan, Jiangxi, Anhui, Jiangsu, Sichuan, and Yunnan. During the period from 1990 through 2004, snails, *S. japonicum* infections, and local control activities were monitored in sentinel sites, which provided valuable data for understanding the changes of schistosomiasis epidemiology and scientifically evaluating the effectiveness of the control activities (Zhao et al. 2005). Since 2005, a total of 80 sentinel sites were assigned in 10 provinces of Hubei, Hunan, Jiangxi, Anhui, Jiangsu, Sichuan, Yunnan, Zhejiang, Shanghai, and Chongqing, and natural and social factors associated with schistosomiasis transmission were also included for monitoring (Zhu et al. 2011).

The Chinese schistosomiasis surveillance network provides plenty of valuable data for decision-makers to develop and adjust the national schistosomiasis control strategy; however, the current surveillance network suffers from problems of underdeveloped monitoring techniques, low data availability, and lack of effective surveillance-response system (Zhang et al. 2016c). Surveillance should therefore be conceived and developed as an intervention approach and a sensitive surveillance-response system should be established to tackle the potential re-emerging schistosomiasis in China (Zhou et al. 2013b; Bergquist et al. 2015; Tambo et al. 2014).

2.8 Novel Tools Employed for Schistosomiasis Control

2.8.1 Geographic Information Systems (GIS) and Remote Sensing (RS)

The GIS and RS techniques were introduced in vector-borne diseases to describe digital map data generated by earth observing satellite sensors for spatial and temporal environment analyses (Hay 2000; Brooker et al. 2002; Leonardo et al. 2005; Stensgaard et al. 2005; Malone 2005). These methods have been used for mapping and modeling transmission of schistosomiasis japonica, identifying high-risk areas of intermediate host snails, and evaluating the ecological transformation and climate change in China (Yang et al. 2005a, b; Zhang et al. 2005, 2009; Zhou et al. 2002; Peng et al. 2010). The GIS/RS technologies had been proved extremely useful for assessing and monitoring the effect of ecological transformation on schistosomiasis transmission in China (Zhou et al. 2001).

2.8.2 Google Earth

In the field schistosomiasis control, the rapid release and sharing of the monitoring information is the prerequisite to the rapid response (Bergquist et al. 2015). Recently, Google Earth was introduced to schistosomiasis control in China, and a

Fig. 2.6 A Google Earth-based surveillance-response system

surveillance-response system of schistosomiasis was developed based on Google Earth (Fig. 2.6), which effectively enhances the use of the monitoring information and achieves the synchronous visualization of the monitoring information (Sun et al. 2011b; Yang et al. 2012). This system displays the graphs and texts directly and clearly and is easy and simple to perform, suggesting that Google Earth is a user-friendly, effective, and feasible tool for schistosomiasis control (Wang et al. 2016).

2.8.3 *Detection of* S. japonicum *DNA in* O. hupensis *Snails*

To achieve early detection of *S. japonicum* infection in snails, a rapid extraction of snail genomic DNA combined with LAMP assay was developed (Yu et al. 2008), which greatly reduces the identification of infected snails from 60 days (dissection of snails) to approximately 1 week (Xiong et al. 2014). As compared to currently available commercial imported reagents, this kit (Fig. 2.7) has comparable detection efficiency but showing an over 50% reduction in cost (Xiong et al. 2014), which has been integrated into the national schistosomiasis control programs of China. During the process toward transmission interruption and elimination of schistosomiasis, this assay, which greatly improves the sensitivity for detection of *S. japonicum* infection in snails in relation to conventional microscopic approaches, may provide an effective approach for the rapid identification and timely elimination of the risk of schistosomiasis transmission (Qin et al. 2018).

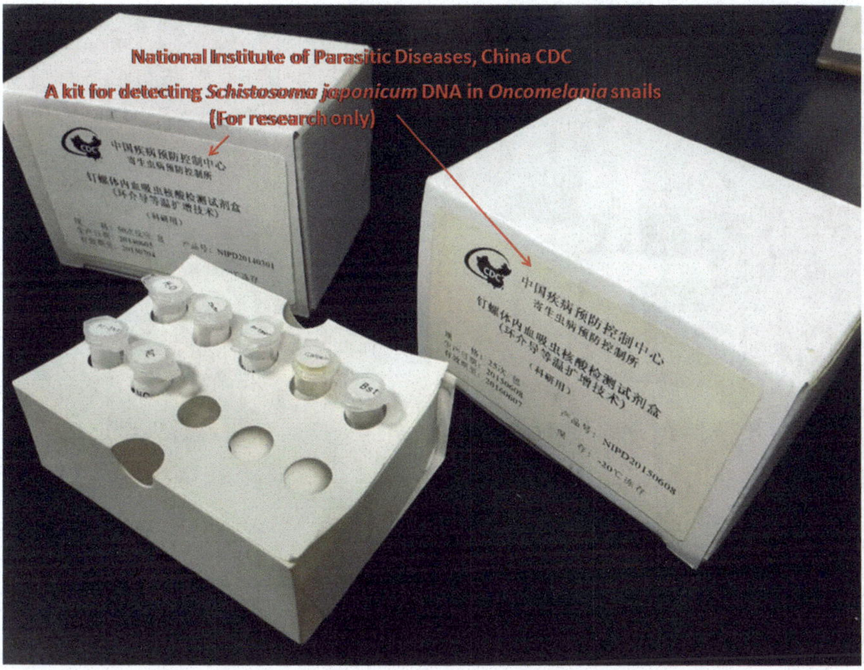

Fig. 2.7 A kit to detect *Schistosoma japonicum* infection in *Oncomelania hupensis* snails

2.8.4 *An Intelligent Device for Detecting* S. japonicum-*Infested Water*

According to the biological feature of *S. japonicum* cercariae that float on water surface and cannot migrate actively (Haas et al. 1987), an intelligent device was developed to identify *S. japonicum*-infested water with a mouse bioassay (Fig. 2.8). This device increases the possibility of detection of *S. japonicum* cercariae through the remote-controlled movement in the water body. Field test showed that the detector reduced the detection from 8 h to 1 h, and increase*d S. japonicum* infection from 15% to 40% in sentinel mice, and the intensity of infection (worm burden) from 0.25 to 2.55 worms per mouse (Yang et al. 2013). The intelligent detector greatly enhances the efficiency for the field detection of the infested water, and has played a critical role in the surveillance-response system for schistosomiasis along the lower reaches of the Yangtze River basin (Liang et al. 2012).

Fig. 2.8 An intelligent device for detecting *Schistosoma japonicum*-infested water

2.9 Conclusions and Perspectives

Experiences and lessons from national schistosomiasis control programs of China during the past six decades demonstrate that strong political will, rational control strategy, and adequate financial support would accelerate the progress toward elimination of schistosomiasis japonica in the country (Zhou et al. 2012; Wang et al. 2014c; Chen et al. 2018).

China has achieved remarkable success in the control of schistosomiasis transmission, and it is now moving toward the progress of elimination of schistosomiasis japonica (Lei and Zhou 2015). However, there are still many factors challenging the progress toward schistosomiasis elimination in China, including huge snail habitats in endemic foci, reduced financial support, difficulty in the management of livestock infections, potential risk of *O. hupensis* snail spread, and increase in the number of imported cases with schistosomiasis (Jing et al. 2018; Shi-Zhu et al. 2019; Jian-Feng et al. 2019; Song et al. 2018; Cui et al. 2018; Wang et al. 2013; Jiao-Jiao 2019). Therefore, the integrated strategy with emphasis on control of the source of *S. japonicum* infection should be further intensified under government leadership and through multi-sectoral collaboration according to *The Three-year Tough Action Plan for Endemic Diseases Control (2018–2020)*, and joint, precision control interventions are implemented to achieve the goal of schistosomiasis elimination in China (Shi-Zhu et al. 2018).

Currently, China is transferring schistosomiasis control knowledge and techniques to African countries and aiding the elimination of schistosomiasis hemotobia in African mainland (Xu et al. 2016b, c; Wang et al. 2019). Let's expect the coming of the miracle "Farewell to God of Plague" as soon as possible in both China and across the world.

Acknowledgements This study was funded by the supports from Jiangsu Provincial Health and Family Planning Commission (grant No. QNRC2016621 and H2018097) and Wuxi Municipal Bureau of Science and Technology (grant No. CSE31N1730).

Conflict of Interests None.

References

Anon (1977) Recent progress in schistosomiasis control. Chin Med J 3:95–99

Bergquist R, Yang GJ, Knopp S, Utzinger J, Tanner M (2015) Surveillance and response: tools and approaches for the elimination stage of neglected tropical diseases. Acta Trop 141:229–234

Brooker S, Hay SI, Bundy DA (2002) Tools from ecology: useful for evaluating infection risk models? Trends Parasitol 18:70–74

Cai YC, Guo J, Chen SH, Tian LG, Steinmann P, Chen MX, Li H, Ai L, Chen JX (2012) Chicken egg yolk antibodies (IgY) for detecting circulating antigens of *Schistosoma japonicum*. Parasitol Int 61:385–390

Carroll DG (1946) Cerebral involvement in schistosomiasis japonica. Bull Johns Hopkins Hosp 78:219–234

Catto J (1905) *Schistosoma cattoi*, a new blood fluke of man. Br Med J 1:11–13, 26

Cecil R, Robertson MC (1940) Schistosomiasis in the Tali-Fu region of Yunnan Province. Chin Med J 57:358–363

Changsong S, Binggui Y, Hongyi L, Yuhai D, Xu X, Huiguo Z, Yong J (2002) Achievement of the World Bank loan project on schistosomiasis control (1992–2000) in Hubei province and the challenge in the future. Acta Trop 82:169–174

Chen MG (1984) New developments in the clinical study of praziquantel. Ji Sheng Chong Xue Yu Ji Sheng Chong Bing Za Zhi 2:193–195

Chen GZ (1986) Schistosomiasis control in hilly regions of Fujian Province. Chin Med J 99:311–316

Chen MG (1989a) Progress and problems in schistosomiasis control in China. Trop Med Parasitol 40:174–176

Chen MG (1989b) Schistosomiasis control program in the People's Republic of China: a review. Southeast Asian J Trop Med Public Health 20:511–517

Chen MG (2005) Use of praziquantel for clinical treatment and morbidity control of schistosomiasis japonica in China: a review of 30 years' experience. Acta Trop 96:168–176

Chen MG, Fu S, Hua XJ, Wu HM (1983) A retrospective survey on side effects of praziquantel among 25,693 cases of schistosomiasis japonica. Southeast Asian J Trop Med Public Health 14:495–500

Chen YY, Liu JB, Huang XB, Cai SX, Su ZM, Zhong R, Zou L, Miao XP (2014) New integrated strategy emphasizing infection source control to curb schistosomiasis japonica in a marshland area of Hubei Province, China: findings from an eight-year longitudinal survey. PLoS One 9: e89779

Chen J, Xu J, Bergquist R, Li SZ, Zhou XN (2018) "Farewell to the God of Plague": the importance of political commitment towards the elimination of schistosomiasis. Trop Med Infect Dis 3: E108

Chu CF (1937) Schistosomiasis japonica in Nanking. Chin Med J 52:651–664

Colley DG, Bustinduy AL, Secor WE, King CH (2014) Human schistosomiasis. Lancet 383:2253–2264

Collins C, Xu J, Tang SL (2012) Schistosomiasis control and the health system in P.R. China. Infect Dis Poverty 1:8

Cui J, Jiang P, Song YY, Zhang X, Wang ZQ (2018) Imported African schistosomiasis and the potential risk of transmission in China. Pathog Glob Health 112:101–105

Dan-Dan L, Yi-Wen L, Xiao-Qing T, Dong-Yun A, Qing-Wu J (2018) Experience and insights of schistosomiasis elimination in Yujiang County of Jiangxi Province-commemoration of the 60th anniversary of publishing Chairman Mao Zedong's two poems "Farewell to the God of Plague". Zhongguo Xue Xi Chong Bing Fang Zhi Za Zhi 30:489–492

Dang H, Xu J, Li SZ, Cao ZG, Huang YX, Wu CG, Tu ZW, Zhou XN (2014) Monitoring the transmission of *Schistosoma japonicum* in potential risk regions of China, 2008–2012. Int J Environ Res Public Health 11:2278–2287

Faust EC (1924) Schistosomiasis in China: biological and practical aspects. Proc R Soc Med 17:31–43

Faust EC, Meleney HE (1924) Studies on schistosomasis japonica. Am J Hyg Monogr Ser 3:339

Fung MS, Xiao N, Wang S, Carlton EJ (2012) Field evaluation of a PCR test for *Schistosoma japonicum* egg detection in low-prevalence regions of China. Am J Trop Med Hyg 87:1053–1058

Gray DJ, Ross AG, Li YS, McManus DP (2011) Diagnosis and management of schistosomiasis. BMJ 342:d2651

Haas W, Granzer M, Garcia EG (1987) Host identification by *Schistosoma japonicum* cercariae. J Parasitol 73:568–577

Hao Y, Zheng H, Zhu R, Guo JG, Wu XH, Wang LY, Chen Z, Zhou XN (2008) Schistosomiasis status in People's Republic of China in 2008. Zhongguo Xue Xi Chong Bing Fang Zhi Za Zhi 20:451–456

Hay SI (2000) An overview of remote sensing and geodesy for epidemiology and public health applications. Adv Parasitol 47:2–27

He P, Gordon CA, Williams GM, Li Y, Wang Y, Hu J, Gray DJ, Ross AG, Harn D, McManus DP (2018) Real-time PCR diagnosis of *Schistosoma japonicum* in low transmission areas of China. Infect Dis Poverty 7:8

Hong QB, Yang K, Huang YX, Sun LP, Yang GJ, Gao Y, Gao Y, Zhang LH, Zhou M, Steinmann P, Liang YS (2011) Effectiveness of a comprehensive schistosomiasis japonica control program in Jiangsu province, China, from 2005 to 2008. Acta Trop 120:S151–S167

Hong XC, Xu XJ, Chen X, Li YS, Yu CH, Yuan Y, Chen YY, Li RD, Qiu J, Liu ZC, Yi P, Ren GH, He HB (2013) Assessing the effect of an integrated control strategy for schistosomiasis japonica emphasizing bovines in a marshland area of Hubei Province, China: a cluster randomized trial. PLoS Negl Trop Dis 7:e2122

Hou XY, McManus DP, Gray DJ, Balen J, Luo XS, He YK, Ellis M, Williams GM, Li YS (2008) A randomized, double-blind, placebo-controlled trial of safety and efficacy of combined praziquantel and artemether treatment for acute *Schistosomiasis japonica* in China. Bull World Health Organ 86:788–795

Hua HY, Liang YS, Zhang Y, Wei JF, Guo HX (2010) The sensitivity of artesunate against *Schistosoma japonicum* decreased after 10 years of use in China. Parasitol Res 107:873–878

Jian-Feng Z, Li-Yong W, Jing X, You-Sheng L, Xiao-Lan Y, Guang-Hui R, Tie-Wu J, Wei W, Xiao-Nong Z (2019) Current status and transmission risks of oversea imported schistosomiasis in China. Zhongguo Xue Xi Chong Bing Fang Zhi Za Zhi 31:26–32

Jiao-Jiao L (2019) Endemic status and control of animal schistosomiasis in China. Zhongguo Xue Xi Chong Bing Fang Zhi Za Zhi 31:40–46

Jing X, Shan L, Chun-Li C, Shi-Zhu L, Xiao-Nong Z (2018) Progress and challenges of schistosomiasis elimination in China. Zhongguo Xue Xi Chong Bing Fang Zhi Za Zhi 30:605–609

Kan HC, Kung JC (1936) Incidence of schistosomiasis japonica in an endemic area in Chekiang. Chin Med J 25:449–456

Kan HC, Yao YT (1934) Some notes on the anti-schistosomiasis japonica campaign in Chih-huai-pan, Kaihua, Chekiang. Chin Med J 23:323–336

Keiser J, Chollet J, Xiao SH, Mei JY, Jiao PY, Utzinger J, Tanner M (2009) Mefloquine--an aminoalcohol with promising antischistosomal properties in mice. PLoS Negl Trop Dis 3:e350

King CH (2010) Parasites and poverty: the case of schistosomiasis. Acta Trop 113:95–104

King CH, Dickman K, Tisch DJ (2005) Reassessment of the cost of chronic helmintic infection: a meta-analysis of disability-related outcomes in endemic schistosomiasis. Lancet 365:1561–1569

Kuo SC (1946) Further studies on distribution of schistosomiasis japonica in Szechwan province, China. J Parasitol 32:367–368

Lei ZL, Zhou XN (2015) Eradication of schistosomiasis: a new target and a new task for the National Schistosomiasis Control Programme in the People's Republic of China in 2015. Zhongguo Xue Xi Chong Bing Fang Zhi Za Zhi 27:1–4

Leonardo LR, Rivera PT, Crisostomo BA, Sarol JN, Bantayan NC, Tiu WU, Bergquist NR (2005) A study of the environmental determinants of malaria and schistosomiasis in the Philippines using remote sensing and geographic information systems. Parassitologia 47:105–114

Li YS, Zhao ZY, Ellis M, McManus DP (2005a) Applications and outcomes of periodic epidemiological surveys for schistosomiasis and related economic evaluation in the People's Republic of China. Acta Trop 96:266–275

Li YS, Chen HG, He HB, Hou XY, Ellis M, McManus DP (2005b) A double-blind field trial on the effects of artemether on *Schistosoma japonicum* infection in a highly endemic focus in southern China. Acta Trop 96:184–190

Li HJ, Wang W, Qu GL, Tao YH, Xing YT, Li YZ, Wei JY, Dai JR, Liang YS (2011a) In-vivo activity of dihydroartemisinin against *Schistosoma japonicum*. Ann Trop Med Parasitol 105:181–185

Li HJ, Wang W, Li YZ, Qu GL, Xing YT, Tao YH, Wei JY, Dai JR, Liang YS (2011b) Effects of artemether, artesunate and dihydroartemisinin administered orally at multiple doses or combination in treatment of mice infected with *Schistosoma japonicum*. Parasitol Res 109:515–519

Li HJ, Wang W, Tao YH, Qu GL, Xing YT, Li YZ, Wei JY, Dai JR, Liang YS (2011c) Dihydroartemisinin-praziquantel combinations and multiple doses of dihydroartemisinin in the treatment of *Schistosoma japonicum* in experimentally infected mice. Ann Trop Med Parasitol 105:329–333

Liang S, Yang C, Zhong B, Qiu D (2006) Re-emerging schistosomiasis in hilly and mountainous areas of Sichuan, China. Bull World Health Organ 84:139–144

Liang YS, Huang YX, Hong QB, Yang K, Sun LP, Dai JR, Yang GJ, Yu CX, Tian ZX, Ge J, Wu HH, Gao Q (2012) Novel strategies and technologies to achieve the transmission control of schistosomiasis in Jiangsu Province. Zhongguo Xue Xi Chong Bing Fang Zhi Za Zhi 24:11

Lier T, Simonsen GS, Wang T, Lu D, Haukland HH, Vennervald BJ, Hegstad J, Johansen MV (2009) Real-time polymerase chain reaction for detection of low-intensity *Schistosoma japonicum* infections in China. Am J Trop Med Hyg 81:428–432

Li-Juan Z, Zhi-Min X, Si-Min D, Hui D, Shan L, Jing X1, Shi-Zhu L, Xiao-Nong Z (2018) Endemic status of schistosomiasis in People's Republic of China in 2017. Zhongguo Xue Xi Chong Bing Fang Zhi Za Zhi 30:481–488

Lin DD, Liu JX, Liu YM, Hu F, Zhang YY, Xu JM, Li JY, Ji MJ, Bergquist R, Wu GL, Wu HW (2008a) Routine Kato-Katz technique underestimates the prevalence of *Schistosoma japonicum*: a case study in an endemic area of the People's Republic of China. Parasitol Int 57:281–286

Lin DD, Xu JM, Zhang YY, Liu YM, Hu F, Xu XL, Li JY, Gao ZL, Wu HW, Kurtis J, Wu GL (2008b) Evaluation of IgG-ELISA for the diagnosis of *Schistosoma japonicum* in a high prevalence, low intensity endemic area of China. Acta Trop 107:128–133

Liu R, Dong HF, Guo Y, Zhao QP, Jiang MS (2011) Efficacy of praziquantel and artemisinin derivatives for the treatment and prevention of human schistosomiasis: a systematic review and meta-analysis. Parasit Vectors 4:201

Liu R, Dong HF, Jiang MS (2012) Artemisinin: the gifts from traditional Chinese medicine not only for malaria control but also for schistosomiasis control. Parasitol Res 110:2071–2074

Liu YX, Wu W, Liang YJ, Jie ZL, Wang H, Wang W, Huang YX (2014) New uses for old drugs: the tale of artemisinin derivatives in the elimination of *Schistosomiasis japonica* in China. Molecules 19:15058–15074

Liu Y, Zhong B, Wu ZS, Liang S, Qiu DC, Ma X (2017) Interruption of schistosomiasis transmission in mountainous and hilly regions with an integrated strategy: a longitudinal case study in Sichuan, China. Infect Dis Poverty 6:79

Logan OT (1905) A case of dysentery in Hunan province, caused by the trematoda, *Schistosoma japonicum*. Chin Med J 19:243–245

Malone JB (2005) Biology-based mapping of vector-borne parasites by geographic information systems and remote sensing. Parassitologia 47:27–50

Mao CP (1948) A review of the epidemiology of schistosomiasis japonica in China. Am J Trop Med Hyg 28:659–672

Mao SB (1986) Recent progress in the control of schistosomiasis in China. Chin Med J 99:439–443

Mao SP (1987) Epidemiology and control of schistosomiasis in the People's Republic of China. Mem Inst Oswaldo Cruz 82:77–82

McManus DP, Dunne DW, Sacko M, Utzinger J, Vennervald BJ, Zhou XN (2018) Schistosomiasis. Nat Rev Dis Primers 4:13

Mehlhorn H, Wu ZD, Ye B (2014) Treatment of human parasitosis in traditional Chinese medicine. Springer, Basel, pp 141–154

Minggang C, Zheng F (1999) Schistosomiasis control in China. Parasitol Int 48:11–19

Office of Endemic Diseases Control, Ministry of Health, Expert Advisory Committee on Schistosomiasis, Ministry of Health, Institute of Parasitic Diseases, Chinese Academy of Preventive Medicine (1998) An introduction of a nationwide sampling survey on schistosomiasis in China. Zhongguo Ji Sheng Chong Xue Yu Ji Sheng Chong Bing Za Zhi 16:84–88

Peng WX, Tao B, Clements A, Jiang QL, Zhang ZJ, Zhou YB, Jiang QW (2010) Identifying high-risk areas of schistosomiasis and associated risk factors in the Poyang Lake region, China. Parasitology 137:1099–1107

Qian YJ, Li SZ, Xu J, Yang K, Huang YX, Cao ZG, Miu F, Dang H, Zhang LJ, Zhang L, Wang Q, Bergquist R, Zhou XN (2015) Potential schistosomiasis foci in China: a prospective study for schistosomiasis surveillance and response. Acta Trop 141:342–348

Qian C, Zhang Y, Zhang X, Yuan C, Gao Z, Yuan H, Zhong J (2018) Effectiveness of the new integrated strategy to control the transmission of *Schistosoma japonicum* in China: a systematic review and meta-analysis. Parasite 25:54

Qin ZQ, Xu J, Feng T, Lv S, Qian YJ, Zhang LJ, Li YL, Lv C, Bergquist R, Li SZ, Zhou XN (2018) Field evaluation of a loop-mediated isothermal amplification (LAMP) platform for the detection of *Schistosoma japonicum* infection in *Oncomelania hupensis* snails. Trop Med Infect Dis 3: E124

Qing-Wu J, Li-Ying W, Jia-Gang G, Ming-Gang C, Xiao-Nong Z, Engels D (2002) Morbidity control of schistosomiasis in China. Acta Trop 82:115–125

Ross AG, Olveda RM, Chy D, Olveda DU, Li Y, Harn DA, Gray DJ, McManus DP, Tallo V, Chau TN, Williams GM (2015) Can mass drug administration lead to the sustainable control of schistosomiasis? J Infect Dis 211:283–289

Sandbach FR (1977) Farewell to the god of plague--the control of schistosomiasis in China. Soc Sci Med 11:27–33

Seto EY, Wong BK, Lu D, Zhong B (2011) Human schistosomiasis resistance to praziquantel in China: should we be worried? Am J Trop Med Hyg 85:74–82

Sewell RB (1919) The possible occurrence of *Schistosoma japonicum*, Katsurada in India. Rec Ind Mus 16:425–430

Shi-Zhu L, Jing X, Shan L, Chun-Li C, Qiang W, Hong-Liang Q, Jun Y, Xiao-Nong Z (2018) Interpretation of the three - year tough action plan for endemic diseases control (2018–2020): schistosomiasis. Zhongguo Xue Xi Chong Bing Fang Zhi Za Zhi 30:601–604

Shi-Zhu L, Jing X, Tian-Ping W, Li-Yong W, Kun Y, Wei W, Shan L, Chun-Li C, Xiao-Nong Z (2019) Upholding Chinese spirit on schistosomiasis control in the new era to accelerate the progress towards schistosomiasis elimination in China. Zhongguo Xue Xi Chong Bing Fang Zhi Za Zhi 31:1–13

Sleigh A, Li X, Jackson S, Huang K (1998) Eradication of schistosomiasis in Guangxi, China. Part 1: setting, strategies, operations, and outcomes, 1953–92. Bull World Health Organ 76:361–372

Song LG, Zeng XD, Li YX, Zhang BB, Wu XY, Yuan DJ, He A, Wu ZD (2018) Imported parasitic diseases in mainland China: current status and perspectives for better control and prevention. Infect Dis Poverty 7:78

Spiridion JT (1936) Schistosomiasis japonica: an account of an outbreak. J Trop Med Hyg 39:161–164

Steinmann P, Keiser J, Bos R, Tanner M, Utzinger J (2006) Schistosomiasis and water resources development: systematic review, meta-analysis, and estimates of people at risk. Lancet Infect Dis 6:411–425

Stensgaard A, Jørgensen A, Kabatereine NB, Malone JB, Kristensen TK (2005) Modeling the distribution of *Schistosoma mansoni* and host snails in Uganda using satellite sensor data and geographical information systems. Parassitologia 47:115–125

Stoll NR (1947) This wormy world. J Parasitol 33:1–18

Sun LP, Wang W, Liang YS, Tian ZX, Hong QB, Yang K, Yang GJ, Dai JR, Gao Y (2011a) Effect of an integrated control strategy for schistosomiasis japonica in the lower reaches of the Yangtze River, China: an evaluation from 2005 to 2008. Parasit Vectors 4:243

Sun LP, Liang YS, Wu HH, Tian ZX, Dai JR, Yang K, Hong QB, Zhou XN, Yang GJ (2011b) A Google Earth-based surveillance system for schistosomiasis japonica implemented in the lower reaches of the Yangtze River, China. Parasit Vectors 4:223

Sun LP, Wang W, Zuo YP, Hong QB, Du GL, Ma YC, Wang J, Yang GJ, Zhu DJ, Liang YS (2017) A multidisciplinary, integrated approach for the elimination of schistosomiasis: a longitudinal study in a historically hyper-endemic region in the lower reaches of the Yangtze River, China from 2005 to 2014. Infect Dis Poverty 6:56

Tambo E, Ai L, Zhou X, Chen JH, Hu W, Bergquist R, Guo JG, Utzinger J, Tanner M, Zhou XN (2014) Surveillance-response systems: the key to elimination of tropical diseases. Infect Dis Poverty 3:17

Tang CC (1936) Schistosomiasis japonica in Fukien with special reference to the intermediate host. Chin Med J 50:1585–1590

Tang CC (1939) Further investigation on schistosomiasis japonica in Futsing, Fukien Province. Chin Med J 56:452–473

Wang LD (2006) Experiences and perspectives of schistosomiasis control in China. People' Medical Publishing House, Beijing, pp 1–2

Wang W, Liang Y (2015) Mass drug administration (MDA) for schistosomiasis. J Infect Dis 211:848–849

Wang L, Utzinger J, Zhou XN (2008) Schistosomiasis control: experiences and lessons from China. Lancet 372:1793–1795

Wang W, Dai JR, Liang YS, Huang YX, Coles GC (2009a) Impact of the South-to-North Water Diversion Project on the transmission of *Schistosoma japonicum* in China. Ann Trop Med Parasitol 103:17–29

Wang LD, Chen HG, Guo JG, Zeng XJ, Hong XL, Xiong JJ, Wu XH, Wang XH, Wang LY, Xia G, Hao Y, Chin DP, Zhou XN (2009b) A strategy to control transmission of *Schistosoma japonicum* in China. N Engl J Med 360:121–128

Wang LD, Guo JG, Wu XH, Chen HG, Wang TP, Zhu SP, Zhang ZH, Steinmann P, Yang GJ, Wang SP, Wu ZD, Wang LY, Hao Y, Bergquist R, Utzinger J, Zhou XN (2009c) China's new strategy to block *Schistosoma japonicum* transmission: experiences and impact beyond schistosomiasis. Tropical Med Int Health 14:1475–1483

Wang W, Dai JR, Li HJ, Shen XH, Liang YS (2010) Is there reduced susceptibility to praziquantel in *Schistosoma japonicum*? Evidence from China. Parasitology 137:1905–1912

Wang C, Chen L, Yin X, Hua W, Hou M, Ji M, Yu C, Wu G (2011) Application of DNA-based diagnostics in detection of schistosomal DNA in early infection and after drug treatment. Parasit Vectors 4:164

Wang W, Li Y, Li H, Xing Y, Qu G, Dai J, Liang Y (2012a) Immunodiagnostic efficacy of detection of *Schistosoma japonicum* human infections in China: a meta analysis. Asian Pac J Trop Med 5:15–23

Wang W, Wang L, Liang YS (2012b) Susceptibility or resistance of praziquantel in human schistosomiasis: a review. Parasitol Res 111:1871–1877

Wang W, Dai JR, Li HJ, Shen XH, Liang YS (2012c) The sensitivity of *Schistosoma japonicum* to praziquantel: a field evaluation in areas with low endemicity of China. Am J Trop Med Hyg 86:834–836

Wang W, Liang YS, Hong QB, Dai JR (2013) African schistosomiasis in mainland China: risk of transmission and countermeasures to tackle the risk. Parasit Vectors 6:249

Wang W, Li HJ, Qu GL, Xing YT, Yang ZK, Dai JR, Liang YS (2014a) Is there a reduced sensitivity of dihydroartemisinin against praziquantel-resistant *Schistosoma japonicum*? Parasitol Res 113:223–228

Wang W, Li TY, Ji Y, Qu GL, Qian YL, Li HJ, Dai JR, Liang YS (2014b) Efficacy of artemether and artesunate in mice infected with praziquantel non-susceptible isolate of *Schistosoma japonicum*. Parasitol Res 113:925–931

Wang W, Dai JR, Liang YS (2014c) Apropos: factors impacting on progress towards elimination of transmission of *Schistosomiasis japonica* in China. Parasit Vectors 7:408

Wang XY, He J, Yang K, Liang S (2016) Applications of spatial technology in schistosomiasis control programme in the People's Republic of China. Adv Parasitol 92:143–163

Wang X, Wang W, Wang P (2017) Long-term effectiveness of the integrated schistosomiasis control strategy with emphasis on infectious source control in China: a 10-year evaluation from 2005 to 2014. Parasitol Res 116:521–528

Wang XY, He J, Juma S, Kabole F, Guo JG, Dai JR, Li W, Yang K (2019) Efficacy of China-made praziquantel for treatment of *Schistosomiasis haematobium* in Africa: a randomized controlled trial. PLoS Negl Trop Dis 13:e0007238

Weerakoon KG, Gordon CA, Gobert GN, Cai P, McManus DP (2016) Optimisation of a droplet digital PCR assay for the diagnosis of *Schistosoma japonicum* infection: a duplex approach with DNA binding dye chemistry. J Microbiol Methods 125:19–27

Weerakoon KG, Gordon CA, Williams GM, Cai P, Gobert GN, Olveda RM, Ross AG, Olveda DU, McManus DP (2017) Droplet digital PCR diagnosis of human schistosomiasis: parasite cell-free DNA detection in diverse clinical samples. J Infect Dis 216:1611–1622

Wen LY, Chen JH, Ding JZ, Zhang JF, Lu SH, Yu LL, Shen LY, Wu GL, Zhou XN, Zheng J (2005) Evaluation on the applied value of the dot immunogold filtration assay (DIGFA) for rapid detection of anti-*Schistosoma japonicum* antibody. Acta Trop 96:142–147

Wen LY, Yan XL, Zhang JF, Li LS, Lin LJ, Huang SY, Li L, Liang YS (2011) Strategy of solidification and surveillance for schistosomiasis in transmission-interrupted provinces in China. Zhongguo Xue Xi Chong Bing Fang Zhi Za Zhi 23:18–21, 31

Wright WH (1950) Bilharziasis as a public health problem in the Pacific. Bull World Health Organ 2:581–595

Wu W, Huang Y (2013) Application of praziquantel in schistosomiasis japonica control strategies in China. Parasitol Res 112:909–915

Wu XH, Chen MG, Zheng J (2005) Surveillance of schistosomiasis in five provinces of China which have reached the national criteria for elimination of the disease. Acta Trop 96:276–281

Wu XH, Zhang SQ, Xu XJ, Huang YX, Steinmann P, Utzinger J, Wang TP, Xu J, Zheng J, Zhou XN (2008) Effect of floods on the transmission of schistosomiasis in the Yangtze River valley, People's Republic of China. Parasitol Int 57:271–276

Xia CM, Rong R, Lu ZX, Shi CJ, Xu J, Zhang HQ, Gong W, Luo W (2009) *Schistosoma japonicum*: a PCR assay for the early detection and evaluation of treatment in a rabbit model. Exp Parasitol 121:175–179

Xianyi C, Liying W, Jiming C, Xiaonong Z, Jiang Z, Jiagang G, Xiaohua W, Engels D, Minggang C (2005) Schistosomiasis control in China: the impact of a 10-year World Bank loan project (1992–2001). Bull World Health Organ 83:43–48

Xiao SH (2005) Development of antischistosomal drugs in China, with particular consideration to praziquantel and the artemisinins. Acta Trop 96:153–167

Xiao SH (2013) Mefloquine, a new type of compound against schistosomes and other helminthes in experimental studies. Parasitol Res 112:3723–3740

Xiao S, Tanner M, N'Goran EK, Utzinger J, Chollet J, Bergquist R, Chen M, Zheng J (2005) Recent investigations of artemether, a novel agent for the prevention of schistosomiasis japonica, mansoni and haematobia. Acta Trop 96:153–167

Xiao SH, Chollet J, Utzinger J, Mei JY, Jiao PY, Keiser J, Tanner M (2009a) Effect of single-dose oral mefloquine on the morphology of adult *Schistosoma japonicum* in mice. Parasitol Res 105:853–861

Xiao SH, Mei JY, Jiao PY (2009b) Further study on mefloquine concerning several aspects in experimental treatment of mice and hamsters infected with *Schistosoma japonicum*. Parasitol Res 106:131–138

Xiao SH, Keiser J, Chen MG, Tanner M, Utzinger J (2010) Research and development of antischistosomal drugs in the People's Republic of China a 60-year review. Adv Parasitol 73:231–295

Xiao-Nong Z, Shi-Zhu L, Qing-Biao H, Kun Y, Shan L, Jing X (2018) Remain true to our original aspiration for farewell to the God of Plague, compose the new chapter for the national schistosomiasis control programme scientifically-commemoration of 60th anniversary of publishing Chairman Mao Zedong's two poems "Farewell to the God of Plague". Zhongguo Xue Xi Chong Bing Fang Zhi Za Zhi 30:1–4

Xiong CR, Yin XR, Song LJ, Shen S, Wang J, Gao H, Liu Q, Yu CX, Yang K (2014) Comparison of the effectiveness of loop-mediated isothermal DNA amplification (LAMP) and microscopic dissection at detecting snails infected with *Schistosoma japonicum*. Chin J Pathogen Biol 9:1084–1087

Xu J, Rong R, Zhang HQ, Shi CJ, Zhu XQ, Xia CM (2010) Sensitive and rapid detection of *Schistosoma japonicum* DNA by loop-mediated isothermal amplification LAMP. Int J Parasitol 40:327–331

Xu J, Feng T, Lin DD, Wang QZ, Tang L, Wu XH, Guo JG, Peeling RW, Zhou XN (2011a) Performance of a dipstick dye immunoassay for rapid screening of *Schistosoma japonicum* infection in areas of low endemicity. Parasit Vectors 4:87

Xu J, Peeling RW, Chen JX, Wu XH, Wu ZD, Wang SP, Feng T, Chen SH, Li H, Guo JG, Zhou XN (2011b) Evaluation of immunoassays for the diagnosis of *Schistosoma japonicum* infection using archived sera. PLoS Negl Trop Dis 5:e949

Xu J, Guan ZX, Zhao B, Wang YY, Cao Y, Zhang HQ, Zhu XQ, He YK, Xia CM (2015) DNA detection of *Schistosoma japonicum*: diagnostic validity of a LAMP assay for low-intensity infection and effects of chemotherapy in humans. PLoS Negl Trop Dis 9:e0003668

Xu J, Steinman P, Maybe D, Zhou XN, Lv S, Li SZ, Peeling R (2016a) Evolution of the national schistosomiasis control programmes in the People's Republic of China. Adv Parasitol 92:1–38

Xu J, Bergquist R, Qian YJ, Wang Q, Yu Q, Peeling R, Croft S, Guo JG, Zhou XN (2016b) China-Africa and China-Asia collaboration on schistosomiasis control: a SWOT analysis. Adv Parasitol 92:435–466

Xu J, Yu Q, Tchuenté LA, Bergquist R, Sacko M, Utzinger J, Lin DD, Yang K, Zhang LJ, Wang Q, Li SZ, Guo JG, Zhou XN (2016c) Enhancing collaboration between China and African countries for schistosomiasis control. Lancet Infect Dis 16:376–383

Yang GJ, Vounatsou P, Zhou XN, Tanner M, Utzinger J (2005a) A potential impact of climate change and water resource development on the transmission of *Schistosoma japonicum* in China. Parassitologia 47:127–134

Yang GJ, Vounatsou P, Zhou XN, Utzinger J, Tanner M (2005b) A review of geographic information system and remote sensing with applications to the epidemiology and control of schistosomiasis in China. Acta Trop 96:117–129

Yang K, Sun LP, Huang YX, Yang GJ, Wu F, Hang DR, Li W, Zhang JF, Liang YS, Zhou XN (2012) A real-time platform for monitoring schistosomiasis transmission supported by Google Earth and a web-based geographical information system. Geospat Health 6:195–203

Yang K, Sun LP, Liang YS, Wu F, Li W, Zhang JF, Huang YX, Hang DR, Liang S, Bergquist R, Zhou XN (2013) *Schistosoma japonicum* risk in Jiangsu province, People's Republic of China: identification of a spatio-temporal risk pattern along the Yangtze River. Geospat Health 8:133–142

Yao YT (1936) Report on the investigation of schistosomiasis japonica. Chin Med J 50:1667–1669

Yu JM, de Vlas SJ, Jiang QW, Gryseels B (2007) Comparison of the Kato-Katz technique, hatching test and indirect hemagglutination assay (IHA) for the diagnosis of *Schistosoma japonicum* infection in China. Parasitol Int 56:45–49

Yu CX, Yin XR, Hua WQ, Gao Q (2008) Establishment of loop mediated isothermal DNA amplification for identifying *Oncomelania* snails infected with *Schistosoma japonicum*. Chin J Pathogen Biol 3:661–664

Yuan H, Jiagang G, Bergquist R, Tanner M, Xianyi C, Huanzeng W (2000) The 1992–1999 World Bank schistosomiasis research initiative in China: outcome and perspectives. Parasitol Int 49:195–207

Zhang W, Wong CM (2003) Evaluation of the 1992–1999 World Bank schistosomiasis control project in China. Acta Trop 85:303–313

Zhang ZY, Xu DZ, Zhou XN, Zhou Y, Liu SJ (2005) Remote sensing and spatial statistical analysis to predict the distribution of *Oncomelania hupensis* in the marshlands of China. Acta Trop 96:205–212

Zhang ZJ, Carpenter TE, Lynn HS, Chen Y, Bivand R, Clark AB, Hui FM, Peng WX, Zhou YB, Zhao GM, Jiang QW (2009) Location of active transmission sites of *Schistosoma japonicum* in lake and marshland regions in China. Parasitology 136:737–746

Zhang Z, Zhu R, Ward MP, Xu W, Zhang L, Guo J, Zhao F, Jiang Q (2012) Long-term impact of the World Bank loan project for schistosomiasis control: a comparison of the spatial distribution of schistosomiasis risk in China. PLoS Negl Trop Dis 6:e1620

Zhang XG, Li GX, Zhao SS, Xu FL, Wang YH, Wang W (2014) A review of dihydroartemisinin as another gift from traditional Chinese medicine not only for malaria control but also for schistosomiasis control. Parasitol Res 113:1769–1773

Zhang LJ, Xu ZM, Qian YJ, Dang H, Lv S, Xu J, Li SZ, Zhou XN (2016a) Endemic status of schistosomiasis in People's Republic of China in 2015. Zhongguo Xue Xi Chong Bing Fang Zhi Za Zhi 28:611–617

Zhang JF, Xu J, Bergquist R, Yu LL, Yan XL, Zhu HQ, Wen LY (2016b) Development and application of diagnostics in the national schistosomiasis control programme in the People's Republic of China. Adv Parasitol 92:409–434

Zhang LJ, Li SZ, Wen LY, Lin DD, Abe EM, Zhu R, Du Y, Lv S, Xu J, Webster BL, Rollinson D, Zhou XN (2016c) The establishment and function of schistosomiasis surveillance system towards elimination in the People's Republic of China. Adv Parasitol 92:117–141

Zhao GM, Zhao Q, Jiang QW, Chen XY, Wang LY, Yuan HC (2005) Surveillance for schistosomiasis japonica in China from 2000 to 2003. Acta Trop 96:288–295

Zhen J (1993) A brief introduction to a nation-wide sampling survey on schistosomiasis. Chin Med J 106:569–575

Zheng J, Gu XG, Xu YL, Ge JH, Yang XX, He CH, Tang C, Cai KP, Jiang QW, Liang YS, Wang TP, Xu XJ, Zhong JH, Yuan HC, Zhou XN (2002) Relationship between the transmission of schistosomiasis japonica and the construction of the Three Gorge Reservoir. Acta Trop 82:147–156

Zhou D, Li Y, Yang X (1994) Schistosomiasis control in China. World Health Forum 15:387–389

Zhou XN, Malone JB, Kristensen TK, Bergquist RN (2001) Application of geographic information systems and remote sensing to schistosomiasis control in China. Acta Trop 79:97–106

Zhou X, Dandan L, Huiming Y et al (2002) Use of Landsat TM satellite surveillance data to measure the impact of the 1998 flood on snail intermediate host dispersal in the lower Yangtze River basin. Acta Trop 82:199–205

Zhou XN, Wang LY, Chen MG, Wu XH, Jiang QW, Chen XY, Zheng J, Utzinger J (2005a) The public health significance and control of schistosomiasis in China--then and now. Acta Trop 96:97–105

Zhou XN, Wang LY, Chen MG, Wang TP, Guo JG, Wu XH, Jiang QW, Zheng J, Chen XY (2005b) An economic evaluation of the national schistosomiasis control programme in China from 1992 to 2000. Acta Trop 96:255–265

Zhou XN, Jiang QW, Sun LP, Wang TP, Hong QB, Zhao GM, Wen LY, Yin ZC, Wu XH, Lin DD (2005c) Schistosomiasis control and surveillance in China. Zhongguo Xue Xi Chong Bing Fang Zhi Za Zhi 17:161–165

Zhou XN, Guo JG, Wu XH, Jiang QW, Zheng J, Dang H, Wang XH, Xu J, Zhu HQ, Wu GL, Li YS, Xu XJ, Chen HG, Wang TP, Zhu YC, Qiu DC, Dong XQ, Zhao GM, Zhang SJ, Zhao NQ, Xia G, Wang LY, Zhang SQ, Lin DD, Chen MG, Hao Y (2007a) Epidemiology of schistosomiasis in the People's Republic of China, 2004. Emerg Infect Dis 13:1470–1476

Zhou YB, Yang MX, Wang QZ, Zhao GM, Wei JG, Peng WX, Jiang QW (2007b) Field comparison of immunodiagnostic and parasitological techniques for the detection of schistosomiasis japonica in the People's Republic of China. Am J Trop Med Hyg 76:1138–1143

Zhou XN, Bergquist R, Leonardo L, Yang GJ, Yang K, Sudomo M, Olveda R (2010) Schistosomiasis japonica control and research needs. Adv Parasitol 72:145–178

Zhou YB, Liang S, Chen GX, Rea C, He ZG, Zhang ZJ, Wei JG, Zhao GM, Jiang QW (2011) An integrated strategy for transmission control of *Schistosoma japonicum* in a marshland area of China: findings from a five-year longitudinal survey and mathematical modeling. Am J Trop Med Hyg 85:83–88

Zhou YB, Liang S, Jiang QW (2012) Factors impacting on progress towards elimination of transmission of schistosomiasis japonica in China. Parasit Vectors 5:275

Zhou YB, Liang S, Chen GX, Rea C, Han SM, He ZG, Li YP, Wei JG, Zhao GM, Jiang QW (2013a) Spatial-temporal variations of *Schistosoma japonicum* distribution after an integrated national control strategy: a cohort in a marshland area of China. BMC Public Health 13:297

Zhou XN, Bergquist R, Tanner M (2013b) Elimination of tropical disease through surveillance and response. Infect Dis Poverty 2:1

Zhu YC (2005) Immunodiagnosis and its role in schistosomiasis control in China: a review. Acta Trop 96:130–136

Zhu Y, He W, Liang Y, Xu M, Yu C, Hua W, Chao G (2002) Development of a rapid, simple dipstick dye immunoassay for schistosomiasis diagnosis. J Immunol Methods 266:1–5

Zhu R, Zhao GM, Li HZ, Guo JG (2011) Development and prospect of surveillance network of schistosomiasis in China. Zhongguo Xue Xi Chong Bing Fang Zhi Za Zhi 23:14–17

Zhu HQ, Xu J, Zhu R, Cao CL, Bao ZP, Yu Q, Zhang LJ, Xu XL, Feng Z, Guo JG (2014) Comparison of the miracidium hatching test and modified Kato-Katz method for detecting *Schistosoma japonicum* in low prevalence areas of China. Southeast Asian J Trop Med Public Health 45:20–25

Chapter 3
Cholera (Blue Skin Disease) and Its History

Heinz Mehlhorn

Abstract This chapter reports details on the detection and spreading of the cholera on earth until today.

Keywords Cholera · Blue skin disease · Worldwide epidemics

3.1 Agent of Disease

Bacterium *Vibrio cholerae*. The genus name *Vibrio* goes back to the description of the German scientist Otto Friedrich Müller in the year 1772, when he observed "vibrating" small organisms at the surface of contaminated water of lakes. In total 12 out of 78 *Vibrio* species are known to be pathogenic. Infectious *Vibrio cholerae* bacteria appear in the intestinal mucus of humans as comma-shaped, curved stages measuring 1.5–3 μm in length. Their slow movements are induced by a single terminal flagellum (Fig. 3.4).

3.2 First Description

In the year 1893, the Noble Prize winner Robert Koch, Berlin (Germany), proved that *Vibrio cholerae* is the agent of a disease leading to diarrhea and vomiting (Koch 1884).

H. Mehlhorn (✉)
Department of Parasitology, Heinrich Heine University, Düsseldorf, Germany
e-mail: mehlhorn@uni-duesseldorf.de

H. Mehlhorn, S. Klimpel (eds.), *Parasite and Disease Spread by Major Rivers on Earth*, Parasitology Research Monographs 12,
https://doi.org/10.1007/978-3-030-29061-0_3

3.3 Cholera Epidemics in the World

3.3.1 Name

The name of this exclusively human disease of the intestinal tract has its source in the Greek term "kholera," which is derived from *khole* = bile and means "flowing yellow bile." This description is based on the reddish-yellow-greenish-blueish color of the fluids vomited and excreted by infected persons (Fig. 3.1). As can be seen in the literature dealing with disappearing societies, this disease has rather often killed exclusively humans especially in times, when efficacious drugs had not been yet

Fig. 3.1 Drawings of the German Dr. Robert Froriep (1831) showing excretions of cholera patients (Verlag des Landes Industrie Comptoir, Weimar). *1*. Vomited fluid. *2*. Rice-water-like intestinal excretions. *3*. Bloody intestinal excretions. *4.–6*. Urine of survivors of cholera

available for treatment. The various subtypes of the disease are described worldwide by different names (e.g., *Cholera asiatica*, *C. orientalis*, *C. epidemica*) being based on the local occurrence or the severeness of the symptoms (*Cholera sicca*).

3.3.2 History

The *first worldwide cholera epidemic* being clearly documented in human history started in the year 1817 along the river *Ganges* close to Calcutta. From there the agents of disease (unknown at this time) were constantly spread until 1824 being dispersed from India all over Southeast Asia and from there to the Middle East countries, to Europe, and also to many countries in Africa as well as to China and Russia. From the present knowledge, it can be concluded that the worldwide spreading of this disease was supported—if not due—to the worldwide activities of the British Army at this time and mainly by wandering pilgrims of various religions as well as by refugees. The *second worldwide epidemic* of the cholera disease lasted from 1826 to 1837 (1841).[1] After starting in Mekka (Saudi Arabia) it hit especially Egypt but also European countries more intensively than the first epidemic. Thus it is not astonishing that this disease occurred undiscovered as a bacterial disease in London, where the early epidemiologist *John Snow* (1813–1858) traced the cholera in Soho at London in 1854. In the year 1832 also the North American continent was hit very intensively. The cholera bacteria were imported and spread by European immigrants, who fled from the numerous wars and from bad harvests in Europe due to decreased temperatures. The *third cholera pandemic* started in the year 1846, reached high spreading rates in many countries beginning in the year 1852 and lasted until 1860. Especially during the Crimean war (1853–1856), at which English troops had been intensively involved, it extended its expansion from Europe again to North America and thus soon after victims also occurred in South America being spread by traveling businessmen. The *fourth cholera pandemic* lasted from 1863 until 1876 spreading again from harbors in India to Naples (Italy) and to Spain and to some other European regions, but also infected people in North Africa and South America who died in especially high numbers. The *fifth cholera pandemic* occurred from 1881 (1883)[2]–1896. Coming from India it spread in Egypt, all over Europe and in many countries in Asia and in South and North America [Robert Koch proved the bacterial nature of the cholera when studying the excretions of patients in Egypt (1883) and India (1884)] and thus opened means to avoid infections and further spreading of this disease. The *sixth cholera pandemic* followed in the year 1899 and lasted very long until 1923. Arriving very probably by ship from Asia it mainly hit Egypt, the countries of the Arabian peninsula, Persia, India and the Philippines, but also killed many people in

[1]Some authors see enlargement of this pandemic.

[2]Some sources claim that this epidemic started only in the year 1883.

Russia and in several European countries. The effects of this epidemic were less severe, since the German Nobel Prize winner Robert Koch (1843–1950) had discovered the bacterial agents of disease in the year 1893 (published in 1894), so that from thereon different methods and means to control the spreading and to cure the symptoms of the disease had been developed.

The 7th—since 1961 until today ongoing—*pandemic* due to infections with *Vibrio cholerae* had its origin in Indonesia and persists (actually spreading) especially in developing countries (e.g., especially in those south of the Sahara desert), but is also constantly imported in single cases into many other countries by business people or by tourists, who mainly have visited third world countries without avoiding contaminated food.

Until the discovery of the agent of disease—the bacterium *Vibrio cholerae*—published by Robert Koch (1884), the reason of the disease had been unknown. Some people claimed that this disease occurs as punishment sent from God for sinners. The German medical professor *Maximilian von Pettenkofer* (1818–1901) (see insert) was a supporter of the *miasma theory* (diseases due to bad air). To prove this theory, he drank a glass of water containing cultured stages of the cholera bacillus in order to prove their non-importance as agent of cholera diarrheas. Since he developed no or only very low-graded diarrheic symptoms, he and others remained convinced that this bacillus is not the reason of the worldwide-occurring death cases. However, he did not know that the infectious dose of different bacteria may vary considerably. For example, the infectious dose of *Shigella* stages needs just 10 organisms, that of *Salmonella* stages 10^4–10^5 specimens, while that of cholera, however, needs around 10^9 infectious stages depending on the strain in order to induce really significant symptoms.

3.4 Final Discovery of the Agents of the Cholera

The decisive detection was done during an expedition to Egypt starting on August 16, 1883, when Robert Koch and his colleagues Treskow, Fischer, and Gaffky (Fig. 3.2) entered the ship *Mongolia*, which served regularly the route Brindisi (Italy) to Bombay (India) via Port Said (Egypt) at the Suez channel, which had just been opened some years before (November 16, 1869). Koch and colleagues arrived in Port Said (Egypt) on August 23, 1883, and traveled by another boat to Alexandria, where they examined in a clinic a dead patient and his feces, from which he established cultures, wherein he succeeded to recognize a comma-shaped bacterium as an agent of cholera (Fig. 3.3). At the same time the young, just 27-year-old French scientist *Louis Thuillier* (a co-worker of Pettenkofer in Munich and Pasteur in Paris) died there due to cholera. He had described enlarged platelets and had kept them as the agents of cholera (Grüntzig and Mehlhorn 2010). Since the cholera epidemic started to run slowly down in Egypt, Robert Koch and his co-workers decided to follow their studies in the "motherland" of the cholera (India). Thus he and his co-workers left Egypt on November 13, 1883, by ship and arrived in Calcutta

Fig. 3.2 Doctoral members of the Egypt-India cholera expedition (1883–1884). From left: Koch (sitting), Treskow, Fischer and Gaffky (Robert Koch Institute, Berlin)

on December 11, 1883 (Figs. 3.4 and 3.5), where the English surgeon general Dr. *Douglas Cunningham* (1843–1944) supported their work (Grüntzig and Mehlhorn 2010). They succeeded in a very short time to prove that all of the investigated 22 death bodies and 21 survivors of the infection contained the comma-like bacteria (Fig. 3.3) thus verifying their preceding findings in Egypt. All these findings together were the basis of the publication which was written by Robert Koch after their return from their 4 months long expedition (1883–1884).

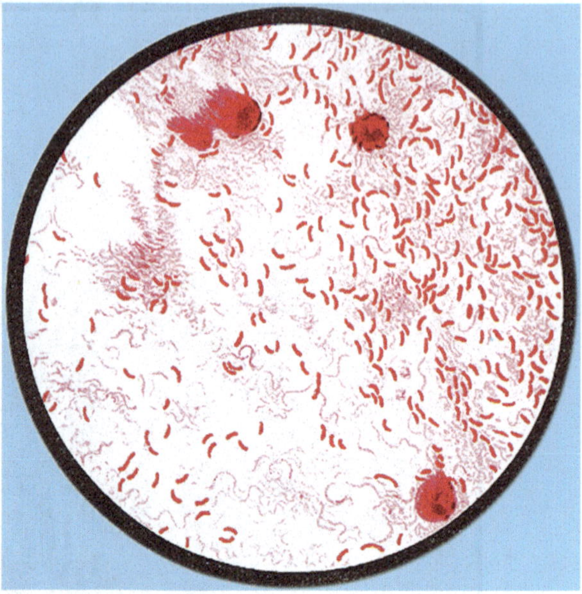

Fig. 3.3 Micrograph from a presentation table of Bernhard Nocht Institute Hamburg from the year 1909 showing the comma-like bacteria of *Vibrio cholerae*

3.5 Symptoms of Disease

The oral uptake of the cholera bacterium (*Vibrio cholerae*) induces only in 15% of the cases significant symptoms of disease. But in general there are three phases: (1) cholera diarrhea, (2) collapse (algidium stage); (3) reaction. The incubation period takes 1–5 days. During early slight reactions the infected persons excrete ~1 liter of watery feces per day accompanied by abdominal cramps (Gyr and Steffen 2010). However, ongoing infections are eventually characterized by daily excretion of 15 or even more liters of watery diarrheas and by intense vomiting (Fig. 3.1). Both effects lead quickly to severe dehydration (exsiccosis) and may induce collapse and electrolyte imbalance often being characterized by sunken eyes, cold skin, and reduced skin elasticity. Thus the skin may look blueish due to the dehydration, so in early times this disease was also called *blue disease* (Fig. 3.6) due to cyanosis. Common symptoms during this phase are hypovolemia, hemoconcentration, oliguria, and anuria in addition to metabolic acidosis and loss of amounts of potassium, followed by collapse and giddiness. However, fever is in general rare or not very high, but may occur as follow-up of a secondary bacterial or viral infection due to the weakened immune system (Gyr and Steffen 2010).

3.6 Start of Symptoms (Incubation Period)

Symptoms start very quickly or delayed after an oral infection (e.g., 2 h–5 days). Common are rice-water-like appearing diarrheas and intense vomiting of clear fluids (Fig. 3.1). If not treated the loss of fluid may kill infected persons very quickly

Fig. 3.4 Map showing the Indian towns, where Robert Koch and co-workers staged in India during the years 1883–1884 on their cholera expedition

because of exsiccosis, hypokalemia, and hypoglycemia. Thus sudden death cases may occur as the German poet Heinrich Heine reported in April 1832 out of Paris, where he was correspondent of the German newspaper *Augsburger Allgemeine Zeitung*, from a dancing festival in a hotel, where people wore head masks and danced intensely. Many of the dancers suddenly dropped down onto the floor and died. When the masks were taken away, the faces of the dead persons had a purple color—a sign of the cholera, which apparently had reached Paris.

Fig. 3.5 Photo of pilgrims at the Ganges in Benares (India) showing how easy cholera bacteria could have been transmitted by oral uptake during praying and bathing in contaminated water

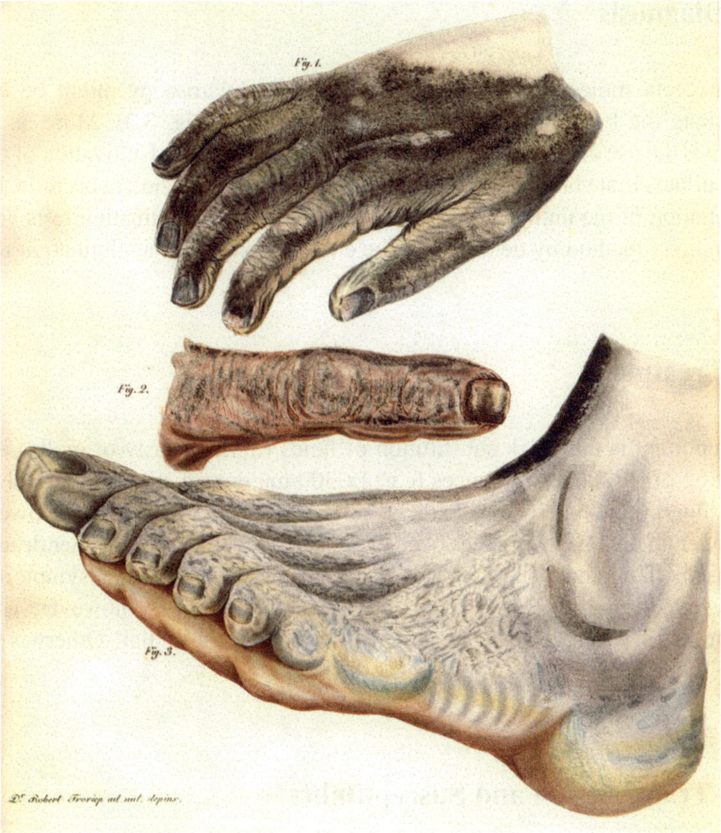

Fig. 3.6 Dr. Froriep's drawings showing the drying effects on the hand, finger, and foot of cholera patients in India in the year 1832 (Verlag des LandesIndustrieComptoirs, Weimar)

3.7 Duration of Symptoms

Infections of healthy adult people may induce only a few, slight symptoms for some days, if sufficient fluid is ingested. This occurs, however, mainly in cases with bacterial strains of low toxicity. However, children are much higher endangered in cases of high-graded loss of water. If untreated the disease may kill up to 60% of the patients or even more in cases of weak or old persons. These fatal cases are mainly due to severe loss of water (desiccosis). If therapy starts quickly accompanied by intense water replacement, death rate may drop down to 1% (Gyr and Steffen 2010).

3.8 Diagnosis

Besides recent molecular biological methods light microscopy might be used to differentiate the bacteria contained in the fecal fluids (Fig. 3.3). More and more common is the use of the following tests as methods of choice: Cultivation of isolates on thiosulfate citrate bile sucrose agar, which is selective for cholera bacteria. For the differentiation of the until now known serotype mainly agglutination tests are used or the indirect method by detection of the cholera toxin in an agglutination test.

3.9 Treatment

Most important is the quick substitution of fluids (intravenously or orally) and the application of antibiotic substances (e.g. 1x 300 mg doxycycline respectively tetracycline, fluoroquinolones, ciprofloxacin, ceftazidime, cefotaxime, etc.). However, in some endemic regions resistances occur against tetracycline so that other drugs have to be used. If treatment is done shortly after occurrence of first symptoms, the lethality rate drops to 1% or even less. Most important is, however, constant replacement of water loss and treatment of children is essential! Otherwise death cases will occur.

3.10 Transmission and Susceptibility

Human infections mainly occur due to contact with products contaminated by human feces. Thus infections may also occur by oral uptake of bacteria-contaminated water, fruits, or vegetables. However, also raw or undercooked fish and especially shellfish may contain the *Vibrio* stages, since they are often reproduced inside planktonic microorganisms, which belong to the food of a broad spectrum of fish and shellfish species. However, when consumed by humans, large amounts of bacteria must be taken up in order to induce the typical severe symptoms of a patent cholera disease. On the other side, children, elder people, and individuals with lowered immunity (e.g., AIDS patients, etc.) may suffer considerably more intensive from such an infection, and fatal cases occur consequently more often even at very low doses of ingested bacteria (Gyr and Steffen 2010).

Once the cholera bacteria have survived the passage of the acid fluids of the stomach and have reached the intestine, they do no longer need their single flagellum to move. Thus they stop to produce the protein flagellin and express several different surface proteins, which help to survive due to camouflage (=they are not detected by the host's defense system). Then they start to produce toxic proteins, which lead to the induction of several watery diarrheas per day, which set free infectious bacteria in the host's surroundings and thus enable the quick transfer of the bacteria to other human hosts, especially in cases of reduced hygienic conditions. These facts make it understandable that especially children or adults with a reduced immune system and people in rural regions and in locations with low hygienic conditions are especially endangered and their death rates reach highest levels during epidemics, especially when people like refugees or immigrants live in close contact to each other inside small rooms and use commonly a small number of toilets without sufficient handwashing and disinfection systems. Furthermore, cholera symptoms increase, when people suffer from the lack of sufficient drinking and safe water. This explains why especially many cases of cholera occurred among immigrants while transportation on ships on the rivers of the USA in the years following 1832, 1849, and 1866 (Pyle 2010; Rosenberg 1962).

Since the cholera disease (see features Robert Koch and Pettenkofer) is induced (as it is known today) by many types of *Vibrio cholerae*, it is notable that some strains induce severe symptoms and others lower ones or even none. Therefore, epidemics occurred in the past leading to extreme high numbers of death cases and others with much lower effects. Nevertheless, cholera became widespread in the nineteenth century and has killed at least several tens of millions humans worldwide. For example, India alone lost between 1900 and 1920 8–9 million people, or Russia claimed death cases of at least 1.5 million people in only 4 years (1847–1851).

Still today the *seventh pandemic cholera* affects about three to five million humans worldwide and leads to the death of 29,000–143,000 deaths per year. These numbers are more or less estimated by WHO, since many countries do not really count or do not diagnose the correct disease. In the latter cases, the risk of death may reach up to 60%, while, e.g., in Europe and the USA, the diagnosed death rates range today between 1 and 5%, especially since some nontoxic strains induce only mild or even no symptoms. However, it was also observed that under not very clear circumstances, "nontoxic" strains may become toxic. Thus public health officers should be aware especially after summer holidays that epidemics might occur.

Robert Koch: A Life for Science

1843	Born on December 11.
1862	Start of studies in Human Medicine at Göttingen University, Germany.
1866	Passes medical examinations.
1870	Serves in the German Army during the French-German War (1870–1871).
1872	Becomes District Physician in Wollstein (today Poland)—starts research on *anthrax*.
1876	Koch's famous *paper on anthrax* is published.
1877	He perfects techniques for microscopy of bacteria—first photomicrographs of bacteria appear.
1879	*Koch* moves to Breslau (today Wroczlaw, Poland) to become District Physician.

(continued)

1881	The famous culture techniques for bacteria (plate technique) are published on August 18th, then he starts his research on tuberculosis.
1882	Koch's paper on tubercle bacillus is published. – *Ehrlich's* paper on acid-fast staining is published. – Fannie and Walter *Hesse* develop the agar method for bacterial cultivation. – First confrontations between the "microbiology popes" *Louis Pasteur* and *Robert Koch.* – *Pasteur's* anthrax vaccine is used in Germany.
1883	*Koch's* cholera expedition to Egypt (Aug. 16) and India (Dec. 11). *Thullier* from the French cholera-team died in Alexandria, Egypt. Koch's "postulates" for bacterial transmissions are published first.
1883/1884	*Koch* isolated the *cholera agent* (*Vibrio cholerae*) and kept it in culture. Return to Berlin from India. Results prove that cholera is not due to "miasms" as claimed by *Pettenkofer* and others.
1885	*Koch* appointed as Professor of Hygiene at the University of Berlin and Director of the Hygiene Institute. World Cholera Conference is held in Berlin.
1890	– The pharmaceutical product *tuberculin* is announced as first cure for tuberculosis. – Massive controversies on tuberculin efficacy; *Koch* resigns his professorship to spend full time in tuberculin development. –Koch receives leadership of the new institute "Institute of Infectious Diseases." – December 4th *Koch's* disciples *Paul Ehrlich* and *Shibasaburo Kitasato* publish their epoch-making papers on the antitoxins of diphtheria and tetanus.
1891	The *Institute for Infectious Diseases* opens at Berlin; *Koch* publishes details on tuberculin.
1892	Severe cholera outbreaks occur in London and Hamburg—startup to control the cholera and typhoid disease by water filtration technique.
1893	*Emil von Behring* develops an antitoxin to diphtheria.
1894	*Behring* leaves Koch's institute to open a new institute and later a company (the famous Behring Werke at Marburg, now GSK, Glaxo Smith Kline).
1896–1900	*Koch* starts research expeditions to Africa studying *theileriosis* (=a tick-transmitted cattle disease), to India to study *plague* and to Italy to study *malaria.*
1901	*Koch's* disciple *Behring* is awarded the first *Nobel Prize for Medicine.*

(continued)

1902	Research of *typhoid fever* in Trier (Germany) leads to *Koch's* insights into the importance of human "silent carrier status." Expedition to Rhodesia (now Zimbabwe) to study an outbreak of "Rinderpest" (*cattle plague*).
1904	Expedition to Africa to study several tropical diseases.
1905	December 12 *Koch* awarded the *fourth Nobel Prize for Medicine* (for his research on tuberculosis)—after *Behring* (1901—diphtheria antitoxin), *Ross* (1902—malaria transmission) and *Finsen* (1903) and *Pawlow* (1904) (-light therapy respectively for conventional reflexes).
1906–1907	Koch's *sleeping sickness* (due to trypanosomes) *expedition* to East Africa. Use of the first trypanocidal drug (Germanin = Suramin) developed by the Bayer company in Germany.
1908	– *Koch* at congress in New York, receives honors. – *Koch* in Japan, overwhelming honors by the Emperor for *Koch* and his famous disciple *Kitasato*, who detected parallelly to *Pasteur's* coworker *Yersin* the agent of plague during the outbreak in Hong Kong: the bacterium *Yersinia (Pasteurella) pestis*. Participation at the International *Tuberculosis congress* in Washington to discuss the controversy on the identity of bovine and human tuberculosis.
1909	*Koch* suffers from severe health problems.
1910	May 27th *Koch* dies at an age of only 67 years in Baden-Baden (Germany). His body is buried in a "shrine" inside the governmental *Robert Koch Institute* at Berlin (Germany).

Literature

- Brock T (1999) Robert Koch, a life in medicine and bacteriology. ASM Press, Washington
- Grüntzig J, Heinz Mehlhorn M (2010) Robert Koch: Seuchenjäger und Nobelpreisträger. Spektrum Akad. Verlag, Heidelberg
- Grüntzig J, Mehlhorn H (2010) Expeditions into the empire of plagues. Düsseldorf University Press, Düsseldorf
- Koch R (1884) Über das Cholera-Bakterium (on the cholera bacterium). DtschMedWschr 10:725–728

Max Joseph von Pettenkofer (1818–1901)

This highly educated scientist (born 1818 in the Bavarian kingdom in South Germany) was a well-known chemist and hygienist. After studies in the institute of the famous German chemistry professor *Justus von Liebig* (1803–1878) in Gießen (Germany), he was appointed as a chemist at the Munich University in the year 1845, where he reached a full professorship in the year 1865. During these times he recognized that in the growing German towns large hygienic problems were arising. After his application the Bavarian government established three departments of the new scientific field *hygiene* in the university towns Munich, Würzburg, and Erlangen, and *hygiene* became an obligatory study subject for medical students in order to increase and protect public health.

Von Pettenkofer and *Koch* conflicted—in a today unusual intensity—on the identification of the agent (respectively cause) of the symptoms of a disease, which later at this time and still today is named *cholera* and may lead to death—even today if not treated.

Von Pettenkofer was one of the promoters of the so-called miasma theory, which claimed at this time that bad air (which occurred increasingly in large towns) is responsible for the life-threatening diarrhea that occurred at this time around the world (see chapters *cholera*). He was as deeply convinced by his theory that he consumed in presence of several witnesses on the seventh of October 1882 a fluid containing cultured agents of cholera. He suffered only from very mild (if at all) symptoms of cholera, while at this time hundred thousands of humans died in Asia and Europe during the *fifth cholera world-wide epidemic*. The reasons why he did not get severe symptoms are surely based on the fact that he ingested probably only a low number of the bacteria and that he probably had been already preinfected during the years before as effect of the ongoing fifth pandemic of cholera (see chapter 3.3.2).

However, his great merits are surely based on the fact that he established basic methods in the scientific field of *hygiene* in towns, which still today save the life of millions of people and animals. He died shooting himself on February 10, 1901, in a phase of mental depression.

Literature

- Grüntzig J, Heinz Mehlhorn M (2010) Robert Koch: Seuchenjäger und Nobelpreisträger. Spektrum Akad. Verlag, Heidelberg
- Grüntzig J, Mehlhorn H (2010) Expeditions into the empire of plagues. Düsseldorf University Press, Düsseldorf
- Locher WG (2007) Max von Pettenkofer (1818–1901) as a pioneer of modern hygiene and preventive medicine. Environ Health Prevent Med 12:238–245
- Von Pettenkofer (1841) The value of health to a city (Baltimore). John Hopkins Press, Baltimore

During the actually running *seventh pandemic of cholera* many serogroups of *Vibrio cholerae* have been diagnosed and shown to occur in many countries. However, only two of them (01 and 0139) caused and cause considerable outbreaks. The type 01, however, caused all severe outbreaks worldwide, while the type 0139 induced outbreaks only in the past after being first detected in Bangladesh. However, both types induced the same severe degree of symptoms of disease (Gyr and Steffen 2010).

3.11 Obligatory Announcements to Health Authorities

The occurrence of the cholera in a family, in schools, and in other institutions has to be announced immediately to the health authorities, which will start measurements to avoid further spreading.

3.12 Treatment

The first very important step of treatment is prompt administration of amounts of safe drinking water to replace the loss of water due to the diarrheas. The *WHO/ UNICEF* recommends the use of the so-called *oral rehydration solution*, which is distributed by WHO (1 sachet of powder has to be diluted in 1 L of clean water). This uptake should be repeated at least six times on the first days. If persons are already in a shock stage, they should be treated by appropriate antibiotics to diminish the diarrhea and by intravenous injection of fluids. Children under 5 years should be treated—according to WHO—by application of a zinc solution or by a solution containing glucose (13.5 g/L), potassium (1.5 g/L), NaCl (2.6 g/L), and sodium citrate (2.9 g/L). *Antibiotic treatment* for 1–3 days shortens the length of the disease and reduces its severity. *Doxycycline* (1 × 300 mg for adults) is recommended as first line, but also *cotrimoxazole*, *erythromycin*, *tetracycline*, *furazolidone*, and *chloramphenicole* are active, but sometimes it is not fully successful.

3.13 Vaccination

The first anti-cholera vaccine was developed by the Russian scientist *Waldemar Mordechai Wolf Haffkine* (1860–1930) at the *Pasteur Institute* in Paris. He himself was the first test person when he started to test it in 1896 in the Indian town Mumbai. Today persons, who travel to or work in endangered regions of potential cholera outbreaks, should use one of the three WHO prequalified oral cholera vaccines:

- *Dukoral*® (has to be administered with a buffer solution/150 mL water),
- *Euvichol-Plus*®, does not need a buffer solution,

- *Shanchol*™, does not need a buffer solution.

3.14 Hope

WHO hopes that cholera outbreaks should end on earth around the year 2030 based on its program *Ending Cholera*. However, a hope remains a hope, especially in times of "hurly-burly" in the human populations around the world.

Bibliography

Aberth J (2012) Plagues in world history. Rowman & Littlefield, Lanham

Azman AS, Rudolph KE, Cummings DA, Lessler J (2013) The incubation period of cholera: a systematic review. J Infect 66:432–438

Bell J, Condie DF (1832) Epidemic cholera. Rep Coll Physicians to the Board of Health. Th. DeSilver, Philadelphia

Blake PA (1993) Epidemiology of cholera in the Americas. Gastroenterol Clin North Am 22:639–660

Byrne JP (2008) Encyclopedia of pestilence and pandemics: their impact on human history. ABC-Clio

CDC (2015) Cholera treatment

Committee of International Health and City Physicians (1849) Asiatic cholera. Boston

Council of Hygiene, Public Health of New York (1865) Report on cholera epidemic. Sanford, New York

Daly WJ (2008) The black cholera comes to the central valley of America in the 19th century-1832, 1849, and later. Trans Am Clin Climatol Assoc 119:143–152

Drake D (1832) Practical treatise on the history, prevention and treatment of epidemic cholera. Casey and Fairbanks

Drasar BS, Forrest BD (2012) Cholera and the ecology of *Vibrio cholerae*. Springer, Heidelberg

Fotedar R (2001) Vector potential of houseflies (*Musca domestica*) in the transmission of *Vibrio cholerae* in India. Acta Trop 78:31–34

Gestmann F, Förster M, Mehlhorn H et al (2012) Flies as vectors of microorganisms potentially inducing severe diseases in humans and animals. In: Mehlhorn H (ed) Arthropods as vectors of emerging diseases. Parasitology research monographs, vol 3. Springer, Heidelberg

Greenough WB (2008) The blue death disease and the water we drink. J Clin Investig 118:8–16

Grüntzig J, Mehlhorn H (2010) Expeditions into the empire of plagues. Düsseldorf University Press, Düsseldorf

Gyr NE, Steffen R (2010) Cholera. In: Löscher T, Burchard GD (eds) Tropenmedizin in Klinik und Praxis, 4th edn. Thieme, Stuttgart, pp 425–431

Hawgood BJ (2007) Waldemar Mordechai Haffine (1860–1930): prophylactic vaccination against cholera and bubonic plague in British India. J Med Biography 15:9–19

Hempel S (2006) The medical detective: John Snow: Cholera and the mystery of the Broad Street pump. Granta Books. ISBN 1862078424

Koch R (1884) Über das Cholera-Bakterium (on the cholera bacterium). Dtsch Med Wschr 10:725–728

Krishna BV, Patil AB, Chandrasekhar MR (2006) Fluoroquinolone-resistant *Vibrio cholerae* isolated during a cholera outbreak in India. Trans R Soc Trop Med Hyg 100:224–226

Lacey SW (1995) Cholera calamitous past, ominous future. Clin Inf Dis 20:1409–1419

Lan R, Reeves PR (2002) Pandemic spread of cholera: genetic diversity and relationships within the seventh pandemic clone of *Vibrio cholerae* determined by amplified fragment length polymorphism. J Clin Microbiol 40:172–181
Mehlhorn H (ed) (2016) Encyclopedia of parasitology, 3 vols, 4th edn. Springer, New York
Nelson EJ, Harris JB, Morris JG Jr, Calderwood SB, Camilli A (2009) Cholera transmission: the host, pathogen and bacteriophage dynamic. Nat Rev Microbiol 7:693–702
Paine M (1832) Letters on the cholera asphyxia as it appears in New York. Collins and Hanny, New York
Pettenkofer M (1887) Zum gegenwärtigen Stand der Cholerafrage (On the present status oft he cholera question). R. Oldenbourg, München
Pyle GF (2010) The diffusion of cholera in the United States in the nineteenth century. Geograph Anal 1:58–75
Ramunitha T (2008) Antibiotic resistance in *Vibrio cholera*. In: Faruque SM, Nair GB (eds) Genomic and molecular biology. Caister Academic, Poole
Rosenberg CE (1962) The cholera years of the USA 1832, 1849 and 1866. University of Chicago Press, Chicago
Sack DA, Sack RB, Chaignat CL (2006) Getting serious about cholera. New Engl J Med 355:649–651
Sinclair D, Abba K, Zaman K, Quadri F, Graves PMC (2011) Oral vaccines for preventing cholera. Cochrane Database Syst Rev 3:CD008603
Sutton GA (1853) A report to the Indiana State Medical Society in the Asiatic cholera in the state, 1853. Proc Indiana Med Soc 3–54
US Surgeons General's Office (1875) Epidemic of 1873 in the United States. Executive Document No. 95. Washington: Gov Printing Office
WHO (2017) Cholera vaccines: WHO position paper. Weekly Epidemiol Rec 34:477–500
WHO (2019) Cholera – key facts. Datasheet of recent status, Geneva

Chapter 4
Spreading of Cholera Along Rivers in Early Times of the USA Before Detection of the Agents of Disease

Heinz Mehlhorn

Abstract This chapter shows how three epidemics of cholera are spread in the USA by immigration of Europeans and spreading along the rivers.

Keywords Cholera in the USA · Riverboats as spreading tools

The name of the extremely virulent intestinal disease *cholera* has its origin in the Greek term *khole*, which means "bile," since the fluid fecal excretions of infected persons have a yellowish-green-bluish color. There exist different names (variations) of this disease: *Cholera asiatica*, *C. indica* (the classic one), *C. orientalis*, *C. epidemica*, and *C. infantum*. Up to now, *seven pandemics* of the bacterium *Vibrio cholerae* have been noted. All of them started in regions along the river *Ganges* in India, although at first larger amounts of cases had been noted elsewhere. From there these bacteria took their way around the world affecting or even killing until today many millions of people. Even today, when drugs for treatment and active vaccines are worldwide available, many people still die due to the present version of *V. cholerae*, which represents the since 1961 ongoing *seventh pandemic*, which had reached at first Africa coming from India in the year 1971 and which spread in the Americas starting in the year 1981 (see Chap. 3). Detailed information can be found in publications ranging from 1832 until today (Ali et al. 2015; Azman et al. 2013; Bell and Condie 1832; Brock 1999; Drake 1832; Koch 1884; Mehlhorn 2016; Paine 1832; Rosenberg 1962; US Surgeons General Office 1875; WHO 2017).

Looking backward into the first worldwide documented spreading periods, it can be guessed that these bacteria had always been present in their Asian home countries. However, the population in these countries had apparently reached a resistant status at a certain degree, so that not very many death cases occurred and thus the problem might have been underestimated. This supposition underlines the fact that patent

H. Mehlhorn (✉)
Heinrich Heine University, Düsseldorf, Germany
e-mail: mehlhorn@uni-duesseldorf.de

H. Mehlhorn, S. Klimpel (eds.), *Parasite and Disease Spread by Major Rivers on Earth*, Parasitology Research Monographs 12,
https://doi.org/10.1007/978-3-030-29061-0_4

infections with these bacteria only occur, if large numbers up to millions of them are ingested at once. This is an extremely high level compared to the high-graded infectious stages of some *Shigella* or *Salmonella* strains, "where only 10 (!), *Shigella*, respectively, 10^4 *Salmonella* bacteria are able to establish severe patent infections in humans with fatal outcome (Daly 2008).

The *first cholera pandemic* started in the year *1817* in the Indian Bengal region along the Ganges river and lasted until *1824* dispersing from the Indian Ganges regions to many countries in Southeast Asia, the Middle East, Europe, and Eastern Africa. It is believed that especially pilgrims wandering to holy sites (e.g., to Mekka, Saudi Arabia, of recent times) but also members of the British Navy and Army as well as from other armies, which had been heavily active in these regions, have transported this agent of disease to Europe, which at this time was a great battlefield, onto which the armies of the great powers France, England, Austria, Russia, and Prussia fought for supremacy of their countries ruled by related noble families. Even after the *battle of Waterloo* (June 18, 1815), where Prussian (German) and English troops defeated finally Napoleon's French troops, it took years to reestablish a sufficient food production. Thus the population of many countries suffered from the lack of food due to a very long cold period in Europe, due to destroyed fields by war actions and due to the fact that young farmers had been forced compulsatory to leave their farms, to become soldiers and often died in action. In addition, the crisscross through Europe marching troops spread agents of disease due to open air defecation, lack of clean drinking water, and fly-transmitted agents of disease. In addition to the reasons described above there occurred an Europe-wide disease of the potatoes due to the spreading of a fungus named *Phytophthora infestans*. All these events introduced severe long-lasting famine so that hundred thousands of death cases occurred among the European population, and thus large numbers (especially) of German, English, and Irish people were forced to leave their home countries and to emigrate to North America, where still today (2019) around 42% of the USA families have at least one member with German roots, 30% with Irish or English roots. The despair of these emigrants must have been really extraordinary, since at these times the present stable state USA of our days did not yet exist. Fights between Mexican, Southern, and North America troops and altogether against the original native population (erroneously called Indians) afforded many hundred thousand death cases besides the victims of the "normal diseases" of today, which during these times also in Europe killed large numbers of victims of tuberculosis, scarlet fever, etc., since at this time efficacious drugs did not yet exist.

All this together was severe enough and hard to survive, but it became even worse, when the *second worldwide pandemic of cholera* (1826–1837) reached in the years 1830–1832 the not yet fully established United States with its large and constant growing amount of immigrating European people. At this time the agent (*Vibrio cholerae*) of the spreading intestinal disease (Fig. 4.1) was not yet known and thus there existed no ideas and means, how to avoid infections. The agent of disease was later discovered by the German *Robert Koch* in the year 1883 (Figs. 4.2 and 4.3) and thus there was no treatment available. This disease, which at this time was spreading in Europe (coming from India) arrived in the USA in the years

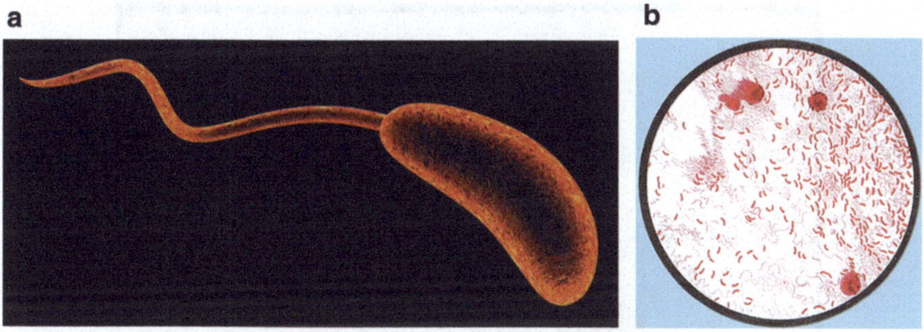

Fig. 4.1 (**a**) Light micrograph of a *Vibrio cholerae* bacterium. Note that the stage has a flagellum-like protrusion. https://www.cdc.gov/cholera/images/cholera-banner.jpg. (**b**) Photo of a smear preparation of hundreds of cholera vibriones shown on a teaching table of the Hamburg Institute for Tropical Diseases

Fig. 4.2 Photo of Robert Koch, who received the Nobel Prize for medicine in 1905

between 1830 and 1832 (Fig. 4.4). Its quick onset among healthy-appearing persons and the quickly following death of infected victims led to many types of discrimination of infected persons. Death due to this disease was, e.g., considered as divine punishment for the progressing sinfall of humans. The few medical doctors were unaware of the agents of most infectious diseases, believed in the existence of local

Fig. 4.3 Robert Koch (sitting) and his coworkers (from left: Treskow, Fischer and Gaffky). All of them had been participants at the cholera expeditions to Egypt and India

miasmas and thus fled together with other inhabitants from affected towns/villages. Thus a quick spreading of the disease in the Americas occurred from Eastern to Western regions, since people fled from affected towns to others and thus spread the agents of disease.

However, several medical circles in the USA started examinations, how this disease might be distributed on the North American continent (Daly 2008; Rosenberg 1962; Pyle 1969), while in Europe medical researchers like the German Robert Koch (1843–1910) and the French Louis Pasteur (1822–1895) and his group looked

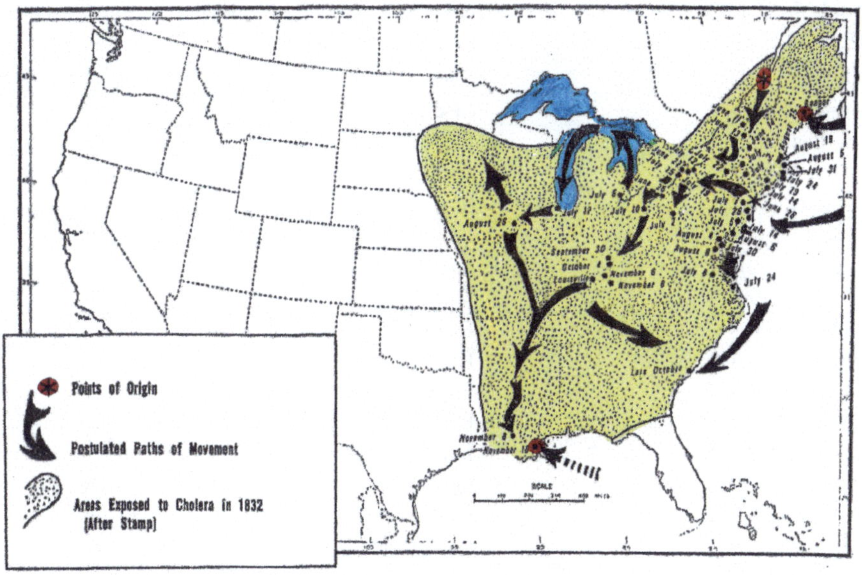

Fig. 4.4 Diagrammatic representation of the pathways of the cholera epidemic in the year 1832 (according to Pyle 1969). Note that the cholera arrived from Europe and spread southwards along the ocean, the Mississippi and its confluents

for potential agents of diseases, which were finally discovered by the German Nobel Prize winner Robert Koch in the year 1893 (published in the year 1894) after participating at an expedition to endemic regions in Egypt and India (Figs. 4.2, 4.3 and Table 4.1). However, he received the Noble Prize in the year 1905 *not* for the detection of the *Vibrio cholerae* bacterium, but for his results in the detection process of the agent of the tuberculosis (*Bacillus tuberculosis*) which he had published already in the year 1882.

Koch's finding of the defined agents of the *cholera disease* in the year 1883 explained why this disease had posed extremely high problems to the American physicians, who tried to control the three repeating cholera epidemics starting in the USA in the years 1832, 1849, and 1866. When looking at the arrival of the diseased persons in certain towns, these doctors noted that apparently the postulated (yet unknown) agents of disease arrived with immigrants on ships in the New York harbor or in ports of regions of the Canada of our days, from where further transportations of apparently infected Europeans occurred either by ships along the American coast or on rivers (Figs. 4.4, 4.5 and 4.6). The region of New Orleans of present days was also one of the starting points of the epidemics later called cholera (Figs. 4.5 and 4.6). From there infected (of course at first symptomless) persons were transported northwards and helped to spread this disease into the inner regions of North America. As the American author Pyle (1969) showed, the medical doctors in the USA finally became aware that agents of the disease had been imported by European immigrants and were distributed by not visibly infected people on ships

Table 4.1 Officially recognized entry dates for cholera of the 1832 epidemic showing the quick process

City	Date	City inhabitants
Plattsburg, NY	June 11	4913
Burlington, VT	June 13	3526
White Hall, NY	June 14	2888
Niagara, NY	June 22	1401
Erie, PA	June 26	1329
New York City	June 26	202,589
Pittsburg, PA	July 2	12,542
Albany, NY	July 3	24,238
Philadelphia, PA	July 5	161,410
Richmond, VA	July 6	16,060
Detroit, MI	July 6	2222
Newark, NY	July 7	10,953
Cleveland, OH	July 10	1076
Schenectady, NY	July 12	11,405
Rochester, NY	July 12	9269
Chicago, IL	July 12	4470
New Haven, CT	July 14	10,180
New Brunswick, NJ	July 14	7830
Buffalo, NY	July 15	8653
Lockport, NY	July 19	2121
Hartford, CT	July 19	7074
Frankfort, NY	July 22	2620
Newport, RI	July 24	8010
Norfolk, VA	July 24	14,998
Jersey City, NJ	July 26	12,568
Annapolis, MD	July 30	2623
Providence, RI	July 31	16,836
Baltimore, MD	August 4	80,625
Boston, MA	August 5	61,393
New Castle, DE	August 6	2463
Washington, DC	August 8	18,827
Andover, MA	August 18	4540
Haverhill, MA	August 25	3912
Rock Island, IL	August 26	320
Cincinnati, OH	September 30	24,831
Covington, KY	October 4	743
Charleston, SC	Late October	30,286
Lexington, KY	November 6	6104
Frankfort, KY	November 6	1680
Baton Rouge, LA	November 8	9809
New Orleans, LA	November 18	46,310
Bangor, ME	December 26	2863

Taken from Pyle (1969)

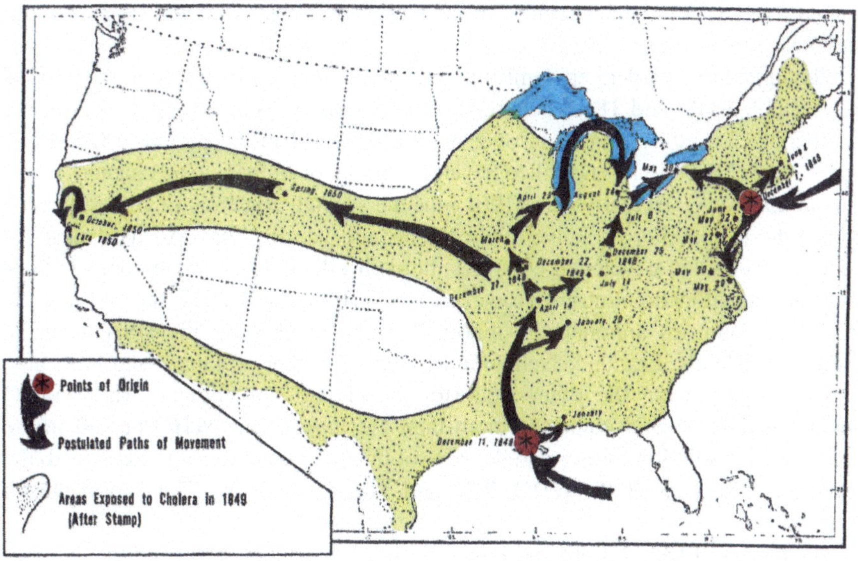

Fig. 4.5 Diagrammatic representation of the pathways of the cholera epidemic in the year 1849 (according to Pyle 1969). Note that the cholera arrived apparently from Europe in New York and New Orleans and was spread along rivers, but also by tracks to western regions

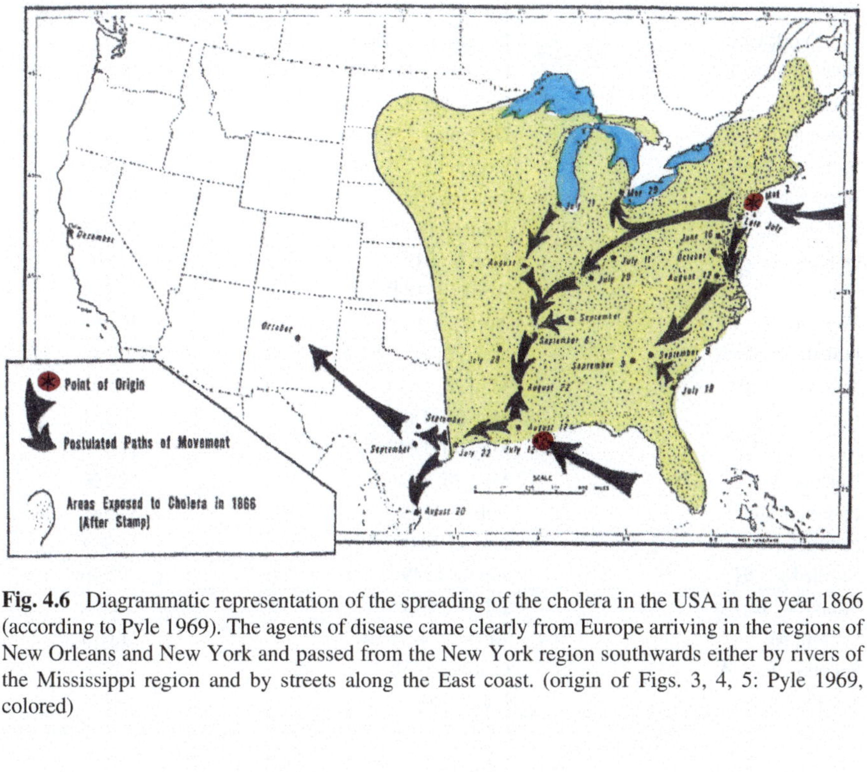

Fig. 4.6 Diagrammatic representation of the spreading of the cholera in the USA in the year 1866 (according to Pyle 1969). The agents of disease came clearly from Europe arriving in the regions of New Orleans and New York and passed from the New York region southwards either by rivers of the Mississippi region and by streets along the East coast. (origin of Figs. 3, 4, 5: Pyle 1969, colored)

when traveling in the inner regions of the North American continent (Figs. 4.4, 4.5 and 4.6).

The three large cholera epidemics in the United States, which started there in the years 1832, 1849, and 1866, belonged to the worldwide cholera epidemics of the numbers 2, 3, and 4, which ran over the whole world in the periods 1826–1837, 1846–1860, 1863–1875 and killed several millions of humans (see chapter "Cholera" in this book). Actually since 1961 runs the seventh epidemic of cholera, which due to the availability of active medicaments and vaccines has lost its danger for people in so-called developed countries, but still kills today many humans in poor and underdeveloped countries. However, the WHO map of 2017 showed that the following countries announced "imported" cases of cholera: the USA, Canada, Caribbean islands, England, Netherlands, Germany, Korea, Japan, and Australia. Central Africa is on the other side still the "hoard" of cholera, where WHO documented in several countries more than 100 cases. The WHO passed in the year 2008 a resolution to promote the control of cholera and to try to extinguish the cholera by the year 2030 (WHO 2017 and May 2018 at the 71st World Health Assembly).

The papers of Daly (2008) and Pyle (1969) followed the arrival and the distribution of an until this time unknown disease spreading in a country, which had in most

Table 4.2 Officially recognized entry dates for cholera of the 1849 epidemic (Pyle 1969)

City	Date	City inhabitants
New York City	December 1, 1848	515,547
New Orleans, LA	December 11, 1848	116,375
Louisville, KY	December 22, 1848	46,194
Cincinnati, OH	December 25, 1848	115,436
St. Louis, MO	December 27, 1848	77,860
Nashville, TN	January 20, 1849	10,165
Mobile, AL	January 1849	20,515
Quincy, IL	March 1849	6902
Cairo, IL	April 14, 1849	242
Chicago, IL	April 29, 1849	29,963
Philadelphia, PA	May 22, 1849	340,045
Baltimore, MD	May 22, 1849	169,054
Buffalo, NY	May 30, 1849	42,261
Richmond, VA	May 30, 1849	27,570
Norfolk, VA	May 30, 1849	14,326
Boston, MA	June 4, 1849	136,881
Newark, NJ	June 4, 1849	38,894
Sandusky, OH	July 8, 1849	1040
Frankfort, KY	July 14, 1849	3308
Detroit, MI	August 14, 1849	21,019
Sacramento, CA	October 1850	6820
San Francisco, CA	Late 1850	34,775

places no or only insufficient medical services. Facts had been that masses of new inhabitants entered in America as refugees from Europe fleeing from permanent wars, lack of food, and overcrowded large cities at the beginning of industrialization. They arrived exhausted in a country with an unknown future and in constant wars with the original inhabitants and other powers like Mexico, Spain, etc. Pyle (1969) describes that in 1832 the growing but not yet established USA "was barely more than a frontier country" and that "most major cities of the time periodically drowned in mud" and "that pigs roamed the streets of New York." On the other side, water transportation of people started at this time using the Ohio-Mississippi system and the Hudson River and Erie Canal. Also Great lakes and the Atlantic coast from New York to New Orleans was intensively used to transport people arriving from Europe and goods of any type. Thus it is completely clear that the agents of the three epidemics of the cholera became quickly spread by sea and river pathways, respectively, all over the growing United States. Tables 4.1, 4.2, and 4.3 and the plates (Figs. 4.4, 4.5, and 4.6) published by Pyle (1969) showing the process of the spreading of the cholera in the years 1832, 1849, and 1866 give an impression

Table 4.3 Officially recognized entry dates for cholera of the 1866 epidemic (Pyle 1969)

City	Date	City inhabitants
New York City	May 2	942,292
Detroit, MI	May 29	79,577
Baltimore, MD	May 29	267,354
Cincinnati, OH	June 16	216,239
New Orleans, LA	July 11	191,418
Savannah, GA	July 12	28,235
Chicago, IL	July 18	298,977
Galveston, TX	July 21	13,818
Little Rock, AR	July 22	12,380
Louisville, KY	July 28	100,753
Philadelphia, PA	July 29	674,022
Richmond, VA	Late July	51,038
Baton Rouge, LA	August 12	6498
Brownsville, TX	August 17	4905
Vicksburg, MS	August 20	12,443
St. Louis, MO	August 22	310,864
Nashville, TN	August	25,865
Memphis, TN	September 2	40,226
Augusta, GA	September 6	15,389
Atlanta, GA	September 9	21,789
San Antonio, TX	September 9	12,256
Austin, TX	September	4428
Washington DC	October	109,199
Albuquerque, NM	October	1307
San Francisco, CA	December	149,473

how quick this in the USA unknown disease was spread and how many people died and how many had been sick in addition.

The unknown reasons of the origin of this disease overwhelmed poor people having left their homes in Europe looking for a better future. They feared punishment by god and the end of the world. Thus invisibly infected people often left their homes to move away from a town of "sinners" and thus transported the much later detected virus to other towns. Therefore, it is not astonishing that the second USA epidemics reached quickly the West coast of the USA—supported by the then finished railway connection between the East and West coast of the growing USA (Sutton 1853).

References

Ali M, Nelson AR, Lopez AL, Sack D (2015) Updated global burden of cholera in endemic countries. PLoS Negl Trop Dis 9:e0003832

Azman AS, Rudolph KE, Cummings DA, Lessler J (2013) The incubation period of cholera: a systematic review. J Infect 66:432–843

Bell J, Condie DF (1832) Epidemic cholera. Report of the College of Physicians to the Board of Health. Th. DeSilver, Philadelphia

Brock T (1999) Robert Koch, a life in medicine and bacteriology. ASM Press, Washington

Daly WJ (2008) The black cholera comes to the central valley of America in the 19th century – 1832, 1849, and later. Trans Am Clin Climatol Assoc 119:143–152

Drake D (1832) Practical treatise on the history, prevention and treatment of epidemic cholera. Casey and Fairbanks

Koch R (1884) Über das Cholera-Bakterium (on the Cholera bacterium). Dtsch Med Wschr 10:725–728 (detected in 1883)

Mehlhorn H (ed) (2016) Encyclopedia of parasitology, 3 vols, 4th ed. Springer, New York

Paine M (1832) Letters on the cholera asphyxia as it appears in New York. Collins and Hanny, New York

Pyle GF (1969) The diffusion of cholera in the United States in the nineteenth century. Geograph Anal 1:58–75

Rosenberg CE (1962) The cholera years of the USA 1832, 1849 and 1866. University of Chicago Press, Chicago

Sutton GA (1853) A report to the Indiana State Medical Society in the Asiatic cholera in the state, 1853. Proc Indiana Med Soc 3–54

US Surgeons General's Office (1875) The cholera of 1873 in the US. Government Printing Office

WHO (2017) Cholera vaccines: WHO position paper. Weekly Epidemiol Rec 34:477–500

Chapter 5
Human Cercarial Dermatitis (HCD) or Swimmer's Itch Along Rivers and Lakes

Heinz Mehlhorn

Abstract This chapter describes the often unexplained occurrence of skin itching of persons that have taken a bath in a river or lake or have worked with related water.

Keywords Swimmer's itch · Cercarial dermatitis · Flukes · Schistosomes

The so-called blood flukes of the family Schistosomatidae contain at least 14 genera of medical or veterinary importance (Loker and Brant 2006; Jouet et al. 2009; Mehlhorn 2016). Nine of them parasitize as adults in birds; however, their larvae (cercariae) (Fig. 5.1) enter the skin of humans and induce slight to very severe (up to life-threatening) symptoms of disease in the case of allergic reactions of the immune system either due to the penetration of extremely large numbers of cercariae or due to a high sensitivity of this person, while final hosts (ducks, other aquatic birds) show low or even no symptoms even in cases of high-graded infections.

5.1 Life Cycle

The life cycle of the bird schistosomes runs in principle in the same way as that of the *Schistosoma* species of humans or vertebrates, where these final hosts excrete larva-containing or larva-developing eggs (Fig. 5.2). The larval stages (miracidia) (Fig. 5.3) hatch from the eggs and swim by help of their superficial layer of cilia to snails, wherein cercariae (Fig. 5.1) are developed, which finally leave the snails, search in the water for endothermic hosts and thus enter also the skin of humans besides that of many animals. In the case of bird schistosomes they enter mainly the

H. Mehlhorn (✉)
Heinrich Heine University, Düsseldorf, Germany
e-mail: mehlhorn@uni-duesseldorf.de

H. Mehlhorn, S. Klimpel (eds.), *Parasite and Disease Spread by Major Rivers on Earth*, Parasitology Research Monographs 12,
https://doi.org/10.1007/978-3-030-29061-0_5

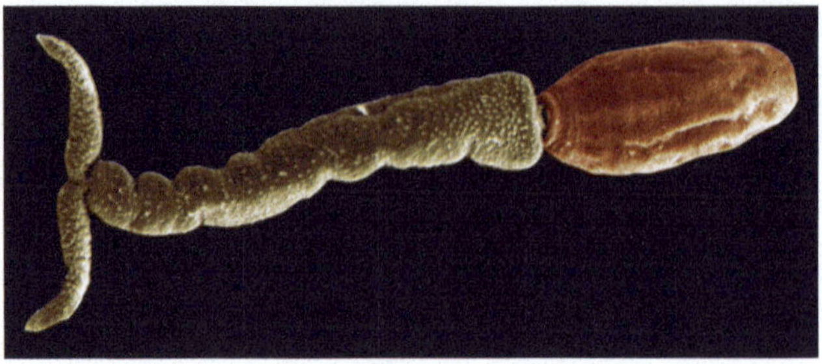

Fig. 5.1 Scanning electron micrograph showing the principal organization of a so-called bifurcated cercarial stage, which contains the anterior worm body and a motile tail with two propeller-like protrusions, which induces the aspect of bifurcation

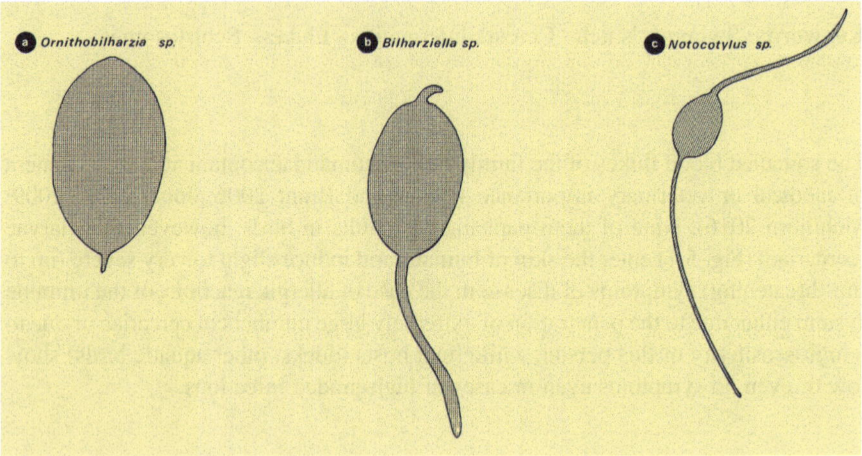

Fig. 5.2 Diagrammatic representation of the eggs of three very common genera of bird schistosomes. The protrusion(s) may help to become fixed at plants thus avoiding to be driven away by streaming water

skin of birds and pass from there to the blood vessel system. However, in case the cercariae enter the skin of humans, while the latter are bathing or working in the water of rivers, lakes or rice fields, etc., the cercariae are unable to proceed their development to adult worms, but are attacked by the human immune system. This may lead to skin reactions, which might be slight (itching), but also to life-threatening shock reactions (Table 5.1) (Aldhoun et al. (2012), Ebbs et al. (2016), Hörweg et al. (2006), Jouet et al. (2015), Kolářová et al. (2013), Lawton et al. (2014), Mahdavi et al. (2013), Pinto et al. (2017), Prüter et al. (2017), Santoro et al. (2011), Yakhchali et al. (2013)).

How common digenean worms are in nature is shown by the paper of Rząd et al. (2013), who studied the worm burden of tufted ducks (*Aythya fuligula*) and greater scaups (*Aythya marila*) during their overwintering in the North-Western region of Poland. These birds contained in high numbers the adults of the following

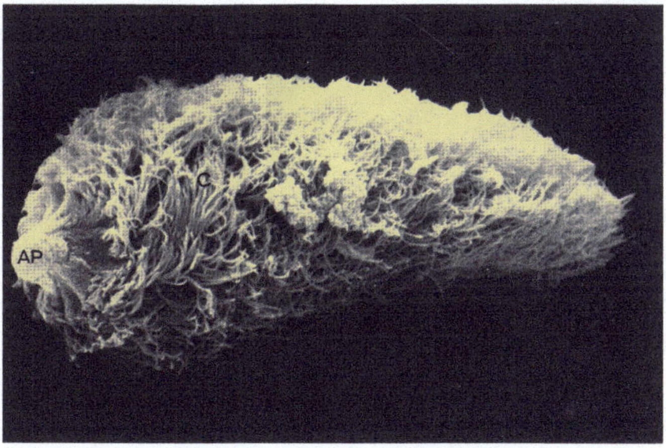

Fig. 5.3 Scanning electron micrograph of a miracidium larva, which hatches from the eggs of trematodes, swims around before entering a snail, develops the sporocyst stage, wherein finally the infectious cercariae are developed, which after release may enter other final hosts

Table 5.1 Examples of bird schistosomes, the larvae of which induce the symptoms of swimmer's itch = cercarial dermatitis in humans

Species	Length of adults (mm)	Intermediate hosts (snails)	Final hosts
Bilharziella polonica	♂ ♀ 2	*Planorbis planorbis, Plenobarius corneus, Anisus vortex, Radix* sp.	Geese, ducks, several further aquatic birds
Trichobilharzia szidati syn. *T. ocellata* (Europe/Asia)	♂ 3–4 ♀ 5–7	*Lymnaea* species, *Stagniola palustris*	Ducks, aquatic birds
Trichobilharzia franki	♂ 4 ♀ 5	*Radix auriculata*	Ducks, aquatic birds
Trichobilharzia regenti	♀ 7	*Radix* species	Ducks (*Anseri anseri*)
Ornithobilharzia canaliculata	♂ ♀ 5	?	Mediterranean gull
Ornithobilharzia turkestanium	♂ ♀ ~ 4	*Lymnaea* sp., *L. gedrosiana*	Ruminants, rodents, deer
Trichobilharzia anseri	♂ ♀~ 3.5–5	*Radix balthica* (syn. *R. peregra*)	Geese
Trichobilharzia querquedulae	♂ ♀~ 4	*Physa* species	*Anas* species

trematodes *Amblosoma exile, Australopatemon minor, Cotylurus cornutus, Cyathocotyle prussica, Echinoparyphium recurvatum, Echinostoma revolutum,* and *Notocotylus alternatus*. In addition, the tufted ducks hosted *Bilharziella*

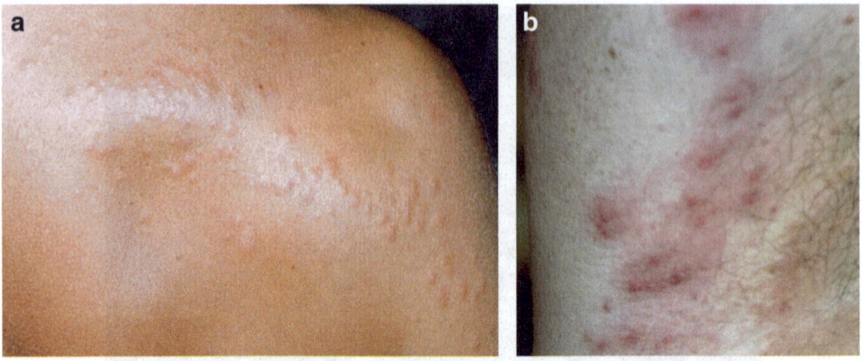

Fig. 5.4 (**a**, **b**) Skin symptoms after penetration of cercariae into human skin. (**a**) Quick reaction after minutes. (**b**) Intense reaction on the following days

polonica, *Cryptocotyle concava,* and *Echinostoma nordiana*. These facts make it clear that cercariae will penetrate in large numbers into the skin of humans and animals when entering the contaminated waters in this region.

5.2 Symptoms of Cercarial Dermatitis

At the site, where the cercariae had penetrated into the skin of humans, slight to severe reactions occur depending on an existing sensibilization or not, which would have occurred after proceeding penetrations. In the latter case, the reaction may occur within a few minutes leading to intense itching followed by an erythemateous, maculeous, or papuleous exanthema, which are accompanied by skin reddening and strong itching (Fig. 5.4a, b). In case of infections with large numbers of cercariae, sensitive persons may suffer from collapse. In the case of sensitive persons, protection from penetration might be done by application onto the skin of a rather thick sunscreen with an insect/tick repellent (e.g., Viticks).

References

Aldhoun JA, Podhorský M, Holická M, Horák P (2012) Bird schistosomes in planorbid snails in the Czech Republic. Parasitol Int 61:250–259

Ebbs ET, Loker ES, Davis NE, Flores V, Veleizan A, Brant SV (2016) Schistosomes with wings: how host phylogeny and ecology shape the global distribution of *Trichobilharzia querquedulae* (Schistosomatidae). Int J Parasitol 46:669–677

Hörweg C, Saltmann H, Auer H (2006) Cercarial dermatitis in Austria: questionnaires as useful tools to estimate risk factors. Wiener Klin Wschr (Central Eur J Med) 118:77–80

Jouet D, Ferté H, Hologne C, Kaltenbach ML, Depaquit J (2009) Avian schistosomes in French aquatic birds: a molecular approach. J Helminthol 83:181–189

Jouet D, Kolářová L, Patrelle C, Ferté H, Skírnisson K (2015) *Trichobilharzia anseri* n. sp. (Schistosomatidae: Digenea), a new visceral species of avian schistosomes isolated from greylag goose (*Anser anser* L.) in Iceland and France. Infect Genet Evol 34:298–306

Kolářová L, Skírnisson K, Ferté H, Jouet D (2013) *Trichobilharzia mergi* sp. nov. (Trematoda: Digenea: Schistosomatidae), a visceral schistosome of *Mergus serrator* (L.) (Aves: Anatidae). Parasitol Int 62:300–308

Lawton SP, Lim RM, Dukes JP, Cook RT, Walker AJ, Kirk RS (2014) Identification of a major causative agent of human cercarial dermatitis, *Trichobilharzia franki* (Müller and Kimmig 1994), in southern England and its evolutionary relationships with other European populations. Parasit Vectors 7:277

Loker ES, Brant SV (2006) Diversification, dioecy and dimorphism in schistosomes. Trends Parasitol 22:521–528

Mahdavi S, Farahnak A, Mobedi I, Rad MM, Azadeh H (2013) Survey of migratory birds (Anatidae: *Anas platyrhynchos*) for schistosome parasites from Mazandaran Province, Northern Iran. Iran J Parasitol 8:333–336

Mehlhorn H (ed) (2016) Encyclopedia of parasitology, 3 vols, 4th ed. Springer, New York

Pinto HA, Pulido-Murillo EA, de Melo AL, Brant SV (2017) Putative new genera and species of avian schistosomes potentially involved in human cercarial dermatitis in the Americas, Europe and Africa. Acta Trop 176:415–420

Prüter H, Sitko J, Krone O (2017) Having bird schistosomes in mind-the first detection of *Bilharziella polonica* (Kowalewski 1895) in the bird neural system. Parasitol Res 116:865–870

Rząd I, Sitko J, Kavetska K, Kalisińska E, Panicz R (2013) Digenean communities in the tufted duck [*Aythya fuligula* (L., 1758)] and greater scaup [*A. marila* (L., 1761)] wintering in the north-west of Poland. J Helminthol 87:230–239

Santoro M, Mattiucci S, Kinsella JM, Aznar FJ, Giordano D, Castagna F, Pellegrino F, Nascetti G (2011) Helminth community structure of the Mediterranean gull (*Ichthyaetus melanocephalus*) in Southern Italy. J Parasitol 97:364–366

Yakhchali M, Mirrajei SY, Malekzadeh-Viayeh R (2013) Detection of infection with larval stages of *Ornithobilharzia turkestanicum* using PCR in field-collected snails of *Lymnaea gedrosiana* from Northwestern Iran. Iran J Parasitol 8:627–633

Chapter 6
Blood-Sucking Leeches: Not Only Nasty but Also Dangerous and Helpful

Heinz Mehlhorn

Abstract This chapter considers the effects and dangers due to blood-sucking leeches attacking human skin.

Keywords Leeches · Blood suckers · Hirudinea · Leeches as therapeutics

6.1 Introduction

Leeches belong to the systematical group (phylum) of *Annelida*, which got their name from their external aspect showing ringlike sections (Figs. 6.1 and 6.2). In contrast to earthworms, which belong to the same animal phylum, many leeches have developed a parasitic lifestyle attacking. Depending on their specialization, they attack a large variety of endothermic animals as well as humans around the globe (Tables 6.1 and 6.2). They lure for warm-blooded prey such as humans and many species of animals. Some species of leeches attack also several fish species.

Especially members of the species belonging to the genus *Hirudo*—the Latin word for leech—are known worldwide since the first description had been done already two thousand years ago by physicians like Nicandros, Themison, Gala, and Avicenna (Mory et al. 2000) followed by the more detailed description of Linné in the year 1758 covering *Hirudo medicinalis* followed by that of *H. troctina* (1816), *H. verbana* (1820), *Macrobdella decora* (Say 1824), *Hirudinaria manillensis* (1878), and *H. orientalis* (2015) (Grau et al. 2018; Kutschera 2012; Whitaker et al. 2014). Leeches are mostly temporarily blood-sucking ectoparasites on mammals (including humans, monkeys, horses, etc.), but some species attack additionally also frogs, tadpoles, turtles, and even various fish species (see Table 6.1). Thus they are able to digest different types of host blood.

H. Mehlhorn (✉)
Heinrich Heine University, Düsseldorf, Germany
e-mail: mehlhorn@uni-duesseldorf.de

H. Mehlhorn, S. Klimpel (eds.), *Parasite and Disease Spread by Major Rivers on Earth*, Parasitology Research Monographs 12,
https://doi.org/10.1007/978-3-030-29061-0_6

Fig. 6.1 Macrophoto of the dorsal side of a *Hirudo verbana* leech

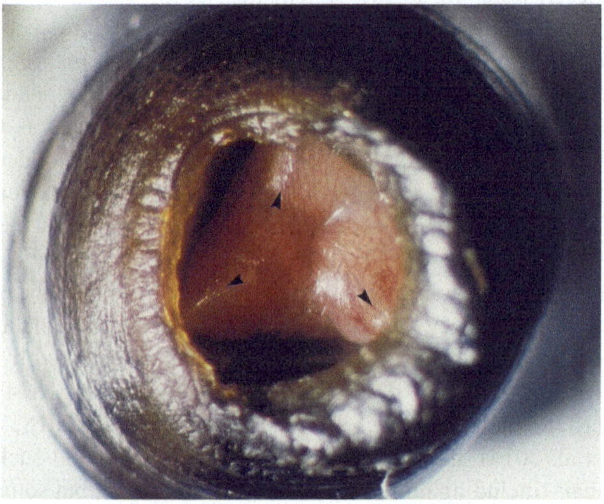

Fig. 6.2 Macrophoto of the anterior sucker of a *Hirudo medicinalis* leech showing three jaws, which cut the skin of attacked hosts

Modern molecular biological investigations showed that in the meantime since the first descriptions large numbers of leeches have been developed worldwide, which differ not only in their genes but also in their colors. For example, Kutschera described in 2007 dark stages as *H. verbana nigra* and a bit later (2012) those with a median dark dorsal line as *H. verbana monostriata*.

Table 6.1 Common leeches

Family/species	Length (mm)	Main hosts/habitat
Acanthobdellidae		
Acanthobdella peledina	35	Salmonid fish/skin
Rhynchobdellidae		
Piscicola geometra	70	Freshwater fish/skin
Theromyzon tessulatum	50	Waterbirds/nose, pharynx, trachea
Haementeria ghilianii	500	Many vertebrates/skin
Glossiphonia spp.	20	Snails, oligochaetes/skin
Gnathobdellidae		
Hirudo medicinalis	170	Mammals, *humans*/skin
Haemadipsa spp.	70	*Humans*, mammals, birds/skin
Xerobdella spp.	50	Amphibia/skin
Limnatis nilotica	120	Horses, mammals/nose, pharynx
Haemopis sanguisuga	100	Feeds on oligochaetes
Pharyngobdellidae		
Erpobdella octoculata	20	Predacious leeches, feed on insect larvae

Table 6.2 Leeches of the genus *Hirudo* (family Gnathobdellidae) and related genera

Species	Distribution	Length
Hirudo medicinalis = *H. officinalis*	Europe	Up to 20 cm when stretched
H. verbana syn. *H. officinalis*	Mediterranean regions, Hungary	Up to 10 cm
H. orientalis	Caucasian regions, Asia	Up to 10.8 cm
H. troctina	African regions	~10 cm
Macrobdella decora	North America	5–20 cm
H. nipponia	Japan	3–6 cm
Hirudinaria manillensis	Mexico, Central America	~9 cm
Haemadipsa sp.	Asia	5–20 cm

6.2 Biology and Morphology of Leeches

Most of the members of the Hirudinea (Tables 6.1 and 6.2), which are endowed with a complete intestinal tract, look like platyhelminths and are dorsoventrally flattened monoecious animals (Hermaphrodites). In general they possess 32 inner segments (except for 29 in *Acanthobdella*), wherefrom each corresponds to 2–14 outer annuli. Some authors, however, count 34 segments instead of 32. Except for members of the genus *Acanthobdella*, the species of the Hirudinea (Figs. 6.1, 6.2, 6.3, 6.4, 6.5, and 6.6), parasitic or not, are characterized by an *oral sucker* and a posterior disk-like *acetabulum*, which both act as strong holdfast organs supporting movements on the surface of animal hosts, plants, etc. (Fig. 6.6). Bloodsucking leeches produce enzymes (e.g., hirudin) preventing coagulation of the host blood during feeding. Closed circulatory vessels and a definitive respiratory system are lacking in most

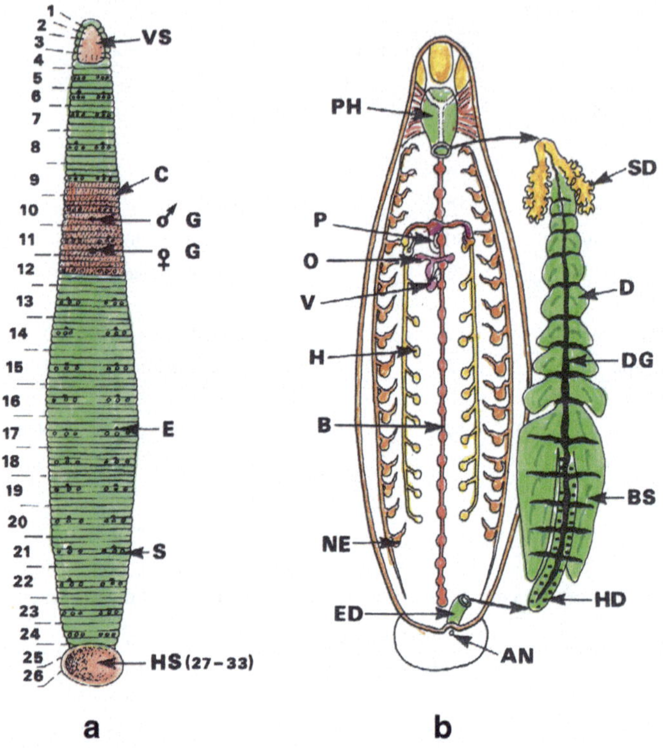

Fig. 6.3 Diagrammatic representations of *Hirudo medicinalis*. (**a**) Ventral aspect; the outer annulation does not correspond to the inner segments (1–32). (**b**) Inner structure. After copulation and transfer of so-called double spermatophores, fertilized eggs are packed together into slime (cocoon) deriving from the clitellum and are deposited on plants. After a period (depending on the temperature of the water), the young worms hatch from the eggs and reach maturity within 5–6 months (*AN* = anus; *B* = ventral nerve cord; *BS* = last intestinal cecum; *C* = clitellum; *D* = intestinal ceca; *DG* = dorsal blood vein; *E* = nephropores = openings of the 17 pairs of metanephridia; *ED* = rectum; *G* = genital pore; *H* = testis; *HD* = hind gut; *HS* = posterior sucker formed by fusions of segments 27–33; *NE* = nephridium; *O* = ovary; *P* = penis; *PH* = pharynx; *S* = sensory papillae; *SD* = salivary glands; *V* = vagina; *VS* = apical sucker)

species, and the coelom cavities are restricted to a few blood-transporting lacunae. The ontogenesis of the Hirudinea proceeds directly, without involving larval stages such as the trochophora larva, which is characteristic for other annelids. Thus, juvenile worms hatch immediately from eggs, which are joined in packs of 1–200 into a cocoon. The latter is fixed at water plants or stones 2 days to months after copulation of two individuals by their clitellum region, which is situated at segments 9–11 (Fig. 6.3) (Mehlhorn 2016; Utevsky and Trontelj (2005).

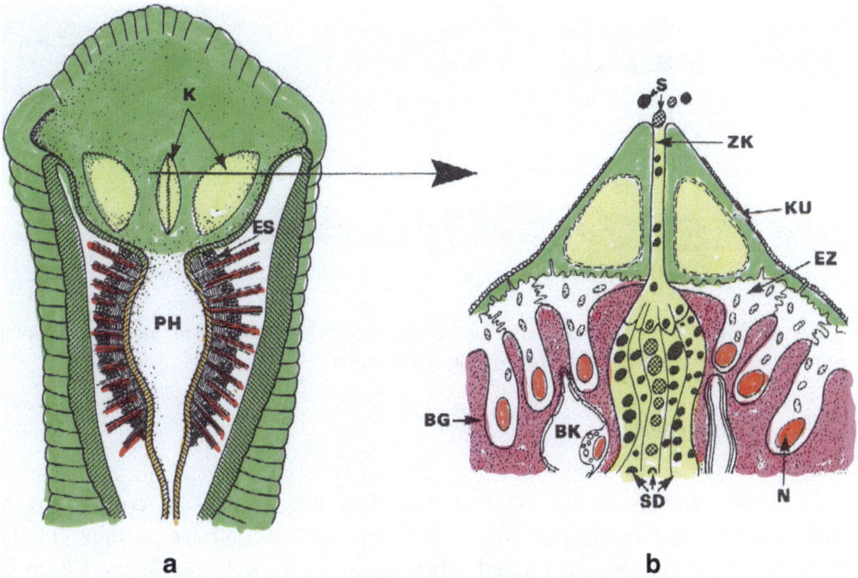

Fig. 6.4 *Hirudo medicinalis.* Diagrammatic representations. (**a**) Longitudinal section through the anterior end showing the cutoff muscular pharynx (*PH*) and the terminal ends of the salivary glands. (**b**) Cross section through one of the three jaws (along the *arrow line*)—according to Westheide in Siewing. (*BG* = connective tissue; *BK* = blood capillary; *ES* = terminal portion of the unicellular salivary glands; *EZ* = epithelium cell; *K* = jaw; *KU* = cuticle; *N* = nucleus; *PH* = pharynx; *S* = drop of saliva; *SD* = salivary gland with secret granules; *ZK* = opening channel of the unicellular salivary glands)

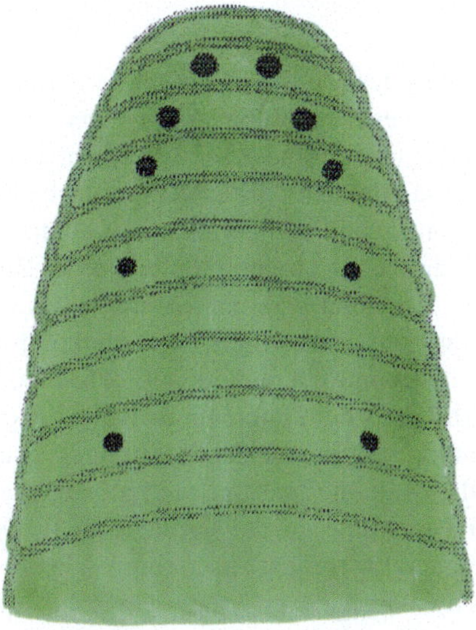

Fig. 6.5 Diagrammatic representation of the location of the eyes of *Hirudo medicinalis* in its anterior segments

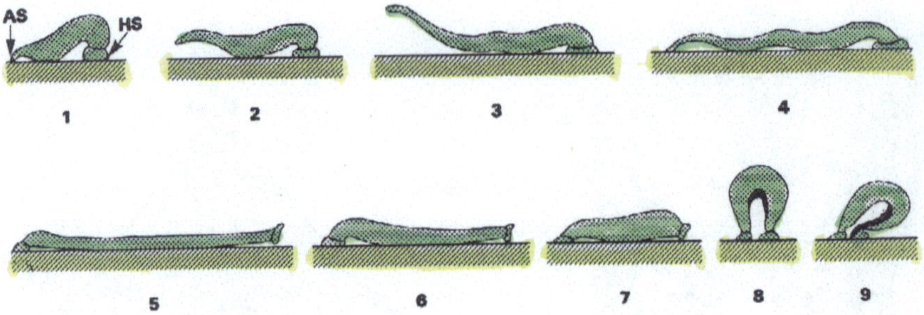

Fig. 6.6 Diagrammatical demostration of the movements of *Hirudo medicinalis* along the skin of a host or on soil (*AS* = anterior sucker; *HS* = posterior sucker)

6.2.1 *Integument*

As in other members of the phylum Annelida, the leeches are covered by a unilayered, cellular epidermis which excretes onto its surface a thin cuticle consisting of a proteinaceous ground substance including collagen fibers. Chitin is only present on the rigid bristles (chaeta) which are, however, absent in most hirudineans except for *Acanthobdella species*. This cuticle remains very elastic and is molted several times indefinite intervals in the case of Hirudinea, but not in other annelids.

6.2.2 *Intestine*

The intestine of parasitic leeches consists of a frequently armed mouth, a pharynx, an esophagus, a stomach, an intestine with 6–7 large paired diverticles, a hindgut, and the rectum ending with the anus. In the case of the medicinal leech *H. medicinalis*, the teeth of the three jaws are pored by the excretion canaliculi of unicellular glands, which produce besides anesthetic and vasodilatory compounds an anticoagulant substance (hirudin) which acts as an antithrombokinase. Due to this fact, a large leech may ingest about 15 g blood within a few minutes (i.e., tenfold of their own body weight). This phenomenon was used for centuries in human medicine as a general medicinal remedy to control blood pressure. The blood cells are stored in the intestinal diverticles, while the fluid compounds are excreted rather quickly. This ingested large amount of blood allows the leeches to survive starving periods of up to 2 years. The blood cells inside the intestine are retained due to the activity of the hirudin in full shape for a long time and are digested only bit by bit due to the activity of a particular, permanently present bacterium (*Pseudomonas [Aeromonas] hirudinis*) instead of its own digestive enzymes (which, however, are not completely absent). The bacteria are packed by the adult worm on its eggs and are thus become transmitted to the next generation. Investigations of our group showed a very long (half year) survival of many agents of diseases (viruses, bacteria, protozoan parasites) inside the stored blood. Furthermore, the antibodies against HIV and hepatitis B viruses were found in African leeches, so that it is proven that leeches pose a considerable danger for their hosts when becoming attached to their skin.

6.2.3 *Excretory System*

The excretory system of leeches consists of 10–17 pairs of metanephridia (e.g., in *Hirudo medicinalis*; Fig. 6.3), which each opens by a porus at the ventral body surface. These metanephridia start with a ciliated funnel that opens into the body cavity or into a spherical nephridial capsule (depending on the species) and thus are able to collect materials, which must be excreted.

6.2.4 *Nervous System*

The nervous system of leeches (e.g., *Hirudo medicinalis*; Fig. 6.3) corresponds in principle to that of other annelids and thus is composed of a ventral chord system with a typical broad fused ganglion in each segment. From there 3–4 ringlike running pairs of nerves extend to the periphery. The ganglia of the segments of the posterior sucker are fused to form a large terminal ganglion. Anteriorly, a supraesophageal ganglion has developed innervating sensillae of various types. This epi- = supraesophageal ganglion is connected to the subesophageal ganglion which innervates the mouth and the anterior portions of the intestine.

6.3 Use of Leeches for Therapy

Since centuries until today the sucking activity of leeches is used for therapeutical purposes (enhancing the venous outflow by removing stagnant blood during plastic and reconstructive microsurgery, etc.), because these leeches have developed a system to keep blood fluid, which is based on a group of bioactive substances (such as small molecules, peptides, and proteins). For example, the so-called *hirudin* is produced and stored until its use in their salivary gland cells (Figs. 6.3 and 6.4) (Müller et al. 2016; Hildebrandt and Lemke 2011; Beka et al. 2018). This activity is surely very useful and thus was commonly and often used in former times, when related medicaments had not yet been developed as they are now available for recent physicians and their patients.

While in former times leeches were mostly collected just prior to use in rivers and lakes, it is today only allowed to use exclusively cultivated and thus pathogen-free leeches for attaching them onto the skin of human or animal patients. The need to do

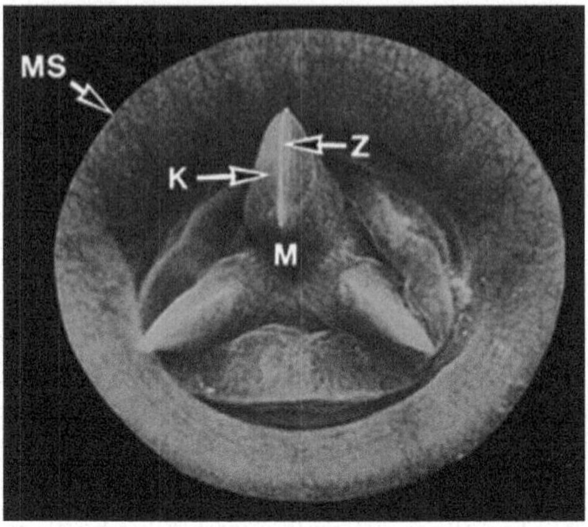

Fig. 6.7 Scanning electron micrograph of the mouth sucker (MS) of the leech *Hirudo medicinalis* showing the mouth opening (M), three cutting systems (K) the centers of which are lined by a row of tiny teeth

this was experimentally shown by transmission experiments of Nehili et al. (1994) and findings of other groups (Shope 1957; Snower et al. 1989).

The experiments of Nehili et al. (1994) have clearly proven that leeches can easily be infected by parasites, bacteria, and viruses, which then remain and survive for a long time inside the intestine of the leeches. In details the test leeches sucked successfully on animals that had been experimentally infected with bacteria (*Aeromonas hydrophila* = *Pseudomonas hydrophila*) and protozoans such as *Toxoplasma gondii*, *Plasmodium berghei*, and *Trypanosoma brucei*, which all did reproduce inside these mice for at least 5–6 weeks until the experiments were stopped. These experiments showed that leeches from ponds or rivers may transmit severe agents of diseases. Then further experiments were undertaken to check the presence of viruses in leeches captured in rivers in Cameroon (Africa). It was documented that these African leeches contained blood, which was serologically positive for human immunodeficiency virus (HIV) and agents of hepatitis B. For further experiments leeches were bought in German pharmacies. It was found that their blood contained up to 11 different species of bacteria. These bacteria had been infective even after storage of the leeches for 6 months at 3 °C as well as in leeches kept at room temperature in aquaria (Nehili et al. 1994).

These experiments clearly demonstrate that contacts to living leeches in endemic regions must be avoided by both: local inhabitants as well as tourists, since otherwise eventually deadly agents of diseases might be transmitted.

References

Beka L, Fullmer MS, Colston SM, Nelson MC, Talagrand-Reboul E, Walker P, Ford B, Whitaker IS, Lamy B, Gogarten JP, Graf J (2018) Low-level antimicrobials in the medicinal leech select for resistant pathogens that spread to patients. MBio 9:e01328–e01318

Grau D, Masson R, Villiet M, Lamy B (2018) Leech management before application on patient: a nationwide survey of practices in French university hospitals. Antimicrob Resist Infect Control 7:19

Hildebrandt JP, Lemke S (2011) Small bite, large impact-saliva and salivary molecules in the medicinal leech, *Hirudo medicinalis*. Naturwissenschaften 98:995–1008

Kutschera U (2012) The *Hirudo medicinalis* species complex. Naturwissenschaften 99:433–434

Mehlhorn H (ed) (2016) Encyclopedia of parasitology, 3 vols, 4th ed. Springer, New York

Mory RN, Mindell D, Bloom DA (2000) The leech and the physician: biology, etymology, and medical practice with *Hirudinea medicinalis*. World J Surg 24:878–883

Müller C, Mescke K, Liebig S, Mahfoud H, Lemke S, Hildebrandt JP (2016) More than just one: multiplicity of hirudins and hirudin-like factors in the medicinal leech, *Hirudo medicinalis*. Mol Gen Genomics 291:227–240

Nehili M, Ilk C, Mehlhorn H, Ruhnau K, Dick W, Njayou M (1994) Experiments on the possible role of leeches as vectors of animal and human pathogens: a light and electron microscopy study. Parasitol Res 80:277–290

Shope RE (1957) The leech as a potential virus reservoir. J Exp Med 105:373–382

Snower DP, Ruef C, Kuritza AP, Edberg SC (1989) *Aeromonas hydrophila* infection associated with the use of medicinal leeches. J Clin Microbiol 27:1421–1422

Utevsky SY, Trontelj P (2005) A new species of the medicinal leech (Oligochaeta, Hirudinida, *Hirudo*) from Transcaucasia and an identification key for the genus *Hirudo*. Parasitol Res 98:61–66

Whitaker IS, Maltz M, Siddall ME, Graf J (2014) Characterization of the digestive tract microbiota of *Hirudo orientalis* (medicinal leech) and antibiotic resistance profile. Plast Reconstr Surg 133:408e–418e

Chapter 7
Important Foodborne Trematodiases in the Lower Mekong River Basin

Oranuch Sanpool, Tongjit Thanchomnang, Wanchai Maleewong, and Pewpan M. Intapan

Abstract This chapter reviews the important foodborne trematodiases that affect public health in the lower Greater Mekong Subregion (GMS), including the Kingdom of Cambodia (Cambodia), Lao People's Democratic Republic (Lao PDR), Myanmar, Vietnam, and Thailand. Highly diverse environments, the economy, and political factors in these areas affect the epidemiology and outcomes of human parasite control initiatives. We explore the current scientific understanding of the epidemiology of important foodborne trematodiases. Based on this data, future research should be designed to achieve a more accurate understanding of the epidemiology and transmission to prevent and control these medically important flukes.

Keywords Foodborne trematodiases · Lower greater Mekong subregion · Opisthorchiasis · Clonorchiasis · Paragonimiasis · Fascioliasis · Trematodes · Snails · *Opisthorchis* · *Paragonimus* · *Fasciola*

List of Abbreviations

AEC	ASEAN Economic Community
Cambodia	Kingdom of Cambodia
GMS	Greater Mekong Subregion
Lao PDR	Lao People's Democratic Republic

O. Sanpool · W. Maleewong · P. M. Intapan (✉)
Faculty of Medicine and Research and Diagnostic Center for Emerging Infectious Diseases, Department of Parasitology, Khon Kaen University, Khon Kaen, Thailand
e-mail: pewpan@kku.ac.th

T. Thanchomnang
Faculty of Medicine, Mahasarakham University, Mahasarakham, Thailand

H. Mehlhorn, S. Klimpel (eds.), *Parasite and Disease Spread by Major Rivers on Earth*, Parasitology Research Monographs 12,
https://doi.org/10.1007/978-3-030-29061-0_7

7.1 Introduction

The Greater Mekong Subregion (GMS) encompasses six countries: the Kingdom of Thailand, Union of Myanmar, Lao People's Democratic Republic (Lao PDR), Kingdom of Cambodia, Socialist Republic of Vietnam, and the Yunnan province of the People's Republic of China. These countries share the benefits of being Mekong river countries in the lower Mekong sub-region, including increased political and economic significance and the opportunity to become one of the fastest growing regions in the world. Economic growth and an open-borders execution that began in 2015, which expanded migration among the ASEAN Economic Community (AEC) countries (ASEAN Economic Community Blueprint 2008), have caused environmental problems that affect public health (Asian Development Bank 2018). This chapter aims to review the important foodborne trematodiases that affect the public health of people in the lower GMS countries.

7.2 Important Foodborne Trematodiases

7.2.1 *Opisthorchiasis*

Opisthorchiasis is a fish-borne disease caused by a human liver fluke in Genus *Opisthorchis*. *Opisthorchis viverrini* is one of the main public health problems in the Southeast Asian region, especially along the Mekong River (Cambodia, Lao People's Democratic Republic, Vietnam, and Thailand), where there is a culture of eating raw or undercooked fish (Fig. 7.1) that contain encysted metacercariae of the liver fluke (Hung et al. 2013; Sripa et al. 2011; Sripa and Pairojkul 2008).

The life cycle of *O. viverrini* needs a single snail genus; the first intermediate host includes *Bithynia funiculata, Bithynia siamensis goniomphalos*, and *Bithynia siamensis siamensis* in Southeast Asia (Fig. 7.2). When free-swimming cercariae emerge from snails, they infect the second intermediate host, cyprinoid fishes (Fig. 7.3a) and develop as metacercariae (Fig. 7.3b). The final host (human, dog, cat, etc.) is infected when it consumes raw or semi-cooked fish containing infective metacercariae. In the definitive host, the metacercariae are excysted in the duodenum and move to the bile ducts toward the liver. They become sexually mature in the bile ducts, where they start laying eggs. Sexual maturity is reached after several weeks. Adults are long-lived and eat on epithelial cells of the intrahepatic bile ducts (Fig. 7.4).

Although most infected cases are asymptomatic, complications can include hepatobiliary diseases, such as hepatomegaly, cholangitis, cholecystitis, peri-ductal fibrosis, and gallstones. Severe chronic infection is a strong risk factor for cholangiocarcinoma (Honjo et al. 2005; IARC 1994; Zheng et al. 2017). In the chronic stage, most complications of human opisthorchiasis are hepatobiliary diseases, including cholangitis, cholecystitis, peri-ductal fibrosis gallstones, and hepatomegaly. An association with cholangiocarcinoma has also been found (Honjo et al. 2005; IARC 1994; Ayé et al. 2015).

Fig. 7.1 The ingredients for the preparation of traditional dishes made from raw fish (Koi-pla) (**a**) and other raw dishes (Koi-pla) (**b**)

Opisthorchiasis is widespread in or near the basin of the lower Mekong River. More than 10 million people are estimated to be infected with *O. viverrini*: nearly 8 million in Thailand and 2 million in Lao People's Democratic Republic (Sripa et al. 2010). High prevalence rates are found in North and Northeast Thailand, especially in rural populations (Sripa et al. 2011; Jongsuksuntigul and Imsomboon 2003; Sithithaworn et al. 2012), and in the neighboring southern and central regions of Lao People's Democratic Republic (Forrer et al. 2012; Sayasone et al. 2009). Human infections with *O. viverrini* and the presence of metacercariae in intermediate hosts have been reported in several provinces of Cambodia (Chai et al. 2014; Miyamoto et al. 2014; Sohn et al. 2012; Yong et al. 2014). Miyamoto et al. (2014) found *O. viverrini* eggs in stool samples from five provinces of Cambodia, among which 15 villages had an egg-positive rate of >10%. Molecular identification confirmed the presence of *O. viverrini* in 30 samples from 5 villages.

The parasite is likewise endemic in southern and central parts of Vietnam (Dao et al. 2016; Dung et al. 2014). A 2015 study stated that the prevalence of *O. viverrini* infection was 11.4% in central Vietnam (Dao et al. 2016). In Myanmar, *O. viverrini* was found in 34 of 364 (9.3%) participants in rural areas in central regions by stool examination using a modified formalin-ether concentration technique and confirmation by molecular methods (Aung et al. 2017). *O. viverrini* metacercariae were also found in the cyprinoid fish (Fig. 7.5), *Puntius brevis,* in the Bago Region of Central Myanmar (Sanpool et al. 2018). New records and confirmation of *O. viverrini* adults in experimental hamsters were revealed by morphological and molecular analyses (Sanpool et al. 2018).

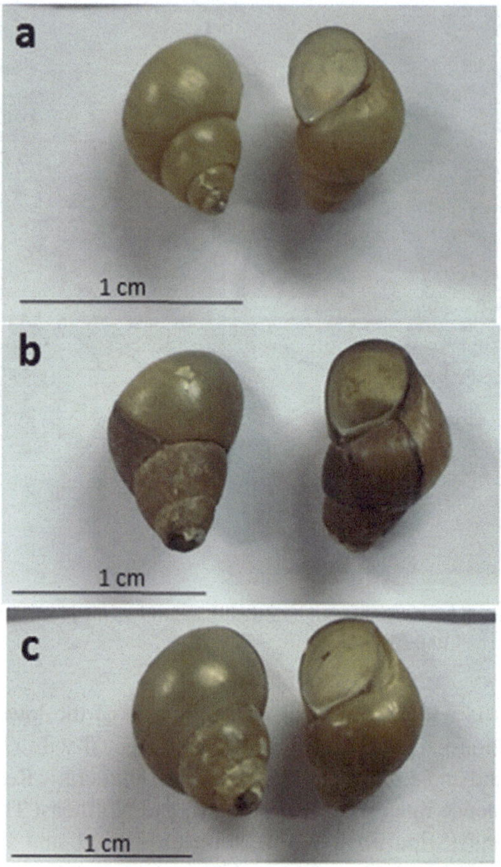

Fig. 7.2 *Bithynia* snails are the first intermediate host of *Opisthorchis viverrini*: (**a**) *Bithynia siamensis siamensis*, (**b**) *Bithynia siamensis goniomphalos*, and (**c**) *Bithynia funiculata*

Fig. 7.3 (**a**) Cyprinid fish are the second intermediate host of *Opisthorchis viverrini*. (**b**) *O. viverrini* metacercariae were found in from cyprinid fish

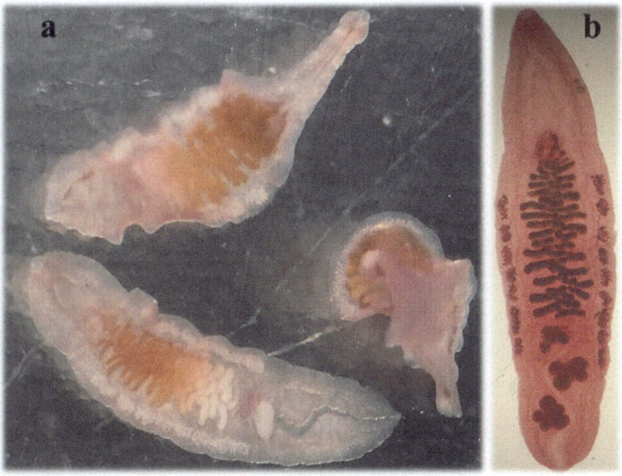

Fig. 7.4 The adult stage of *Opisthorchis viverrini*: (**a**) the living adults and (**b**) a worm stained with carmine stain

Fig. 7.5 A female street vendor and other freshwater fish at the morning market in the Bago Region, Central Myanmar (located 91 km northeast of Yangon)

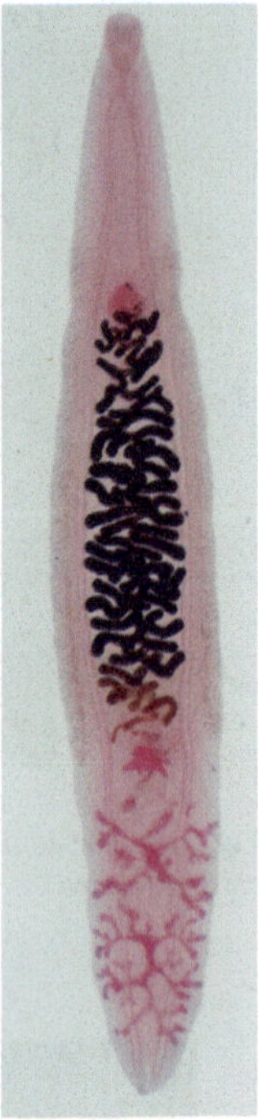

Fig. 7.6 Adult-stage *Clonorchis sinensis* with carmine staining

7.2.2 *Clonorchiasis*

Clonorchiasis is a disease caused by Chinese liver flukes of the genus *Clonorchis*; *Clonorchis sinensis* is the only species in mammals. *C. sinensis* adults are lanceolate-shaped digenetic trematodes (Fig. 7.6) belonging to class Trematoda, phylum Platyhelminthes. This liver fluke is the most important species of foodborne zoonotic parasites in East Asia, including China (apart from the northwestern area), Korea, the far-eastern part of Russia, Taiwan, and the lower Mekong subregion of northern Vietnam, where it is still actively endemic (Hong and Fang 2012).

The life cycle requires both sexual and asexual reproductions in snail, fish, and mammal hosts (Qian et al. 2016). Embryonated eggs are placed by adult worms and discharged in the biliary ducts and stools of final hosts. The discharged eggs are eaten by a snail intermediate host and release miracidia, which develop to sporocysts, rediae, and cercariae. After developed cercariae shed into water and encounter the second intermediate cyprinid fish, they invade the mucous skin and develop into encysted metacercariae in the subcutaneous tissues or muscles. When a final mammal host (cat, dog, human, etc.) eats raw or undercooked infected fish, the larvae excyst in the duodenum; then, the excysted juvenile flukes move to the intrahepatic bile duct through the ampulla of Vater and grow into adult flukes (Qian et al. 2016).

Human parasitic infection is asymptomatic or has non-specific symptoms, such as asthenia, nausea, indigestion, headache, dizziness, vertigo, abdominal discomfort, diarrhea, or abdominal pain. Heavy and chronic infection causes several symptoms in the liver and biliary systems, primarily cholelithiasis, cholangitis, and cholecystitis (Hong and Fang 2012). This fluke is presently categorized as a definite Group 1 biological agent (carcinogen) by the International Agency of Research on Cancer (IARC) based on plenty of evidence in humans (IARC 1994, 2012).

In Vietnam, clonorchiasis is endemic in human populations in northern provinces, whereas cats and dogs are the reservoir hosts (Doanh and Nawa 2016). The prevalence rates were 0.2 to 37.5%, with infection hotspots (26.0–37.5%) being reported in Nam Dinh province and Ninh Binh province (23.5–31.0%) (Doanh and Nawa 2016).

In Thailand, Traub and coworkers (2009) used molecular-based technology to detect *C. sinensis* eggs in human stools collected from a rural community in eastern Thailand, Chachoengsao province. Of these, 23% (5/22) were found by microscopy to be positive for *Opisthorchis*-like eggs.

7.2.3 Paragonimiasis

Paragonimiasis, a major food-borne zoonosis in Asian countries, is caused by infection with lung flukes of the genus *Paragonimus*. More than 50 species of genus *Paragonimus* have been identified in Asia, America, and Africa, but most of them occur in eastern Asia (Blair et al.1999; Kong and Doanh 2015). Molecular phylogenetic analyses have shown that *Paragonimus* species in Asia can be separated into four major species complexes—*Paragonimus heterotremus*, *Paragonimus skrjabini*, *Paragonimus ohirai,* and *Paragonimus westermani* complexes—and two independent species, *Paragonimus vietnamensis* and *Paragonimus macrorchis* (Nawa et al. 2014). Although *P. vietnamensis* was reported only in northern Vietnam (Doanh et al. 2007), *P. macrorchis* was firstly found in China by Chen (1962, 1964) and then sporadically in that country (Wei et al. 1993), Sri Lanka (Kannangara 1969) and Thailand (Miyazaki and Vajrasthira 1967; Waikagul et al. 1989).

Humans are infected by ingesting raw or undercooked freshwater crabs containing infectious metacercariae (Fig. 7.7). Approximately 200 million people have been exposed and 20 million individuals possibly infected worldwide (WHO 1995). Although *Paragonimus westermani* is the main human pathogen in China, Korea, and Japan (Liu et al. 2008; Blair et al. 1999, 2007; Nawa et al. 2014), *P. heterotremus* is the major causative agent of human paragonimiasis in south region of China (Liu et al. 2008), Lao PDR (Yahiro et al. 2008; Vidamaly et al. 2009), Vietnam (De et al. 2000; Le et al. 2006; Doanh et al. 2011), India (Devi et al. 2007), and Thailand (Vanijanonta et al. 1981; Maleewong 2003). *Paragonimus pseudoheterotremus* has also been reported as a human pathogen in Thailand (Intapan et al. 2012).

Paragonimus is a digenean trematode of carnivorous mammals. Generally, living adult worms are shaped like coffee beans with a pinkish brown color (Fig. 7.8a, b). Operculated eggs are excreted in stool (Fig. 7.8c) and mature to miracidia in

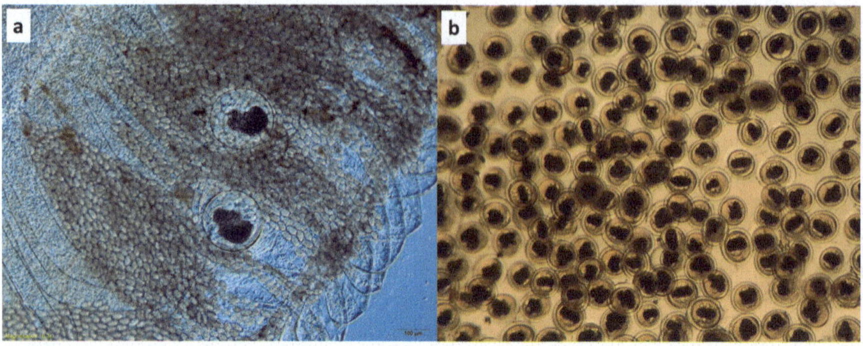

Fig. 7.7 *Paragonimus heterotremus* metacercariae in mountainous crab gill (**a**) and isolated *P. heterotremus* metacercariae (**b**) recovered from mountainous crabs in Thailand

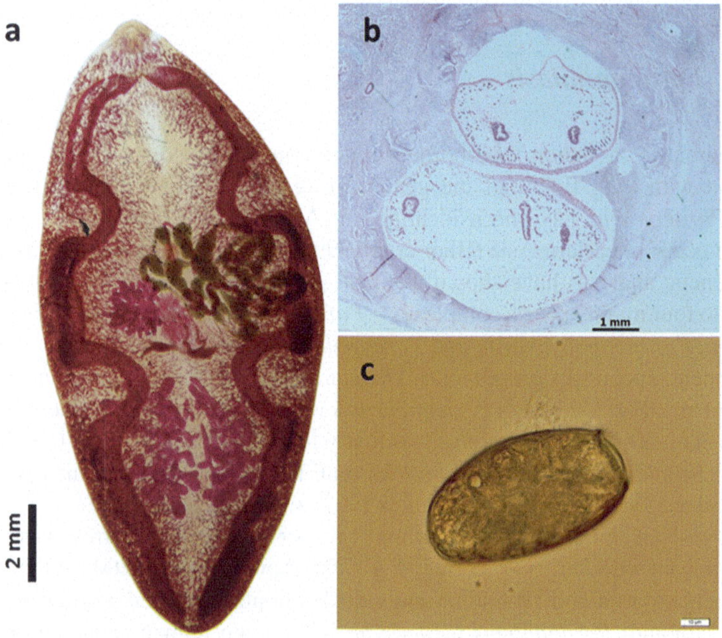

Fig. 7.8 *Paragonimus heterotremus:* (**a**) adult stained with carmine stain, (**b**) the cross-section of two adult worms in the lung of the final host, and (**c**) an egg

freshwater. They infect the first intermediate hosts of freshwater snails, in which they develop three generations of asexual reproduction: redia, daughter redia, and cercaria. Microcercus cercariae are released into freshwater and infect the gills and muscles of the second intermediate crustacean hosts (Fig. 7.9a, b), freshwater crabs (*Eriocheir*, *Geothelphusa*, and *Potamon* species) and crayfish (*Cambaroides* and *Procambarus* species). The second intermediate hosts can be infected through the

Fig. 7.9 Natural water resources in Tak province, northern Thailand: (**a**) a habitat of mountainous crabs, (**b**) mountainous crabs as the second intermediate host of *Paragonimus heterotremus,* and (**c**) a popular dish in Thailand, papaya salad mixed with raw mountainous crabs (possibly contaminated with *P. heterotremus* metacercariae) supplemented with Thai morning-glory (*Ipomoea aquatica*) and yellow velvetleaf (*Limnocharis flava*) (possibly contaminated with *Fasciola* spp. metacercariae in the lower greater sub-regions)

ingestion of cercariae-infected snails and become metacercariae, which are infective to the definitive hosts. Various mammals may have different susceptibilities to different species; however carnivorous mammals, felines, and canines, are the most important definitive hosts. Wild rodents and boars are paratenic (transport) hosts of *P. westermani*. Human infection is correlated to the food habits (Fig. 7.9c) of people in several endemic areas (Kong and Doanh 2015).

7.2.4 Epidemiology

In Thailand, human paragonimiasis was reported in Saraburi Province, Central Thailand (Yoonuan et al. 2008), Nan Province (Watthanakulpanich et al. 2005), and Phitsanulok Province (Waree et al. 2001). In Vietnam, *Paragonimus* and paragonimiasis were recently reviewed (Doanh et al. 2013). *P. heterotremus* has been revealed to infect humans in some northern provinces and local people in Lai Chau and Yen Bai; the northern provinces are still distressed from *P. heterotremus* infection (Doanh et al. 2013). The sero-prevalence rate was 12.7% (28/220) in the Sin Ho district of Lai Chau Province (Doanh et al. 2011).

In Lao PDR, the clinical and radiological characteristics of 50 Laotian pleuropulmonary paragonimiasis patients were observed between 2003 and 2007 (Kanpittaya et al. 2010); the chest radiographs revealed abnormal findings (49/50; 98%). All 49 patients showed pulmonary parenchymal abnormalities, while pleural effusion was also reported (11/49; 22%). General intraparenchymal findings include multiple small cysts (90%), irregular linear densities (68%), and nodular opacities (40%) (Kanpittaya et al. 2010).

In Myanmar, *Paragonimus heterotremus* metacercariae collected from Myawaddy Province, Myanmar were identified based on ITS2 and cox1 sequences (Sanpool et al. 2013). People living in this area may be at risk for infection with this lung fluke and should be monitored for human paragonimiasis.

7.2.5 Fascioliasis

Fascioliasis is a major public health problem in many countries caused by liver fluke infection. *Fasciola hepatica* is recognized as "the common liver fluke" or "the sheep liver fluke." A related fluke is *Fasciola gigantica* (Mas-Coma et al. 2005). *Fasciola hepatica* has a global distribution but is predominant in temperate zones, whereas *F. gigantica* is also reported on most continents but mostly in tropical regions of Asia, Southeast Asia, and Africa (Andrews 1999). Humans are frequently infected by the consumption of aquatic plants (Figs. 7.9c and 7.10) that have infective metacercariae (Mas-Coma et al. 1999). The estimated number of infected individuals is in the range of 2.4–17 million (WHO 1995; Mas-Coma et al. 2015) and the estimated at-risk population is 80 million (Keiser and Utzinger 2005).

The adult worm has a leaf-shaped body, with a broadly pointed posterior end. A cone-like anterior extension of the body has two suckers. The intestinal ceca is long with a large number of lateral branches, protruding the posterior end of the body (Fig. 7.11). The size of *Fasciola gigantica* size is approximately double the length of *F. hepatica*. *Fasciola gigantica* has a more elongated and narrower shape. In *Fasciola gigantica*, the ceca are more branched, especially those to the midline of the body, and the branches of the ovary are longer and more numerous. The two species overlap; hybrid specimens may show intermediate forms in endemic areas (Mas-Coma et al. 2015).

Fig. 7.10 Aquatic plants: (**a**) water caltrops (*Trapa natans*) and (**b**) water chestnut (*Eleocharis dulcis*) (possibly contaminated with *Fasciola* spp. metacercariae in lower greater subregions)

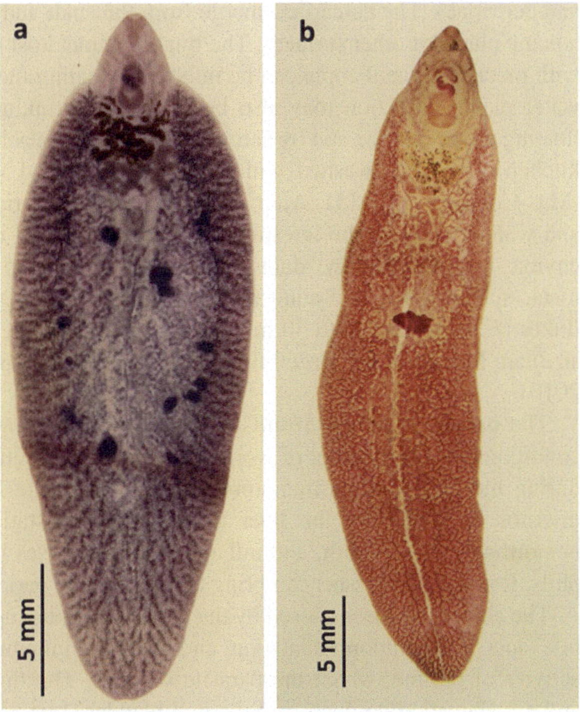

Fig. 7.11 The adult stage of *Fasciola* spp.: (**a**) *Fasciola hepatica* and (**b**) *Fasciola gigantica* stained with carmine stain

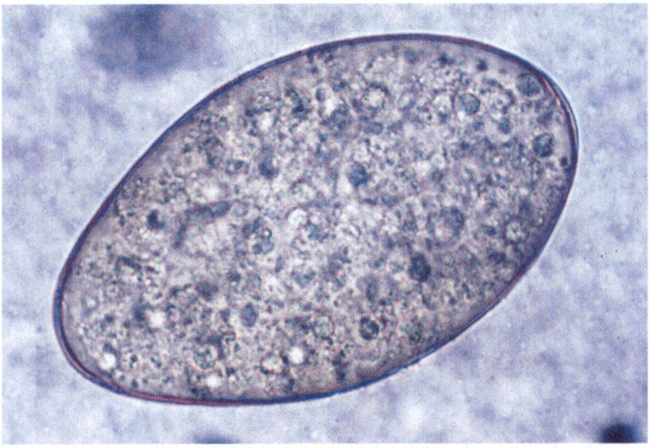

Fig. 7.12 The egg of *Fasciola* spp.

Fascioliasis is a zoonosis that is transmitted from animals to humans. Susceptible reservoir hosts are sheep, pigs, buffaloes, donkeys, horses, goats, dromedaries, camels, rabbits, and rodents. Immature fluke eggs are excreted in the feces of infected animals or humans and water in the environment (Fig. 7.12). The eggs become embryonated in water; the miracidia release from the eggs and enter a suitable snail intermediate host, comprising the genera *Galba*, *Fossaria,* and *Pseudosuccinea*. The parasites grow in the snail to several stages (sporocysts, rediae and cercariae). The cercariae emerge from the snail and encyst as metacercariae on aquatic plants or other surfaces. The final mammal host is infected by aquatic plants with metacercariae. People can be infected by eating metacercariae-attaching freshwater plants. Infection may also be caused by drinking water contaminated with floating metacercariae and by eating metacercariae attached to the surface of food or kitchen appliances washed with water contaminated with floating metacercariae (Mas-Coma et al. 2015). After eating, the metacercariae excyst in the duodenum and wander through the intestinal wall, the peritoneal cavity, and the liver parenchyma into the biliary ducts, where they develop into adults. In humans, metacercariae grow into adult flukes within approximately 3–4 months. The adult flukes (*F. hepatica*: up to 30 mm by 13 mm in size; *F. gigantica*: up to 75 mm in size) are located in the large biliary ducts of the final mammalian host (Sripa et al. 2010).

The parasite infection results in death for sheep and cattle. Human disease is conditional on the number of worms and stage of infection. The acute phase occurs during the migration of immature flukes through the liver. Pathology shows the invasion of worms in the liver tissues and devastation of parenchymal tissue, hemorrhage, worm death, and inflammatory responses largely mediated by eosinophils. Repair mechanisms can bring about massive periportal fibrosis.

The chronic phase is caused by the mechanical obstruction of the ducts, including bile duct proliferation, dilatation and fibrosis, inflammatory responses, and the activity of proline, which the parasite secretes. The flukes are found in the intra- and extrahepatic bile ducts and the gall bladder (Lim et al. 2008). Anemia may be caused by blood loss through bile duct lesions.

Fascioliasis can be symptomatic or asymptomatic (Mas-Coma et al. 2015). Symptomatic cases occur approximately 2 months following the consumption of metacercariae, which results in abdominal pain, hepatomegaly, fever, diarrhea, urticarial vomiting, and eosinophilia during the acute phase due to the migration of the immature fluke through the hepatic parenchyma. In the chronic phase, the symptoms are more distinct and reveal intermittent biliary obstruction and inflammation triggered by the adult fluke inside the bile ducts. Ectopic locations of flukes have been reported, including the intestinal wall, lungs, subcutaneous tissue, and pharyngeal mucosa (Mas-Coma et al. 2015). Abnormal migration by immature *Fasciola* through the skin in human fascioliasis has been found in Vietnam (Le et al. 2007; Xuan et al. 2005).

7.2.6 Epidemiology

In Thailand, a high prevalence of *F. gigantica* has been reported in cattle and buffaloes in areas with large water reservoirs in Thailand (Srihakim and Pholpark 1991) and Cambodia (Suon et al. 2006).

Sporadic cases of human fascioliasis have been reported in Thailand (Kachintorn et al. 1988; Aroonroch et al. 2006). In Vietnam, from 1998 to 2000, approximately

500 cases of human fascioliasis were reported (Tran et al. 2001). The World Health Organization (1995) estimated that 5000 cases have to be treated annually in endemic provinces (WHO 1995). In Laotian villages, a 2.4% prevalence rate of human fascioliasis was found by a stool examination, with a rate of 13.8% found after serological diagnosis (Quang et al. 2008). In Cambodia, two schoolchildren were found to be positive for *Fasciola hepatica* by serology testing; one sample was found to be positive for *Fasciola* spp. by polymerase chain reaction (Bless et al. 2015).

Acknowledgments The authors were supported by a Distinguished Research Professor Grant, Thailand Research Fund (Grant no. DPG6280002).

Bibliography

Introduction

ASEAN Economic Community Blueprint (2008) Declaration on the ASEAN Economic Community blueprint. http://asean.org/wp-content/uploads/archive/5187-10.pdf. Accessed 4 Mar 2017

Asian Development Bank (ADB). Greater Mekong Subregion Environmental Performance Assessment 2006–2016. https://www.adb.org/publications/gms-environmental-performance-assessment-2006-2016. Accessed 14 Dec 2018

Opisthorchiasis

Aung WPP, Htoon TT, Tin HH, Thinn KK, Sanpool O, Jongthawin J et al (2017) First report and molecular identification of *Opisthorchis viverrini* infection in human communities from Lower Myanmar. PLoS One 12(5):e0177130

Ayé SP, Rajpho V, Phongluxa K, Vonghachack Y, Hattendorf J, Hongvanthong B et al (2015) Subtle to severe hepatobiliary morbidity in *Opisthorchis viverrini* endemic settings in southern Laos. Acta Trop 141(Pt B):303–309

Chai JY, Sohn WM, Na BK, Yong TS, Eom KS, Yoon CH et al (2014) Zoonotic trematode metacercariae in fish from Phnom Penh and Pursat, Cambodia. Korean J Parasitol 52:35–40

Dao TTH, Bui TV, Abatih EN, Gabriël S, Nguyen TT, Huynh QH et al (2016) *Opisthorchis viverrini* infections and associated risk factors in a lowland area of Binh Dinh Province, Central Vietnam. Acta Trop 157:151–157

Dung VT, Waikagul J, Thanh BN, Vo DT, Nguyen DN, Murrell K (2014) Endemicity of *Opisthorchis viverrini* liver flukes, Vietnam, 2011–2012. Emerg Infect Dis 20:152–154

Forrer A, Sayasone S, Vounatsou P, Vonghachack Y, Bouakhasith D, Vogt S et al (2012) Spatial distribution of, and risk factors for, *Opisthorchis viverrini* infection in southern Lao PDR. PLoS Negl Trop Dis 6:e1481

Honjo S, Srivatanakul P, Sriplung H, Kikukawa H, Hanai S, Uchida K et al (2005) Genetic and environmental determinants of risk for cholangiocarcinoma via *Opisthorchis viverrini* in a densely infested area in Nakhon Phanom, northeast Thailand. Int J Cancer 117:854–860

Hung NM, Madsen H, Fried B (2013) Global status of fish-borne zoonotic trematodiasis in humans. Acta Parasitol 58:231–258

IARC (1994) Infection with liver flukes (*Opisthorchis viverrini, Opisthorchis felineus* and *Clonorchis sinensis*). IARC Monogr Eval Carcinog Risks Hum 61:121–175
Jongsuksuntigul P, Imsomboon T (2003) Opisthorchiasis control in Thailand. Acta Trop 88:229–232
Miyamoto K, Kirinoki M, Matsuda H, Hayashi N, Chigusa Y, Sinuon M et al (2014) Field survey focused on *Opisthorchis viverrini* infection in five provinces of Cambodia. Parasitol Int 63:366–373
Sanpool O, Aung WPP, Rodpai R, Maleewong W, Intapan PM (2018) Human liver fluke *Opisthorchis viverrini* (Trematoda, Opisthorchiidae) in Central Myanmar: New records of adults and metacercariae identified by morphology and molecular analysis. Acta Trop 185:149–155
Sayasone S, Vonghajack Y, Vanmany M, Rasphone O, Tesana S, Utzinger J et al (2009) Diversity of human intestinal helminthiasis in Lao PDR. Trans R Soc Trop Med Hyg 103:247–254
Sithithaworn P, Andrews RH, Nguyen VD, Wongsaroj T, Sinuon M, Odermatt P et al (2012) The current status of opisthorchiasis and clonorchiasis in the Mekong Basin. Parasitol Int 61:10–16
Sohn W, Yong TS, Eom KS, Pyo KH, Lee MY, Lim H et al (2012) Prevalence of *Opisthorchis viverrini* infection in humans and fish in Kratie Province, Cambodia. Acta Trop 124:215–220
Sripa B, Pairojkul C (2008) Cholangiocarcinoma: lessons from Thailand. Curr Opin Gastroenterol 24:349–356
Sripa B, Kaewkes S, Intapan PM, Maleewong W, Brindley PJ (2010) Food-borne trematodiases in Southeast Asia. Epidemiology, pathology, clinical manifestation and control. Adv Parasitol 72:305–350
Sripa B, Bethony JM, Sithithaworn P, Kaewkes S, Mairiang E, Loukas A et al (2011) Opisthorchiasis and *Opisthorchis*-associated cholangiocarcinoma in Thailand and Laos. Acta Trop 120(Suppl 1):S158–S168
Yong TS, Chai JY, Sohn WM, Eom KS, Jeoung HG, Hoang EH et al (2014) Prevalence of intestinal helminths among inhabitants of Cambodia (2006-2011). Korean J Parasitol 52:661–666
Zheng S, Zhu Y, Zhao Z, Wu Z, Okanurak K, Lv Z (2017) Liver fluke infection and cholangiocarcinoma: a review. Parasitol Res 116:11–19

Clonorchiasis

Doanh PN, Nawa Y (2016) *Clonorchis sinensis* and *Opisthorchis* spp. in Vietnam: current status and prospects. Trans R Soc Trop Med Hyg 110:13–20
Hong ST, Fang Y (2012) *Clonorchis sinensis* and clonorchiasis, an update. Parasitol Int 61:17–24
IARC (1994) Infection with liver flukes (*Opisthorchis viverrini, Opisthorchis felineus* and *Clonorchis sinensis*). IARC Monogr Eval Carcinog Risks Hum 61:121–175
IARC (2012) Working Group on the Evaluation of Carcinogenic Risks to Humans. Biological agents. Volume 100 B. A review of human carcinogens. IARC Monogr Eval Carcinog Risks Hum 100(Pt B):1–441
Qian MB, Utzinger J, Keiser J, Zhou XN (2016) Clonorchiasis. Lancet 387:800–810
Traub RJ, Macaranas J, Mungthin M, Leelayoova S, Cribb T, Murrell KD et al (2009) A new PCR-based approach indicates the range of *Clonorchis sinensis* now extends to Central Thailand. PLoS Negl Trop Dis 3:e367

Paragonimiasis

Blair D, Xu ZB, Agatsuma T (1999) Paragonimiasis and the genus *Paragonimus*. Adv Parasitol 42:113–222

Blair D, Agatsuma T, Wang W (2007) Paragonimiasis. In: Murrell KD, Fried B (eds) Food-borne parasitic zoonoses. World class parasites, vol 11. Springer, New York, pp 117–150

Chen HT (1962) Some new Chinese paragonimid trematodes: with notes on certain technical problems in the study of the group. Acta Sci Natur Univ Sunyatseni 3:58–64. (In Chinese)

Chen HT (1964) Morphological studies on the adult *Paragonimus* (M.) *macrorchis* Chen, 1962, with a consideration on the taxonomy of Paragonimidae. Acta Zool Sin 16:381–392. (In Chinese)

De NV, Cong LD, Kino H, Son DT, Vien HV (2000) Epidemiology, symptoms and treatment of paragonimiasis in Sin Ho district, Lai Chau province, Vietnam. Southeast Asian J Trop Med Public Health 31(Suppl 1):26–30

Devi KR, Narain K, Bhattacharya S, Negmu K, Agatsuma T, Blair D et al (2007) Pleuropulmonary paragonimiasis due to *Paragonimus heterotremus*: molecular diagnosis, prevalence of infection and clinicoradiological features in an endemic area of northeastern India. Trans R Soc Trop Med Hyg 101:786–792

Doanh PN, Shinohara A, Horii Y, Habe S, Nawa Y, Le NT (2007) Description of a new lung fluke species, *Paragonimus vietnamensis* sp. nov. (Trematoda, Paragonimidae), found in northern Vietnam. Parasitol Res 101:1495–1501

Doanh PN, Dung do T, Thach DT, Horii Y, Shinohara A, Nawa Y (2011) Human paragonimiasis in Viet Nam: epidemiological survey and identification of the responsible species by DNA sequencing of eggs in patients' sputum. Parasitol Int 60:534–537

Doanh PN, Horii Y, Nawa Y (2013) *Paragonimus* and paragonimiasis in Vietnam: an update. Korean J Parasitol 51:621–627

Intapan PM, Sanpool O, Thanchomnang T, Imtawil K, Pongchaiyakul C, Nawa Y, Maleewong W (2012) Molecular identification of a case of *Paragonimus pseudoheterotremus* infection in Thailand. Am J Trop Med Hyg. 87:706–709

Kannangara DWW (1969) Occurrence of the lung fluke *Paragonimus macrorchis* Chen, 1962 in Ceylon. Ceylon J Med Sci 18:33–38

Kanpittaya J, Sawanyawisuth K, Vannavong A, Intapan PM, Maleewong W, Zhang W et al (2010) Different chest radiographic findings of pulmonary paragonimiasis in two endemic countries. Am J Trop Med Hyg 83:924–926

Kong Y, Doanh PN (2015) NawaY. *Paragonimus*. In: Xiao L, Ryan U, Feng Y (eds) Biology of Foodborne Parasites. CRC press, Taylor & Francis Group, New York, pp 445–462

Le TH, Van De N, Blair D, McManus DP, Kino H, Agatsuma T (2006) *Paragonimus heterotremus* Chen and Hsia (1964), in Vietnam: a molecular identification and relationships of isolates from different hosts and geographical origins. Acta Trop 98:25–33

Liu Q, Wei F, Liu W, Yang S, Zhang X (2008) Paragonimiasis: an important food-borne zoonosis in China. Trends Parasitol 24:318–323

Maleewong W (2003) *Paragonimus* species. In: Miliotis MD, Bier JW (eds) International handbook of foodborne pathogens. New York. Marcel Dekker Inc, New York, pp 601–611

Miyazaki I, Vajrasthira S (1967) Occurrence of the lung fluke *Paragonimus macrorchis* Chen, 1962 in Thailand. J Parasitol 53:894–895

Nawa Y, Thaenkham U, Doanh PN, Blair D (2014) Helminth-Trematode: *Paragonimus westermani* and *Paragonimus* Species. In: Motarjemi Y, Moy G, Todd E (eds) Encyclopedia of Food Safety. Elsevier Inc, Oxford, pp 179–188

Sanpool O, Intapan PM, Thanchomnang T, Janwan P, Nawa Y, Blair D et al (2013) Molecular variation in the *Paragonimus heterotremus* complex in Thailand and Myanmar. Korean J Parasitol 51:677–681

Vanijanonta S, Radomyos P, Bunnag D, Harinasuta T (1981) Pulmonary paragonimiasis with expectoration of worms: a case report. Southeast Asian J Trop Med Public Health 12:104–106

Vidamaly S, Choumlivong K, Keolouangkhot V, Vannavong N, Kanpittaya J, Strobel M (2009) Paragonimiasis: a common cause of persistent pleural effusion in Lao PDR. Trans R Soc Trop Med Hyg 103:1019–1023

Waikagul J, Setasuban P, Vajrasthira S (1989) On a species of *Paragonimus* lung fluke metacercaria in Thailand. Southeast Asian J Trop Med Public Health 20:495–497

Waree P, Polseela P, Pannarunothai S, Pipitgool V (2001) The present situation of paragonimiasis in endemic area in Phitsanulok Province. Southeast Asian J Trop Med Public Health 32(Suppl 2):51–54

Watthanakulpanich D, Waikagul J, Dekumyoy P, Muangkhum P, Praevanit R, Mongkhonmu S (2005) Paragonimiasis in Nan Province, northern Thailand. Southeast Asian J Trop Med Public Health 36:853–857

Wei MB, He G, Shang SM et al (1993) Discovery of human *Pagumogonimus macrorchis* infection and its distribution. Chin J Zoonoses 4-6(In Chinese):9

WHO (1995) Control of foodborne trematode infections. World Health Organization Technical Report Series, Geneva, vol 849, pp 1–157

Yahiro S, Habe S, Duong V, Odermatt P, Barennes H, Strobel M, Nakamura S (2008) Identification of the human paragonimiasis causative agent in Lao People's Democratic Republic. J Parasitol 94:1176–1177

Yoonuan T, Vanvanitchai Y, Dekumyoy P, Komalamisra C, Kojima S, Waikagul J (2008) Paragonimiasis prevalences in Saraburi Province, Thailand, measured 20 years apart. Southeast Asian J Trop Med Public Health 39:593–600

Fascioliasis

Andrews S (1999) The Life cycle of *Fasciola hepatica*. In: Dalton JP (ed) Fasciolosis. Oxon, UK, CABI Publishing, Wallingford, pp 1–29

Aroonroch R, Worawichawong S, Nitiyanant P, Kanchanapitak A, Bunyaratvej S (2006) Hepatic fascioliasis due to *Fasciola hepatica*: a two-case report. J Med Assoc Thail 89:1770–1774

Bless PJ, Schär F, Khieu V, Kramme S, Muth S, Marti H et al (2015) High prevalence of large trematode eggs in schoolchildren in Cambodia. Acta Trop 141(Pt B):295–302

Kachintorn U, Atisook K, Tesjaroen S, Lertakyamanee N, Plengvanit U (1988) *Fasciola gigantica*: the second case of human infection in Thailand. J Med Assoc Thail 71:451–455

Keiser J, Utzinger J (2005) Emerging foodborne trematodiasis. Emerg Infect Dis 11:1507–1514

Le TH, De NV, Agatsuma T, Blair D, Vercruysse J, Dorny P et al (2007) Molecular confirmation that *Fasciola gigantica* can undertake aberrant migrations in human hosts. J Clin Microbiol 45:648–650

Lim JH, Mairiang E, Ahn GH (2008) Biliary parasitic diseases including clonorchiasis, opisthorchiasis and fascioliasis. Abdom Imaging 33:157–165

Mas-Coma S, Bargues MD, Esteban JG (1999) Human fascioliasis. In: Dalton JP (ed) Fasciolosis. CABI Publishing, Wallingford, pp 411–434

Mas-Coma S, Bargues MD, Valero MA (2005) Fascioliasis and other plant-borne trematode zoonoses. Int J Parasitol 35:1255–1278

Mas-Coma S, Valero A, Bargues MD (2015) *Fasciola* and *Fasciolopsis*. In: Xiao L, Ryan U, Feng Y (eds) Biology of Foodborne Parasites. CRC press, Taylor & Francis Group, New York, pp 371–404

Quang TD, Duong TH, Richard-Lenoble D, Odermatt P, Khammanivong K (2008) Emergence in humans of fascioliasis (from *Fasciola gigantica*) and intestinal distomatosis (from *Fasciolopsis buski*) in Laos. Sante 18:119–124

Srihakim S, Pholpark M (1991) Problem of fascioliasis in animal husbandry in Thailand. Southeast Asian J Trop Med Public Health 22:352–355

Sripa B, Kaewkes S, Intapan PM, Maleewong W, Brindley PJ (2010) Food-borne trematodiases in Southeast Asia. Epidemiology, pathology, clinical manifestation and control. Adv Parasitol 72:305–350

Suon S, Hol D, Siek S, McLean M, Copeman B (2006) Seasonal differences in the incidence of infection with *Fasciola gigantica* in Cambodian cattle. Trop Anim Health Prod 38:23–28

Tran VH, Tran TK, Nguye HC, Pham HD, Pham TH (2001) Fascioliasis in Vietnam. Southeast Asian J Trop Med Public Health 32(Suppl 2):48–50

WHO (1995) Control of foodborne trematode infections. World Health Organization Technical Report Series, Geneva, vol 849

Xuan le T, Hung NT, Waikagul J (2005) Cutaneous fascioliasis: a case report in Vietnam. Am J Trop Med Hyg 72:508–509

Chapter 8
Important Protozoan Diseases in the Lower Mekong River Basin

Tongjit Thanchomnang, Oranuch Sanpool, Pewpan M. Intapan, and Wanchai Maleewong

Abstract The Greater Mekong Subregion (GMS) encompasses the resources of the Mekong River in six areas: the Kingdom of Cambodia, the Yunnan Province of the People's Republic of China (PRC), Lao People's Democratic Republic (Lao PDR), Union of Myanmar, Kingdom of Thailand, and the Socialist Republic of Vietnam. The lower Mekong River Basin is now one of the world's fastest growing economies. A number of factors in the lower GMS affect the epidemiology and outcomes of human parasite control initiatives, including the environment, the economy, and politics. In this chapter, we review the important protozoa diseases that affect the public health of people in the lower GMS, including malaria, leishmaniasis amoebiasis, giardiasis, cryptosporidiosis, and toxoplasmosis.

Keywords Lower Mekong River Basin · Malaria · Leishmaniasis · Amoebiasis · Giardiasis · Cryptosporidiosis · Toxoplasmosis · *Entamoeba* · *Leishmania* · *Giardia* · HIV · AIDS

List of Abbreviations

AIDS	Acquired immunodeficiency syndrome
CL	Cutaneous leishmaniasis
GMS	Greater Mekong Subregion
HIV	Concomitant human immunodeficiency virus
Lao PDR	Lao People's Democratic Republic
MCL	Mucocutaneous leishmaniasis
PRC	People's Republic of China
VL	Visceral leishmaniasis

T. Thanchomnang
Faculty of Medicine, Mahasarakham University, Mahasarakham, Thailand

O. Sanpool · P. M. Intapan · W. Maleewong (✉)
Department of Parasitology, Faculty of Medicine and Research and Diagnostic Center for Emerging Infectious Diseases, Khon Kaen University, Khon Kaen, Thailand
e-mail: wanch_ma@kku.ac.th

H. Mehlhorn, S. Klimpel (eds.), *Parasite and Disease Spread by Major Rivers on Earth*, Parasitology Research Monographs 12,
https://doi.org/10.1007/978-3-030-29061-0_8

8.1 Introduction

The Greater Mekong Subregion (GMS) is the area encompassing the resources of the Mekong River in six areas: the Kingdom of Cambodia, the Yunnan Province of the People's Republic of China (PRC), Lao People's Democratic Republic (Lao PDR), Union of Myanmar, Kingdom of Thailand, and the Socialist Republic of Vietnam. The GMS is 2.6 million km^2 and has a total population of approximate 328 million (ADB 2018). The Mekong River is 4909 km in length from its origin on the Tibetan Plateau in the PRC to its end in the Vietnam delta. The basin has an area of 795,000 km^2, extending over six countries: the PRC (21%), Myanmar (3%), the Lao PDR (25%), Thailand (23%), Cambodia (20%), and (8%). The population of the lower Mekong River Basin was approximately 65 million in 2016 (ADB 2018).

The lower Mekong River Basin is now one of the world's fastest growing economies. However, this spectacular economic growth has resulted in environmental problems, such as reductions in natural resources, environmental quality (including air pollution, biodiversity and habitats, climate change, fish, forests, land degradation, solid waste management, and water resources), ecosystem services, and public health. Such negative impacts subvert sustainable development in the GMS by affecting future economic growth and poverty curtailing in these countries (ADB 2018).

A number of factors in the lower GMS affects the epidemiology and outcomes of human parasite control initiatives, including the environment, the economy, and politics. This has led to new challenges in the fight against human parasitosis. Thus, this chapter aims to review the important protozoan diseases that affect the public health of people in the lower GMS.

8.2 Important Vector-Borne Protozoa

8.2.1 *Malaria*

Malaria is caused by *Plasmodium* parasites. The parasites are spread to people through the bites of infected female *Anopheles* mosquitoes. There are five parasite species that cause malaria in humans: *Plasmodium falciparum*, *Plasmodium vivax*, *Plasmodium malariae*, *Plasmodium ovale*, and *Plasmodium knowlesi* (Jongwutiwes et al. 2011; Saeung 2012). The life cycle of *Plasmodium* involves several distinct stages in the malaria vectors and human hosts (Phillips et al. 2017).

The World Health Organization estimated that 216 million cases of malaria occurred worldwide in 2016 and about 445,000 people died from the disease (WHO 2017a). In recent years, countries of the GMS have accelerated their efforts to prevent, diagnose, and treat malaria. Furthermore, resistance of *P. falciparum* to artemisinin and other antimalarial medicines has reached alarming levels in certain areas of the GMS. The goals of the regional strategy are to eliminate malaria by 2030 in all GMS countries and, considering the urgent action required against multidrug resistance in the GMS, to eliminate *P. falciparum* malaria by 2025 (WHO 2017b).

Vector control is the main way to prevent and reduce malaria transmission. Two forms of vector control are effective in a wide range of circumstances: long-lasting insecticide-treated nets and indoor residual spraying. Indoor residual spraying with insecticide is a powerful way to rapidly reduce malaria transmission (WHO 2017a).

In the years 2010 and 2016, the number of reported malaria cases confirmed by microscopy in GMS countries were as follows: Lao PDR, 1597 and 13,223; Thailand, 11,301 and 7921; Cambodia, 3695 and 14,277; Myanmar, 6717 and 103,285; Vietnam, 4161 and 19,638; and China, 2921 and 4990, respectively (WHO 2017a).

Between 2012 and 2016, the reported number of malaria cases in the GMS fell by 74% because these countries have enacted measures to prevent, diagnose, and treat malaria in addition to implementing enhanced surveillance systems. Malaria deaths decreased by 91% in the same period. However, since 2008, the presence of multidrug-resistant malaria parasites in the subregion has been a growing concern (WHO 2017b).

In Thailand Malaria remains a health problem in Thailand. Ten provinces in Thailand—Tak, Yala Srisaket, Mae Hong Son, Narathiwat, Ubon Ratchathani, Songkhla, Kanchanaburi, Chanthaburi, and Prachinburi—recorded the highest prevalence of malaria cases (n = 10,965), with *P. vivax* (82.36%) and *P. falciparum* (12.03%) being responsible for many of the cases reported in 2017 (Bureau of Vector Borne Diseases, Ministry of Public Health, Thailand 2017). The distribution of malaria in Thailand displays significant geographical heterogeneity, with most malaria cases concentrated along the Myanmar and Cambodia borders (Cui et al. 2012; Corbel et al. 2013). Imported cases in the border regions with Myanmar, Cambodia, and Lao PDR are a major hurdle to overcome in Thailand. In 2016, out of a total of 18,758 reported malaria cases, 41% were attributed to indigenous transmission within villages located along Thailand's borders with neighboring countries. Nine percent of cases were attributed to transmission from outside Thailand's borders (WHO 2017a). *Plasmodium falciparum* is becoming resistant to artemisinin and causing outbreaks of malaria in areas along the Thai-Cambodia border.

Among the 77 *Anopheles* species recognized in Thailand (Harbach 2018), members of the Dirus Complex, the Minimus Complex, and the Maculatus Group are regarded as the primary malaria vectors. *Anopheles aconitus*, *An. annularis*, *An. barbirostris* s.l., *An. epiroticus* (distributed in coastal area), *An. Pseudowillmori,* and *An. sawadwongporni* are considered to be secondary vectors (Rattanarithikul et al. 1996, 2006; Linton et al. 2005; Sallum et al. 2005; Tainchum et al. 2015; Sriwichai et al. 2016). Additionally, *An. philippinensis* is a suspected vectors based on positive oocysts in the midgut (Rattanarithikul et al. 2006).

In Myanmar Areas with community health workers providing malaria diagnosis and treatment services experienced declines in *P. falciparum* and *P. vivax* malaria incidence of 70% (95% CI 66–73%) and 64% (59–68%), respectively, in each year of their presence. Rapid diagnostic test positivity rates declined similarly, with decreases of 70% (95% CI 66–73%) for *P. falciparum* and 65% (95% CI 61–69%) for *P. vivax* with each year of operation (McLean et al. 2018).

In Cambodia In Cambodia, *P. falciparum* has markedly decreased since 2009. This may be due to the distribution of insecticide-treated bed nets and the presence of village malaria workers, who perform early diagnosis and treatment of malaria for their villagers. Malaria cases have significantly decreased, especially in western Cambodia (i.e., Pailin and Battambang Provinces) (Maude et al. 2014). Malaria transmission in Cambodia has been low recently; however, 21 of the 25 provinces are still endemic, and more than half of the population (approximately 8.6 million people) is still at risk of malaria. The risk of malaria transmission is highest in the forest or forest fringe areas in the northeastern part of the country (MoH 2016). Four provinces in northeastern Cambodia (Rattanakiri, Mondulkiri, Steung Treng, and Kratie Provinces) account for only 3% of the national population; however, they were reported to have had 43% and 25% of falciparum and vivax malaria cases in the country in 2013, respectively (Maude et al. 2014).

In Lao PDR Malaria transmission varies across the different ecological zones of Lao PDR. High transmission areas have been reported in hilly, forested areas in the southern part of the country (Kounnavong et al. 2017). Lower transmission rates have been recorded in the plains along the Mekong River and in high-altitude areas. Between 2002 and 2012, the number of suspected (showing clinical symptoms) and confirmed malaria cases increased by 46% and the number of malaria-related deaths increased by 77% (Kounnavong et al. 2017).

In Vietnam Since the 1990s, malaria has been a major cause of morbidity and mortality in Vietnam. As socio-economic conditions, governmental investments, and community-based monitoring have improved, the prevalence of malaria in Vietnam decreased (Thang et al. 2009). By integrating control procedures (i.e., insecticide-treated mosquito nets, early diagnosis and treatment of symptomatic patients; Cui et al. 2012), the number of malaria cases has decreased (Bui et al. 2011). Similarly to the situation in Lao PDR, malaria in Vietnam is widespread in the central and southern parts and found mostly in the forests or forest edges (Cui et al. 2012). *Plasmodium falciparum* is the most prevalent species; however, infections with the monkey malaria parasite, *P. knowlesi*, have also been found in children in the central region (Van den Eede et al. 2009).

8.2.2 Leishmaniasis

Leishmaniasis is a a vector-borne disease caused by flagellated protozoans of the genus *Leishmania* (Fig. 8.1). The disease is prevalent in tropical and subtropical areas (Steverding 2017). Transmission occurs via a bite from the phlebotomine

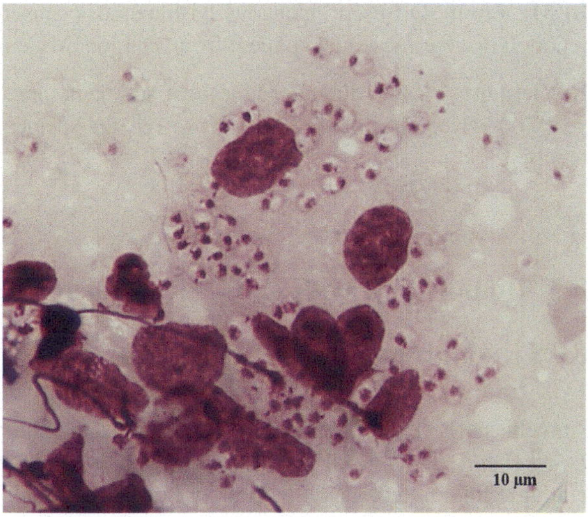

Fig. 8.1 *Leishmania tropica* amastigotes form in a Giemsa-stained smear of an ulcer sample

female sandfly. *Leishmania* is classified in two subgenera, *Leishmania* (*Leishmania*) and *Leishmania* (*Viannia*) (Lainson et al. 1977).

The three major clinical forms of leishmaniasis are cutaneous leishmaniasis (CL), mucocutaneous leishmaniasis (MCL), and visceral leishmaniasis (VL). VL or kala-azar, the most severe form, is caused by parasite multiplication in the reticuloendothelial system, principally the liver, spleen, lymph node, and bone marrow. VL causes a broad range of symptoms from asymptomatic to a fatal outcome (Belo et al. 2014; Bañuls et al. 2011). Concomitant human immunodeficiency virus (HIV)/acquired immunodeficiency syndrome (AIDS) enhances the risk of active VL and presents in poor responses to treatment (Cota et al. 2011; Alemayehu et al. 2016).

At least 20 species of *Leishmania* have been reported as human pathogens. The primary causative agents of VL are *Leishmania donovani* complex species, including *L. donovani* and *L. infantum* in the Old World and *L. chagasi* (synonym *L. infantum*) in the New World. *L. tropica* and *L. amazonensis* primarily cause CL, but could be viscerotropic in some cases (Grimaldi et al. 1989; Akhoundi et al. 2016).

In Thailand Previously, leishmaniasis cases in Thailand were mainly imported. However, emerging autochthonous leishmaniasis sporadic cases have now been found. Two causative agents, *L. martiniquensis* and *L. siamensis*, have been confirmed (Leelayoova et al. 2017). CL and VL have been reported among Thai patients. Disseminated CL and VL were also reported in some HIV/AIDS patients (Leelayoova et al. 2017; Supsrisunjai et al. 2017; Chiewchanvit et al. 2015). Positive polymerase chain reaction tests for *Leishmania* DNA were significantly associated with the amount of viral load, whereas non-injection drug use and CD4+ levels of <500 cells/μL were potential risk factors for seropositivity of *Leishmania* (Manomat et al. 2017).

In Myanmar Diffused cutaneous leishmaniasis caused by *L. martiniquensis* was reported in a Burmese male patient from Yangon, Myanmar (Noppakun et al. 2014).

In Vietnam VL was diagnosed for the first time in three HIV-positive women in the year 2002 (Keenihan et al. 2002). However, the species of causative agent was not reported.

8.3 Important Food- and Water-Borne Protozoa

8.3.1 Amoebiasis

Amoebiasis is a disease caused by the protozoa *Entamoeba coli*, *E. hartmanni*, *E. gingivalis*, *E. polecki*, *E. chattoni*, and *E. histolytica*. An invasive intestinal or extraintestinal infection can occur, with approximately 100,000 deaths reported annually (Sandipan 2015). Various *Entamoeba* species are found in humans, including *Entamoeba coli*, *E. hartmanni*, *E. gingivalis*, *E. gingivalis*, *E. polecki*, *E. chattoni*, and *E. polecki* (Clark and Diamond 1997). *Entamoeba histolytica* (Fig. 8.2) is the harmful agent responsible for amoebic dysentery and amoebic liver abscesses. Although harmless species such as *E. coli* and *E. dispar* are also found, *Entamoeba moshkovskii* is prevalent in *E. histolytica* endemic areas (Sandipan 2015).

Amoebiasis is found worldwide, especially in the tropics. The disease is an important water- and food-borne disease, affecting approximately 12% of the world's population; it is also a major cause of traveler's diarrhea (Sandipan 2015). Drinking unhygienic water causes contamination with *Entamoeba* infective cysts. Poor hygiene and insufficient knowledge of water sterilization in populations, especially in developing countries, can lead to widespread pathogenic *E. histolytica* infection.

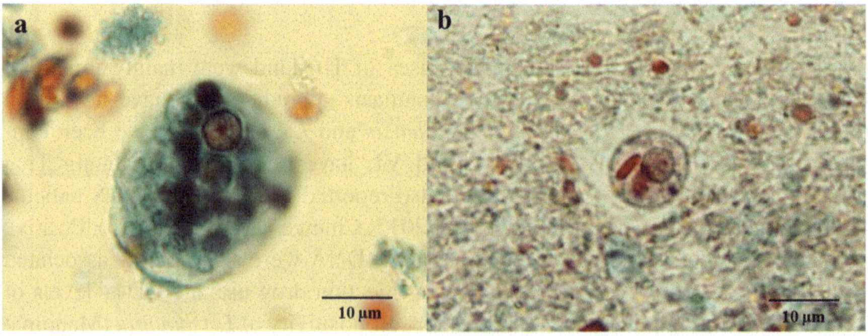

Fig. 8.2 *Entamoeba histolytica* trophozoite (**a**) and cyst (**b**) stained with trichrome

In Thailand From 1975 to 1983, a total of 418 patients with amoebic liver abscesses were admitted to Siriraj Hospital, Bangkok (Boonyapisit et al. 1993). In 1988, outbreaks of diarrheal disease caused by *E. histolytica* were reported in travelers returning from Thailand (de Lalla et al. 1988). A survey of a suburban area at Simum subdistrict, Mueang district, Nakhon Ratchasima Province in 2007 found a prevalence of 11.7% (Kitvatanachai et al. 2008). The prevalence in villages in Khon Kaen Province, northeastern Thailand in 2013 was 0.8% (Boonjaraspinyo et al. 2013). The prevalence of *E. histolytica*/*E. dispar* in children in suburban public primary schools in Saraburi, the central region of Thailand, in 2016 was 1.1% (Assavapongpaiboon et al. 2018).

In Lao PDR The prevalence of *Entamoeba histolytica*/*E. dispar* in of Lak Sip village, Luang Prabang district, Luang Prabang Province of northern Lao PDR in 2012 was 2.62 (95% confidence interval: 0.8–4.4) (Ribas et al. 2017).

In Cambodia A survey in Bat Dambang, Cambodia in 2004 found an *E. histolytica* infection rate of 0.8% (Park et al. 2004). The prevalence of *E. histolytica*/*E. dispar* in schoolchildren from Battambang Province, northwestern Cambodia in 2015 was 17.5% (Liao et al. 2017).

In Myanmar The prevalence of *E. histoytica* infection among schoolchildren in suburban areas near Yangon, Myanmar was 1.2% (Kim et al. 2016). The prevalance of *E. histolytica*/*E. dispar* amongst Myanmar migrant workers in Malaysia was reported to be 11.6% (Sahimin et al. 2016).

In Vietnam During 2005–2013, infection rates of 2.0–11.2% were reported for *E. histolytica* (Pham-Duc et al. 2013; Pearson and Singh 2010; Uga et al. 2005). A potential risk factor is the contaminated hand of population with *E. histolytica* cysts (Pham-Duc et al. 2011).

8.3.2 Giardiasis

Giardiasis is a disease caused by the flagellate protozoan *Giardia duodenalis* (synonyms *G. lamblia* and *G. intestinalis*) (Fig. 8.3), which is prevalent worldwide. Human transmission occurs both directly and indirectly by the fecal-oral route through the consumption of food or water contaminated with infected cysts. Human infection is caused by two genetically distinct groups of *G. duodenalis*—namely, assemblages A and B. The remaining six assemblages (C–H) infect other mammals (Cacciò and Lalle 2015). Giardiasis is the main water-borne diarrhea-causing disease in poor and urban slum areas because of lack of good hygiene, health education, and appropriate surveillance (Dib et al. 2008).

In Thailand *Giardia* sp. was found in untreated water samples from Thailand (22.58%; 21/93) (Kumar et al. 2016), with the prevalence of giardiasis varying from 3.5 to 6.0% (Yanola et al. 2018; Boontanom et al. 2014; Saksirisampant

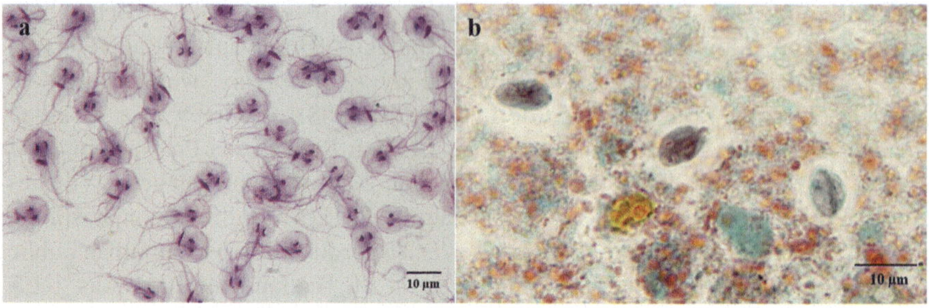

Fig. 8.3 *Giardia duodenalis* trophozoites (**a**) stained with Giemsa's and cysts (**b**) stained with trichrome

et al. 2012). The genetic characterization of *Giardia duodenalis* isolated in hilltribe children from northern Thailand was performed using polymerase chain reaction methods specific for small subunit ribosomal rRNA (SSU-rRNA) and glutamate dehydrogenase (gdh) gene. This study revealed that the distribution of *Giardia* assemblages varied in these two populations. Assemblage BIV occurred primarily in children from the Hod District, whereas assemblage AII was more often found in children from the Mae-Chaem District (Saksirisampant et al. 2012). The genotypes of *G. duodenalis* in infected preschool children at Sanamchaiket, Chachoengsao Province, central Thailand, were reported to be assemblage AII (30.0%), BIII (10.0%), and BIV (60.0%) (Boontanom et al. 2011). The genotypes of *G. duodenalis* in orphans living in suburban area outside Bangkok, central Thailand were genotypic assemblage A subtype AII (47.2%) and assemblage B subtype BIV (27.8%) (Boontanom et al. 2014). Moreover, the molecular identification of *Giardia intestinalis* samples collected from rivers, tributaries, and a major irrigation canal in northern Thailand were assemblages B (zoonotic) and E (non-zoonotic) (Chuah et al. 2016).

In Cambodia *Giardia duodenalis* cysts were detected in 18.9% (94/498) of symptomatic Cambodian children by microscopy and 27.7% (138/498) by molecular tests. Using a multiplex real-time polymerase chain reaction targeting assemblages A and B, assemblage B (106; 76.8% of all *Giardia* positive samples) was found to be the most predominant, followed by assemblage A (12.3%) and mixed infections (5.1%) (Moore et al. 2016). Between 2006 and 2011, the prevalence of giardiasis in symptomatic Cambodian children attending a provincial hospital in Siem Reap, Cambodia was 8.0% (Moore et al. 2012). In 2012, the prevalence in Dong village, Rovieng district, Preah Vihear province (bordering Thailand and Lao PDR) was 22.0% (Schär et al. 2014).

In Lao PDR The prevalence of *Giardia* infection among children under 15 years old in two villages in Khammouane Province, southeastern Lao PDR was 8.6% (Kobayashi et al. 1996). In 1999, two parasitological surveys in Xai Udom, a small fishing community on the Nam Ngum reservoir, revealed the prevalence of *Giardia* infection to be 4.9–5.2% (Sitthithaworn et al. 2006). In addition, the prevalence of

Giardia infection in inhabitants of Lak Sip village, Luang Prabang district, Luang Prabang Province, in 2012 was 6.56% (95% confidence interval: 3.8–9.3) (Ribas et al. 2017).

In Myanmar The prevalence of *Giardia* sp. in migrant workers in Malaysia was reported to be 10.8% (Sahimin et al. 2016). Among immigrant children in an environmental survey of child daycare centers and Myanmar migrant workers in Thailand, the prevalance was 10.2–14.1% (Sagnuankiat et al. 2016; Nuchprayoon et al. 2009).

In Vietnam In mountainous northern-western Vietnam, the adjusted prevalence of *G. lamblia* infection (4.1%) among six ethnic groups (Muong, Kinh, Dao, Thai, Tay, and Hmong) was estimated at 3.2%, which is low in comparison with other countries (Verle et al. 2003; Dib et al. 2008).

8.3.3 Cryptosporidiosis

Cryptosporidiosis has the highest impact in developed nations through food-borne, water-borne, person-to-person, and zoonotic outbreaks of diarrhea (Yoder and Beach 2010). In developing countries, cryptosporidiosis is also a major cause of diarrhea and malnutrition in children and individuals with AIDS (Xiao and Ryan 2015). *Cryptosporidium* spp. (Fig. 8.4) are obligate protozoans classified in the family Cryptosporidiidae, order Eimeriida, class Coccidea, phylum Apicomplexa. At present, approximately 26 *Cryptosporidium* species are reported to infect several hosts, including *Cryptosporidium hominis* and *C. viatorum* in humans and *C. parvum* in ruminants and humans (Xiao and Ryan 2015).

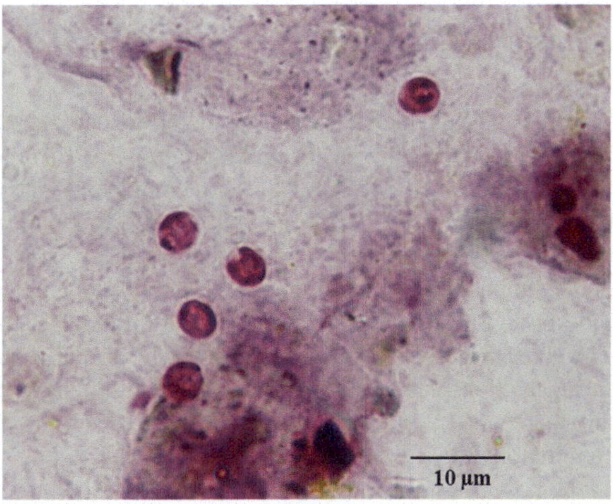

Fig. 8.4 *Cryptosporidium* spp. oocysts in a fecal sample are stained red with a modified acid-fast stain

Cryptosporidium spp. lives in the brush borders of the gastrointestinal, respiratory, and renal epithelium of different vertebrates, causing enterocolitis, diarrhea, and cholangiopathy in humans (Chalmers and Davies 2010). Short-term illnesses such as watery diarrhea, nausea, vomiting, and weight loss occur in infected immunocompetent children and adults. Infection can be prolonged and life-threatening in immunocompromised persons, especially in patients with AIDS/HIV (Xiao and Ryan 2015).

In Thailand Cryptosporidiosis is an important infectious disease among HIV/AIDS patients in Thailand, with a prevalence rate between 2.5% and 25% (Srisuphanunt et al. 2011); a rate of 3.7% was found in children with acute diarrhea (Thamlikitkul et al. 1987). Animals are a reservoir for a zoonotic *Cryptosporidium* infection in humans, with a prevalence of 31.2% in dogs (Tangtrongsup et al. 2017), 5–7.6% in dairy cows (Inpankaew et al. 2017), 1% in long-tailed macaques (Sricharern et al. 2016), and 5.7–8.7% in buffaloes (Inpankaew et al. 2014). *Cryptosporidium* oocysts were detected in 26.7% of treated water samples from Thailand (Kumar et al. 2016).

In Cambodia The prevalence of infection in symptomatic Cambodian children was 2.2–7.7% (Moore et al. 2016). A cryptosporidiosis prevalence of 13% was found among patients presenting for care in Phnom Penh, Cambodia (Kong et al. 2007).

In Myanmar A total of 3.4% of infants between 2 and 11 months of age were found to be passing cryptosporidium oocysts (Aye et al. 1994).

In Vietnam Animals are reservoir hosts, with a prevalence of 14.5% in pigs (Nguyen et al. 2013) and 18.9% in native beef calves (Nguyen et al. 2012).

In Lao PDR *Cryptosporidium* infection was found in 6.6% of HIV-infected patients in Lao PDR (Paboriboune et al. 2014).

8.3.4 Toxoplasmosis

Toxoplasmosis, which is caused by the protozoan *Toxoplasma gondii*, continues to be a significant public health problem globally. *Toxoplasma gondii* is a coccidian parasite classified in phylum Apicomplexa, class Sporozoasida, subclass Coccidiasina, order Eimeriorina, and family Toxoplasmatidae. It has an unusually broad range of intermediate hosts, such as sheep, goats, pigs, chickens, and many game animal species. Felids (i.e. cats, tigers) serve as definitive hosts and produce the oocyst stage in the environment. The protozoa have complicated life cycles. *Toxoplasma gondii* is transmitted by the fecal-oral route, consumption of infected meat, and transplacental transfer from mother to fetus (Dubey and Beattie 1988; Hill and Dubey 2015). Toxoplasmosis can cause death in patients with AIDS (Luft and Remington 1992).

Although the disease is the one of the most common protozoan infections in humans, its prevalence varies from place to place. Trade exchanges and human migration affect the dispersion of *Toxoplasma* genotypes across countries or continents (Mercier et al. 2010). At least 43 *Toxoplasma* characterized strains have been reported in tropical South East Asian countries. Type I strains and atypical genotypes with most Type I alleles are mainly found in wild animals (e.g., wild boars in Malaysia, bats in Myanmar). The ToxoDB#9 genotype was found in Vietnam (Mercier et al. 2010).

In Thailand The prevalence is reported to range from 2.8 to 9.6% in healthy persons in various parts of Thailand (Morakote et al. 1984; Maleewong et al. 1989; Nishigawa et al. 1989; Wongkamchai et al. 1995; Pinlaor et al. 2000), with rates of 2.1–28.3% among pregnant women (Morakote et al. 1984; Maleewong et al. 1989; Nissapatorn et al. 2011; Andiappan et al. 2014a) and 2.6% among non-pregnant women (Sakae et al. 2013). In HIV-positive individuals, 23.2–53.7% were reported to be positive for toxoplasma IgG antibody, while 5.3–29.5% of HIV-negative individuals were positive (Sukthana et al. 2000; Wanachiwanawin et al. 2001).

In Vietnam Prevalence rates in Duyen Thai village (a suburban agricultural area of Hanoi) and the villages of Mai Chau district (in a mountainous valley near Laos border) were 15.7–24.3% (Sery et al. 1988). A seroprevalence study of *T. gondii* antibodies in Vietnamese villagers performed using the Sabin-Feldman dye test found rates of 4.19% (95% confidence interval: 1.78–4.62) (Udonsom et al. 2008).

In Cambodia The weighted national estimated prevalence for toxoplasmosis is 5.8% (Priest et al. 2016).

In Lao PDR The seroprevalence of *Toxoplasma* antibodies confirmed using a complement fixation test was 15.3% (Catár et al. 1992).

In Myanmar Anti-*Toxoplasma* IgG and IgM antibodies were screened using standard commercial ELISA kits. The overall prevalence of *Toxoplasma* infection in Myanmar pregnant women was 30.70% (95% confidence interval: 27.92–37.16) (Andiappan et al. 2014b).

Acknowledgments The authors were supported by a Distinguished Research Professor Grant, Thailand Research Fund.

References

Akhoundi M, Kuhls K, Cannet A, Votýpka J, Marty P, Delaunay P et al (2016) A historical overview of the classification, evolution, and dispersion of *Leishmania* parasites and sandflies. PLoS Negl Trop Dis 10:e0004349

Alemayehu M, Wubshet M, Mesfin N (2016) Magnitude of visceral leishmaniasis and poor treatment outcome among HIV patients: meta-analysis and systematic review. HIV AIDS (Auckl) 8:75–81

Andiappan H, Nissapatorn V, Sawangjaroen N, Chemoh W, Lau YL, Kumar T et al (2014a) *Toxoplasma* infection in pregnant women: a current status in Songklanagarind hospital, southern Thailand. Parasit Vectors 7:239

Andiappan H, Nissapatorn V, Sawangjaroen N, Nyunt MH, Lau YL, Khaing SL et al (2014b) Comparative study on *Toxoplasma* infection between Malaysian and Myanmar pregnant women. Parasit Vectors 7:564

Asian Development Bank (ADB) (2018) Greater Mekong subregion environmental performance assessment 2006–2016. https://www.adb.org/publications/gms-environmental-performance-assessment-2006-2016. Accessed 14 Dec 2018

Assavapongpaiboon B, Bunkasem U, Sanprasert V, Nuchprayoon S (2018) Cross-sectional study on intestinal parasitic infections in children in suburban public primary schools, Saraburi, the central region of Thailand. Am J Trop Med Hyg 98(3):763–767

Aye T, Moe K, Nyein MM, Swe T (1994) Cryptosporidiosis in Myanmar infants with acute diarrhea. Southeast Asian J Trop Med Public Health 25:654–656

Bañuls AL, Bastien P, Pomares C, Arevalo J, Fisa R, Hide M (2011) Clinical pleiomorphism in human leishmaniases, with special mention of asymptomatic infection. Clin Microbiol Infect 17 (10):1451–1461

Belo VS, Struchiner CJ, Barbosa DS, Nascimento BW, Horta MA, da Silva ES et al (2014) Risk factors for adverse prognosis and death in American visceral leishmaniasis: a meta-analysis. PLoS Negl Trop Dis 8(7):e2982

Boonjaraspinyo S, Boonmars T, Kaewsamut B, Ekobol N, Laummaunwai P, Aukkanimart R et al (2013) A cross-sectional study on intestinal parasitic infections in rural communities, northeast Thailand. Korean J Parasitol 51:727–734

Boontanom P, Mungthin M, Tan-Ariya P, Naaglor T, Leelayoova S (2011) Epidemiology of giardiasis and genotypic characterization of *Giardia duodenalis* in preschool children of a rural community, central Thailand. Trop Biomed 28:32–39

Boontanom P, Pipatsatitpong D, Tan-Ariya P, Mungthin M, Siripattanapipong S, Naaglor T, Leelayoova S (2014) Incidence and risk factors of *Giardia duodenalis* infection in an orphanage, Thailand. Trop Biomed 31:525–533

Boonyapisit S, Chinapak O, Plengvanit U (1993) Amoebic liver abscess in Thailand, clinical analysis of 418 cases. J Med Assoc Thail 76:243–246

Bui HM, Clements AC, Nguyen QT, Nguyen MH, Le XH, Hay SI et al (2011) Social and environmental determinants of malaria in space and time in Viet Nam. Int J Parasitol 41:109–116

Bureau of Vector Borne Diseases (2017) Annual report 2017, Bureau of vector borne diseases, Ministry of Public Health

Cacciò SM, Lalle M (2015) Giardia. In: Xiao L, Ryan U, Feng Y (eds) Biology of foodborne parasites. CRC press, New York, pp 175–193

Catár G, Giboda M, Gutvirth J, Hongvanthong B (1992) Seroepidemiological study of toxoplasmosis in Laos. Southeast Asian J Trop Med Public Health 23:491–492

Chalmers RM, Davies AP (2010) Minireview: clinical cryptosporidiosis. Exp Parasitol 124:138–146

Chiewchanvit S, Tovanabutra N, Jariyapan N, Bates MD, Mahanupab P, Chuamanochan M et al (2015) Chronic generalized fibrotic skin lesions from disseminated leishmaniasis caused by *Leishmania martiniquensis* in two patients from northern Thailand infected with HIV. Br J Dermatol 173:663–670

Chuah CJ, Mukhaidin N, Choy SH, Smith GJD, Mendenhall IH, Lim YAL et al (2016) Prevalence of *Cryptosporidium* and *Giardia* in the water resources of the Kuang River catchment, Northern Thailand. Sci Total Environ 562:701–713

Clark CG, Diamond LS (1997) Intraspecific variation and phylogenetic relationships in the genus *Entamoeba* as revealed by riboprinting. J Eukaryot Microbiol 44:142–154

Corbel V, Nosten F, Thanispong K, Luxemburger C, Kongmee M, Chareonviriyaphap T (2013) Challenges and prospects for dengue and malaria control in Thailand, Southeast Asia. Trends Parasitol 29:623–633

Cota GF, de Sousa MR, Rabello A (2011) Predictors of visceral leishmaniasis relapse in HIV-infected patients: a systematic review. PLoS Negl Trop Dis 5:e1153

Cui L, Yan G, Sattabongkot J, Cao Y, Chen B, Chen X et al (2012) Malaria in the greater Mekong subregion: heterogeneity and complexity. Acta Trop 121:227–239

de Lalla F, Rizzardini G, Cairoli GA, Rinaldi E, Santoro D, Ostinelli A (1988) Outbreak of amoebiasis in tourists returning from Thailand. Lancet 2:847

Dib HH, Lu SQ, Wen SF (2008) Prevalence of *Giardia lamblia* with or without diarrhea in South East, South East Asia and the Far East. Parasitol Res 103:239–251

Dubey JP, Beattie CP (1988) Toxoplasmosis of animals and man. CRC Press, Boca Raton, FL

Grimaldi G Jr, Tesh RB, McMahon-Pratt D (1989) A review of the geographic distribution and epidemiology of leishmaniasis in the New World. Am J Trop Med Hyg 41:687–725

Harbach RE (2018) *Anopheles* classification. Mosquito taxonomic inventory. http://mosquito-taxonomic-inventory.info. Accessed 5 May 2018

Hill DE, Dubey JP (2015) *Toxoplasma gondii*. In: Xiao L, Ryan U, Feng Y (eds) Biology of foodborne parasites. CRC Press, New York, pp 209–222

Inpankaew T, Jiyipong T, Wongpanit K, Pinyopanuwat N, Chimnoi W, Kengradomkij C et al (2014) Molecular detection of *Cryptosporidium* spp. infections in water buffaloes from North-east Thailand. Trop Anim Health Prod 46:487–490

Inpankaew T, Jiyipong T, Sunanta C, Kengradomkij C, Pinyopanuwat N, Jittapalapong S (2017) Prevalence and molecular characterization of bovine *Cryptosporidium* from dairy cows in Northern Thailand. Acta Parasitol 62:772–774

Jongwutiwes S, Buppan P, Kosuvin R, Seethamchai S, Pattanawong U, Sirichaisinthop J et al (2011) *Plasmodium knowlesi* malaria in humans and macaques, Thailand. Emerg Infect Dis 17:1799–1806

Keenihan SH, Schetters T, Carey JV, Taverne J (2002) Visceral leishmaniasis in Vietnam. Trends Parasitol 18:152

Kim MJ, Jung BK, Cho J, Kim DG, Song H, Lee KH et al (2016) Prevalence of intestinal protozoans among schoolchildren in suburban areas near Yangon, Myanmar. Korean J Parasitol 54(3):345–348

Kitvatanachai S, Boonslip S, Watanasatitarpa S (2008) Intestinal parasitic infections in Srimum suburban area of Nakhon Ratchasima Province, Thailand. Trop Biomed 25:237–242

Kobayashi J, Vannachone B, Xeutvongsa A, Manivang K, Ogawa S, Sato Y et al (1996) Prevalence of intestinal parasitic infection among children in two villages in Lao PDR. Southeast Asian J Trop Med Public Health 27:562–565

Kong BN, Harwell JI, Suos P, Lynen L, Mohiuddin S, Reinert S et al (2007) Opportunistic infections and HIV clinical disease stage among patients presenting for care in Phnom Penh, Cambodia. Southeast Asian J Trop Med Public Health 38:62–68

Kounnavong S, Gopinath D, Hongvanthong B, Khamkong C, Sichanthongthip O (2017) Malaria elimination in Lao PDR: the challenges associated with population mobility. Infect Dis Poverty 6:81

Kumar T, Abd Majid MA, Onichandran S, Jaturas N, Andiappan H, Salibay CC et al (2016) Presence of *Cryptosporidium parvum* and *Giardia lamblia* in water samples from Southeast Asia: towards an integrated water detection system. Infect Dis Poverty 5:3

Lainson R, Ward RD, Shaw JJ (1977) *Leishmania* in phlebotomid sandflies: VI. Importance of hindgut development in distinguishing between parasites of the *Leishmania mexicana* and *L. braziliensis* complexes. Proc R Soc Lond B Biol Sci 199:309–320

Leelayoova S, Siripattanapipong S, Manomat J, Piyaraj P, Tan-Ariya P, Bualert L et al (2017) Leishmaniasis in Thailand: a review of causative agents and situations. Am J Trop Med Hyg 96:534–542
Liao CW, Chiu KC, Chiang IC, Cheng PC, Chuang TW, Kuo JH et al (2017) Prevalence and risk factors for intestinal parasitic infection in schoolchildren in Battambang, Cambodia. Am J Trop Med Hyg 96:583–588
Linton YM, Dusfour I, Howard TM, Ruiz F, Manh ND, Dinh TH et al (2005) Anopheles (Cellia) epiroticus (Diptera:Culicidae), a new malaria vector species in the southeast Asian Sundaicus complex. Bull Entomol Res 95:329–339
Luft BJ, Remington JS (1992) Toxoplasmic encephalitis in AIDS. Clin Infect Dis 15:211–222
Maleewong W, Lulitanond V, Pipitgool V, Auwijitaroon Y, Kuttsarejariya S, Morakote N (1989) Prevalence of toxoplasma antibodies in blood donors and pregnant women in Khon Kaen Province. J Med Assoc Thail 72:256–259
Manomat J, Leelayoova S, Bualert L, Tan-Ariya P, Siripattanapipong S, Mungthin M et al (2017) Prevalence and risk factors associated with *Leishmania* infection in Trang Province, southern Thailand. PLoS Negl Trop Dis 11(11):e0006095
Maude RJ, Nguon C, Ly P, Bunkea T, Ngor P, de la Canavati TSE et al (2014) patial and temporal epidemiology of clinical malaria in Cambodia 2004–2013. Malar J 13:385
McLean ARD, Wai HP, Thu AM, Khant ZS, Indrasuta C, Ashley EA et al (2018) Malaria elimination in remote communities requires integration of malaria control activities into general health care: an observational study and interrupted time series analysis in Myanmar. BMC Med 16:183
Mercier A, Devillard S, Ngoubangoye B, Bonnabau H, Bañuls AL, Durand P et al (2010) Additional haplogroups of *Toxoplasma gondii* out of Africa: population structure and mouse-virulence of strains from Gabon. PLoS Negl Trop Dis 4(11):e876
MoH (2016) Cambodia malaria elimination action framework 2016–2020. Ministry of Health, Phnom Penh, p 7
Moore CE, Hor PC, Soeng S, Sun S, Lee SJ, Parry CM et al (2012) Changing patterns of gastrointestinal parasite infections in Cambodian children: 2006–2011. J Trop Pediatr 58:509–512
Moore CE, Elwin K, Phot N, Seng C, Mao S, Suy K et al (2016) Molecular characterization of *Cryptosporidium* species and *Giardia duodenalis* from symptomatic Cambodian children. PLoS Negl Trop Dis 10(7):e0004822
Morakote N, Thamasonthi W, Charuchinda K, Khamboonruang C (1984) Prevalence of *Toxoplasma* antibodies in Chiang Mai population. Southeast Asian J Trop Med Public Health 15:80–85
Nguyen ST, Fukuda Y, Tada C, Sato R, Duong B, Nguyen DT et al (2012) Molecular characterization of *Cryptosporidium* in native beef calves in central Vietnam. Parasitol Res 111:1817–1820
Nguyen ST, Fukuda Y, Tada C, Sato R, Huynh VV, Nguyen DT et al (2013) Molecular characterization of *Cryptosporidium* in pigs in Central Vietnam. Parasitol Res 112:187–192
Nishigawa H, Tuntasuvan D, Triwanatham N (1989) Preliminary survey of toxoplasmosis in dog, cat and human. J Trop Med Parasitol 12:53–59
Nissapatorn V, Suwanrath C, Sawangjaroen N, Ling LY, Chandeying V (2011) Toxoplasmosis-serological evidence and associated risk factors among pregnant women in southern Thailand. Am J Trop Med Hyg 85:243–247
Noppakun N, Kraivichian K, Siriyasatien P (2014) Disseminated dermal leishmaniasis caused by *Leishmania siamensis* in a systemic steroid therapy patient. Am J Trop Med Hyg 91:869–870
Nuchprayoon S, Sanprasert V, Kaewzaithim S, Saksirisampant W (2009) Screening for intestinal parasitic infections among Myanmar migrant workers in Thai food industry: a high-risk transmission. J Immigr Minor Health 11:115–121
Paboriboune P, Phoumindr N, Borel E, Sourinphoumy K, Phaxayaseng S, Luangkhot E et al (2014) Intestinal parasitic infections in HIV-infected patients, Lao People's Democratic Republic. PLoS One 9(3):e91452

Park SK, Kim DH, Deung YK, Kim HJ, Yang EJ, Lim SJ et al (2004) Status of intestinal parasite infections among children in Bat Dambang, Cambodia. Korean J Parasitol 42:201–203

Pearson RJ, Singh U (2010) Approaches to characterizing *Entamoeba histolytica* transcriptional regulation. Cell Microbiol 12:1681–1690

Pham-Duc P, Nguyen-Viet H, Hattendorf J, Zinsstag J, Dac Cam P, Odermatt P (2011) Risk factors for *Entamoeba histolytica* infection in an agricultural community in Hanam province, Vietnam. Parasit Vectors 4:102

Pham-Duc P, Nguyen-Viet H, Hattendorf J, Zinsstag J, Phung-Dac C, Zurbrügg C et al (2013) *Ascaris lumbricoides* and *Trichuris trichiura* infections associated with wastewater and human excreta use in agriculture in Vietnam. Parasitol Int 62:172–180

Phillips MA, Burrows JN, Manyando C, van Huijsduijnen RH, Van Voorhis WC, Wells TNC (2017) Malaria. Nat Rev Dis Primers 3:17050

Pinlaor S, Ieamviteevanich K, Pinlaor P, Maleewong W, Pipitgool V (2000) Seroprevalence of specific total immunoglobulin (Ig), IgG and IgM antibodies to *Toxoplasma gondii* in blood donors from Loei Province, Northeast Thailand. Southeast Asian J Trop Med Public Health 31:123–127

Priest JW, Jenks MH, Moss DM, Mao B, Buth S, Wannemuehler K et al (2016) Integration of multiplex bead assays for parasitic diseases into a national, population-based serosurvey of women 15–39 years of age in Cambodia. PLoS Negl Trop Dis 10(5):e0004699

Rattanarithikul R, Konishi E, Linthicum KJ (1996) Detection of *Plasmodium vivax* and *Plasmodium falciparum* circumsporozoites antigen in anopheline mosquitoes collected in southern Thailand. Am J Trop Med Hyg 54:114–121

Rattanarithikul R, Harrison BA, Harbach RE, Panthusiri P, Coleman RE, Panthusiri P (2006) Illustrated keys to the mosquitoes of Thailand. IV. Anopheles. Southeast Asian J Trop Med Public Health 37(Suppl 2):1–128

Ribas A, Jollivet C, Morand S, Thongmalayvong B, Somphavong S, Siew CC et al (2017) Intestinal parasitic infections and environmental water contamination in a rural village of northern Lao PDR. Korean J Parasitol 55:523–525

Saeung A (2012) *Anopheles* (Diptera: Culicidae) species complex in Thailand: identification, distribution, bionomics and malaria-vector importance. Int J Parasitol Res 4:75–82

Sagnuankiat S, Wanichsuwan M, Bhunnachet E, Jungarat N, Panraksa K, Komalamisra C et al (2016) Health status of immigrant children and environmental survey of child daycare centers in Samut Sakhon Province, Thailand. J Immigr Minor Health 18:21–27

Sahimin N, Lim YA, Ariffin F, Behnke JM, Lewis JW, Mohd Zain SN (2016) Migrant workers in Malaysia: current implications of sociodemographic and environmental characteristics in the transmission of intestinal parasitic infections. PLoS Negl Trop Dis 10(11):e0005110

Sakae C, Natphopsuk S, Settheetham-Ishida W, Ishida T (2013) Low prevalence of *Toxoplasma gondii* infection among women in northeastern Thailand. J Parasitol 99:172–173

Saksirisampant W, Boontanom P, Mungthin M, Tan-Ariya P, Lamchuan D, Siripattanapipong S et al (2012) Prevalence of giardiasis and genotypic characterization of *Giardia duodenalis* in hilltribe children, Northern Thailand. Trop Biomed 29:331–338

Sallum MAM, Peyton EL, Wilkerson RC (2005) Six new species of the *Anopheles leucosphyrus* group, reinterpretation of *An. elegans* and vector implications. Med Vet Entomol 9:158–199

Sandipan G (2015) Entamoeba. In: Xiao L, Ryan U, Feng Y (eds) Biology of foodborne parasites. CRC press, New York, pp 131–148

Schär F, Inpankaew T, Traub RJ, Khieu V, Dalsgaard A, Chimnoi W et al (2014) The prevalence and diversity of intestinal parasitic infections in humans and domestic animals in a rural Cambodian village. Parasitol Int 63:597–603

Sery V, Zastěra M, Prokopec J, Radkovsky J, The PH, Canh DT (1988) To the problem of toxoplasmosis in Vietnam. Bull Inst Marit Trop Med Gdynia 39:181–185

Sithithaworn P, Sukavat K, Vannachone B, Sophonphong K, Ben-Embarek P, Petney T et al (2006) Epidemiology of food-borne trematodes and other parasite infections in a fishing community on the Nam Ngum reservoir, Lao PDR. Southeast Asian J Trop Med Public Health 37:1083–1090

Sricharern W, Inpankaew T, Keawmongkol S, Supanam J, Stich RW, Jittapalapong S (2016) Molecular detection and prevalence of *Giardia duodenalis* and *Cryptosporidium* spp. among long-tailed macaques (*Macaca fascicularis*) in Thailand. Infect Genet Evol 40:310–314

Srisuphanunt M, Saksirisampant W, Karanis P (2011) Prevalence and genotyping of *Cryptosporidium* isolated from HIV/AIDS patients in urban areas of Thailand. Ann Trop Med Parasitol 105:463–468

Sriwichai P, Samung Y, Sumruayphol S, Kiattibutr K, Kumpitak C, Payakkapol A et al (2016) Natural human *Plasmodium* infections in major *Anopheles* mosquitoes in western Thailand. Parasit Vectors 9:17

Steverding D (2017) The history of leishmaniasis. Parasit Vectors 10:82

Sukthana Y, Chintana T, Lekkla A (2000) *Toxoplasma gondii* antibody in HIV-infected persons. J Med Assoc Thail 83(6):681–684

Supsrisunjai C, Kootiratrakarn T, Puangpet P, Bunnag T, Chaowalit P, Wessagowit V (2017) Disseminated autochthonous dermal leishmaniasis caused by *Leishmania siamensis* (PCM2 Trang) in a patient from Central Thailand infected with human immunodeficiency virus. Am J Trop Med Hyg 96:1160–1163

Tainchum K, Kongmee M, Manguin S, Bangs MJ, Chareonviriyaphap T (2015) *Anopheles* species diversity and distribution of the malaria vectors of Thailand. Trends Parasitol 31:109–119

Tangtrongsup S, Scorza AV, Reif JS, Ballweber LR, Lappin MR, Salman MD (2017) Prevalence and multilocus genotyping analysis of *Cryptosporidium* and *Giardia* isolates from dogs in Chiang Mai, Thailand. Vet Sci 4(2):E26

Thamlikitkul V, Tepmongkol M, Lamon C, Sripochang S, Rungnapawate W, Suvajeejarun T (1987) Cryptosporidiosis in Siriraj Hospital, Bangkok, Thailand. Southeast Asian J Trop Med Public Health 18:229–232

Thang ND, Erhart A, le Hung X, le Thuan K, Xa NX, Thanh NN et al (2009) Rapid decrease of malaria morbidity following the introduction of community-based monitoring in a rural area of central Vietnam. Malar J 8:3

Udonsom R, Lekkla A, Chung PT, Cam PD, Sukthana Y (2008) Seroprevalence of *Toxoplasma gondii* antibody in Vietnamese villagers. Southeast Asian J Trop Med Public Health 39:14–18

Uga S, Hoa NT, le Thuan K, Noda S, Fujimaki Y (2005) Intestinal parasitic infections in schoolchildren in a suburban area of Hanoi, Vietnam. Southeast Asian J Trop Med Public Health 36:1407–1411

Van den Eede P, Van HN, Van Overmeir C, Vythilingam I, Duc TN, le Hung X et al (2009) Human *Plasmodium knowlesi* infections in young children in central Vietnam. Malar J 8:249

Verle P, Kongs A, De NV, Thieu NQ, Depraetere K, Kim HT et al (2003) Prevalence of intestinal parasitic infections in northern Vietnam. Tropical Med Int Health 8:961–964

Wanachiwanawin D, Sutthent R, Chokephaibulkit K, Mahakittikun V, Ongrotchanakun J, Monkong N (2001) *Toxoplasma gondii* antibodies in HIV and non-HIV infected Thai pregnant women. Asian Pac J Allergy Immunol 19:291–293

WHO (2017a) World Malaria Report 2017. WHO Global Malaria Programme, p 196

WHO (2017b) The Mekong malaria elimination programme: countries of the Greater Mekong making inroads in the race towards elimination, p 8

Wongkamchai S, Rungpitaransi B, Wongbunnate S, Sittapairochana C (1995) *Toxoplasma* infection in healthy persons and in patients with HIV or ocular disease. Southeast Asian J Trop Med Public Health 26:655–658

Xiao L, Ryan U (2015) Cryptosporidium. In: Xiao L, Ryan U, Feng Y (eds) Biology of foodborne parasites. CRC press, New York, pp 77–95

Yanola J, Nachaiwieng W, Duangmano S, Prasannarong M, Somboon P, Pornprasert S (2018) Current prevalence of intestinal parasitic infections and their impact on hematological and nutritional status among Karen hill tribe children in Omkoi District, Chiang Mai Province, Thailand. Acta Trop 180:1–6

Yoder JS, Beach MJ (2010) *Cryptosporidium* surveillance and risk factors in the United States. Exp Parasitol 124:31–39

Chapter 9
Free-Living Amoebae and Associated Drug-Resistant Bacteria in Surface Waters: A Public Health Concern?

P. Muchesa, C. Bartie, and T. G. Barnard

Abstract Protozoan parasites are a diverse group of microorganisms. Among the protozoa, free-living amoeba can reside in freshwater that includes surface waters and pose risk of opportunistic infections caused by some species. In addition, interactions of these free-living amoebae with drug-resistant bacteria in surface waters can pose a public health risk as they promote the survival and proliferation of these drug-resistant bacteria.

Keywords Free-living amoeba · Drug-resistant bacteria · Surface waters · Public health · Drug-resistant bacteria

9.1 Introduction

The availability of potable water in poor countries is a challenge as it causes serious health problems. Polluted water sources such as surface waters that include rivers and lakes can contribute to waterborne infections which may result in morbidities and mortalities. Microbial contamination of these surface waters with organisms such as *Vibrio*, *Campylobacter*, *Shigella*, and *Salmonella* has been widely reported to cause cholera, campylobacteriosis, shigellosis, and salmonellosis, respectively, in most rural communities. However, the occurrence of other pathogenic waterborne parasites such as free-living amoebae (FLA) in surface waters has been scarcely reported. FLA have gained greater prominence in the last decades as the focus of studies in a wide variety of environments due to their rare but often serious infections (Schuster 2002) and their interactions with other microorganisms (Brown et al. 2007). FLA and associated bacteria, in particular, are part of environmental microbial communities of water, soil, and aquatic environments to which humans are

P. Muchesa (✉) · C. Bartie · T. G. Barnard
Water and Health Research Centre, Faculty of Health Sciences, University of Johannesburg, Doornfontein, South Africa

H. Mehlhorn, S. Klimpel (eds.), *Parasite and Disease Spread by Major Rivers on Earth*, Parasitology Research Monographs 12,
https://doi.org/10.1007/978-3-030-29061-0_9

increasingly being exposed. Therefore, environmental contamination and environmental transmission of FLA can have important implications on general public health. It is therefore important to know the different species of FLA (pathogenic and non-pathogenic), their interactions with other drug-resistant pathogens, and the impacts of the interactions in environments associated with human activity such as rivers.

9.2 Free-Living Amoebae (FLA)

Free-living amoebae are unicellular heterotrophic protozoa that consist of several genera with amoebal species that are categorized based on morphological and behavioral characteristics (Loret and Greub 2010). FLA are grouped under the class Sarcodina of the four classes of parasitic protozoa that have a significance for health. The other three are Flagellata, Sporozoa, and Infusoria (Omarova et al. 2018). FLA encompass a wide range of phylogenetically distant microorganisms with most FLA characterized by at least two developmental stages in their life cycle, a trophozoite and a cyst, with some FLA such as *Mayorella* and *Amoeba* not forming cysts. Some like *Naegleria* have an additional temporary motile flagellate stage, while other FLA like *Dictyostelium discoideum* form complex multi-individual structures that mimic multicellular organisms (Greub and Raoult 2004; Fritz-Laylin and Cande 2010; Kang et al. 2017). Trophozoite is the metabolic active stage which feeds on bacteria, fungi, viruses, yeast, algae, and other small organic particles in the environment and reproduces by binary fission (Gonzalez-Robles et al. 2001; Marciano-Cabral and Cabral 2003; Khan 2008). This contributes to nutrient recycling by assimilating nutrients such as carbon, nitrogen, and phosphate and releasing them into the environment by excretion or death (Rodriguez-Zaragoza 1994; Siddiqui and Khan 2012).

Exposure of trophozoites to adverse environmental such as adverse pH, temperature, osmotic pressure, or presence of biocides and nutrient depletions leads to formation of cysts. Amoebal cyst walls contain cellulose and/or chitin that forms a physical barrier which protects the FLA from adverse physical, chemical, and environmental conditions (Winiecka-Krusnell et al. 2002; McDonnell 2007; Fouque et al. 2012). Most amoebal cyst walls are made up of two layers, the inner endocyst and the outer ectocyst. *Balamuthia mandrillaris* cyst wall has three layers: an outer thin and irregular ectocyst, an inner thick endocyst, and a middle amorphous fibrillar mesocyst (Schuster et al. 2003; Dunnebacke et al. 2004). Amoebal cysts have been shown to resist desiccation of up to 21 years, acidic pH of up to 2.0, radiation treatments of 250 rads γ irradiation, and temperatures of 60 °C (Kilvington 1989; Aksozek et al. 2002; Sriram et al. 2008; Coulon et al. 2010). Studies have also shown amoebal cysts to resist oxidizing agents such as chlorine, used in water treatment, indicating their widespread distribution in municipal water networks (Storey et al. 2004; Chang et al. 2009; Thomas et al. 2010; Thomas and Ashbolt 2011).

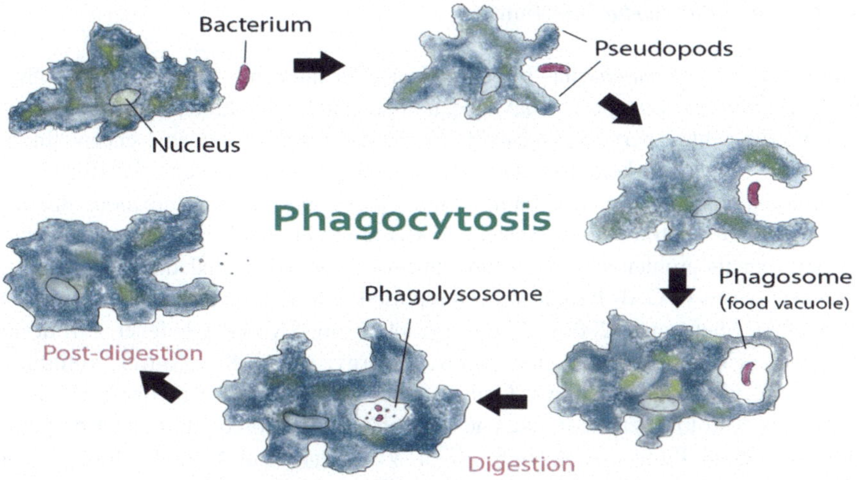

Fig. 9.1 Phagocytic uptake and digestion of bacteria by free-living amoebae (Siddiqui and Khan 2012)

Acanthamoeba cysts have also been shown to be resistant to 2% (v/v) glutaraldehyde solution, a commonly used disinfectant in hospital settings (Coulon et al. 2010).

Free-living amoebae feed by phagocytosis in which actin-based cytoskeleton rearrangements of pseudopods enable them to encircle and engulf food particles or microorganisms, e.g., bacteria (Clarke and Maddera 2006). After uptake, the bacteria are internalized in phagosomes before digestion can take place. The phagosomes will then fuse with lysosomes to form phagolysosomes. Digestion of ingested microorganisms occurs by catabolism action of lysosomal acid hydrolases (Sun-Wada et al. 2009). Bactericidal superoxide radicals produced by b-type cytochrome and NADPH oxidase in the lysosome have also been used by some FLA such as *Acanthamoeba* to kill internalized microorganisms. After digestion, degradation products are expelled by exocytosis (Siddiqui and Khan 2012; McLean 2014) (Fig. 9.1).

9.2.1 Pathogenic FLA

Although mostly non-pathogenic, some genera of FLA such as *Acanthamoeba*, *Naegleria*, *Balamuthia*, and *Sappinia* are responsible for opportunistic and non-opportunistic infections in humans. The taxonomy for these four genera that is based on molecular phylogenetics has been proposed which comprises the supergroup, first rank, and second rank. *Acanthamoeba*, *Balamuthia*, and *Sappinia* are derived from the *Amoebozoa* supergroup, and *Naegleria* is derived from the *Excavata* supergroup which consists of unicellular organisms belonging to the domain *Eukaryota*.

9.2.1.1 *Acanthamoeba* Infections

The genus *Acanthamoeba* has been implicated in more infections than any other FLA. *Acanthamoeba* species have been associated with human infections that involve the central nervous system (CNS) (granulomatous amoebic encephalitis), eyes (*Acanthamoeba* keratitis), and skin (cutaneous acanthamoebiasis) (Fig. 9.1). Granulomatous amoebic encephalitis (GAE) is an opportunistic infectious disease that affects the central nervous system (CNS) of immune-deficient individuals with chronic health problems such as tuberculosis, HIV/AIDS, and diabetic patients. However, cases of GAE have also been reported in healthy individuals (Visvesvara et al. 2007; Sheng et al. 2009). Clinical symptoms of GAE include fever, stiff neck, headache, seizures, dizziness meningitis, lethargy, visual disturbances, vomiting, coma, and visual abnormalities. Granulomatous amoebic encephalitis in the later stage may lead to seizures, coma, and death if not diagnosed and treated properly (Environmental Protection Agency 2003; Visvesvara and Maguire 2006; Khan 2008). *Acanthamoeba* keratitis (AK) is a painful, vision threatening corneal infection that may cause blindness in healthy individuals following corneal abrasions or corneal trauma due to injury. Clinical symptoms of AK start with eye pain, eye redness and tearing, epiphora, and photophobia which develops into corneal epithelium ulceration and blindness if not diagnosed and treated early (Wilhelmus et al. 2008; Trabelsi et al. 2012). The infection is difficult to treat and has been diagnosed mainly in people who wear contact lenses (Chong and Dana 2007; Visvesvara et al. 2007). Improper maintenance and poor sanitary conditions, like rinsing and storing of lenses in tap water or non-sterile saline solutions and swimming in contaminated water, increase the risk of AK in these individuals (Seal 2003; Trabelsi et al. 2012). Another rare opportunistic infection of FLA is cutaneous acanthamoebiasis. This infection is associated with immune-compromised individuals, particularly HIV-infected patients, where the first case reported was in 1986 (Visvesvara et al. 2007). This amoebal infection is characterized by lesions (mostly in the face and trunk), pustules, papules, and skin ulcerations that are colonized with *Acanthamoeba*. Cutaneous acanthamoebiasis has also been reported in few cases of immune-competent individuals. The route of exposure to *Acanthamoeba* is the skin, before entry into the bloodstream.

9.2.1.2 Other FLA Infections

Naegleria fowleri causes a central nervous system (CNS) infection called primary amoebic meningoencephalitis, a rare but rapidly fatal opportunistic infection which can result in death in 3–7 days. The thermophilic character of *N. fowleri* allows it to proliferate in hot summer months of the year where temperatures are likely to be high. Risk of exposure is therefore high in warm freshwater bodies such as rivers, lakes, ponds, and swimming pools where children and young adults participate in recreational activities (Visvesvara et al. 2007; Heggie 2010). Inhalation of the

N. fowleri trophozoites, flagellates, or cysts acts as a route of infection through the nasal passages where it attaches to the nasal mucosa before migrating to the brain. In the brain, it divides rapidly causing death within a week or less following exposure if not diagnosed and treated early (John 1993; Rojas-Hernandez et al. 2004; Visvesvara and Maguire 2006; Didier et al. 2009). *Balamuthia* amoebic encephalitis (BAE) is also a rare, fatal CNS disease with an incubation period between a few weeks to a few years (Bravo et al. 2006). The disease occurs particularly in immunocompromised individuals such as cancer and HIV patients. However, unlike most cases of *Acanthamoeba* GAE, BAE has been reported in children and older individuals with no known history of immunodeficiency (Schuster et al. 2004; Visvesvara and Maguire 2006; Tavares et al. 2006). The only case of *Sappinia* amoebic encephalitis (SAE) was described by Gelman et al. (2001) in which an immune-competent young male survived the infection. *Sappinia diploidea* was originally identified as the causative amoebae before molecular techniques identified it as *S. pedata* (Brown et al. 2007; Qvarnstrom et al. 2009). In this single case of SAE, a sinus infection was observed before onset of symptoms such as nausea, vomiting, photophobia, bifrontal headache, blurry vision, and loss of consciousness for a brief period. Inhalation of *S. pedata* through the nasopharynx before spreading to the CNS is thought to be the route of infection (Gelman et al. 2001).

9.3 Occurrence of Pathogenic FLA in Surface Waters

The ability of FLA to withstand adverse environmental conditions has allowed them to colonize a wide range of habitats. Free-living amoebae have been isolated from air, soil, and water environments where their proliferation and diversity largely depends on environmental factors such as temperature, pH, moisture, and availability of nutrients (Rodriguez-Zaragoza 1994; Bass and Bischoff 2001; Thomas et al. 2006). In aquatic environments, FLA parasites are found in natural environments such as rivers, lakes, and hot springs (Loret and Greub 2010). Free-living amoebae parasites widely spread in a variety of aquatic natural habitats that include surface water and groundwater (Loret and Greub 2010). Table 9.1 summarizes the occurrence of pathogenic FLA and associated diseases from surface water resources, with some of them being used for drinking water production.

The prevalence of FLA in surface waters such as rivers, lakes, ponds, and marine waters has been described worldwide, in the USA (Ettinger et al. 2003), Mexico (Bonilla-Lemus et al. 2010), Iran (Mahmoudi et al. 2015), Uganda (Sente et al. 2016), Switzerland (Gianinazzi et al. 2010), Bulgaria (Tsvetkova et al. 2004), Poland (Łanocha et al. 2009), and Turkey (Özçelik et al. 2012). Different physicochemical parameters such as annual temperature fluctuation, dissolved organic matter, and pH determine the occurrence and composition of FLA in these environments. For example, the prevalence of *Naegleria* and *Acanthamoeba* has been found to be more in spring and early to midsummer, whereas *Vannella*, *Vahlkampfia*, and *Hartmannella* were more prevalent in late summer and spring (Ettinger et al.

Table 9.1 Pathogenic FLA isolated in surface waters

FLA	Disease	Source	References
Acanthamoeba species	Granulomatous amoebic encephalitis, amoebic keratitis; cutaneous acanthamoebiasis	Lakes	Immunocompetent (for keratitis) and immunocompromised (for encephalitis)
Naegleria fowleri	Primary amoebic meningoencephalitis	Lakes, hot springs, rivers	Immunocompetent children and young adults
Balamuthia mandrillaris	*Balamuthia* amoebic encephalitis	Soil, air, and stagnant recreational water	Immunocompetent children or elderly and immunocompromised individuals
Sappinia species	*Sappinia* amoebic encephalitis	Rivers	Isolated once from an immunocompetent adult

2003). The presence of FLA in surface and groundwaters has also been associated with high levels of dissolved organic matter (Loret et al. 2008; Valster et al. 2009). Contamination of natural surface water sources, particularly those associated with human activity, is an important risk factor for amoebic infections and a possibility of a rise of human diseases. In addition, these surface waters, potentially contaminated with FLA, are usually used in drinking water production systems and institutional and domestic water systems where humans are naturally exposed. It is therefore important to study the ecology, pathogenicity, and interactions associated with FLA in rivers, lakes, ponds, and marine waters as they can be sources of FLA parasitic infections.

9.4 Free-Living Amoeba Interactions with Drug-Resistant Bacteria

In addition to their involvement in human disease, FLA can also interact and act as hosts of other pathogenic microorganisms that include bacteria, fungi, and viruses. FLA generally feed on these pathogenic organisms. However, some of these organisms, collectively known as "endocytobionts," are taken by FLA through phagocytosis but resist complete digestion. Some of these endocytobionts do not only survive intracellularly but also multiply inside the amoebal host before being released (Greub and Raoult 2004; Siddiqui and Khan 2012; Scheid 2014). Therefore, FLA may act as reservoirs and/or transmitters of pathogenic microorganisms with a public health and environmental health potential. Furthermore, these organisms are reported to coexist with FLA in the aquatic environments, including surface waters (Muchesa et al. 2017). Studies in situ and in vitro on interactions of drug-resistant bacterial endocytobionts with FLA are important to understand how this can affect humanity.

Bacterial interactions with FLA can be divided into three groups: those which multiply and clearly cytopathic causing lysis of amoebal cells, those which survive

Table 9.2 Examples for clinically relevant bacterial endocytobionts associated with FLA

Bacterial endocytobiont	FLA host	Type of interaction
Acinetobacter baumannii∗∗∗	*Acanthamoeba castellanii* *Acanthamoeba polyphaga*	Intracellular survival
Pseudomonas aeruginosa∗∗∗	*Acanthamoeba polyphaga*	Intracellular multiplication
Escherichia coli∗∗∗	*Acanthamoeba castellanii*	Intracellular survival
Serratia marcescens∗∗∗	*Acanthamoeba* spp.	Intracellular survival, intracellular multiplication
Methicillin-resistant Staphylococcus aureus∗∗	*Acanthamoeba polyphaga*	Intracellular multiplication packaged in vesicles
Helicobacter pylori∗∗	*Acanthamoeba* spp.	Intracellular survival
Campylobacter∗∗	*Acanthamoeba polyphaga*	Intracellular survival
Salmonella spp.∗∗	*Acanthamoeba polyphyga*	Packaged in vesicles
Shigella spp.∗	*Acanthamoeba* spp.	Intracellular survival

WHO priority: ∗∗∗Critical, ∗∗High, ∗Medium

without multiplication, and those which are packaged and expelled in amoeba vesicles (Denoncourt et al. 2014). Some bacterial endocytobionts can belong to more than one of these three groups depending on the species and environmental conditions (Table 9.2). Drug-resistant bacterial endocytobionts reported in the literature on the World Health Organization global priority list of antibiotic-resistant bacteria which can be used to guide research, discovery, and development of new antibiotics include *Acinetobacter baumannii*, *Pseudomonas aeruginosa*, *Escherichia coli*, *Serratia marcescens*, *Staphylococcus aureus*, *Helicobacter pylori*, *Campylobacter*, *Salmonella* spp., and *Shigella* spp. All these bacteria have been reported to be resistant to one or more groups of antibiotics such as carbapenem, cephalosporin, and fluoroquinolone. Infections associated with these multidrug-resistant bacteria include systematic blood, respiratory tract, urinary tract, ocular, gastrointestinal tract, and skin infections to exposed individuals (Rice 2008; Özdemir et al. 2011; Perez et al. 2014).

9.5 Public Health Implications of Interactions

The survival of pathogenic endocytobionts in water sources has increased the clinical significance of FLA because of their role in the transmission of pathogenic bacterial endocytobionts in the environment. The transmission of pathogenic endocytobionts is further enhanced by the resistance of their amoebal hosts to biocides in water systems that use surface water as a source of drinking water. This in turn contributes to the spread of antibiotic-resistant pathogens from the surface waters to the drinking water systems where humans can be exposed.

9.5.1 Antibiotic Resistance

The widespread, increased, and prolonged use of antimicrobial agents has led to the selection of different multidrug-resistant bacteria over the years. In that context, it is interesting to note that the interaction of amoebae with pathogenic bacterial endocytobionts has increased their ability to resist antibiotics. Although the mechanism used by these bacterial endocytobionts to acquire drug resistance is not yet known, some studies have shown that amoebae harboring *L. pneumophila* and *M. avium* were found to enhance decreased antibiotic susceptibility to these bacteria (Barker et al. 1995; Miltner and Bermudez 2000).

9.5.2 Increase in Virulence

The survival of a wide variety of pathogenic bacteria inside amoebae provides a platform in which horizontal transfer of genes can occur between the intracellular bacteria and their amoeba hosts (Albert-Weissenberger et al. 2007; Moliner et al. 2010; Gomez-Valero and Buchrieser 2013). Bacteria grown in amoebae have also been shown to be more virulent to humans than they were prior to FLA infection (Cirillo et al. 1997; Neumeister et al. 2000) as their virulent genes are upregulated (Schmitz-Esser et al. 2008). This selection of virulent traits has been observed in *Legionella* where amoebae-grown bacteria have been shown to be more resistant to adverse environmental conditions such as high temperature and acidity compared to free growing *Legionella*. Increased infectivity of amoebae-grown *Legionella* to mammalian cells has also been observed (Greub and Raoult 2004). Therefore, altered genetic makeup and selection of virulent traits for bacteria associated with amoebae could be a source of new pathogenic bacteria in surface waters, if they are present.

9.5.3 Biocide Resistance

Amoebal cyst walls contain cellulose that forms a physical barrier, which protects the ARB from physical extremes of temperatures (Winiecka-Krusnell and Linder 2001; Storey et al. 2004), ultraviolet radiation (Aksozek et al. 2002), and biocides and disinfectants mainly used to treat water in water systems (Coulon et al. 2010; Dupuy et al. 2014). In our previous study, FLA were isolated in final effluent samples with a residual chlorine concentration of between 0.01 and 1.10 mg/L (Muchesa et al. 2014). The wastewater was discharged into a river, pointing to a possibility of these FLA being spread by the river as it flows. This also increases the risk to the health of communities living in the vicinity of the river that uses the water for multiple purposes which include drinking, agricultural, and recreational purposes. The resistance of FLA to biocidal treatments, chlorine in particular, is therefore of critical importance in the treatment of water in water distribution systems since humans are constantly exposed to these environments (McDonnell 2007; Loret et al. 2008). A study done by Storey et al. (2004) showed that amoebal cysts can survive chlorine concentrations as high as 100 mg/L for 10 min.

9.6 Other Important FLA and Enteric Protozoan Parasites in Surface Waters

Some FLA present in aqueous environments are not directly responsible for human disease; however, their interaction and their role as reservoirs and in protecting other pathogens make them clinically relevant, especially in hospital water distribution systems. The FLA, *Vermamoeba vermiformis*, *Vanella* spp., and *Vahlkampfia* spp. and protozoan parasites like *Tetrahymena pyriformis*, *Cryptosporidium*, and *Giardia* have been described in surface waters (Greub and Raoult 2004; Thomas et al. 2010; Cateau et al. 2014). Thus, studying these FLA is important from a public point of view. In this review particular focus is put on *V. vermiformis*, *D. discoideum*, and *T. pyriformis* as the more prevalent FLA and protozoa in the environment of the non-pathogenic FLA.

9.6.1 Vermamoeba vermiformis

Vermamoeba vermiformis (previously known as *Hartmannella vermiformis*) have been isolated in natural and man-made water systems (Rodriguez-Zaragoza 1994; Ovrutsky et al. 2013; Cateau et al. 2014). In particular, some studies have reported them to be the most common FLA isolated from hospital water systems (Rohr et al. 1998; Thomas et al. 2006; Pagnier et al. 2015). *Vermamoeba vermiformis* like *Acanthamoeba* has two developmental stages: the trophozoites and the cysts. Trophozoites, 28–42 μm in length, have an elongate cylindrical motile form of smooth monopodial that protrudes laterally onward the direction of movement. Mitochondria, posterior uroidal filaments, contractile vacuoles (sometimes more numerous), and a single nucleus are observed in the ultrastructure of trophozoites. Cysts (average diameter of 8.7 μm) have a smooth spherical shape with a characteristic double-walled structure which consists of the endocyst and the ectocyst (Smirnov and Michel 1999; Dykova et al. 2005; Fouque et al. 2015).

9.6.2 Dictyostelium discoideum

Dictyostelium discoideum is a unicellular protozoan that belongs to social amoebae. It is a soil amoeba that preys on bacteria and other microorganism (Cosson and Soldati 2008; Müller-Taubenberger et al. 2012). Separate and independent vegetative cells of *Dictyostelium* amoebae grow and divide by simple mitotic divisions. Depletion of nutrients in the environment triggers changes in gene expression and secretion of cAMP by *Dictyostelium* cells. This signal causes cells to form multicellular aggregates which further undergo differentiation and morphogenetic changes. The aggregation of thousands of individual cells results in the formation of a fruiting body that is made up of spore and stalk cells. The stalk cells are dead cells, while spore cells are highly resistant cells that form new *Dictyostelium* cells when conditions become favorable (van Driessche et al. 2002; Maeda and Chida 2013). *Dictyostelium discoideum* motility, phagocytosis, chemotaxis, signal transduction properties, and cell differentiation during development have made it a useful model organism in study of neurodegenerative disorders, autosomal diseases, and innate immunity (Müller-Taubenberger et al. 2012). Besides its uses in biomedical research, *D. discoideum* have also been known to be harbor pathogenic organisms such as *Salmonella typhimurium* (Sillo et al. 2011), *Pseudomonas aeruginosa* (Skriwan et al. 2002), and *Klebsiella pneumoniae* (Benghezal et al. 2006).

9.6.3 Tetrahymena pyriformis

Tetrahymena pyriformis (syn. *Glaucoma pyriformis, Tetrahymena geleii*) is a unicellular ciliated protozoan that is often isolated in freshwater environments. *Tetrahymena pyriformis* morphology is characterized by an ovoid, pear-shaped body that is uniformly ciliated with a roughly triangular mouth. The body is 40–60 μm in length and has approximately 600 cilia that are used for locomotion and ingestion (Brennen and Winet 1977; Kim et al. 2011). The ultrastructure of *T. pyriformis* has a central, spherically shaped macronucleus with a micronucleus and a contractile vacuole near the posterior end. The absence of a cell wall makes it highly sensitive to environmental contamination resulting in a fast cellular response. This has made *T. pyriformis* a useful eukaryotic model organism in environmental and industrial pollutants and for toxicological studies at the genetic level (Gutiérrez et al. 2003; Lee and Gutell 2012). However, some studies have also shown that pathogens such as *E. coli*, *Legionella* spp., and *Mycobacterium avium* can resist digestion processes, persist, and even multiply within *T. pyriformis* (Steele and McLennan 1996; Strahl et al. 2001; Smith et al. 2012).

9.6.4 Cryptosporidium *and* Giardia

Cryptosporidium is a protozoan parasite commonly found in sewage and surface waters and occasionally in treated water. The life cycle of *Cryptosporidium* begins with sporulated oocysts which are round, thick-walled, environmentally stable and 4–6 μm in diameter. The life cycle consists of excystation, where sporozoites are released from an excysting oocyst; schizogony (syn. merogony), where asexual reproduction takes place; gametogony, the stage at which gametes are formed; four fertilization stages of the macrogametocyte by a microgamete to form a zygote; oocyst wall formation; and sporogony, where sporozoites form within the oocyst (Thompson and Monis 2004; Rosales et al. 2005). *Giardia* is a flagellated protozoan parasite that inhabits the small intestines of humans and other animals. The life cycle of *Giardia* species is characterized by two developmental stages: the trophozoite or feeding stage and the cyst form. The size of trophozoites ranges from 9 to 21 μm long depending on the species. Trophozoites have a pear-shaped body that contains a large ventral sucking disc (cytostome), two nuclei, two slender median rods, and eight flagella in four pairs. Trophozoites feed by attaching on surface of the intestinal villi. *Giardia* cysts, which are smaller (8–14) than trophozoites, are formed when trophozoites detach from intestinal villi. The ovoid cysts contain two or four nuclei and visible remnants of organelles (Xiao et al. 2004; Smith et al. 2007). *Cryptosporidium* and *Giardia* are important causes of moderate-to-severe diarrhea accompanied by headaches, fever, cramping, loss of appetite, weight loss, and nausea, with *Cryptosporidium* being reported as the most common diarrhea-causing protozoan parasite worldwide (WHO 2009; Xiao 2010). Surface waters are contaminated in two major ways: effluent from wastewater treatment plants with *Cryptosporidium* or *Giardia* spores or oocysts and runoff from certain agricultural operations or human feces that may contain the spores (Hunter and Thompson 2005).

9.7 Conclusions

Some species of FLA are known to cause opportunistic infections in humans. These species with non-pathogenic FLA can interact with drug-resistant bacteria in habitats such as surface waters. The interactions can help the bacteria to survive and/or multiply using FLA as an environmental host. Surface waters therefore are important sources of disease-causing FLA and associated bacterial endocytobionts. This has implications in public health as humans can be exposed to contaminated water. Thus town authorities are obliged to control rigorously the drinking water in and around towns.

References

Aksozek A, McClellan K, Howard K, Niederkorn JY, Alizadeh H (2002) Resistance of *Acanthamoeba castellanii* cysts to physical, chemical, and radiological conditions. J Parasitol 88(3):621–623

Albert-Weissenberger C, Cazalet C, Buchrieser C (2007) *Legionella pneumophila* - a human pathogen that co-evolved with fresh water protozoa. Cell Mol Life Sci 64(4):432–448

Barker J, Scaife H, Brown MRW (1995) Intraphagocytic growth induces an antibiotic-resistant phenotype of *Legionella pneumophila*. Antimicrob Agents Chemother 39:2684–2688

Bass P, Bischoff PJ (2001) Seasonal variability in abundance and diversity of soil gymnamoebae along a short transect in south eastern USA. J Eukaryot Microbiol 48:475–479

Benghezal M, Fauvarque MO, Tournebize R, Froquet R, Marchetti A, Bergeret E, Lardy B, Klein G, Sansonetti P, Charette SJ, Cosson P (2006) Specific host genes required for the killing of *Klebsiella* bacteria by phagocytes. Cell Microbiol 8:139–148

Bonilla-Lemus P, Ramírez-Bautista GA, Zamora-Muñoz C, Ibarra-Montes MDR, Ramírez-Flores E, Hernández-Martínez MD (2010) *Acanthamoeba* spp. in domestic tap water in houses of contact lens wearers in the metropolitan area of Mexico City. Exp Parasitol 126(1):54–58

Bravo FG, Cabrera J, Gottuzo E, Visvesvara GS (2006) Cutaneous manifestations of infection by free-living amebas. In: Tyring SK, Lupi O, Hengge UR (eds) Tropical dermatology. Elsevier, Philadelphia, pp 49–55

Brennen C, Winet H (1977) Fluid mechanics of propulsion by cilia and flagella. Annu Rev Fluid Mech 9(1):339–398

Brown MW, Spiegel FW, Silberman JD (2007) Amoeba at attention: phylogenetic affinity of *Sappinia pedata*. J Eukaryot Microbiol 54(6):511–519

Cateau E, Delafont V, Hechard Y, Rodier MH (2014) Free-living amoebae: what part do they play in healthcare-associated infections? J Hosp Infect 87:131–140

Chang CW, Kao CH, Liu YF (2009) Heterogeneity in chlorine susceptibility for *Legionella pneumophila* released from *Acanthamoeba* and *Hartmannella*. J Appl Microbiol 106:97–105

Chong EM, Dana MR (2007) *Acanthamoeba* keratitis. Int Ophthalmol Clin 47:33–46

Cirillo JD, Falkow S, Tompkin LS, Bermudez L (1997) Interaction of *Mycobacterium avium* with environmental amoebae enhances virulence. Infect Immun 65:3759–3767

Clarke M, Maddera L (2006) Phagocyte meets prey: uptake, internalization, and killing of bacteria by *Dictyostelium* amoebae. Eur J Cell Biol 85:1001–1010

Cosson P, Soldati T (2008) Eat, kill or die: when amoeba meets bacteria. Curr Opin Microbiol 11:271–276

Coulon C, Collignon A, McDonnell G, Thomas V (2010) Resistance of *Acanthamoeba* cysts to disinfection treatments used in health care settings. J Clin Microbiol 48:2689–2697

Denoncourt AM, Paquet VE, Charette SJ (2014) Potential role of bacteria packaging by protozoa in the persistence and transmission of pathogenic bacteria. Front Microbiol 5:240

Didier ES, Weiss LM, Cali A, Marciano-cabra LF (2009) Overview of the presentations on microsporidia and free-living amoebae at the 10th international workshops on opportunistic Protists. Eukaryot Cell 8:441–445

Dunnebacke TH, Schuster FL, Yagi S, Booton GC (2004) *Balamuthia mandrillaris* from soil samples. Microbiology 150:2837–2842

Dupuy M, Berne F, Herbelin P, Binet M, Berthelot N, Rodier MH, Soreau S, Héchard Y (2014) Sensitivity of free-living amoeba trophozoites and cysts to water disinfectants. Int J Hyg Environ Health 217:335–339

Dyková I, Nowak BF, Crosbie PBB, Fiala I, Pecková H, Adams MB, Macháčková B, Dvořáková H (2005) *Neoparamoeba branchiphila* n. sp., and related species of the genus *Neoparamoeba* Page, 1987: morphological and molecular characterization of selected strains. J Fish Dis 28:49–64

Environmental Protection Agency (2003) Health effects support document for *Acanthamoeba*, Office of Water (4304T) Health and Ecological Criteria Division, Washington, DC 20460

Ettinger MR, Webb R, Harris SA, Mclninch SP, Garman CG, Brown BL (2003) Distribution of free-living amoebae in James River, Virginia, USA. Parasitol Res 89:6–15

Fouque E, Trouilhe MC, Thomas V, Hartemann P, Rodier MH, Hechard Y (2012) Cellular, biochemical, and molecular changes during encystment of free-living amoebae. Eukaryot Cell 11:382–387

Fouque E, Héchard Y, Hartemann P, Humeau P, Trouilhé MC (2015) Sensitivity of *Vermamoeba* (*Hartmannella*) *vermiformis* cysts to conventional disinfectants and protease. J Water Health 13:302

Fritz-Laylin LK, Cande WZ (2010) Ancestral centriole and flagella proteins identified by analysis of *Naegleria* differentiation. J Cell Sci 23:4024–4031

Gelman BB, Rauf SJ, Nader R, Popov V, Borkowski J, Chaljub G, Nauta HW, Visvesvara GS (2001) Amoebic encephalitis due to *Sappinia diploidea*. J Am Med Assoc 285(19):2450–2451

Gianinazzi C, Schild M, Zumkehr B, Wuthrich F, Nuesch I, Ryter R, Schurch N, Gottstein B, Muller N (2010) Screening of Swiss hot spring resorts for potentially pathogenic free-living amoebae. Exp Parasitol 126(1):45–53

Gomez-Valero L, Buchrieser C (2013) Genome dynamics in *Legionella*: the basis of versatility and adaptation to intracellular replication. Cold Spring Harb Perspect Med 3(6):a009993. https://doi.org/10.1101/cshperspect.a009993

Gonzalez-Robles A, Flores-Langarica A, Omana-Molina M, Shibayama M (2001) *Acanthamoeba castellanii*: ultrastructure of trophozoites using fast free fixation. J Electron Microsc 50:423–427

Greub G, Raoul D (2004) Microorganisms resistant to free living amoebae. Clin Microbiol Rev 17:413–433

Gutierrez JC, Martin-Gonzalez A, Diaz S, Ortega R (2003) Ciliates as potential source of cellular and molecular biomarkers/biosensors for heavy metal pollution. Eur J Protistol 39:461–467

Heggie TW (2010) Swimming with death: *Naegleria fowleri* infections in recreational waters. Travel Med Infect Dis 8:201–206

Hunter PR, Thompson RCA (2005) The zoonotic transmission of *Giardia* and *Cryptosporidium*. Int J Parasitol 5:1181–1190

John DT (1993) Oppoturnistically pathogenic free-living amoebae. In: Kreier JP, Baker JR (eds) Parasitic Protozoa, vol 3, 2nd edn. Academic, San Diego, CA, pp 143–246

Kang S, Tice AK, Spiegel FW et al (2017) Between a pod and a hard test: the deep evolution of amoebae. Mol Biol Evol 34:2258–2270

Khan NA (2008) *Acanthamoeba* and the blood-brain barrier: the breakthrough. Int J Med Microbiol 57:1051–1057

Kilvington S (1989) Moist-heat disinfection of pathogenic *Acanthamoeba* cysts. Lett Appl Microbiol 9:187–189

Kim DH, Brigandi S, Julius AA, Kim MJ (2011) Real-time feedback control using artificial magnetotaxis with rapidly exploring random tree (RRT) for *Tetrahymena pyriformis* as a microbiorobot. In: IEEE international conference on robotics and automation (ICRA'11), Shanghai, 9–13. IEEE Press, Piscataway, pp 3183–3188

Lanocha N, Kosik-Bogacka D, Maciejewska A, Sawczuk M, Wilk A, Kuzna-Grygiel W (2009) The occurrence *Acanthamoeba* (free living amoeba) in environmental and respiratory samples in Poland. Acta Protozool 48:271–279

Lee JC, Gutell RR (2012) A comparison of the crystal structures of eukaryotic and bacterial SSU ribosomal RNAs reveals common structural features in the hypervariable regions. PLoS One 7: e38203

Loret JF, Greub G (2010) Free-living amoebae: biological by-passes in water treatment. Int J Hyg Environ Health 213:167–175

Loret JF, Jousset M, Robert S, Anselme C, Sauced G, Ribas F, Martinez L, Catalan V (2008) Elimination of free-living amoebae by drinking water treatment processes. Eur J Water Qual 39:37–50

Maeda Y, Chida J (2013) Control of cell differentiation by mitochondria, typically evidenced in *Dictyostelium* development. Biomol Ther 3(4):943–966

Mahmoudi MR, Kazemi B, Haghighi A, Karanis P (2015) Detection of *Acanthamoeba* and *Toxoplasma* in river water samples by molecular methods in Iran. Iran J Parasitol 10(2):250–257
Marciano-Cabral F, Cabral GA (2003) *Acanthamoeba* spp. as agents of disease in humans. Clin Microbiol Rev 16:273–307
Mcdonnell G (2007) Antisepsis, disinfection and sterilization. Types, action and resistance. ASM Press, Washington, DC
McLean J (2014) Developing methodologies for the investigation of free-living amoeba as a tool for pathogen surveillance on dairy farms and aquaculture. A thesis presented to the University of Guelph. Master of Science in Molecular and Cellular Biology
Miltner EC, Bermudez LE (2000) *Mycobacterium avium* grown in *Acanthamoeba castellanii* is protected from the effects of antimicrobials. Antimicrob Agents Chemother 44(7):1990–1994
Moliner C, Fournier PE, Raoult D (2010) Genome analysis of microorganisms living in amoebae reveals a melting pot of evolution. FEMS Microbiol Rev 34(3):281–294
Muchesa P, Mwamba O, Barnard TG, Bartie C (2014) Detection of free-living amoebae using amoebal enrichment in a wastewater treatment plant of Gauteng province, South Africa. Biomed Res Int 2014:575297
Muchesa P, Leifels M, Jurzik L, Barnard TG, Bartie C (2017) Coexistence of free-living amoebae and bacteria in selected South African hospital water distribution systems. Parasitol Res 116:155–165
Müller-Taubenberger A, Kortholt A, Eichinger LL (2012) Simple system – substantial share: the use of *Dictyostelium* in cell biology and molecular medicine. J Eukaryot Microbiol 92(2):45–53
Neumeister B, Reiff G, Faigle M, Dietz K, Northoff H, Lang F (2000) Influence of *Acanthamoeba castellanii* on intracellular growth of different *Legionella* species in human monocytes. Appl Environ Microbiol 66:914–919
Omarova A, Tussupova K, Berndtsson R, Kalishev M, Sharapatova K (2018) Protozoan parasites in drinking water: a system approach for improved water, sanitation and hygiene in developing countries. Int J Environ Res Public Health 15(3):E495. https://doi.org/10.3390/ijerph15030495
Ovrutsky AR, Chan ED, Kartalija M, Bai X, Jackson M, Gibbs S, Falkinham JO, Iseman MD, Reynolds PR, McDonnell G (2013) Co-occurrence of free-living amoebae and nontuberculous *Mycobacteria* in hospital water networks, and preferential growth of *Mycobacterium avium* in *Acanthamoeba lenticulata*. Appl Environ Microbiol 79:3185–3192
Özçelik S, Coşkun KA, Yünlü Ö, Alim A, Malatyalı E (2012) The prevalence, isolation and morphotyping of potentially pathogenic free-living amoebae from tap water and environmental water sources in Sivas. Türk Parazitol Derg 36:198–203
Özdemir M, Pekcan S, Demircili ME, Taşbent FE, Feyzioğlu B, Pirinç S, Baykan M (2011) A rare cause of bacteremia in a pediatric patient with down syndrome: *Sphingomonas paucimobilis*. Int J Med Sci 8:537–539
Pagnier I, Yutin N, Croce O, Makarova KS, Wolf YI, Benamar S, Raoult D, Koonin EV, La Scola B (2015) *Babela massiliensis*, a representative of a widespread bacterial phylum with unusual adaptations to parasitism in amoebae. Biol Direct 10:13
Perez PN, Ramirez M, Fernandez JA, De Guevara LL (2014) A patient presenting with cholangitis due to *Stenotrophomonas maltophilia* and *Pseudomonas aeruginosa* successfully treated with intrabiliary colistin. Curr Infect Dis Rep 6:5147
Qvarnstrom Y, Da Silva AJ, Schuster FL, Gelman BB, Visvesvara GS (2009) Molecular confirmation of *Sappinia pedata* as a causative agent of amoebic encephalitis. J Infect Dis 199 (8):1139–1142
Rice LB (2008) Federal funding for the study of antimicrobial resistance in nosocomial pathogens: no ESKAPE. J Infect Dis 197:1079–1081
Rodriguez-Zaragoza S (1994) Ecology of free-living amoebae. Crit Rev Microbiol 20:225–241
Rohr U, Weber S, Michel R, Selenka F, Wilhelm M (1998) Comparison of free-living amoebae in hot water systems of hospitals with isolates from moist sanitary areas by identifying genera and determining temperature tolerance. Appl Environ Microbiol 64:1822–1824

Rojas-Hernandez S, Rodrıguez-Monroy MA, Lopez-Revilla R, Resendiz-Albor AA, Moreno-Fierros L (2004) Intranasal coadministration of the Cry1Ac protoxin with amoebal lysates increases protection against *Naegleria fowleri* meningoencephalitis. Infect Immun 72 (8):4368–4375

Rosales MJ, Cordón GP, Moreno MS, Sánchez CM (2005) Extracellular like-gregarine stages of *Cryptosporidium parvum*. Acta Trop 95(1):74–78

Scheid P (2014) Relevance of free-living amoebae as hosts for phylogenetically diverse microorganisms. Parasitol Res 113:2407–2414

Schmitz-Esser S, Toenshoff ER, Haider S, Heinz E, Hoenninger VM, Wagner M, Horn M (2008) Diversity of bacterial endosymbionts of environmental *Acanthamoeba* isolates. Appl Environ Microbiol 74:5822–5831

Schuster FL (2002) Cultivation of pathogenic and opportunistic free-living amebas. Clin Microbiol Rev 15:342–335

Schuster FL, Dunnebacke TH, Booton GC, Yagi S, Kohlmeier CK, Glaser C, Vugi D, Bakardjiev A, Azimi P, Maddux-Gonzalez M, Martinez AJ, Visvesvara GS (2003) Environmental isolation of *Balamuthia mandrillaris* associated with a case of amebic encephalitis. J Clin Microbiol 41:3175–3180

Schuster FL, Glaser C, Honarmand S, Maguire JH, Visvesvara GS (2004) *Balamuthia* amebic encephalitis risk. Emerg Inf Dis 10:1510–1512

Seal DV (2003) *Acanthamoeba* keratitis update-incidence, molecular epidemiology and new drugs for treatment. Eye 17:893–905

Sente C, Erume J, Naigaga I, Mulindwa J, Ochwo S, Magambo PK, Namara BG, Kato CD, Sebyatika G, Muwonge K, Ocaido M (2016) Prevalence of pathogenic free-living amoeba and other protozoa in natural and 490 communal piped tap water from Queen Elizabeth protected area, Uganda. Infect Dis Poverty 5:68

Sheng WH, Hung CC, Huang HH, Liang SY, Cheng YJ, Chang SC (2009) First case of granulomatous amebic encephalitis caused by *Acanthamoeba castellanii* in Taiwan. Am J Trop Med Hyg 81:277–279

Siddiqui R, Khan NA (2012) Biology and pathogenesis of *Acanthamoeba*. Parasit Vectors 5:6

Sillo A, Matthias J, Konertz R, Bozzaro S, Eichinger L (2011) *Salmonella typhimurium* is pathogenic for *Dictyostelium* cells and subverts the starvation response. Cell Microbiol 13:1793–1811

Skriwan C, Fajardo M, Hagele S, Horn M, Wagner M, Michel R, Krohne G, Schleicher M, Hacker J, Steinert M (2002) Various bacterial pathogens and symbionts infect the amoeba *Dictyostelium discoideum*. Int J Med Microbiol 291:615–624

Smirnov AV, Michel R (1999) New data on the cyst structure of *Hartmannella vermiformis* page, 1967 (Lobosea, Gymnamoebia). Protistology 1:82–85

Smith HV, Cacciò SM, Cook N, Nichols RA, Tait A (2007) Cryptosporidium and Giardia as foodborne zoonoses. Vet Parasitol 149(1–2):29–40

Smith CD, Berk SG, Brandl MT, Riley LW (2012) Survival characteristics of diarrheagenic *Escherichia coli* pathotypes and *Helicobacter pylori* during passage through the free-living ciliate, *Tetrahymena* sp. FEMS Microbiol Ecol 82:574–583

Sriram R, Shoff M, Botton GC, Fuerst PA, Visvesvara GS (2008) Survival of *Acanthamoeba* cysts after desiccation for more than 20 years. J Clin Microbiol 46(12):4045–4048

Steele TW, McLennan AM (1996) Infection of *Tetrahymena pyriformis* by *Legionella longbeachae* and other *Legionella* species found in potting mixes. Appl Environ Microbiol 62:1081–1083

Storey MV, Winiecka-Krusnell IJ, Ashbolt NJ, Stenstorm TA (2004) The efficacy of heat and chlorine treatment against thermotolerant *Acanthamoebae* and *Legionellae*. Scand J Infect Dis 36:656–662

Strahl ED, Gillaspy GE, Falkinham JO III (2001) Fluorescent acid-fast microscopy for measuring phagocytosis of *Mycobacterium avium*, *Mycobacterium intracellulare* and *Mycobacterium scrofulaceum* by *Tetrahymena pyriformis* and their intracellular growth. Appl Environ Microbiol 67:4432–4439

Sun-Wada GH, Tabata H, Kawamura N, Aoyama M, Wada Y (2009) Direct recruitment of H+-ATPase from lysosomes for phagosomal acidification. J Cell Sci 122:2504–2513

Tavares M, Correia da Costa JM, Carpenter SS, Santos LA, Afonso C, Aguiar A, Pereira J, Cardoso AI, Schuster FL, Yagi S, Sriram R, Visvesvara GS (2006) Diagnosis of first case of *Balamuthia* amoebic encephalitis in Portugal by immunofluorescence and PCR. J Clin Microbiol 44 (7):2660–2663

Thomas JM, Ashbolt NJ (2011) Do free-living amoebae in treated drinking water systems present an emerging health risk. Environ Sci Technol 45:860–869

Thomas V, Herrera-Rimann K, Blannci DS, Greub G (2006) Biodiversity of amoebae and amoeba-resisting bacteria in a hospital water network. Appl Environ Microbiol 72:2428–2438

Thomas V, Mcdonell G, Denyer S, Maillard JY (2010) Free-living amoebae and their intra-cellular pathogenic micro-organisms: risks for water quality. FEMS Microbiol Rev 34:231–259

Thompson RC, Monis PT (2004) Variation in Giardia: implications for taxonomy and epidemiology. Adv Parasitol 58:69–137

Trabelsi H, Dendana F, Sellami A, Sellami H, Cheikhrouhou F, Neji S, Makni F, Ayadi A (2012) Pathogenic free-living amoebae: epidemiology and clinical review. Pathol Biol 60:399–405

Tsvetkova N, Schilld M, Panaiotov S, Kurdova-Mincheva R, Gottsein B, Walochnik J, Aapock H, Lucas MS, Muller N (2004) The identification of free-living environmental isolates of amoebae from Bulgaria. Parasitol Res 92:405–413

Valster RM, Wullings B, Bakker G, Smidt H, Van der Kooij D (2009) Free-living protozoa in two unchlorinated drinking water supplies, identified by phylogenic analysis of 18S rRNA gene sequences. Appl Environ Microbiol 75(14):4736–4746

Van Driessche N, Shaw C, Katoh M, Morio T, Sucgang R, Ibarra M, Kuwayama H, Saito T, Urushihara H, Maeda M, Takeuchi I, Ochiai H, Eaton W, Tollett J, Halter J, Kuspa A, Tanaka Shaulsky G (2002) A transcriptional profile of multicellular development in *Dictyostelium discoideum*. Development 129:1543–1552

Visvesvara GS, Maguire JH (2006) Pathogenic and opportunistic free-living amebas. *Acanthamoeba* spp., *Balamuthia mandrillaris*, *Naegleria fowleri*, and *Sappinia diploidea*. FEMS Immunol Med Microbiol 50(1):1–26

Visvesvara GS, Moura H, Shuster FL (2007) Pathogenic and opportunistic free-living amoebae: *Acanthamoeba spp.*, *Balamuthia mandrillaris*, *Naegleria fowleri* and *Sappinia diploidea*. FEMS Immunol Med Microbiol 50:1–26

Wilhelmus KR, Jones DB, Matoba AY, Hamill MB, Pflugfelder SC, Weikert MP (2008) Bilateral *Acanthamoeba* keratitis. Am J Ophthalmol 145(2):193–197

Winiecka-Krusnell J, Linder E (2001) Bacterial infections of free-living amoebae. Res Microbiol 152(7):613–619

Winiecka-Krusnell J, Wreiber K, Von Euler A, Engstrand L, Linder E (2002) Free-living amoebae promote growth and survival of *Helicobacter pylori*. Scand J Infect Dis 34:253–256

World Health Organization/United Nations International Children's Emergency Fund (WHO/UNICEF) (2009) Geneva/New York. http://apps.who.int/iris/bitstream/10665/44174/1/9789241598415_eng.pdf. Accessed 25 Sep 2016

Xiao L (2010) Molecular epidemiology of cryptosporidiosis: an update. Exp Parasitol 124(1):80–89

Xiao L, Fayer R, Ryan U, Upton SJ (2004) Cryptosporidium taxonomy: recent advances and implications for public health. Clin Microbiol Rev 17(1):72–97

Chapter 10
Nile River, Impact on Human Health in Egypt from Pharaohs Until Now

Fathy Abdel-Ghaffar, Hassan M. Sobhy, Karim S. Metwaly, Sayed AbdEl-Monem, and Heinz Mehlhorn

Abstract This chapter shows the tremendous influence of the Nile on human culture reaching until today.

Keywords Egyptian health history · Waterborne diseases · Waterborne vectors · Malaria · Schistosomiasis

10.1 Introduction

The Nile River has always been the lifeline of Egypt, which provides the country with almost all its water needs. The text of the Greek historian *Herodotus* (~490–430 BC), *Egypt is a gift of the Nile*, is a sign of the high importance of the Nile in the development of one of the most important civilizations in the history of the ancient world.

The silt deposits, which had come and still come today every year in the flood season, formed a fertile soil suitable for the cultivation of many crops, which secured Egypt's economic stability throughout the ancient and middle ages until today.

Egypt covers a land area of approximately 1 million square kilometers; however 95% of the country is a sandy desert today. With an estimated population of 95 million people, the country accounts for one-fourth of the recent Arabic world's population. More than 90% of the Egyptian population lives in only 10% of the entire region along the borders of the Nile River and in the Nile Delta in the northern part of the country.

F. Abdel-Ghaffar (✉) · S. AbdEl-Monem
Faculty of Sciences, Department of Zoology, Cairo University, Giza, Egypt

H. M. Sobhy · K. S. Metwaly
Faculty of African Postgraduate Studies, Cairo University, Giza, Egypt

H. Mehlhorn
Parasitology Institute, Dusseldorf University, Düsseldorf, Germany

H. Mehlhorn, S. Klimpel (eds.), *Parasite and Disease Spread by Major Rivers on Earth*, Parasitology Research Monographs 12,
https://doi.org/10.1007/978-3-030-29061-0_10

10.2 Nile River

The Nile (Fig. 10.1) remains an integral part of the daily life of Egyptians playing an essential role in their habits, traditions, and commercial and economic activities. The Nile is one of the most important rivers in the world and especially in Northeast Africa covering a length of about 6650 km. Together with the Amazonas, it is the longest river on earth. The Nile Basin stretches from the heart of tropical forests in Central Africa to the harshest deserts of the world, covering an area of 3,400,000 km^2 passing 11 countries (Wimberly and Midekisa 2014).

The Nile has two main tributaries, the *White Nile*, which flows from the Great Lakes region of Central Africa, and the *Blue Nile*, which originates from the Lake Tana at the plateau of Ethiopia and is the source of most of the water of the final Nile and its amounts of transported silt (Dumont 2009).

From there the river passes regions with a high diversity of environments along its journey from upstream to downstream offering life for a fabulous animal world and—in addition—is the basis of one of the world's oldest civilizations: the ancient Egyptian civilization. On the other side many of the human diseases being associated with the river have been documented to have affected the Egyptian people throughout its history spanning more than 6000 years BC and do it even until today.

10.3 Nile River in Ancient Egypt

Five million years ago, the Nile began to cross the Sahara Desert in North Africa, playing a vital role in the formation of Egypt and its societies. Beginning 6000 BC the settlements gradually spread along the river and in the delta, marking the beginning of the dawn of ancient Egyptian civilization, in the year 3150 BC, when Egypt became united as national state under a central government. At this time the King Narmer united Upper Egypt and Lower Egypt (Blaha and Blaha 2004).

Although the Nile was the source of life for Egypt, the Egyptians did not worship it especially, but they were thankful to specific gods for blessing them by the Nile and its steady flowing waters. Since they had no name for the river, the Nile was called *Hapy* or *Itrow*, which just means "river" in the ancient Egyptian language (Fig. 10.2) (Hillel 2006).

10.4 Etymology of the Word Nile, Origin and Meaning

The English name Nile and the Arabic names *en-Nîl* and *an-Nîl* are derived from the Latin term *Nilus* and the Ancient Greek term *Neilos* "Νεῖλος" which came from the word "river valley." *Hesiod* at his *Theogony* (700 BC) refers that Nilus (Νεῖλος) was one of the *Potamoi* (river gods) and the son of *Oceanus* and *Tethys*. Another

Fig. 10.1 Map of the Nile River basin

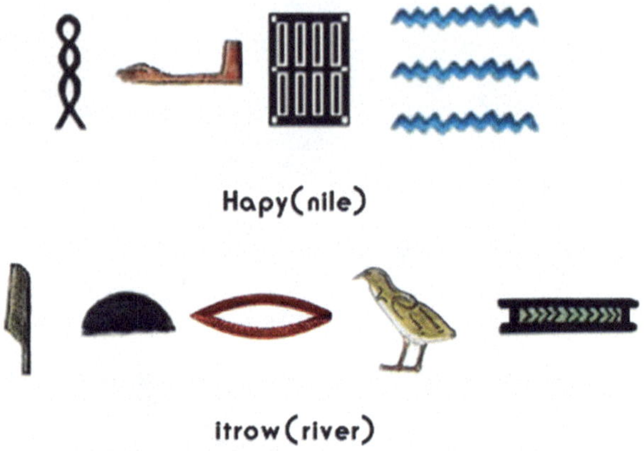

Fig. 10.2 Names of Nile in old Egyptian letters

derivation of Nile might be related to the term Nil (Sanskrit: नील, translit. nila; Egyptian Arabic: نيلة), which refers to *Indigofera tinctoria*, one of the original sources of indigo dye. In Coptic, the words piaro (Sahidic) or phiaro (Bohairic) meaning "the river" (lit. p(h).iar-o "the.canal-great") come from the same ancient name. In the *Odyssey*, the epic poem written by the Greek poet *Homer* (seventh century BC), Aigyptos is the name of the Nile (masculine) as well as that of the country of Egypt (feminine) through which it flows. The Nile in Egypt and Sudan is now called Al-Nīl, Al-Baḥr, and Baḥr Al-Nīl or Nahr Al-Nīl (Smith et al. 1999).

10.5 Nile River in the Life of the Ancient Egyptians

Further on the Nile played an important role in the religious life of the ancient Egyptians: the annual flood god (*Hapi* or *Hapy*) was called (translated in English) "Lord of the Fish and Birds of the Marshes" and "Lord of the River Bringing Vegetation" (Fig. 10.3). Since he appeared in the shape of a person with a large belly and two large breasts, he was a symbol of fertility. *Hapi* did not have temples, but the population celebrated him during the flood season by offering sacrifices and singing hymns, since its water gave new life for plants, animals, and also humans. The god *Osiris* was also closely related to the Nile and the fertility of the landscape. Ancient Egyptians believed that the Nile was the road to life after death on earth. There was a belief that a person would not survive in the "afterlife," if he/she had polluted the waters of the Nile during life on earth (Wilkinson 2003).

Thus the Nile was the source of life for the ancient Egyptians, since they settled and planted along its banks and in the delta different crops such as wheat, flax, papyrus, and others. The Nile was also an important source of fish and various aquatic birds. The Nile was also the most important pathway of transportation of goods for the trade with the different nations living south of Egypt.

Fig. 10.3 Dual depiction of Hapy, the ancient Egyptian god of the Nile and its flood

The ancient year calendar was linked to the three Nile cycles, each of which lasted four months. These seasons were called *Akhet*, a flood season; *Peret*, the growth season; and *Shemu*, the harvest season (Porceddu et al. 2008).

The ancient Egyptians were the first (known) people to practice agriculture on a wide professional scale (Fig. 10.4). There are several clear indications that the beginning of agriculture in Egypt started around 8000 BC, when—depending on the Nile flood—people managed to develop the basic irrigation amazingly, enabling them to grow many different agricultural crops all over the year. This was only possible when the ancient Egyptians had developed a complex of intersecting networks of channels to hold back floodwater to irrigate large amounts of fields (Postel 1999).

In addition, the ancient Egyptians managed it to cultivate orchards, fruit gardens, vineyards, vegetables, medicinal plants, aromatic plants, and various herbs all over the year. This was only possible by permanent irrigation due to the transfer of water throughout the year originating from water wells or directly from the Nile (Wetterstrom and Murray 2001).

Fig. 10.4 (**a**) Plowing with a yoke of horned cattle in ancient Egypt. Painting from the burial chamber of Sennedjem, c. 1200 BC. (**b**) Night hunting and fishing. (**c**) Scene of gardener using a Shaduf, Tomb of Ipuy at Deir el-Medina, west bank of Thebes, TT217. (**d**) Nakht (TT 52), deceased and wife offering, agriculture

Most of the ancient Egyptians worked in agriculture, digging trenches from the Nile to agricultural fields using draw wells and the "Shaduf" (Fig. 10.4c), a primitive machine that lifts water from the Nile to the canals (Postel 1999).

Preservation of the water of the canals was one of the holy activities in ancient Egyptian life. The confession of the dead person to have done it was recognized as declaration of innocence in the courtroom in the afterlife. To have done it is declared within the confession No. 35, in which the dead person admits that he has never obstructed the flow of water in the neighbor's canal. Sophisticated hydraulic means had been developed in the Middle Kingdom (2040–1782 BC) to transport efficiently and to discharge water across the fields (Budge 2016).

10.6 Use of Water in Egypt Throughout History

It is an iron fact that the relationship of the Egyptians with the Nile has not changed through their long history. The obligatory use of the Nile water, either for agriculture, as drinking water, or for other purposes, has remained constant and is still today unchanged. This is what shocked the scientists of the early French expedition campaign when they noted that the living conditions of the Egyptians did not change throughout their history. One of the earliest evidences of the beginning of the practice of agriculture, irrigation, and water management for agriculture was documented by the finding of the mace head of the king scorpion of the "predynastic kings." Ancient Egyptians began to use the Nile waters by constructing canal chains, which allowed the flow of floodwater to sites that were inaccessible. These canals had been designed perpendicular to each other and had been made able to form basins of different sizes (Mays 2010).

During the reign of *King Narmer* (~3000 BC), a large irrigation project was implemented. Later industrial dams with gates were created to control the floodwaters. Artificial irrigation increased the agricultural land area after the flood phase. From the Eighteenth Dynasty, water-lifting operations began using the *Shaduf system*, allowing land to become cultivated close to the river. This area of 0.3 feddans could become irrigated in half a day. The Shaduf system is still used today (Satoh and Aboulroos 2005).

In the year 2650 BC, the ancient Egyptians built the Sadd-el-Kafara dam (Dam of the Pagans) in Wadi Garawi, southeast of Helwan, in the Cairo Governorate. This was done for flood control and offered large-scale water storage. However, the dam failed its function, since it was never fully completed. Finally it collapsed during a large flood. The dam of the Kafara was discovered in 1885 by the German archaeologist *G. Schweinfurth*. During the *Ptolemaic* and *Roman periods*, the methods of irrigation dealing with the waters of the Nile did not change compared to the *Pharaonic era*. However, a new irrigation device was used for irrigation named as the *Tanbur* or the *Archimedes screw*. Saqiya (or waterwheels) were also developed and moved by bulls. This allowed that large tracts of land could be quickly irrigated thus leading to an expansion of agricultural output. The spread of saqiya (waterwheels) during the Ptolemaic and later during Roman times led to the production of summer crops as well as flood crops. This in turn has increased Egypt's wealth (Mays 2010).

The concept of sanitation in ancient Egypt was significantly different from the current situation, since the concept of pathogenic microorganisms was not known at this time. However, the ancient Egyptians built bathrooms inside houses. Some were large, others small depending on the status of the families in the society. In addition, the ancient Egyptian family was responsible for the avoidance of any disposal of garbage and waste in irrigation channels that could be a source of pollution. This was important because the canals and rivers had also been used for bathing purposes. The situation is still the same today in most Egyptian villages, where people stay in close contact with the river as shown in Fig. 10.5 (Filer 2001).

10.7 Parasitic Diseases Along the Nile River

Deep interest in the study of diseases spread among the ancient Egyptians started more than 200 years ago, when the upper-class mummies of ancient Egyptian society were detected and the shroud was removed from the mummies to start autopsy. In 1889, the tissue was first examined using light microscope, and also X-ray technique was first used in the year 1898. Beginning in the year 1993, the overall study of mummies began without focusing on specific mummies. New techniques, especially DNA tests, antibodies, etc., were used (Nunn 2002).

Our knowledge of the diseases of the ancient Egyptians is based on three main sources: human remains, ancient Egyptian art, and medical papyri. There are many indications that the ancient Egyptians suffered from the same diseases that occurred and are studied today. However, it must be taken into account that the average age of the individuals at that time ranged only between 35 and 40 years. It is clear that *epidemics* had been widespread during this period. On the other side, it is a fact that the ancient Egyptians possessed already a great ability to recognize and to diagnose diseases with respect to other countries of their time.

10.7.1 Water-Related Diseases

Infectious diseases associated with water had been and still are the main cause of public morbidity and mortality in many countries. It has become important to study the history of water-related pathogens in order to try to develop a better understanding of the survival of species and their relation to environmental conditions. Water-related diseases are one of the most perilous risks facing humans. They are responsible for the death of about 3.5 million people per year, killing mostly children. These diseases include many species of agents of disease, but they are all directly related to the lack of clean water being free of pathogenic parasites. Water-related infections occur either through exposure to or by drinking of contaminated water as well as due to environmental factors such as heavy rainfall, floods, drought, and temperature change, which may contribute to the emergence and spreading of water-

Fig. 10.5 (**a**, **b**, **c**) Shaduf during history. (**d**, **e**, **g**) Saqiya during history. (**f**) Tanbur. (**h** and **i**) Egyptian farmers

related pathogens. There are four main types of water-related infectious diseases (White et al. 2002; Mehlhorn 2016a–c):

- *Waterborne*

  These diseases result from human and animal wastes that enter and thus contaminate the water with infectious microorganisms, which may lead especially to typhoid fever, cholera, bacillary dysentery, gastroenteritis, leptospirosis, poliomyelitis, aseptic meningitis, infectious hepatitis, amoebic meningoencephalitis, and/or ascariasis. Many waterborne diseases are diarrheal diseases (Ashbolt 2004; Leclerc et al. 2002; Mehlhorn 2016a–c).
- *Water-washed*

  Infections occur due to poor hygiene based on insufficient water availability. These infections occur when infected water touches an eye or skin resulting in serious diseases. Water-washed diseases include shigellosis, trachoma, leprosy, skin infections, ulcers, and conjunctivitis (Braks and de Roda Husman 2013).
- *Water-based*

  The agents of these diseases are parasites (mostly worms) and bacteria, which spend parts of their life cycle in water. The skin is penetrated when contaminated water is used for cleaning or bathing. Agents of disease may also enter the body within contaminated drinking water. Such parasitic diseases include, e.g., schistosomiasis, guinea worm disease, and cysts of *Entamoeba histolytica* (Steinmann et al. 2006).

10.7.2 Water-Related Insect Vectors

These diseases are transmitted by insects that are closely associated with water, since they multiply themselves especially in "standing" water such as lakes, pools, etc. These diseases include among others malaria, filariasis, dengue fever, yellow fever, and river blindness (Braks and de Roda Husman 2013; Mehlhorn 2016a–c).

10.8 Parasitic Diseases of Different Origins

Medical papyrus, temples, and inscriptions of tombs as well as examinations of mummies illustrate the diversity of parasites that have afflicted ancient Egyptians. Many parasites have been identified, such as those of schistosomiasis (bilharziasis), dracunculiasis (guinea worm disease), filariasis, ascariasis, strongyloidiasis, tapeworm diseases, malaria, leprosy, trachoma, and cholera. They have all—as it is known only since about 100 years—different origins, which had been unknown in Old Egypt, so that control and treatment had been (mostly) unknown.

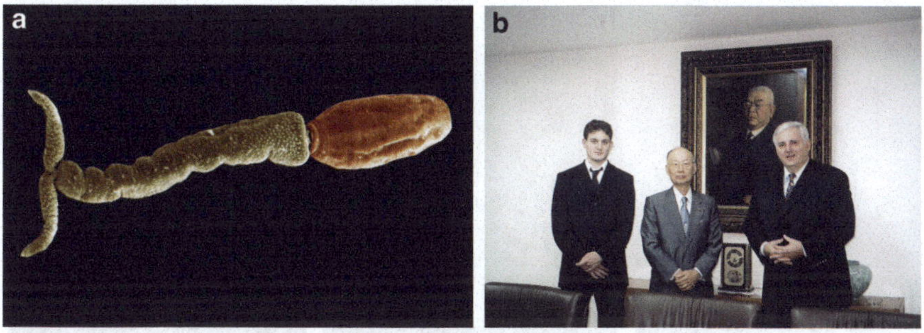

Fig. 10.6 (**a**) Scanning electron micrograph of a typical cercaria of *Schistosoma* spp. with its propelling tail. (**b**) The Nobel Prize winner Prof. Dr. Omura (director of the Kitasato University), Prof. Dr. Heinz Mehlhorn, and Dr. Tim Mehlhorn before the painting of the late Prof. Dr. Kitasato in Tokyo

10.9 Bacterial Diseases

Leprosies due to the bacterium *Mycobacterium leprae*, trachomes due to *Chlamydia trachomatis*, and cholera due to *Vibrio cholerae* have attacked Egyptians through all these times. There had been no means of control—only measurement to soothe the symptoms. Also plague due to infections with *Yersinia* (*Pasteurella*) *pestis* was common along the river, since fleas and their rats harbored these bacteria and transmitted them during their blood meals to humans. The scientists Yersin and Kitasato discovered this transmission at the same time in Hong Kong (Fig. 10.6a). This discovery was honored that one of the famous Japanese universities bears his name. In Egypt the Nile rat was/is the main host. The bacteria came apparently from China (Yunnan province) and hit also Egypt. There are described three major outbreaks: (1) the Justinian plague (sixth and seventh centuries), followed by (2) the epidemic called "Black Death" in the fourteenth and eighteenth centuries, and (3) the third plague pandemics (1855 until 1959) (Lotfy 2015; Panagiotakopulu 2004).

10.10 Typical Parasites

However, several parasitic diseases had been and are still today a major health problem in Egypt requiring considerable numbers of death cases from antiquity to the present. However, it is still today difficult to identify species of parasites just reading old medical papyrus sources, since the direct examination of human tissues remains the best source to identify the different types of parasites. The English scientist *Marc Armand Ruffer* (1959–1917) discovered in 1910 the calcified eggs of *Schistosoma haematobium* in the kidneys of a mummy of the Twentieth Dynasty in the year 1910 only by studying tissue sections.

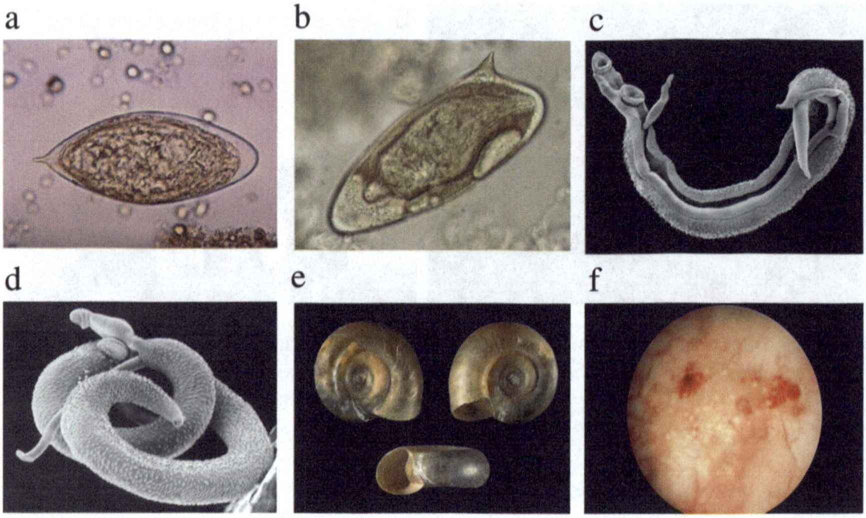

Fig. 10.7 (**a**) *Schistosoma haematobium* egg. (**b**) *Schistosoma mansoni* egg. (**c**, **d**) *Schistosoma mansoni* and *Schistosoma haemotobium*, adult worms. (**e**) *Biomphalaria* freshwater snails that act as an intermediate host for schistosome larvae (**f**) Surface of bladder showing penetrating worm eggs and excreting blood

Many parasitic worms left and leave today clear traces in the feces. Thus it is certain that the ancient Egyptians knew them and gave them special names, but unfortunately the preserved medical papyrus did not provide sufficient information to identify finally the species of these worms. Thus there are transmitted only two names for worms in the medical papyrus: *Pened worms* and *Hefat worms* (Daglio 2005; Nunn 2002).

Schistosomiasis

Schistosomiasis (Figs. 10.6b, 10.7, and 10.8) is one of the most common parasitic infections worldwide, especially in the tropics. There are about 200 million people infected in 76 countries with these trematode agents of disease, and about 800 million people live at risk of an infection even today. The manifestations of the schistosomiasis (bilharziasis) are the incidence of itching and rash, as well as symptoms similar to flu and intestinal disorders (Gönnert 1967; Gönnert and Andrews 1977; Campbell 2008; Roberts et al. 2012; Mehlhorn 2016a–c).

Schistosomiasis was one of the biggest chronic health problems in former Egypt, but is now constantly increasing to the activities of the recent Egyptian government. Worldwide occur three main types of schistosomiasis, two of which are found in Egypt (intestinal schistosomiasis due to *Schistosoma mansoni* and urogenital schistosomiasis due to *Schistosoma haematobium*). In the time of the ancient Egyptians, the identification of worms was done by people with severe myopia due to their ability to distinguish very small objects. The infection of humans by schistosomal worms starts by skin penetration of the swimming *cercariae* (Fig. 10.6a), whereby it

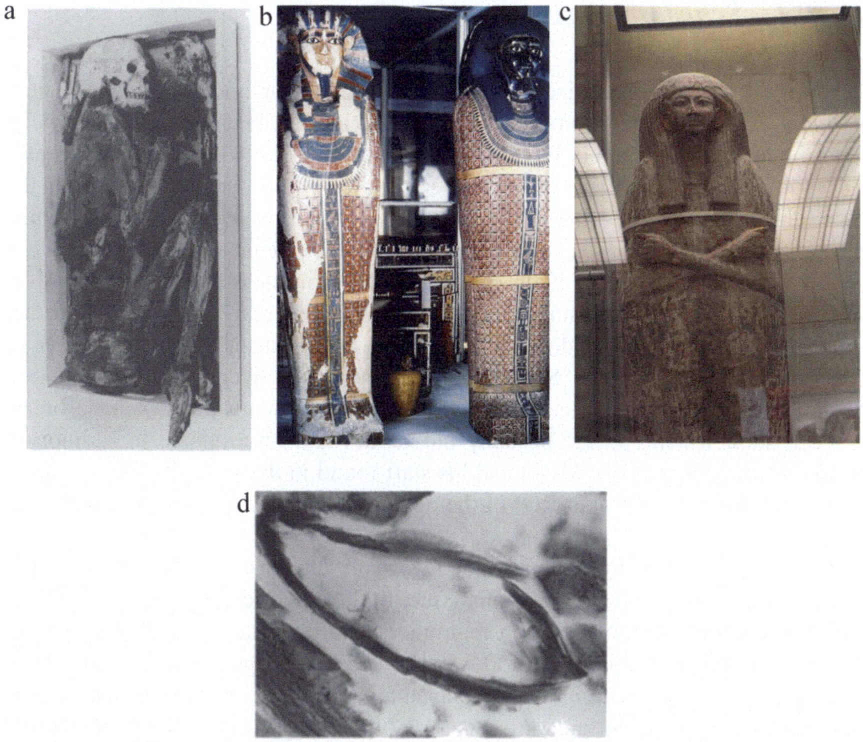

Fig. 10.8 (**a**) Photo of a mummy of late predynastic adolescent (Egypt, approximately 3200 BC) within in which schistosome circulating anodic antigens had been detected. (**b**) Anthropoid coffins originating from the Nekht-Ankh (right) and Khnum-Nakht (left) from the Twelfth Dynasty (1985–1773 BC) as shown in the Manchester Museum. (**c**) Coffin of Nakht, Western Thebes 1190 BC. (**d**) Ovum of *Schistosoma haematobium* with terminal spine (photo taken from a mummy)

releases its propelling tail and is now called *schistosomulum*. After staying some time in the skin and in the lung, these worm larvae wander into the blood vessels: those of *S. mansoni* into those of the liver and those of *S. haematobium* into those of the bladder. There they grow up and become males and females (Fig. 10.7c, d) and become united to persistent couples. The fertilized females excrete typical species-specific structured eggs (Fig. 10.7a, b), which—depending on the species—wander through the walls either of the bladder (*S. haematobium*) or of the intestine (*S. mansoni*), whereby considerable amounts of blood are set free (Fig. 10.7f), which can be diagnosed either in the feces or in the urine of human and animal hosts. If these eggs reach water, they release a so-called *miracidium* larva, which moves inside water by the help of their surface cilia and enters the intermediate host snails (Fig. 10.7e). Therein the *sporocyst* stage is developed, which produces numbers of infectious cercariae, which again may enter human skin.

10.10.1 Schistosomiasis in Ancient Egypt

As shown in the different objects in Fig. 10.8, there is clear evidence that the ancient Egyptian people suffered from schistosomiasis as is described in the *Ebers Papyrus* reporting the phenomenon of bloody urine (Contis and David 1996). There was also a description of apparently infected children who were considered as adults when the blood appeared in their urine, since the old Egyptian physicians thought that this bleeding was a phenomenon similar to that of the first menstruation of girls.

An important symptom of urogenital schistosomiasis is *hematuria* (blood in urine) which causes increasing anemia, wasting of energy, and loss of appetite. Infected persons show in addition to the leading symptoms a reduced resistance to bacterial and viral infections. Schistosomiasis also causes liver cirrhosis resulting in dysfunction of the liver functions and gynecomastia (enlargement of male breast gland tissue), which appears in many old statues and inscriptions. In addition to Ruffer's discovery (1910), Millet and his staff found in the year 1980 *Schistosoma* eggs in the mummy of Nakht-ROM I (belonging to the New Kingdom) (Millet et al. 1980).

In the tissues of many mummies, *Schistosoma* antigens were found, too, especially in the skin as well as in the brain tissues of the bodies dating to the predynastic era and the New Kingdom. Fifteen mummies from the time of the Ballana period (cemetery in Lower Nubia) (350–550 AD) showed also antigens (Miller et al. 1992).

Although modern techniques have played an important a role in the detection of schistosomiasis in ancient mummies, the use of radiology is also important, since a calcified bladder was found using a radiograph in a mummy (Isherwood et al. 1979). Also ELISA tests are very useful (Miller et al. 1992) in the detection of schistosomiasis in the tissues of these mummies. However, immunocytochemistry remains the most effective method in detecting schistosomiasis in ancient samples (Rutherford 2000).

The results of the DNA analysis using PCR confirmed the presence of *S. haematobium* in the liver tissues of the Nekht-Ankh mummy and the presence of *S. mansoni* in the tissues of the intestines of Khnum-Nakht belonging to the Middle Kingdom (~3900 BC) (Matheson et al. 2014). There is a reference to a single prescription in *Ebers Papyrus 265* telling us about the use of Balanites, an anthelminthic in the treatment of schistosomiasis (Campbell 2008). There were observed constant hematurias in feces and urine, respectively, by the French campaign scientists in Egypt (1799–1801), which soon was attributed to schistosomal infections (Sandison and Tapp 1998).

10.10.2 Schistosomiasis in Modern Egypt

1. In 1851 the German parasitologist *Theodor Bilharz* (1825–1862) (Fig. 10.9) discovered schistosomiasis in Egypt during an autopsy of a person infected

Fig. 10.9 Painting of the German medical scientist Theodor Bilharz (1825–1862), who discovered the life cycle of schistosomes during his research stay in Cairo, Egypt

with worms, which caused human urinary schistosomiasis in the Kasr El Ainy Hospital. This finding was later followed by the discovery that between 30% and 40% of Egypt's population suffered from schistosomiasis (Barakat 2013; Roberts et al. 2012; Mehlhorn 2016a–c).

In the year 1937 *Scott* performed the first mass study of bilharzia infections in Egypt. After a comprehensive survey of the data obtained from two million samples from the Egyptian Endemic Diseases Department and examining a sampling of probes of 40,000 people between 1932 and 1934, he confirmed that schistosomiasis was commonly spread in 100 villages. Later it was found that 60% of the population of the northern and eastern delta of the Nile had been infected with *S. haematobium* and *S. mansoni*. In the southern and central delta, 60% of the population contained *S. haematobium* and only 6% had *S. mansoni* (see insert 1). In Upper Egypt *S. mansoni* was not found, whereas *S. haematobium* was common (Scott 1937).

Beginning in the year 1926, the Egyptian government began to take actions to stop spreading of schistosomiasis by establishing control measurements based on a national plan. A two-way action plan was adopted: at first occurred chemotherapeutical treatment of the infected people with a tartar emetic compound followed by control of the snail reproduction by a Bayer product (Barakat 2013).

Between 1935 and 2010 the human infection rate was very high especially among school-age children and a large portion of workers in the agricultural sector. *S. haematobium* was the most prevalent species in Lower Egypt, while *S. mansoni* was most prevalent in the rest of Egypt. Since 1920 the Egyptian government had undertaken great efforts in the fight against schistosomiasis. In the 1930s, the Ministry of Health established a special section to fight endemic diseases. In the 1940s, the country took a number of measures to control snails and examined vulnerable populations. In 1955, the government recognized that schistosomiasis was the major health problem facing Egypt and thus began to develop a comprehensive disease control program. Throughout the years 1960 and 1970, governments have successively strived to combat schistosomiasis using foreign aid for health and economic development (Reich et al. 1998). At the end of the twentieth century, Egypt adopted a strict control program of elimination of schistosomiasis, which was based on two targets: (1) *treatment of humans* with praziquantel (produced at this time by Bayer, Leverkusen, Germany) after development by Prof. Dr. Gönnert (Mehlhorn et al. 1981, 1983) and (2) *elimination of snails* by chemicals applied into water, wherein the vector snails lived. Due to these activities the total occurrence of schistosomiasis in Egypt declined year by year to reach 10% in 1999, 5% in 2000, 3.5% in 2002, and 1.2% in 2006, but increased again to 3–10% in 2010 (Fig. 10.10) (Barakat 2013).

At present (~2019), schistosomiasis has a low prevalence throughout Egypt where infection rates have fallen to a record low of less than 0.2% in all previously infested areas. Recent studies on 30 million people showed less than 50,000 cases in recent times (WHO 2016a).

The success of treatment in Egypt was reached, since in the year 2016 the Ministry of Health and Population decided to establish and adopt a national plan to eliminate schistosomiasis by 2020. The now ongoing strategy is based on four main axes (Abou-El-Naga 2018):

- Large-scale treatment using praziquantel supported through the World Health Organization, whereby the Egyptian government allocated 10 million US dollars to implement this plan (WHO 2017)
- Intensification of the control of the snail
- Health education and behavioral changes of the population
- Expanding complementary public health interventions

10.11 Dracunculiasis (Guinea Worm Disease)

Dracunculiasis (Fig. 10.11) is one of the oldest human diseases, especially in the Middle East and in Africa. This disease is caused by the nematode worm *Dracunculus medinensis* as was already mentioned in many texts of the ancient Greeks, Romans, and Arabs, but first described by Linné (1758). The disease remained widely unknown to western medicine until the Russian scientist *Aleksej Fedchenko* (1844–1873) offered at the end of the nineteenth century (1879) the first detailed description of the worm and its life cycle (Roberts et al. 2012).

The Guinea worm has a life cycle changing between two hosts, humans and "water fleas," e.g., of the genus *Cyclops* (belonging to freshwater copepods). The

Fig. 10.10 (**a**) Nile Delta: view from space. (**b**) Living *Biomphalaria alexandrina* snails collected from Kafr El-Zayat (El Sharazly et al. 2017). (**c**) Exposure of children to canal water in the studied area in Kafr El-Zayat (El Sharazly et al. 2017). (**d**) Discoidal snail species isolated from the studied area. (1) *Biomphalaria alexandrina*. (2) *Planorbis planorbis*. (3) *Helisoma duryi* (El Sharazly et al. 2017). (**e**) Other freshwater snails isolated from the studied area. (4) *Physa acuta*. (5) *Bulinus truncatus*. (6) *Lymnaea natalensis* (El Sharazly et al. 2017)

infection starts when humans drink water containing infected crustaceans. Inside the human intestines the small larvae (500–700 μm) are set free, penetrate the intestinal wall, and wander via lymph vessels to subcutaneous tissues, where they reach maturity (females reach lengths of 70 cm, while males remain small (1–4 mm) and die after copulation) (Paniker et al. 2018; Mehlhorn 2016a–c). Finally the fertilized females break through the skin, e.g., of the leg, and drop down their eggs into the water, where they are ingested by the "water fleas," within which the infectious larva

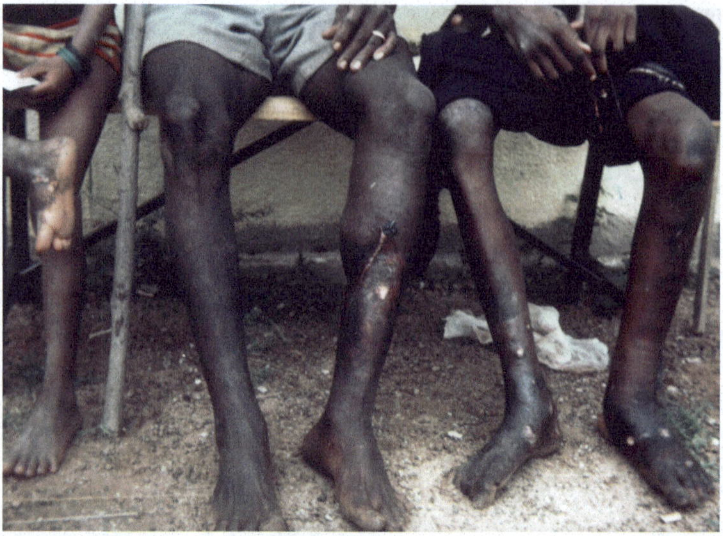

Fig. 10.11 Macrophoto of African patients showing protruded female *Dracunculus* worms in their legs

is developed and stored. The new cycle starts when humans drink contaminated water containing such infected crustaceans (Roberts et al. 2012; Mehlhorn 2016a, b).

Guinea worm infections had been a chronic health problem in ancient Egypt for long until modern times, when treatment stopped propagation. The disease was known since many centuries, and the worm was found in a calcified state in the abdominal wall already of a mummy (mummy 1770) in Manchester (Fig. 10.12), dating back to about 1000 BC. This mummy has an amputation above the knee for the right leg and an amputation below the knee for the left leg, indicating that a failed attempt to extract female worms had led to serious infections and complications (Tapp 1979).

In former times these worms had been extracted from the ankle and wrapped around a small piece of wood and were attracted very slowly and carefully until they are fully extracted from the body. The medical *Ebers Papyrus* contains the diagnosis and description of the process of extracting the Guinea worm from the foot of a person infected with this disease (Nunn 2002).

During the last two centuries, no further cases of dracunculiasis have been reported in Egypt, where the disease has now finally disappeared from the country (Karam and Tayeh 2006). This disease exists actually only in very few countries, and the number of patients has dramatically decreased, since clean drinking water is offered in most small river-bent villages in endemic countries.

Fig. 10.12 (**a**) Coffin of a mummy (1770) in Manchester that contains a *Dracunculus* worm. (**b**) False legs made of wood wrapped in mud and bandages

10.12 Filariasis

10.12.1 Filariasis in Old Egypt

Lymphatic filariasis due to the nematode *Wuchereria bancrofti* (Fig. 10.13) belongs to the oldest and most debilitating human diseases. Mosquitoes are the carriers of the parasites, which are transmitted as larvae from mosquitoes to humans. After completing growth the worms cause lymphatic blockage resulting in swellings and thickening of the skin leading to the so-called elephantiasis, which occurs mainly as large swelling mainly of legs and breasts. About 856 million people in 52 countries around the world remain vulnerable (WHO 2018a).

The agents of this lymphatic filariasis are transmitted by various types of mosquitoes. For example, females of the genera *Anopheles*, *Aedes*, *Culex*, *Ochlerotatus*, and *Mansonia* may act as hosts and vectors (Roberts et al. 2012; Mehlhorn 2016a–c). Tumors due to worms are difficult to detect in mummies. But there are inscriptions on the walls of the graves showing people with inflated genital organs, but swellings of the legs characteristic of the so-called elephant disease had not been recorded on inscriptions (WHO 2018a).

However, filarial worms were found in the mummy of Leeds "Natsef Amun" inside the scrotum skin (Tapp and Wildsmith 1992). Furthermore there are related inscriptions on the walls of the funeral temple of Queen Hatshepsut showing the prince and princess of the country of Punt. In the inscription it is clarified that there are infections with filarial worms (Otsuji 2017) (Fig. 10.14).

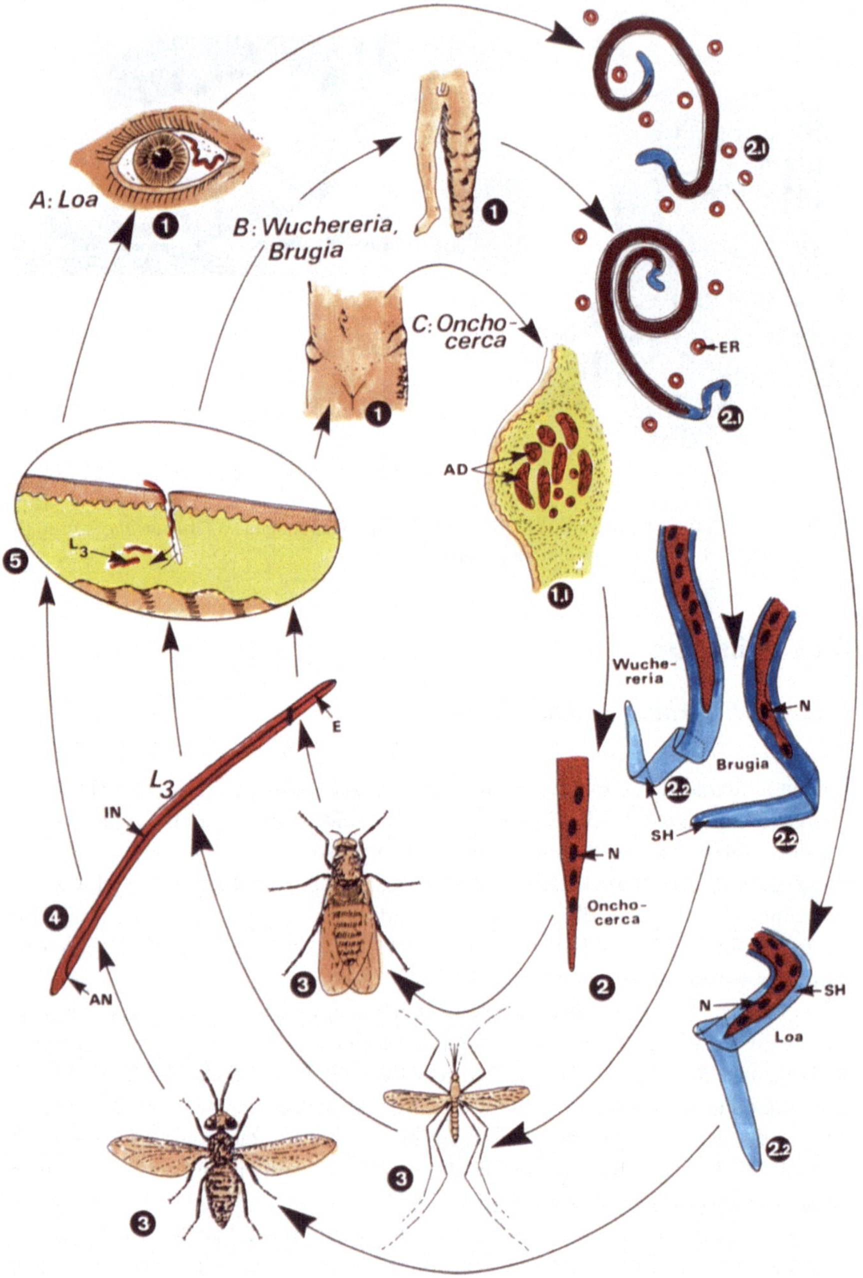

Fig. 10.13 Diagrammatic representation of the life cycles of human filariae. **A** *Loa loa* adult worms (macrofilariae male 3.5 cm, female 7 cm) wander subcutaneously and may pass the anterior chamber of the eye (**1**). **B** *Wuchereria bancrofti* adults (male 4 cm, female 10 cm) and *Brugia malayi* adults (male 3 cm, female 9 cm) live in lymph vessels and lead to a late stage disease called elephantiasis tropica (**1**). **C** *Onchocerca volvulus* adults (male 2–4 cm, female 70 cm) are knotted

10.12.2 Filariasis in Modern Egypt

Egypt is currently completely free of lymphatic filariasis, since the disease has been removed from all parts of the country due to governmental control measurements over 20 years.

The disease had been widespread in rural areas, especially in the eastern delta of the Nile. Control programs had been started in the 1960s relying on the vector control (of *Culex pipiens*) and the treatment of people by use of diethylcarbamazine (DEC). Since 2017 the WHO recommends the use of IDA1 (a combination of ivermectin, DEC, and albendazole) to eliminate filariasis (WHO 2018b).

10.12.3 Strongyloidiasis

Strongyloidiasis (Fig. 10.15) is a chronic parasitic infection of humans caused due to infection by larvae of *Strongyloides stercoralis*. There are approximately 30–100 million people infected worldwide. Infectious stages are transmitted by direct handling of water contaminated with parasites either during agriculture or domestic activities. Children are highly endangered. Infectious larvae penetrate the human skin when touching soil or contaminated water during barefoot walking. After the penetration, the larvae wander via the stomach to the small intestine where they mature and females lay eggs from which the larvae hatch being set free within the

Fig. 10.13 (continued) together in groups in the subcutaneous tissues. Because of host reactions these groups are encapsulated, leading to palpable nodules (**1**). In sections of these nodules coiled adults are seen (**1.1**). Microfilariae may induce blindness. (**1**) Visible signs of disease. (**2**) Microfilariae; the shape (**2.1**), structure (**2, 2.2**), and diurnal occurrence are species specific: they may or may not be sheathed (**2.2**); their terminal nuclei have a species typical appearance (**2, 2.2**); they can be found in blood vessels (*Loa*, *Brugia*, *Wuchereria*) or in lymphatic gaps (*Onchocerca*); their occurrence in the peripheral blood can be periodical (*Loa*, during the day; *Wuchereria* during the night; some subperiodic strains also exist) or may not be (*Onchocerca*, always present, but in lymph vessels). (**3**) Intermediate hosts: depending on the periodic appearance of microfilaria in the host's skin, insects with different biological behavior are involved as vectors. Daytime-feeding vectors (deerflies, *Chrysops* species, blackflies, *Simulium* spp.) transmit *Loa loa* or *Onchocerca volvulus*, whereas night-feeding mosquitoes (*Aedes*, *Culex*, *Anopheles*) may be vectors of the nocturnal strain of *Wuchereria* and *Brugia*. When microfilariae are ingested by intermediate hosts during the blood meal, they penetrate the intestine and enter the abdominal cavity and the thoracic muscles. After a molt the L2 is formed, which has a stumpy shape (sausage stage). Another molt finally leads to the filariform infectious L3. (**4–5**) L3 reach a length of about 1.5 mm and migrate to the proboscis, from which they escape when the vector is feeding. They enter the skin through the wound channel made by the biting insect (**5**, arrow). Inside the final host (man), the larvae migrate until they reach their favorite site of location, where they mature (after another 2 molts) within 1 year (prepatent period). **AD** adult worms (in section), **AN** anus, **E** esophagus, **ER** erythrocyte, **IN** intestine, **L3** third larval stage, **N** nuclei (their arrangement at the poles of microfilariae is species specific), **SH** sheath

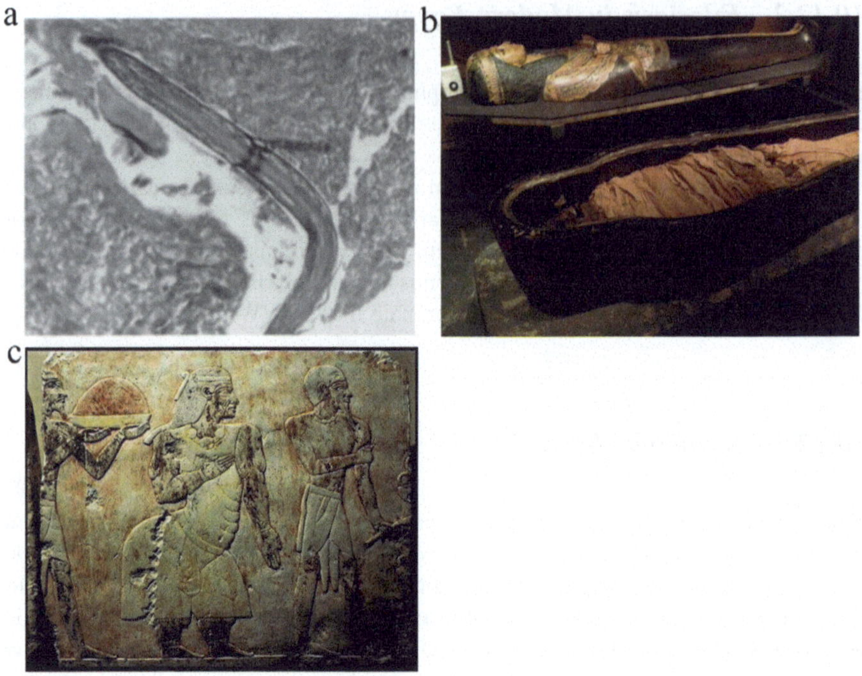

Fig. 10.14 (**a**) A filarial worm from the groin of the Leeds mummy. (**b**) The Leeds mummy. (**c**) A relief depicting the princess of Punt on Pharaoh Hatshepsut's temple at Deir el-Bahri

stool and thus may infect further persons. Sometimes the larvae penetrate the skin already in the anus to restore the same infection. Worms continue their life cycle without needing an intermediary host (WHO 2016b).

Larvae of this species had been found in the intestinal wall of the mummy of Asru (Twenty-Fifth Dynasty) (Fig. 10.16), where the intestine was detected to be located between the legs of the mummy and not in the canopic jar. It should be noted that there is no mention of these worms in medical papyrus (Daglio 2005; Nunn 2002).

10.12.4 Roundworms

Ascariasis (Fig. 10.17) becomes a serious disease of humans only in infections with high numbers of worms leading to belly swelling, shortness of breathing, fever, intestinal pain, and diarrhea. Infection occurs by eating food or drinking water contaminated with *Ascaris* eggs. Inside the intestine the larva hatches from the egg and penetrates the stomach wall and wanders via blood vessels to the lungs and into the alveoli and trachea, wherefrom it may be set free within cough or is swallowed again. Finally it grows up in the intestine and produces eggs which are finally excreted within the feces (Roberts et al. 2012).

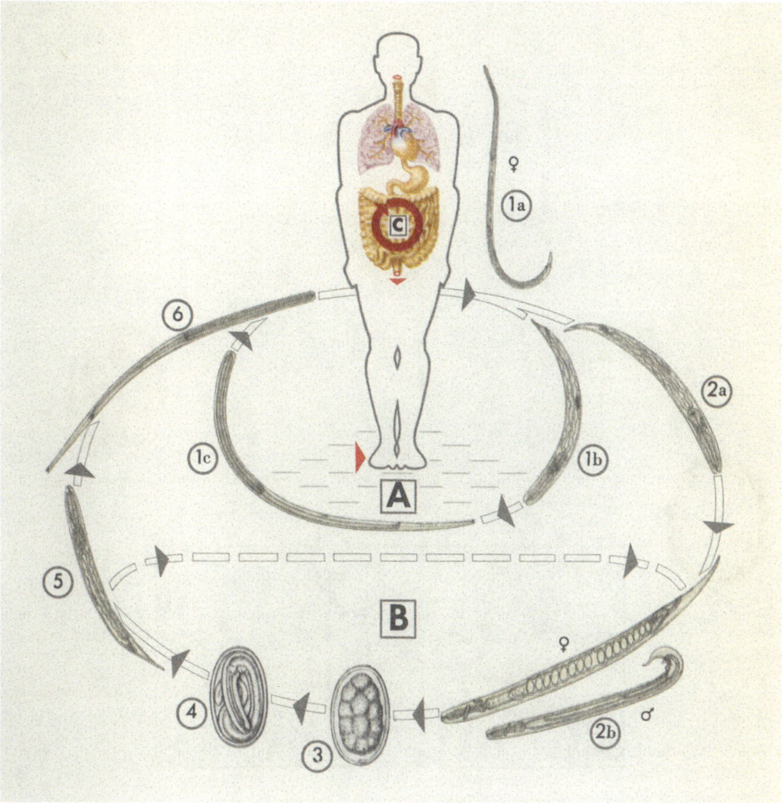

Fig. 10.15 Life cycle of *Strongyloides stercoralis*. **A** Infection. **B** Development outside the body. **C** Development inside the intestine. **1a** Parthenogenetic female inside the intestine. **1b** Larva 1 (rhabditiform) in fresh feces. **1c** Filariform, infectious larva. 2a Larva 1 (rhabditiform) in fresh feces. **2b** Female and male. **3**, **4** Eggs of the sexual generation developing the larva. **5**, **6** Rhabditiform (**5**) and infectious (**6**) larva (From Mehlhorn 2016)

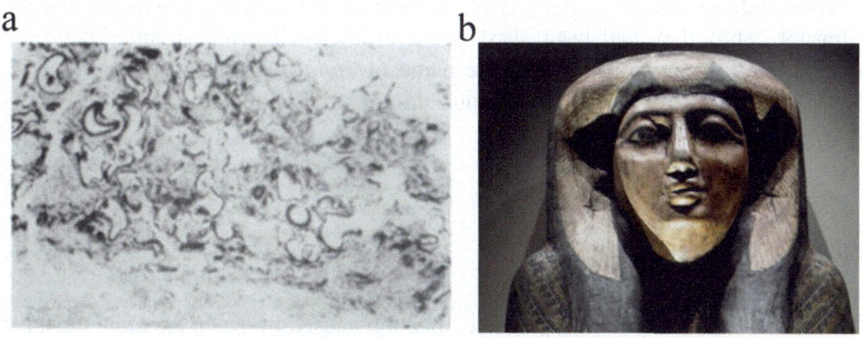

Fig. 10.16 (**a**) Larval forms of the worm *Strongyloides* in the intestines of the mummy Asru. (**b**) The mummy Asru

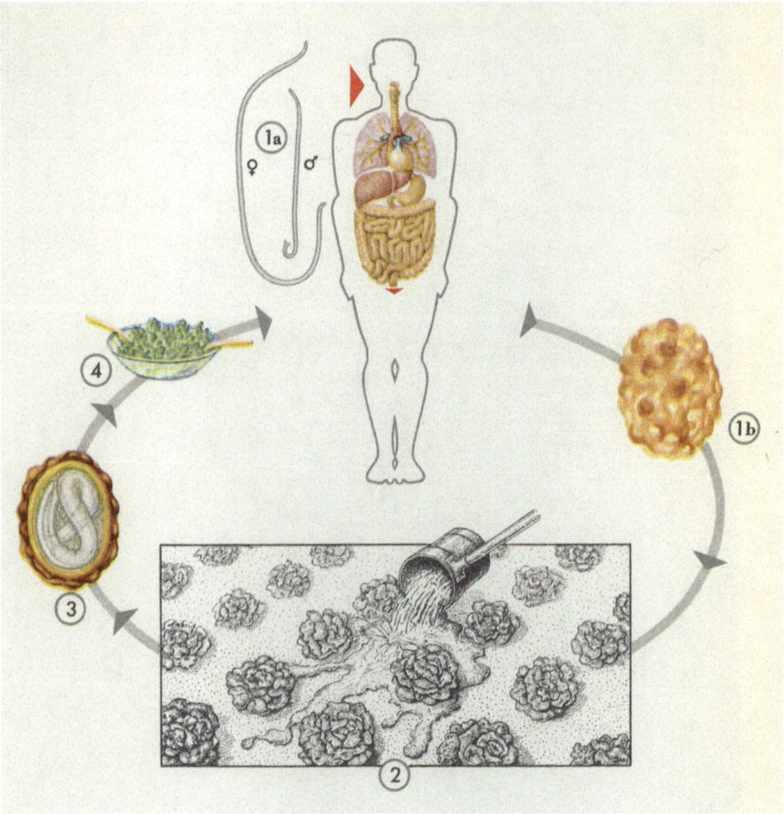

Fig. 10.17 Diagrammatic representation of the life cycle of *Ascaris lumbricoides* (from Piekarski 1962). **1a** Adult worms. **1b** Fertilized egg. **2** Eggs become distributed on salad (etc.) by fecally contaminated water. **3** Egg containing an infectious larva. **4** Infection due to egg-contaminated salad, etc.

Nematodes of the genus *Ascaris* are considered as the worldwide most common parasites. Thus they had been also known to ancient Egyptians, since there are descriptions of similar worms by the name of *Hefat* in a medical papyrus. The eggs of roundworms were also found in the non-wrapped mummy PUM II (Fig. 10.18) (Cockburn et al. 1975).

10.12.5 Tapeworms

Cysticercosis (Fig. 10.19) is a parasitic infection caused by *Taenia solium*, the most dangerous tapeworm, which has a life cycle involving humans and pigs. Human infection occurs by ingesting infected raw meat of pigs and/or by uptake of eggs inside human feces. The larvae of these worms infect the brain, muscles, or other

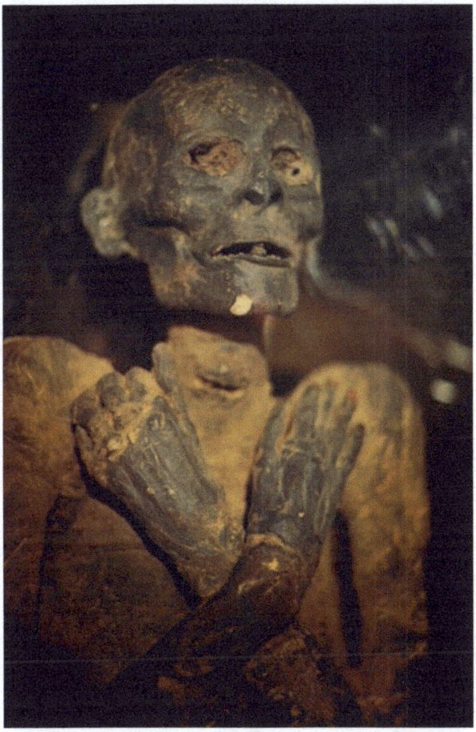

Fig. 10.18 Mummy of Pum II

tissues. Eggs of these worms have been found in the mummy of Nakht showing that such infections had been apparently common in old Egyptian times (Daglio 2005; Nunn 2002).

10.12.6 Malaria

Malaria is a severe infectious disease transmitted by mosquitoes (Figs. 10.21 and 10.22) that carry the sporozoites of the *Plasmodium* parasite. These stages are contained in the mosquito's saliva and are thereby transmitted to human blood, where they enter the red blood cells (Fig. 10.20). Malaria causes symptoms that usually include high fever, fatigue, vomiting, and headache and leads to death in serious cases due to the persisting destruction of the erythrocytes. Humans may become infected by five species of the genus *Plasmodium*; *Plasmodium falciparum* is the most dangerous one. In 2016 an estimated 220 million cases of malaria occurred worldwide and about 445,000 people died, mostly children in the African region (WHO 2019; Cowman et al. 2012) (Fig. 10.21).

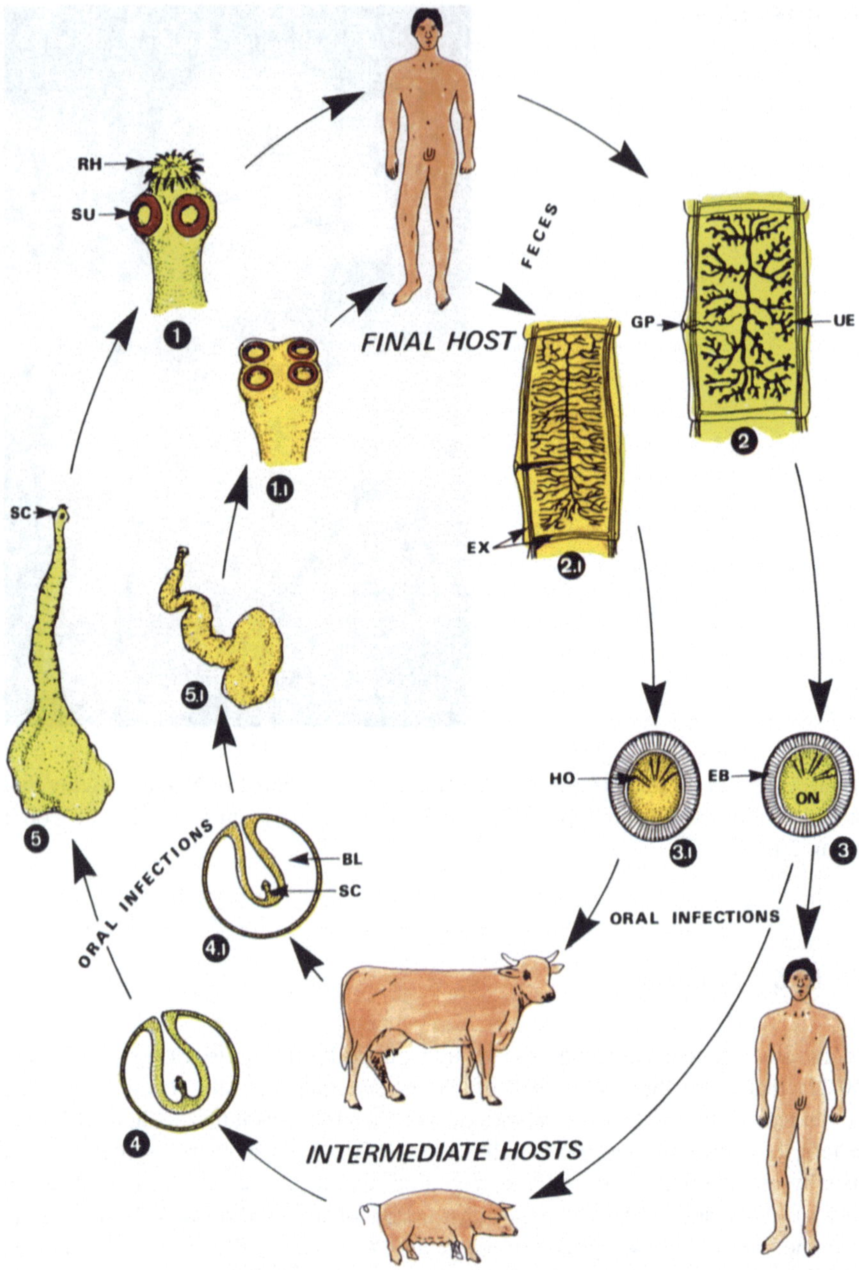

Fig. 10.19 Life cycles of *Taenia solium* (**1–5**) and *T. saginata* (**1.1–5.1**). **1**; **1.1** Scolex of the adult worms (inside human intestine). **2**; **2.1** Mature proglottids (~1 cm in length, whitish to be found in feces; the stretched uterus contains thousands of eggs). **3**; **3.1** After lysis of the proglottids in human feces, the typical eggs are set free. They contain the typical oncosphaera larva (ON). They are ingested by *cattle*, *pigs*, and humans. **4**; **4.1** Inside these hosts, the larva hatches and enters tissues

10.12.6.1 Malaria in Old Egypt

Malaria is known since ancient times and endangered and killed many civilizations. The Greek physician *Hippocrates* (~460–370 BC) studied medicine in Egypt, where he described the symptoms of malaria, which, however, were explained as a bile disease. On the other hand *Herodotus* (2500–2424 BC) stated that Egyptians, especially fishermen, slept under their nets to prevent mosquitoes (Roberts et al. 2012). Thus some relations between malaria fever and mosquitoes should have been known, since there was a link known between marshes and this mysterious fever. However, physicians claimed for long that this disease occurs as a result of inhalation of "bad air" = "mal aria" in Spanish language. This belief remained until the end of the nineteenth century (Hempelmann and Krafts 2013; Mehlhorn 2016a–c) (Fig. 10.22).

Climate and environmental conditions have been favorable for the spread of malaria in ancient Egypt. In some mummies of the prehistoric, modern kingdoms (the Twenty-Fifth Dynasty) and the Nubian Ballana era, malaria infections were detected by immunological tests in some of these mummies. An enlarged spleen (splenomegaly)—a common symptom of malaria—was noted by Ruffer in the year 1913 when examining an infected Coptic mummy.

In the year 1994 *Plasmodium falciparum* antigens were detected in the Nakht mummy (Miller et al. 1994). Samples of the predynastic mummies (3200 BC) were examined in the Marro's Collection at the Anthropological and Ethnographic Museum of Turin (Italy), and 40% of the stored bodies of mummies were found to have been infected with agents of malaria, when paleoimmunological investigation methods were used to detect *Plasmodium* antigens (Massa et al. 2000).

DNA analysis showed that the Pharaoh *Tutankhamun* and his ancestors *Yuya* and *Thuya* had suffered from malaria, too. The analysis showed the appearance of *Plasmodium falciparum* parasites in the samples examined. Malaria and avascular bone necrosis are also discussed to cause the death of Tutankhamun at an early age (Hawass et al. 2010).

10.12.6.2 Malaria in Modern Egypt

In the 1960s malaria was endemic in Egypt with infection rates reaching 20% in some areas (Kenawy 2015). In 1989, Egypt was included in the list of the malaria-endemic countries with national malaria control programs. Between 1999 and 2001, the government managed to keep malaria under strict control. Between 2010 and

Fig. 10.19 (continued) and develops the cysticercus stage. In humans they are also found in the brain. **5; 5.1** When humans ingest such cysticercus larvae in raw meat, the cysticercus grows up to very long adult, bisexual worms (the upper part has male sexual organs, the central one has the female sexual organs, and the terminal portion comprises so-called proglottids filled with infectious eggs (From Mehlhorn 2016a, b, c)

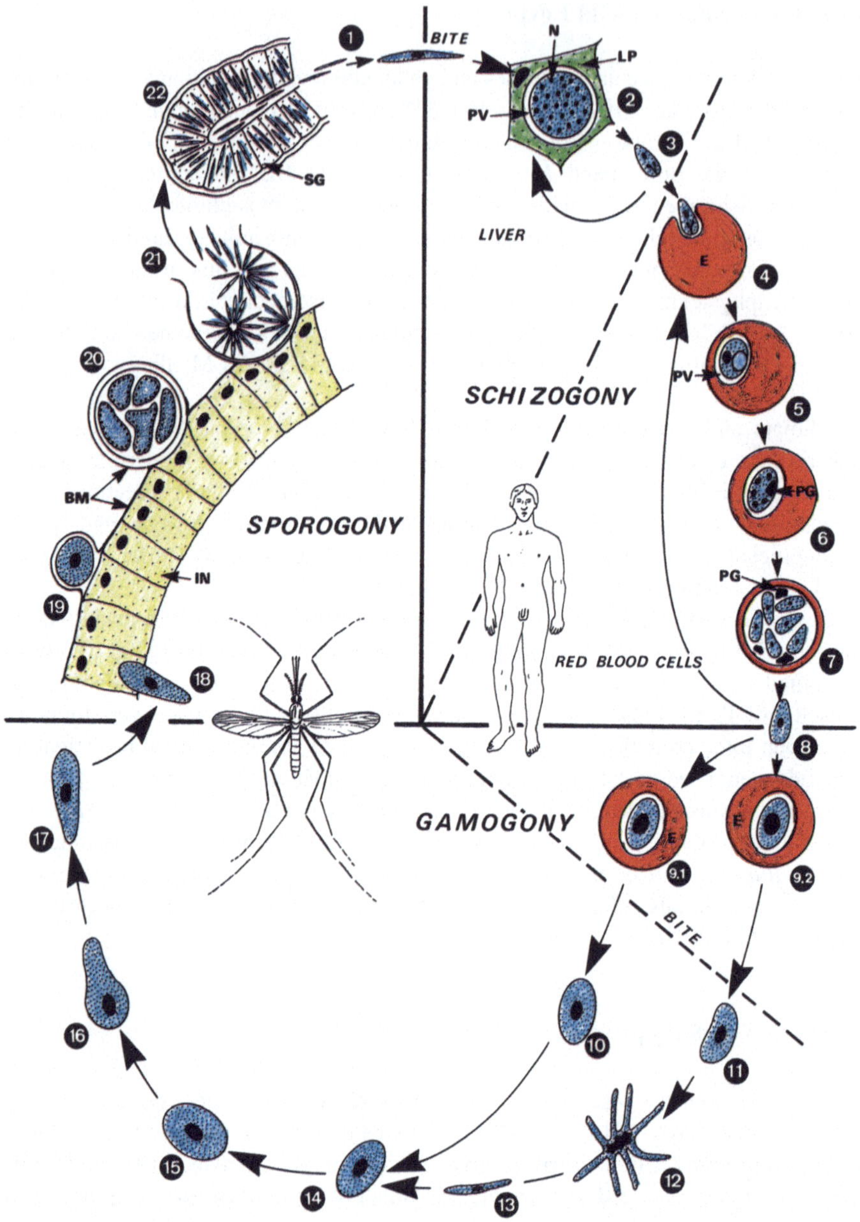

Fig. 10.20 Life cycle of human malaria parasites (*Plasmodium* spp.) without reference to species specific variations. **1** Elongate sporozoites are injected during bite of the female mosquito (*Anopheles* spp.). The sporozoites are distributed by bloodstream and enter liver cells lately within 2 min after infection. **2, 3** Formation of schizonts and merozoites in liver parenchymal cells (exoerythrocytic phase). In some species this cycle may be preserved intracellularly via hypnozoites (dormozoites) for a long time (years) and cause relapses. **4–8** Erythrocytic cycle; liver merozoites enter (after typical prepatent periods) erythrocytes, grow to "signet ring stages," **5** and finally form,

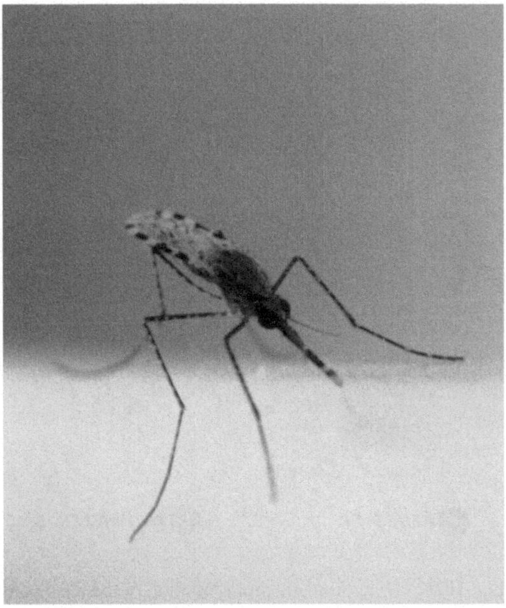

Fig. 10.21 Macroscopical picture of a female *Anopheles stephensi* sucking a sugar solution in the laboratory

2013, Egypt was able to eradicate local malaria almost entirely (Kandeel et al. 2016). However, in other countries malaria may become "imported" by travelers—especially coming from African and Asian countries. Moreover, the environmental changes that have emerged in Southern Egypt, such as the Toshka and El Salam Canal, have led to the emergence of new species of mosquitoes such as *Anopheles multicolor*, *A. sergentii*, and *A. algeriensis*. They are considered as important carriers of *Plasmodium* parasites (El-Bahnasawy et al. 2011).

Fig. 10.20 (continued) as schizonts **6**, several merozoites **7, 8**. During the digestion of hemoglobin, the parasites produce pigment granules (**6, 7**; PG) of hemozoin. The development of such schizonts becomes synchronous and is repeated **4–8** in a 1–3 day cycle (depending on the species). **9** After an indeterminate number of such asexual generations, some merozoites enter erythrocytes and become macro- **9.1** or microgamonts 9.2. The size and shape are species specific (banana shaped in *P. falciparum*). **10–11** When mosquito bites, they ingest erythrocytes containing such gamonts, which are released inside the gut from their enclosing erythrocytes. **12, 13** The microgamonts develop four to eight microgametes within 10–15 min. **14** Fertilization of the macrogamete. **15–19** The resultant zygote quickly elongates and becomes a motile ookinete **17** which penetrates the peritrophic membrane in the mosquito's gut, migrates through the cytoplasm of a gut cell, and begins its transformation into an oocyst (situated between the basal membrane and epithelial cells **19, 20–22**). Formation of multinucleate sporoblasts **20** which give rise to thousands of sporozoites (after 10–14 days). The latter become liberated into the hemocoel (body cavity) and migrate to salivary glands. These slender sporozoites (10–15 × 1 μm), which form a protecting surface coat, are finally injected into a new host at the next feeding act. *BM* basal membrane of intestine, *E* erythrocyte, *IN* intestinal cell, *LP* liver parenchymal cell, *N* nucleus, *PG* pigment, *PV* parasitophorous vacuole, *SG* salivary gland

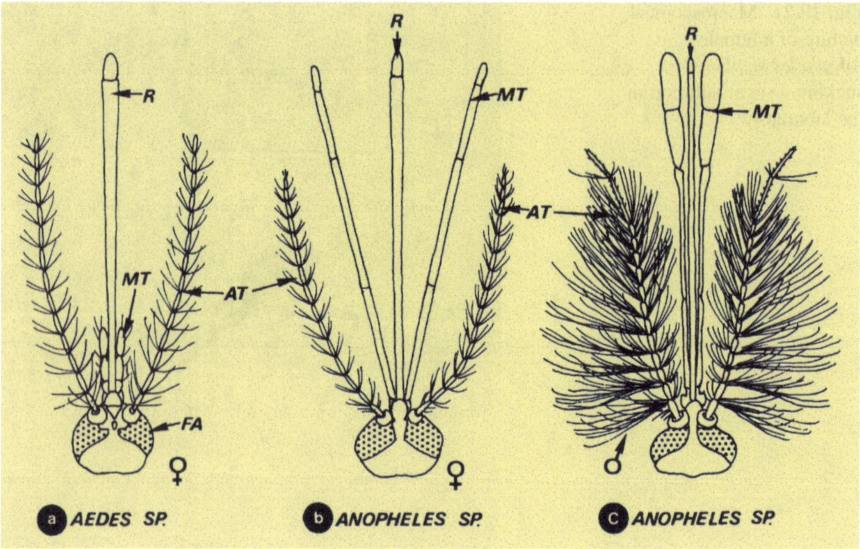

Fig. 10.22 Diagrammatic representation of the heads of the females of three bloodsucking insects. *AT* antennae, *FA* compound eyes, *MT* maxillar palps, R labium (sucking channel)

In 2007, 16 cases of malaria were reported in Egypt. Nine of them were cases of pilgrims, but seven cases had been based on local infections. All cases have been successfully cured using chloroquine (Zaher et al. 2007). Also in the year 2010, 36 cases of malaria occurred, of which 20 persons had been infected during their participation in international peacekeeping activities in Africa. The rest originated from different places in Egypt being apparently imported by travelers (El-Bahnasawy et al. 2010).

Twenty-two malaria cases were recorded in the southern Governorate of Aswan in the year 2014 deriving from *P. falciparum* and *P. vivax* (Kandeel et al. 2016; Kenawy 2015). However, actually Egypt has not experienced a local transmission of malaria in the last 3 years. Thus Egypt is seeking the WHO certification of successful malaria elimination (WHO 2018c).

10.13 Trachoma

The trachoma disease is a chronic granular inflammation of the conjunctiva, caused by the bacterium *Chlamydia trachomatis*. It is a serious eye disease. It is clear that the disease was endemic in ancient Egypt, since papyrus papers refer to a disease called *Nehat* which it is believed to be trachoma. It is possible that the images of blind people engraved into the walls of temples and tombs are potential victims of this malignant disease (Fig. 10.23), which leads to a darkening of the cornea and potential full blindness (Daglio 2005).

Fig. 10.23 Blind musician playing a harp, from the tomb of the ancient Egyptian scribe called Nakht (1422–1411 BCE)

10.14 Leprosy

This disease is a chronic infection being also known as Hansen's disease and caused by one type of the organic bacterium *Mycobacterium leprae*. It usually affects the skin, peripheral nerves, and mucosal surfaces of the respiratory tract as well as the eyes. If leprosy is not treated, it can lead to permanent and increased damage of the skin, nerves, limbs, and eyes. Its impact on some peripheral nerves leads often to distinct patterns of disability.

There is no evidence of leprosy in ancient Egyptian times, but it is believed that it came to Egypt with *Alexander the Great*'s army when returning from India (327–326 BC). But there had been described cases of such infections in the sixth century AD in the area of the El Bigha in Nubia (Sandison and Tapp 1998). Four suspicious cases are also noted in the tombs of the *Ptolemais* in the village of Balat in the oasis of Dakhla (Dzierzykray-Rogalski 1980).

There are also signs of this disease in the medical papyrus called *Ebers Papyrus* describing details of the diagnosis of cases of leprosy. However, the interpretation of such texts is difficult. On the other hand, there are also clear proofs when using modern methods, since Rafi was able to identify leprosy by PCR in the bones of bodies originating from 600 AD.

Arabic doctors in Egypt had an interest in leprosy, which they called *lion's disease*. During the Mameluke era in the tenth century, they began to separate leprosy patients from healthy people in order to avoid suggested transmission (El Dalgamouni 1938).

10.15 Cholera

Cholera (Fig. 10.24) is a waterborne disease caused by the bacterium *Vibrio cholerae*. During periods of the epidemic, the agents of disease are transmitted by contaminated food and water, since these bacteria have the ability to survive in different aquatic environments for long periods. Symptoms of the disease are severe diarrhea and dehydration (Koch 1884; Buchillet 2007).

Cholera has not been known in ancient and medieval times in Egypt. The disease has its home in the Bay of Bengal in South Asia (see Chap. 10.7.1 of this book). However, the disease struck Egypt from the nineteenth century in successive waves. The first epidemic in Egypt started in 1831 due to the arrival of the agents of the second pandemic of cholera. The disease was widely spread in Egypt and killed 150,000 people (6% of the population at this time). Then the disease struck Egypt again in 1834 causing the death of a similar number of humans. In 1848, 1850, and 1855, the cholera struck Egypt again with phases of the third worldwide pandemic in successive attacks, but no specific numbers of victims had been identified. During the fourth pandemic, Egypt has lost about 60,000 victims of cholera between 1881 and 1896. At this time Egypt suffered from a major epidemic. The German professor *Robert Koch* (1843–1910) and his crew of scientists came to Egypt in the year 1883 and detected the bacterium *Vibrio cholerae* as agent of the disease (published in 1884 after further studies in India in this year). Cholera hit the country and resulted in 36,300 deaths in Lower Egypt and 12,170 deaths in Upper Egypt followed by further 20,320 deaths and 17,770 deaths. With the sixth pandemic, Egypt suffered a cholera outbreak in the whole country in 1902, covering 2026 towns and villages, resulting in 40,613 people with 34,595 deaths. In 1947, Egypt was again affected by the epidemic, causing 20,804 severe cases and 10,277 deaths (Shousha 1947). There

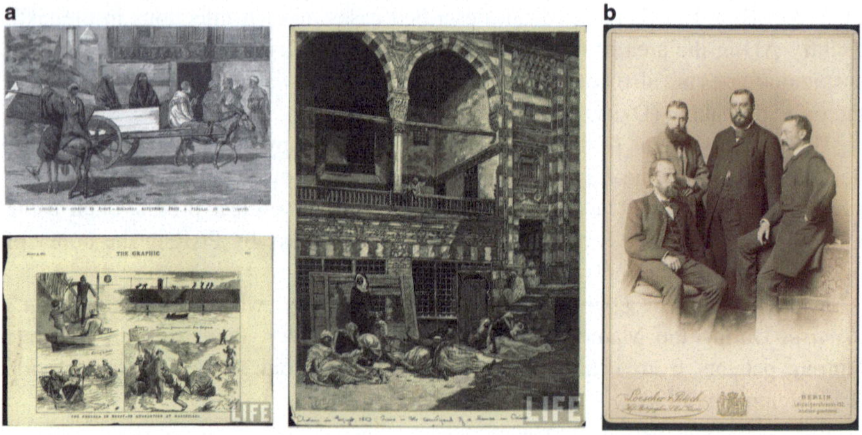

Fig. 10.24 (**a**) Illustrated newspaper show cholera in Egypt during the nineteenth century. (**b**) The German Robert Koch and his crew, which detected the agent of cholera in Egypt 1883

was no spreading of single cases of cholera outbreaks in Egypt after 1947 (Fig. 10.23).

10.16 Other Waterborne Protozoa

The prevalence of parasitic diseases is common in agricultural and rural areas, adversely affecting many areas such as the labor force and wasting economic resources of these regions. Children are also among the most vulnerable persons due to overcrowding of towns and due to unhealthy behavior patterns during these periods (Bakr et al. 2009; Youssef and Uga 2014).

Many of the studies that have been conducted in villages in different parts of Egypt have revealed a high incidence of parasitic diseases and various types of parasites. This can be attributed to unhealthy hygiene habits, especially by water pollution, as well as by direct contact with many animals acting as reservoirs for parasitic diseases. Gastrointestinal protozoan parasites lead to one of the most serious health problems that cause diarrhea in Egypt. There are many types of gastrointestinal protozoan parasites in Egypt that are closely associated with contaminated water (Nazeer et al. 2013; Mehlhorn 2016a–c).

The most important parasite species discovered in Egypt are:

- *Giardia lamblia* (*Giardia intestinalis*), the cause of giardiasis, is one of the major diarrheal diseases around the world. It is transmitted in Egypt due to water contamination. The incidence of an infection in adults ranges between 10% and 34.6%, while in children 11% and 15.4% are infected (Baiomy et al. 2010; Foronda et al. 2008).
- *Cryptosporidium* spp. is one of the main causes of diarrhea in Egypt. Its prevalence ranges depending on the region from 1% to 49% among humans and farm animals (Helmy et al. 2013; Mousa et al. 2010).
- *Heterophyes heterophyes*, this trematode worm affects fish both in brackish and freshwater. The fishermen and those who are directly in contact with water or eat uncooked fish are the most vulnerable group of persons due to this parasite. This parasite is especially concentrated in Egypt in the North Delta region of the Nile. It affects humans, mammals, and birds that feed on fish (Abou-Basha et al. 2000; Lobna et al. 2010).

Many of these species listed above were observed in several areas of Egypt such as El Minya Governorate, Fayoum Governorate (Sakran et al. 2018), Gharbiya Governorate (El-Kowrany et al. 2016), and North Delta (Mohammad and Mohammad 2011) and had apparently been present since the times of the Old Egypt (Fig. 10.24).

Fig. 10.25 *Book of the Dead* of Ani, Weighing of the Heart

10.17 Water Pollution

As already mentioned the Nile has been a blessing for the ancient Egyptians. In the texts describing the trial of the dead people in the afterlife, it is stated that the dead must have respected the Nile and have avoided any contamination of its water by claiming "I have not waded in the water," "I have not stopped the flow of water in the seasons," and "I have not built a dam against flowing water" (Karenga 2003) (Fig. 10.25).

There are many sources of pollution at which the Nile River is exposed due to agricultural and industrial activities discharging their wastes into its water. Although about 4.5 million tons of pollutants are thrown annually into the Nile waters (EOHR 2009), the Nile waters are considered to be reasonably good. Tests of water quality do not show levels of pollution that are high enough to cause alarming health risks. However, it is likely that its water near major cities and industrial centers may have unsafe effects on humans.

A report prepared by the Egyptian Organization for Human Rights (EOHR 2009) lists the pollutants that are thrown into the Nile annually (Figs. 10.26, 10.27, and 10.28):

- 50 tons of very harmful substances
- 25 thousand tons per year of harmful substances
- 35 thousand tons per year of imports of chemical industries
- 270 tons per day of industrial pollutants
- 14 million tons per year of solid waste
- 120 thousand tons per year of pollutants from hospitals

However, drainage canals are considered as most dangerous sources in terms of water pollution. The different pollutants vary considerably. Pathogenic

Fig. 10.26 Sources of Nile pollution (photographer Nasser Nouri)

microorganisms are ranked first, followed by organic compounds, pesticides, heavy metals, and fertilizers. Many villagers in Egypt suffer from many of the symptoms of disease due to direct contact with polluted water. In some areas irrigation water is available still today for domestic use (Mahmoud et al. 2016).

Some studies have shown that almost half of Egypt's drinking water is either polluted or unfit for human consumption. In addition, several human activities are considered to have negative effects on the Nile River, since many Egyptians wash therein clothes and dishes, remove garbage, and throw dead animals into the Nile (Fig. 10.25) (EOHR 2009; Mohamed et al. 2013).

Fig. 10.27 (**a**) Women washing their clothes in the Nile. (**b**) Dead fishes due to local pollution

Fig. 10.28 Garbage along a very scenic riverbank

About 4500 villages out of 7000 villages continue to rely on the drilling of latrines due to the absence of sewage networks, and 76% of the village water is mixed with sewage. It is worth to mention that the pollution hits vegetables, fruits, and crops sold in local markets, since they had been irrigated with wastewater or water contaminated with chemical waste. Thus a large proportion of Nile fish is not suitable for human consumption (EOHR 2009) (Figs. 10.26, 10.27, and 10.28).

Egypt's share of water is about 55 billion cubic meters per year, but there is a loss of about 15 billion cubic meters due to pollution, which costs the state 6% of the total national income due to the need of water purification and treatment of diseases caused by polluted water (Abdel Zaher 2015). According to a number of studies, 100,000 people are suffering annually from kidney failure (EOHR 2009), and the same numbers occur due to liver and pancreatic cancer.

The absence of a legal deterrent is one of the most important reasons for the problem of water pollution, although pollution of the Nile costs a fine of about 200 Egyptian pounds, equivalent to 12 dollars (Abdel Zaher 2015), which is a considerable sum looking at the rather low income of large numbers of Egyptians.

10.18 Chemical Pollutants

Egypt is one of the largest industrial countries on the African continent. Many factories are placed along the Nile River, especially sugar factories, which are concentrated in the south of Upper Egypt, and oil and food factories, which are concentrated in the north of Upper Egypt. These industrial activities consume about 40% of the total Nile water using an average of 638 million cubic meters of water, of which 549 million m^3/year are discharged into the drainage system (El-Sheekh 2009).

Food industries are responsible for 50% of biochemical oxygen demand (BOD), while the chemical industry is responsible for more than 60% of heavy metal discharge. Several studies have shown an important contamination of the Nile water in several areas by heavy metals such as lead and cadmium. Other minerals such as chromium, zinc, copper, iron, and manganese were also detected in the Nile and groundwater as well as in sediments, aquatic plants, and suspended matter. Many heavy metals, especially cadmium, lead, and mercury, have been monitored, as well as trace elements in fish tissue in many areas. However, in most cases only permitted levels have been recorded, but discussion is going on to decrease these "allowed" levels (El-Sheekh 2009; Mohamed et al. 2013).

Agricultural wastewater is one of the most dangerous sources of pollution, including salts, food remnants, pesticide residues, pathogens, and toxic organic and inorganic pollutants. Pesticides are the most dangerous contaminants. Especially organochlorine pesticides have been monitored in many areas along the Nile and were also detected in fish tissues, suspended solids, and sediments (El-Sheekh 2009; Mohamed et al. 2013).

References

Abdel Zaher H (2015) The Nile, a vital source of water, turns into source of disease [WWW Document]. https://thearabweekly.com/nile-vital-source-water-turns-source-disease. Accessed 4 Oct 2019

Abou-Basha LM, Abdel-Fattah M, Orecchia P, Di Cave D, Zaki A (2000) Epidemiological study of heterophyiasis among humans in an area of Egypt. East Mediterr Health J 6:932–938

Abou-El-Naga IF (2018) Towards elimination of schistosomiasis after 5,000 years of endemicity in Egypt. Acta Trop 181:112–121. https://doi.org/10.1016/j.actatropica.2018.02.005

Ashbolt NJ (2004) Microbial contamination of drinking water and disease outcomes in developing regions. Toxicology 198:229–238. https://doi.org/10.1016/j.tox.2004.01.030

Baiomy AMS, Mohamed KAA-H, Ghannam MAM, Shahat SAA-R, Al-Saadawy ASK (2010) Opportunistic parasitic infections among immunocompromised Egyptian patients. J Egypt Soc Parasitol 40:797–808

Bakr IM, Arafa NA, Ahmed MA, Mostafa MEH, Mohamed MK (2009) Prevalence of intestinal parasitosis in a rural population in Egypt, and its relation to socio-demographic characteristics. J Egypt Soc Parasitol 39:371–381

Barakat RMR (2013) Epidemiology of schistosomiasis in Egypt: travel through time: review. J Adv Res 4:425–432. https://doi.org/10.1016/j.jare.2012.07.003

Blaha S, Blaha S (2004) Reconstructing Sub-Saharan, Mayan and other prehistoric civilizations in a mathematical macro – theory of civilizations. Comp Civilizations Rev 50:23–46

Braks MA, de Roda Husman A (2013) Dimensions of effects of climate change on water-transmitted infectious diseases. Air Water Borne Dis 2:1–8. https://doi.org/10.4172/2167-7719.1000109

Buchillet D (2007) Epidemic diseases in the past: history, philosophy and religious thought. In: Tibayrenc M (ed) Encyclopedia of infectious diseases. Wiley, Hoboken, NJ, pp 517–524. https://doi.org/10.1002/9780470114209.ch31

Budge EAW (2016) The ancient Egyptian book of the dead: prayers, incantations, and other texts from the book of the dead. Quarto Publishing Group, New York

Campbell JM (2008) Encyclopedia of infectious diseases. In: David R (ed) Egyptian mummies and modern science. Cambridge University Press, Cambridge, pp 216–233

Cockburn A, Barraco RA, Reyman TA, Peck WH (1975) Autopsy of an Egyptian mummy. Science 187:1155–1160

Contis G, David A (1996) The epidemiology of *Bilharzia* in ancient Egypt: 5,000 years of schistosomiasis. Parasitol Today. https://doi.org/10.1016/0169-4758(96)30007-0

Cowman AF, Berry D, Baum J (2012) The cellular and molecular basis for malaria parasite invasion of the human red blood cell. J Cell Biol 198:961–971. https://doi.org/10.1083/jcb.201206112

Daglio C (2005) La medicina dei faraoni Malattie, ricette e superstizioni dalla farmacopea egizia. Ananke

Dumont H (2009) A description of the Nile basin, and a synopsis of its history, ecology, biogeography, hydrology and natural resources. The Nile: origin, environments, limnology and human use. Monographiae Biologicae, vol 89. Springer, Dordrecht. https://doi.org/10.1007/978-1-4020-9726-3_1

Dzierzykray-Rogalski T (1980) Paleopathology of the Ptolemaic inhabitants of Dakhleh Oasis (Egypt). J Hum Evol 9:71–74. https://doi.org/10.1016/0047-2484(80)90044-5

El Dalgamouni M (1938) The antileprosy campaign in Egypt. Int J Lepr 6:1–10

El Sharazly B, Abou Rayia D, Antonios S, Eissa SH (2017) Current status of *Schistosoma mansoni* infection and its snail host in three rural areas in Gharbia governorate, Egypt. Tanta Med J 44:141. https://doi.org/10.4103/1110-1415.201724

El-Bahnasawy MM, Dabbous HK, Morsy TA (2010) Imported malaria as a threat to Egypt. J Egypt Soc Parasitol 40:773–788

El-Bahnasawy MM, Saleh NMK, Khalil MF, Morsy TA (2011) The impact of three anopheline mosquito species in Toshka, on the introduction of chloroquine resistant *Plasmodium falciparum* to Egypt. J Egypt Soc Parasitol 41:573–592

El-Kowrany SI, El-Zamarany EA, El-Nouby KA, El-Mehy DA, Abo Ali EA, Othman AA, Salah W, El-Ebiary AA (2016) Water pollution in the Middle Nile Delta, Egypt: an environmental study. J Adv Res 7:781–794. https://doi.org/10.1016/j.jare.2015.11.005
El-Sheekh M (2009) River Nile pollutants and their effect on life forms and water quality. In: Dumont HJ (ed) The Nile: origin, environments, limnology and human use. Springer, Dordrecht, pp 395–405. https://doi.org/10.1007/978-1-4020-9726-3_19
EOHR (2009) Water pollution: a time bomb threatening the life of Egyptians. Cairo. http://en.eohr.org/2009/12/17/the-egyptian-organization-for-human-rights-issued-a-qualitative-report-entitled/
Filer J (2001) Hygiene. Oxford Encycl Anc Egypt A-F (vol 2)
Foronda P, Bargues MD, Abreu-Acosta N, Periago MV, Valero MA, Valladares B, Mas-Coma S (2008) Identification of genotypes of *Giardia intestinalis* of human isolates in Egypt. Parasitol Res 103:1177–1181. https://doi.org/10.1007/s00436-008-1113-2
Gönnert R (1967) Experiences in the control of bilharziasis with molluscicides. Ann Soc Belges Med Trop Parasitol Mycol 47:195–203
Gönnert R, Andrews P (1977) Praziquantel, a new broad-spectrum antischistosomal agent. Parasitol Res 52:129–150
Hawass Z, Gad Y, Ismail S, Khairat R, Fathalla D, Hasan N, Ahmed A (2010) Ancestry and pathology in King Tutankhamun's family. Am Med Assoc 303:638–647
Helmy YA, Krucken J, Nockler K, von Samson-Himmelstjerna G, Zessin K-H (2013) Molecular epidemiology of *Cryptosporidium* in livestock animals and humans in the Ismailia province of Egypt. Vet Parasitol 193:15–24. https://doi.org/10.1016/j.vetpar.2012.12.015
Hempelmann E, Krafts K (2013) Bad air, amulets and mosquitoes: 2,000 years of changing perspectives on malaria. Malar J 12:232
Hillel D (2006) The natural history of the bible: an environmental exploration of the Hebrew Scriptures. Columbia University Press, New York
Isherwood I, Jarvis H, Fawcett RA (1979) Radiology of the Manchester mummies. In: David AR (ed) The Manchester Museum mummy project. Manchester University Press, Manchester, pp 25–64
Kandeel A, Haggag AA, El Fetouh MA, Naiel M, Refaey SA, Hassan AH, Ramzy RMR (2016) Control of malaria outbreak due to *Plasmodium vivax* in Aswan Governorate, Egypt. East Mediterr Health J 22:274–279
Karam M, Tayeh A (2006) Dracunculiasis eradication. Bull Soc Pathol Exot 99:377–385
Karenga M (2003) Maat, the moral ideal in ancient Egypt (African studies: history, politics, economics and culture). Routledge, New York
Kenawy M (2015) Review of *Anopheles* mosquitoes and malaria in ancient and modern Egypt. J Mosq Res 5:1–8. https://doi.org/10.5376/jmr.2015.05.0004
Koch R (1884) Über das Cholera-Bakterium (on the cholera bacterium). Dtsch Med Wschr 10:725–728
Leclerc H, Schwartzbrod L, Dei-Cas E (2002) Microbial agents associated with waterborne diseases. Crit Rev Microbiol 28:371–409. https://doi.org/10.1080/1040-840291046768
Linné C (1758) Systema Naturae, 10th edn. Lars Salvius, Stockholm
Lobna SMA, Metawea YF, Elsheikha HM (2010) Prevalence of heterophyiosis in *Tilapia* fish and humans in Northern Egypt. Parasitol Res 107:1029–1034. https://doi.org/10.1007/s00436-010-1976-x
Lotfy WM (2015) Plague in Egypt: disease biology, history and contemporary analysis: a minireview. J Adv Res 6:549–554
Mahmoud MA, Abdelsalam M, Mahdy OA, El Miniawy HMF, Ahmed ZAM, Osman AH, Mohamed HMH, Khattab AM, Zaki Ewiss MA (2016) Infectious bacterial pathogens, parasites and pathological correlations of sewage pollution as an important threat to farmed fishes in Egypt. Environ Pollut 219:939–948. https://doi.org/10.1016/j.envpol.2016.09.044
Massa E, Cerutti N, Savoia D (2000) Malaria in ancient Egypt: paleoimmunological investigation on predynastic mummified remains. Chungará (Arica). https://doi.org/10.4067/S0717-73562000000100003

Matheson C, David R, Spigelman M, Donoghue H (2014) Molecular confirmation of *Schistosoma* and family relationship in two ancient Egyptian mummies. Yearb Mummy Stud 2:39–47

Mays LW (2010) Water technology in ancient Egypt BT. In: Mays L (ed) ancient water technologies. Springer, Dordrecht, pp 53–65. https://doi.org/10.1007/978-90-481-8632-7_3

Mehlhorn H (ed) (2016a) Encyclopedia of parasitology, 3 vols, 4th edn. Springer, New York

Mehlhorn H (2016b) Human parasites, 8th edn. Springer, New York

Mehlhorn H (2016c) Animal parasites, 8th edn. Springer, New York

Mehlhorn H, Becker B, Andrews P, Thomas H, Frenkel JK (1981) In vivo and in vitro experiments on the effects of praziquantel on *Schistosoma mansoni*. A light and electron microscopic study. Drug Res 31:544–554

Mehlhorn H, Kojima S, Rim HJ, Ruenwongsa P, Andrews P, Thomas H, Bunnag B (1983) Ultrastructural investigations on the effects of praziquantel on human trematodes from Asia: *Clonorchis sinensis*, *Metagonimus yokogawai*, *Opisthorchis viverrini*, *Paragonimus westermani* and *Schistosoma japonicum*. Drug Res 33:91–98

Miller RL, Krijger FW, De Jonge N, Armelagos GJ, Deelder AM, Ikram S (1992) Palaeoepidemiology of *Schistosoma* infection in mummies. Br Med J 304:555–556. https://doi.org/10.1136/bmj.304.6826.555

Miller RL, Ikram S, Armelagos GJ, Walker R, Harer WB, Shiff CJ, Baggett D, Carrigan M, Maret SM (1994) Diagnosis of *Plasmodium falciparum* infections in mummies using the rapid manual ParaSight-F test. Trans R Soc Trop Med Hyg 88:31–32

Millet NB, Hart GD, Reyman TA, Zimmerman MR, Lewin PK (1980) ROM1: mummification for the common people. In: Cockburn A, Cockburn E, Reyman TA (eds) Mummies, disease and ancient cultures. Cambridge University Press, Cambridge, pp 91–105

Mohamed A, El Safty A, Siha M (2013) Current situation of water pollution and its effect on aquatic life in Egypt. Egypt J Occup Med 37:95–119

Mohammad KAE-A, Mohammad AAE-A (2011) A survey of *Giardia* and *Cryptosporidium* spp. in rural and urban community in North Delta, Egypt. Zagazig Med J 17:78–85

Mousa KM, Abdel-Tawab AH, Khalil HHM, El-Hussieny NA (2010) Diarrhea due to parasites particularly *Cryptosporidium parvum* in great Cairo, Egypt. J Egypt Soc Parasitol 40:439–450

Nazeer JT, El Sayed Khalifa K, von Thien H, El-Sibaei MM, Abdel-Hamid MY, Tawfik RAS, Tannich E (2013) Use of multiplex real-time PCR for detection of common diarrhea causing protozoan parasites in Egypt. Parasitol Res 112:595–601. https://doi.org/10.1007/s00436-012-3171-8

Nunn J (2002) Ancient Egyptian medicine. University of Oklahoma Press, Norman

Otsuji Y (2017) History, epidemiology and control of filariasis. Trop Med Health 39:3–13. https://doi.org/10.2149/tmh.39-1-suppl_2-3

Panagiotakopulu E (2004) Pharaonic Egypt and the origins of plague. J Biogeogr 31:269–275

Paniker CJ, Ghosh S, Chander J (2018) Paniker's textbook of medical parasitology, 8th edn. Jaypee Brothers Medical, London

Porceddu S, Jetsu L, Markkanen T, Toivari-Viitala J (2008) Evidence of periodicity in ancient Egyptian calendars of lucky and unlucky days. Camb Archaeol J 18:327–339. https://doi.org/10.1017/S095977430S000395

Postel S (1999) Pillar of sand: can the irrigation miracle last? 1st edn. W. W. Norton, New York

Reich MR, Govindaraj R, Dumbaugh K, Yang B, Brinkmann A, El-Saharty S (1998) International strategies for tropical disease treatments: experiences with praziquantel. EDM Res Ser No. 026 1–116

Roberts LS, Janovy JJ, Nadler S (2012) Foundations parasitology, 9th edn. McGraw-Hill Education, New York

Ruffer MA (1910) Note on the presence of "*Bilharzia haematobia*" in Egyptian mummies of the twentieth dynasty [1250-1000 B.C.]. Br Med J 1:16

Rutherford P (2000) The diagnosis of schistosomiasis in modern and ancient tissues by means of immunocytochemistry. Chungará (Arica) 32. https://doi.org/10.4067/S0717-73562000000100021

Sakran T, El-Shahawy G, Shalaby M, Sabry H, Matooq P, Elmallah A (2018) Detection rates of waterborne protozoa in water sources from Fayoum Governorate. Parasitol United J 10:30–38. https://doi.org/10.21608/puj.2017.4734

Sandison AT, Tapp E (1998) Disease in ancient Egypt. In: Cockburn A, Cockburn E, Reyman TA (eds) Mummies, disease and ancient cultures. Cambridge University Press, Cambridge, pp 38–58

Satoh M, Aboulroos S (eds) (2005) Irrigated agriculture in Egypt. Past, present and future. Springer, Cham. https://doi.org/10.1007/978-3-319-30216-4

Scott JA (1937) The incidence and distribution of the human schistosomes in Egypt. Am J Epidemiol 25:566–614. https://doi.org/10.1093/oxfordjournals.aje.a118323

Shousha AT (1947) Cholera epidemic in Egypt. A preliminary report. Bull World Health Organ 1 (2):353–381

Smith CG, El-Kammash MM, Hurst HE (1999) Nile River [WWW Document]. https://www.britannica.com/place/Nile-River. Accessed 4.10.19

Steinmann P, Keiser J, Bos R, Tanner M, Utzinger J (2006) Schistosomiasis and water resources development: systematic review, meta-analysis, and estimates of people at risk. Lancet Infect Dis 6:411–425. https://doi.org/10.1016/S1473-3099(06)70521-7

Tapp E (1979) The unwrapping of a mummy. In: David A (ed) The International ancient Egyptian mummy tissue bank at the Manchester Museum. Manchester Museum Press, Manchester, pp 83–93

Tapp E, Wildsmith K (1992) The autopsy and endoscopy of the Leeds mummy. In: David AR (ed) The mummy's tale. Michael O'Mara, London, pp 132–153

Wetterstrom W, Murray MA (2001) Agriculture. Oxford Encycl Anc Egypt A-F (vol 1)

White GF, Bradley DJ, White AU (2002) Drawers of water: domestic water use in East Africa. 1972. Bull World Health Organ 80:62–63

WHO (2016a) Egypt leverages domestic funding to eliminate schistosomiasis [WWW Document]. https://www.who.int/neglected_diseases/news/Egypt_leverages_domestic_funding_to_eliminate_schistosomiasis/en/

WHO (2016b) Strongyloidiasis [WWW Document]. https://www.who.int/intestinal_worms/epidemiology/strongyloidiasis/en/. Accessed 4.10.19

WHO (2017) Fighting NTDs in Egypt [WWW Document]. https://www.who.int/neglected_diseases/egypt/en/

WHO (2018a) Lymphatic filariasis [WWW Document]. https://www.who.int/news-room/fact-sheets/detail/lymphatic-filariasis. Accessed 4.10.19

WHO (2018b) Egypt: first country in Eastern Mediterranean region to eliminate lymphatic filariasis [WWW Document]. https://www.who.int/neglected_diseases/news/Egypt_first_EMRO_country_eliminate_LF/en/. Accessed 4.10.19

WHO (2018c) Eastern Mediterranean Region [WWW Document]. http://www.emro.who.int/ar/egy/egypt-events/a-review-meeting-about-malaria-elimination-certification-at-who-country-office-for-egypt.html

WHO (2019) Malaria [WWW Document]. https://www.who.int/news-room/fact-sheets/detail/malaria. Accessed 4.10.19

Wilkinson RH (2003) The complete gods and goddesses of ancient Egypt. Choice Reviews Online. Thames & Hudson. https://doi.org/10.5860/choice.41-0039

Wimberly MC, Midekisa AA (2014) Hydro-eépidemiology of the Nile Basin: understanding the complex linkages between water and infectious diseases BT. In: Melesse AM, Abtew W, Setegn SG (eds) Nile River Basin: Ecohydrological challenges, climate change and hydropolitics. Springer International, Cham, pp 219–233. https://doi.org/10.1007/978-3-319-02720-3_12

Youssef AI, Uga S (2014) Review of parasitic zoonoses in Egypt. Trop Med Health 42:3–14. https://doi.org/10.2149/tmh.2013-23

Zaher T, Ahmadi M, Ibrahim A, El-Bahnasawy M, Gouda H, Shahat SA-R (2007) Malaria in Egypt, Saudi Arabia and Yemen: a clinical pilot study. J Egypt Soc Parasitol 37:969–976

Chapter 11
Epidemic Status of Schistosomiasis in Guangdong Province

Zhongdao Wu

Abstract This chapter deals with Chinese efforts to eliminate schistosomiasis.

Keywords Schistosomiasis control · Mass treatment

Schistosomiasis (Figs. 11.1, 11.2, 11.3, and 11.4) used to be a major infectious disease in Guangdong Province that seriously endangered people's health and safety and affected the economic and social development (Huang et al. 2011; He 2011; Health and Family Planning Commission 2016). Because of the warm and humid climate, sufficient precipitation, frequent floods, and strong biological reproduction, schistosomiasis was prevalent in Guangdong Province (a special book on the successful control appears in the series *Parasitology Research Monographs* edited by Liu et al. 2019). Due to the mode of production and living, people there were more probable to contact water containing cercariae and acquire schistosome infections, leading to a decline in rural population, population quality, and labor force, the serious loss of farm animals, and an obvious decline of production, thus greatly jeopardizing the development of rural social economy. Before the founding of New China (before 1949), its popularity was very rampant. After the founding of New China, the epidemiological survey conducted in Guangdong Province (1985) found that schistosomiasis was mainly prevalent in the middle and lower branches of the Dongjiang River and Beijiang River, including 11 counties and 3 farms, e.g., Renhua, Qujiang, and Wengyuan County of Shaoguan City, Yingde and Qingyuan County of Qingyuan City (now called Qingxin District and Qingcheng District, including Qingyuan Overseas Chinese Farm), Huaxian and Zengcheng County (now called Huadu and Zengcheng District) of Guangzhou City, Sihui County of Zhaoqing City (including Dawang Overseas Chinese Farm), and Sanshui (including

Z. Wu (✉)
Department of Parasitology, Zhongshan School of Medicine, Sun Yat-sen University, Guangzhou, China
e-mail: wuzhd@mail.sysu.edu.cn

H. Mehlhorn, S. Klimpel (eds.), *Parasite and Disease Spread by Major Rivers on Earth*, Parasitology Research Monographs 12,
https://doi.org/10.1007/978-3-030-29061-0_11

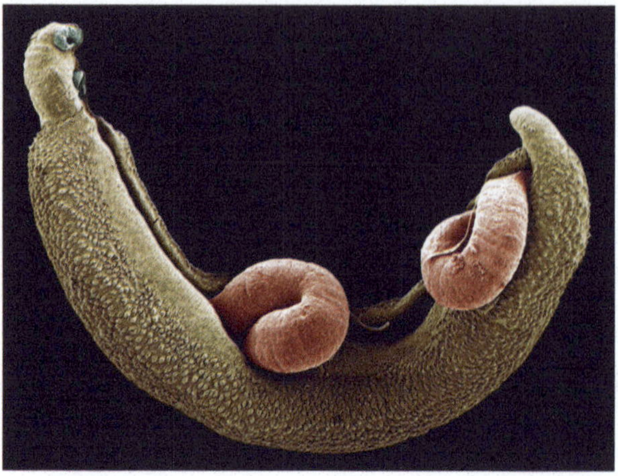

Fig. 11.1 Scanning electron micrograph of a *Schistosoma* couple (the male transports the female)

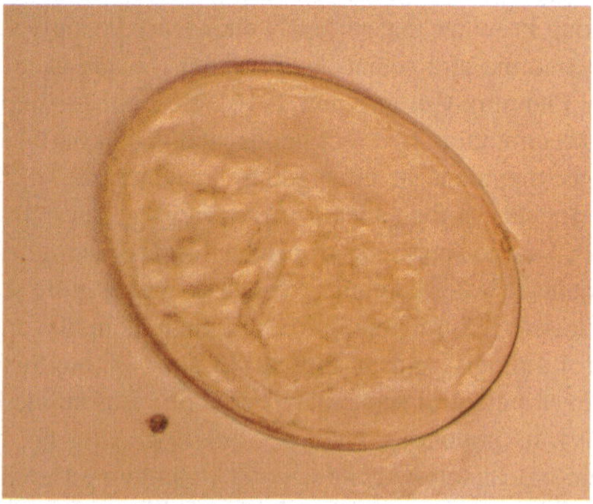

Fig. 11.2 Egg of *Schistosoma japonicum*, which is characterized by a short thorn at the eggshell

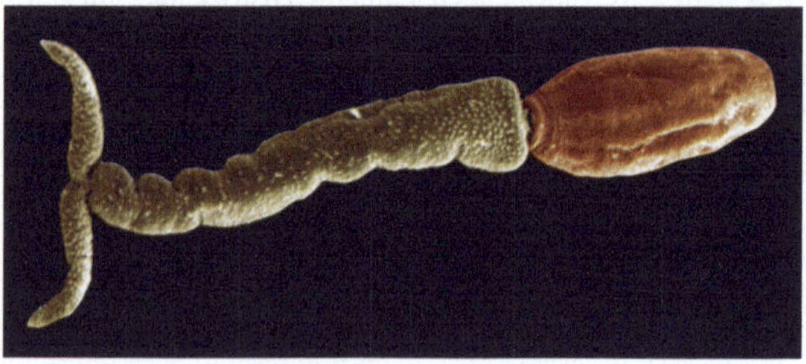

Fig. 11.3 Scanning electron micrograph of a cercaria, the anterior part of which enters the skin of humans and animals and thus starts infection

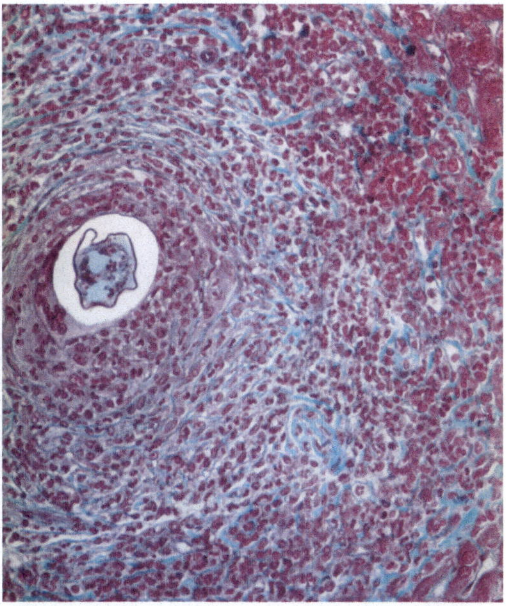

Fig. 11.4 A *Schistosoma* egg in the human liver leading to a granuloma. The gap around the egg is due to tissue drying

Yankou Overseas Chinese Farm), Nanhai, and Shunde County of Foshan City. The snail area was more than 100 million square meters. The average infection rate of people in epidemic areas was 29.04% (1950–1956), and the highest infection rate was 84.21% at administrative village level. There were 84,859 patients with schistosomiasis, including 4432 advanced cases with schistosomiasis (up to 1985).

Under the leadership of the Party Central Committee, the Provincial Party Committee, and the Provincial Government, relevant departments of health, agriculture, water conservancy, and other departments in Guangdong have worked together to formulate and adopt the prevention and control strategy "focusing on snail elimination and accompanied by repeated investigation and treatment of source infections,", implementing the Six Words ("Water (water conservancy), reclaim, planting, elimination (killing snails), treatment (patients and stocks) and management (feces and water)") comprehensive prevention and control technical measures. After more than 30 years of unremitting and painstaking prevention and control in Guangdong, it reached schistosomiasis elimination in 1985 (passed the assessment and reached the Chinese national standards for schistosomiasis elimination). On October 13, 1985, Guangdong was the first to announce the elimination of schistosomiasis in 12 endemic provinces (municipalities and autonomous regions).

Since 1986, the prevention and control of schistosomiasis in Guangdong Province has been transferred to the comprehensive consolidation and surveillance stage. From 1986 to 2015, through systematic monitoring, two outbreaks were discovered and disposed of in time, and effective emergency measures were taken to quickly extinguish the epidemic threats. Since 1993, the province has not found live snails and new local infections for 24 consecutive years. It indicates the schistosomiasis transmission disruption has been effectively consolidated, and great achievements

Fig. 11.5 Snail control by modifying the environment in the 1960s

have been made in the prevention and control of schistosomiasis in Guangdong Province.

In March 2016, the National Health and Family Planning Commission of China (now called the National Health Commission of China) organized experts from the relevant ministries of agriculture, forestry, and water conservancy to form a review team to reassess the schistosomiasis elimination status in areas reaching elimination before. In Guangdong Province, Qujiang District of Shaoguan City and Qingcheng District of Qingyuan City were selected for comprehensive review of schistosomiasis control and monitoring effects, knowledge and skills in prevention and control, epidemic disposal capacity, routine management practices and records, etc. The results suggest that Guangdong Province continues to maintain the state of schistosomiasis elimination, and the above 11 counties and 3 farms in Guangdong Province are no longer recognized as endemic areas for schistosomiasis.

The main problem the province is facing is how to prevent the resurgence of *Oncomelania* snails and the epidemic of imported schistosomiasis from endemic areas including domestic or abroad endemic areas with schistosomiasis (Figs. 11.5 and 11.6).

Fig. 11.6 The Pearl River branch

In the next step, the province will continue to strengthen the monitoring of the snail situation and imported schistosomiasis cases, enhance health education and raise the awareness of schistosomiasis control, strengthen departmental cooperation, and consolidate the province's maintenance of schistosomiasis elimination.

References

He YY (2011) Prevalence and control study of schistosomiasis in Guangdong Province (1949–1985). Guangdong: South China Agricultural University. Dissertation. https://doi.org/10.7666/d.Y2013834

Health and Family Planning Commission of Guangdong Province (2016) Guangdong Province passed the state review to maintain schistosomiasis elimination status. http://zwgk.gd.gov.cn/006940132/201609/t20160926_673829.html

Huang SY, Deng ZH, Zhang QM et al (2011) Endemic situation of schistosomiasis in Guangdong Province from 2004 to 2009. Chin J Schisto Control. 23(2):197–198

Liu Y, Mehlhorn H, Wu Z (eds) (2019) Schistosomiasis control in China – Jiangxi Province as a successful example. Parasitology research monographs, vol 11. Springer International, New York

Chapter 12
Konna: A Help for a Self-Help Project at the Niger River in Mali

Heinz Mehlhorn, Martin Stetzkowski, and Klaus-Christian Withöft

Abstract This chapter shows how fruitful European support can be, when medical and technical support is given as done by a group of German medical students and doctors in the fight against schistosomiasis in an African village.

Keywords Schistosomiasis · Malaria · Onchocercosis · Entamoebiasis · Hookworms

12.1 Introduction

Konna (=this native term means "help me") is a little village in the state of Mali (Table 12.1), which suffered during centuries and suffers even today from big ethic and political problems in addition to persisting ones due to various diseases and problems in healthcare (Table 12.2). This village belongs to the group of bigger ones in Mali comprising houses for about 11,000 inhabitants, who win their food by fishing in the Niger river, plant cultivation, and animal rearing along the river. The location of this village directly at the banks of the Niger river has on one side clear advantages, but also induces several dangers due to enhanced possibilities of transmission of agents of diseases (viruses, bacteria, fungi, parasites; see Table 12.2).

Tables 12.1 and 12.2 show problems in Mali, which today are also present in most of the African countries, since European powers neglected since the eighteenth century for many decades the health problems of the population while beating out the mineral wealth or oil reserves of these countries.

That the help for self-help may work (at least at local levels) shows the German project *Konna e.V.* in Mali induced by some medical doctors already in their times as

H. Mehlhorn (✉)
Department of Parasitology, Heinrich Heine University, Düsseldorf, Germany
e-mail: mehlhorn@uni-duesseldorf.de

M. Stetzkowski · K.-C. Withöft
Konna e.V., Essen, Germany

H. Mehlhorn, S. Klimpel (eds.), *Parasite and Disease Spread by Major Rivers on Earth*, Parasitology Research Monographs 12,
https://doi.org/10.1007/978-3-030-29061-0_12

Table 12.1 Mali: data of a country in West Africa

1. Geography and inhabitants
• *Size*: 1.240 192 km^2
• *Latitude*: 17° 34′ 14.491″ N
• *Longitude*: 3° 59′ 46.198″ W
• *Inhabitants*: 18.54 million people (2017)
• *Birth per woman*: mean ~7.0
• *Official language*: French
• *Local languages*: Bambara, Fulfulde, Malinke, Senoufo, Songhai, Touareg
• *Religions*: 90% Muslims, 9% natural religions, 1% Christians, and others
• *Landscape*: steppe, savannah, desert in northern region
• *Big river*: the upper part of the Niger river passes Mali and thus offers survival of infectious cercariae of schistosomal worms that enter the skin of humans and vertebrate hosts (e.g., cattle)
• *History*: this region was known in the eleventh century as Empire Mali until the fifteenth century. It was followed by the Songhai Empire, which became destructed by nomads; then it became a French colony from 1890 to 1911; during 1911–1958 this region was a part of French Africa. In 1958 it was founded as Republique Soudanaise and formed in 1959–1960 together with Senegal the Mali Federation. After the separation from Senegal in 1960, the state of Mali was founded
• *Big towns along the river*: *Bamako* (capital, ~2 million inhabitants); *Timbuktu* (55,000 inhabitants).
2. Living and health data of large populations in Mali
• Medical healthcare persons: 0.1 per 1000 inhabitants
• Governmental expenses for healthcare: 2% of the budget
• Beds in governmental hospitals: 0.2 per 1000 inhabitants
• Access to clean drinking water: 61–74% (village/town)
• Access to sanitary systems: 58–93% (village/town)
• Lack of sufficient food for children: 43% cases
• Children daily working: 50% cases
• Rate of analphabetic persons: ♀ 83%, ♂ 63%
• Life expectancy in Mali (mean): ♀ 41 years, ♂ 40 years; in Germany: ♀ 84 years, ♂ 79 years
3. Main human parasitic diseases
• Amoebiasis (due to *Entamoeba histolytica*)
• Giardiasis (=lambliasis) (*Giardia duodenalis, G. lamblia*)
• Malaria (due to *Plasmodium* species)
• Human schistosomiasis (locally up to 95%) (*Schistosoma haematobium, S. mansoni*)
• Ascariasis (nematodes, locally up to 95%) (*Ascaris lumbricoides*)
• Enterobiasis (nematodes, locally up to 95%) (*Enterobius vermicularis*)

All these infectious diseases are directly or indirectly based on the lack of clean drinking water and safe removal and storage of human and animal feces

students during the years 1988–1989 and the since years ongoing support of the German Konna e.V. (registered association) in Germany.

Table 12.2 Common important parasitic or other infectious diseases of humans and animals in Mali (examples) and their pathways of transmission

Origin	Agent of disease	Disease	Pathway of disease
Water-borne	Trematodes: *Schistosoma* species	Bilharziasis, schistosomiasis	Entering of cercariae into skin
Water-borne	Protozoans *Entamoeba histolytica*	Entamoebiasis	Ingestion of cysts in/on contaminated food or in drinking water
Water-borne	Viruses	Hepatitis A, E	Drinking water
Water-borne	Bacteria	Typhoid fever, leptospirosis, meningitis	Drinking water, contaminated food
Vector-borne	Protozoans *Plasmodium* species	Malaria	Bites of *Anopheles* mosquitoes
Vector-borne	Protozoans *Trypanosoma* species	Trypanosomiasis	Bites of tsetse flies
Vector-borne	Protozoans *Leishmania* species	Leishmaniasis	Bites of sandflies
Vector-borne	Nematode worms Filariae	Onchocercosis	Bites of mosquitoes
Vector-borne	Viruses	Dengue fever, Yellow fever	Bites of *Aedes* species
Vector-borne	Bacteria	Plague, *Pasteurella (Yersinia) pestis*	Fleas (bites, feces)
Vector-borne	Bacteria	Meningitis, *Neisseria*	Person to person, handshake, flies
Food-borne	Tapeworms *Taenia* species	Taeniasis	Cysts (cysticerci) in raw meat
Food-borne	Tapeworms *Diphyllobothrium* species	Diphyllobothriasis	Stages in raw fish
Fecal-borne	Cestode worms *Taenia* sp., *Echinococcus* sp.	Taeniasis, echinococcosis	Oral uptake of worm eggs
Fecal-borne	Nematode worms *Enterobius vermicularis*	Enterobiasis	Oral uptake of food contaminated with worm eggs or self-infection by fecally contaminated hands
Fecal-borne	Nematode worms *Ascaris* sp.	Ascariasis	Egg-contaminated food
Fecal-borne	Nematode worms, hookworms *Ancylostoma* sp., *Necator* sp.	Hookworm disease	Skin penetration of free-living larvae on soil after their hatching from eggs

12.2 History

At the end of the year 1988, medical students of the Heinrich Heine University in Düsseldorf guided by Martin Stetzkowski and Klaus-Christian Withoeft transported medical equipment to the regional hospital at Sikasso in Mali and stayed in constant personal contact after finishing their medical studies including their promotion to the doctor of medical sciences (Dr. med.). This group of several young doctors founded in the year 1998 the association Konna e.V., which made it possible to acquire personal money and German governmental support for their plans to establish and run a medical laboratory in the village. Already in the year 1999, these young doctors succeeded in close cooperation with the chief and imam of the village Konna (Mr. El Hadj Mamadou Kampo) to establish a small diagnostic laboratory in Konna at the banks of the Niger river (Fig. 12.1). At first simple and inexpensive, but highly efficient diagnostic methods were established already at the very beginning. By increasing the membership of the Konna e.V. in Germany and by obtaining donations from various sides, already in the year 2000 a physician and a laboratory assistant had been employed (nowadays paid by the government of Mali). This first doctor (Mr. Aboubacar Kampo) stayed there from early months in the years 2001 until 2003, when Konna e.V. granted him a 3-month stay at a clinic in Kenya. Then

Fig. 12.1 Map of Mali showing the location of the village of Konna at the Niger river

Table 12.3 Projects and introduction of methods in Konna as consequences of the establishment of a laboratory by Konna e.V.

Status before starting the project: Medical care of Konna and 17 other villages was in the hands of one male nurse and two midwives	
1999	Establishment of a medical laboratory in Konna inside the healthcare station of Konna guided by Mr. Ousmane Kampo. Start of the use of two solar cookers for sterilization of laboratory equipment
2000	Beginning of the ongoing support of the healthcare person Mr. "Olel" Traoré in the fight to diminish new infections with *Schistosoma* worms
2001	Construction of three sewage tanks in order to improve water hygiene. Since this year the salary of a physician in the infirmary is funded
2002	Construction of ten further sewage tanks
2003	Establishment of a technical workshop to construct solar cookers and construction of five more sewage tanks
2004	Support in the fight against the spreading of a cholera epidemic due to *Vibrio cholerae*. New chief of the village is Ibrahim Kampo
2005	The Konna project is supported by the Mali government by paying the salaries of the physician and the laboratory assistant
2006	Konna e.V. supports the establishment of water pumps to get clean drinking water
2007	Since this year Konna e.V. supports financially construction of further sewage pits, large water wells, and water transport systems
2008	Introduction and support of the new medical doctor in the village
2009	Opening of a new very large well for fresh drinking water
2010	Further construction of sewage tanks
2011	Repair of the drinking water systems of the primary schools; sponsoring of new equipment and repair of the current equipment of the infirmary
2012	Construction of further wells and school connections to drinking water
2013	Great disturbances, since the region surrounding Konna was starting point of the military fights between Islamist and French/Mali troops. The organization *Doctors Without Borders* and persons of the infirmary in Konna organized the treatment of injured persons
2014–2018	Normalization of the daily life, ongoing further support of the work at the infirmary by Konna e.V.

he got a grant to study public health at the University of New Orleans, which qualified him to work for UNICEF. From there he was sent to ten different countries. Currently he is the UNICEF representative person in Ivory Coast engaged with the health amelioration of children. The laboratory was always kept in function and regularly supported by manpower during visits of German doctors. This laboratory with its personal was and is highly effective in the fight against diseases by prevention and regular treatment of infected persons (Table 12.3). For example, in the past year 2018, a considerable amount of about 3000 persons had been treated.

12.3 Malaria in Mali

Malaria (Fig. 12.2) is in Mali the main cause of human mortality and morbidity and covers about 42% of the consultations in local health centers, although the occurrence of this disease is dominated by an annual periodicity. In order to get malaria, there is no important impact on the type of urbanization needed—except for the obligatory presence of water as breeding basis for the mosquitoes (Fig. 12.3). Thus there is only the need of evaporation, humidity, and rainfall occurring for 3 months so that sufficient water is present, which can be used by mosquitoes of the genus *Anopheles* species to lay their eggs upon the surface. These conditions, however, are fulfilled along the Niger river as well in big towns like Bamako (Sissoko et al. 2017) as in small and larger settlements along the river throughout the whole year (Figs. 12.4, 12.5, 12.6, 12.7, and 12.8). Thus there remains the need to control mosquitoes close to dwellings to block the entering of mosquitoes into houses by gauze nets, to block mosquito bites by application of repellents onto the skin, and to treat infected persons as soon as possible after diagnosis of malaria. These measurements will help to avoid infection rates of 52% in children below 5 years as had been occurred in the years 2012–2013 in the rather large town Bamako (capital of Mali)

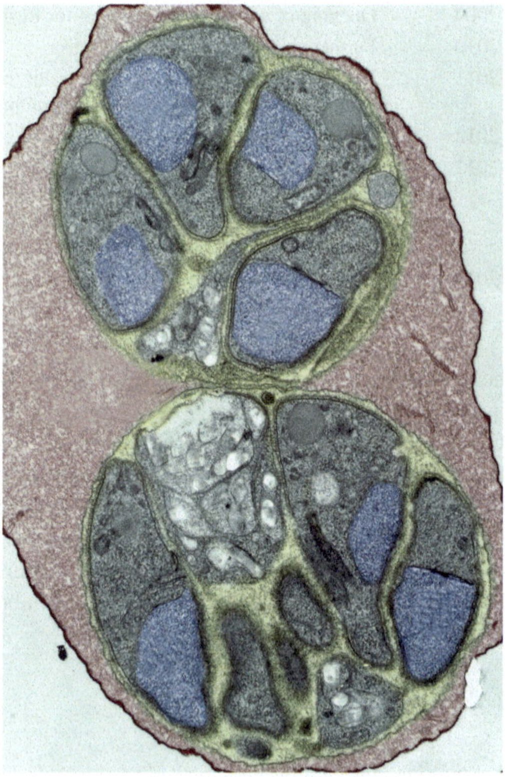

Fig. 12.2 Electron micrograph of a section through a red blood cell containing in two vacuoles a growing schizont of *Plasmodium falciparum*, both developing merozoites. This species induces the life-threatening disease of malaria tropica. Also cristalline remnants can be seen

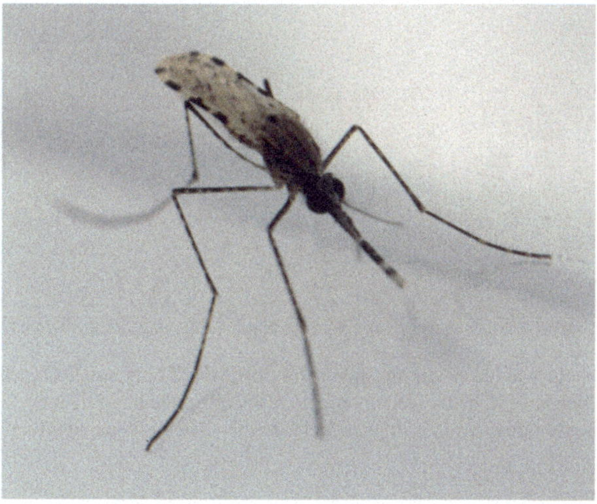

Fig. 12.3 Light micrograph of a female *Anopheles* mosquito (vector of the infectious agents (sporozoites) of malaria)

Fig. 12.4 Bank of the Niger river during dry season showing a high population of humans—all being endangered by malaria and schistosomiasis due to their vector stages (mosquitoes in malaria) respectively cercariae (in schistosomiasis)

Fig. 12.5 This photo was taken during the dry season, where only small branches of the Niger river were left, which have to be crossed in order to reach either the main river or the village Konna. During the rainy season this region is flooded and then the village is situated directly at the banks of the river

Fig. 12.6 The people use the water of the small branches of the Niger river during the dry season also for laundry and dishwashing. The risk of infection with cercariae of *Schistosoma haematobium* is obvious, since the conditions for vector snails are almost as good as in ponds

with about two million inhabitants (Sissoko et al. 2015, 2017; Coulibaly et al. 2013) (Fig. 12.9).

12.4 Schistosomiasis in Mali

This waterborne disease is initiated due to skin penetration of cercariae while persons stay in water. It is the second most important parasitic disease of humans and animals in Mali. After malaria schistosomiasis occurs as intestinal form due to *Schistosoma mansoni* (adult worms live in the mesenterial blood vessels of the intestine) and as urogenital parasites (*S. haematobium*, Figs. 12.10, 12.11, 12.12, and 12.13) (Mehlhorn 2016, 2017). Sacko et al. (2011) investigated the situation of

Fig. 12.7 Children swimming in a pond in the Dogon country. Ponds like this represent a perfect habitat for the vector snails of *Schistosoma haematobium* and offer perfect conditions for the complete life cycle of this worm (These ponds are also breeding sites of mosquitoes). Since the children do not use a proper toilet, *Schistosoma* eggs are released with the urine into the water, miracidia hatch and penetrate snail tissues, within the snails cercariae develop via sporocyst stages, and the cercariae are released from the snails and penetrate human skin. The adult worms develop within the human host and produce eggs, which are released with urine again thus closing the infection cycle

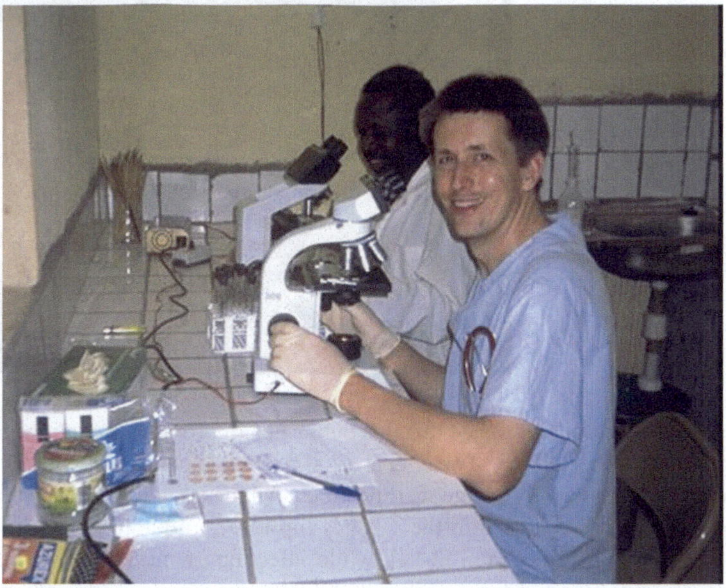

Fig. 12.8 Dr. Martin Stetzkowski (front) (Germany) and Mr. Sangaré (behind) (Mali) examining blood smears in the laboratory for potential malaria infections at the Konna research station

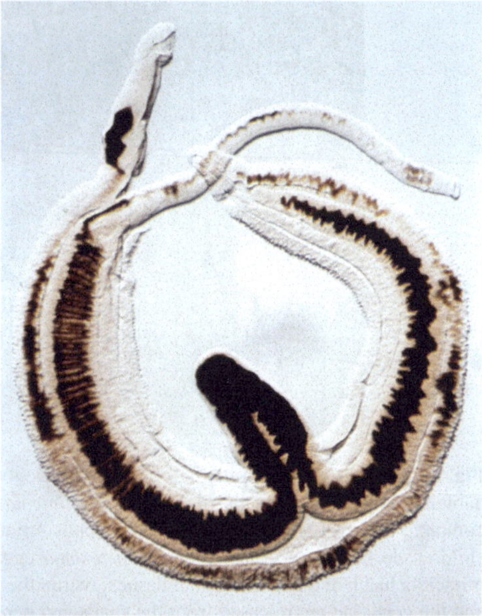

Fig. 12.9 Light micrograph of an adult couple of *Schistosoma mansoni.* The male worm bears the female in a ventral deepening. The intestine of the female is filled with black appearing blood of the host, which has been sucked by its mouth, which is situated in the center of the anterior sucker

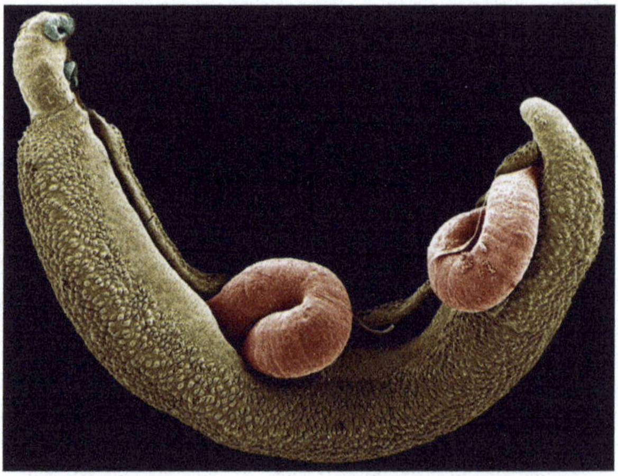

Fig. 12.10 Scanning electron micrograph of an adult couple of *Schistosoma mansoni* showing the female embedded in a ventral folding of the male

schistosomiasis in Mali and stated that this public health problem in Mali, which had been studied over the years by Traoré et al. (2007) and Clements et al. (2009), is still today increasing, especially during warm periods. All these studies showed that very high infestation rates occurred in the villages (up to 97%) due to *S. haematobium* (symptom: blood in urine, Fig. 12.14), while infections with *S. mansoni* reached 12.5% (symptom: blood in feces). Since both infections may lead to death, prevention measurements and medical treatment programs are highly needed. The most important primary measurement is to avoid ongoing repeated infections. This affords that villages do not release feces and urine into the river, that people do not enter

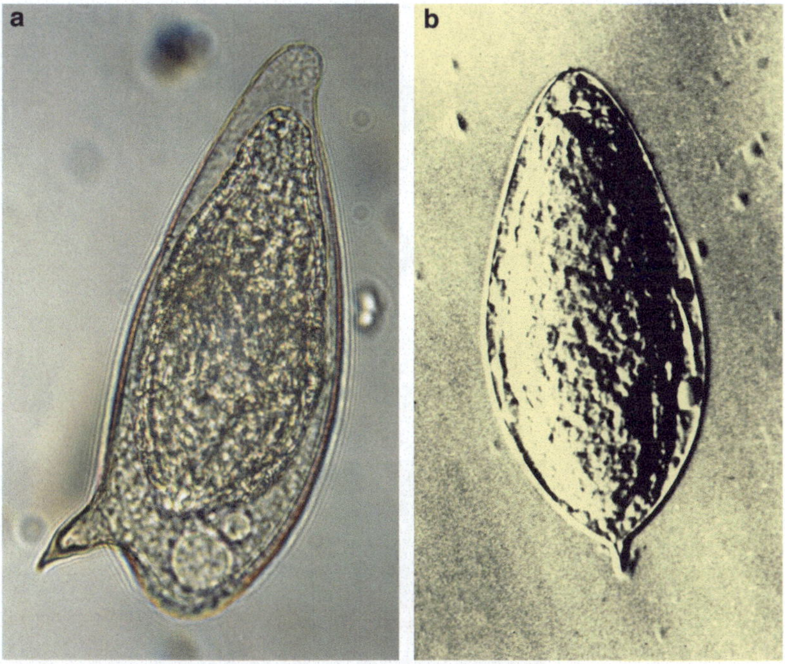

Fig. 12.11 Light micrographs of eggs of *S. mansoni* (left) (in feces) and *S. haematobium* (in urine), which can be clearly distinguished by the location of a protruding thorn at the egg shell

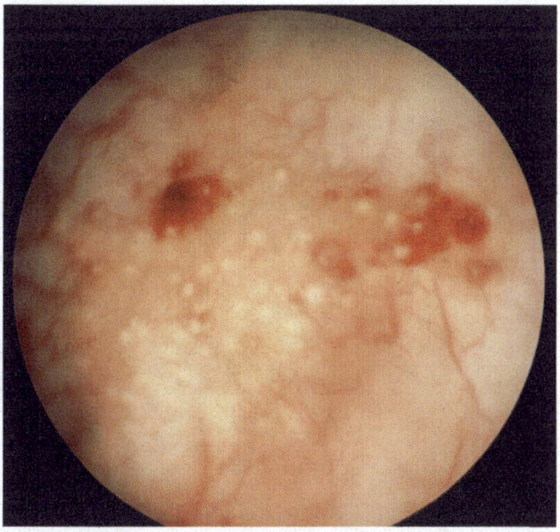

Fig. 12.12 Translucent, whitish appearing *Schistosoma* eggs in the wall of the human bladder inducing intense bleedings (compare Fig. 12.14)

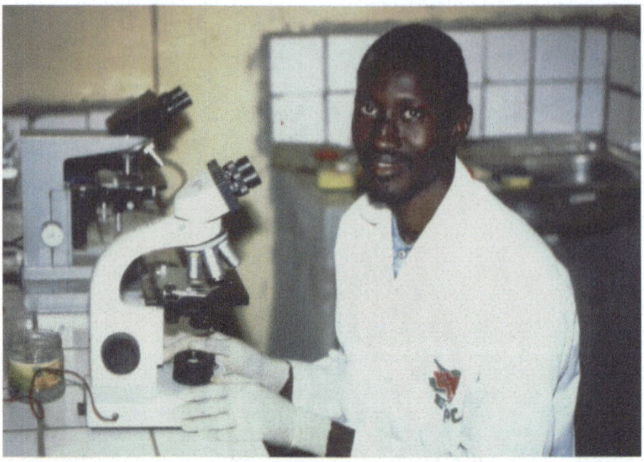

Fig. 12.13 Ousmane Kampo working in the laboratory looking for schistosomal eggs

Fig. 12.14 Urine of *Schistosoma*-infected children in Konna containing different amounts of blood. The different colors from yellow to dark red indicate different intensities of infections depending on the amount of blood being set free when the eggs pass the wall of the bladder. However, all can be treated with praziquantel, if available (!)

Fig. 12.15 Opening of a sewage pit system, which contains material which absorbs contaminated water

water without protecting shoes (boots), that people excreting bloody urine or feces are treated as early as possible with praziquantel tablets, and that control measurements decrease the number of rats along houses, rivers, and lakes.

Construction of Sewage Pits
Since the village Konna had no sewage system for the disposal of *wastewater* (=urine and washing water, but no feces), these fluids traditionally had been discharged just onto the ground. Thus these runlets formed ponds which represented an ideal reservoir for all kinds of pathogens and their potential vectors. In particular, they represent ideal habitats for snails (=intermediate host of the bilharziasis worms). In order to reduce the propagation of such agents of diseases, "soakaways" (sewage pits) have been developed and constructed to collect the wastewater, thus avoiding the formation of ponds (Fig. 12.15). For this purpose a pit had to become a dugout and filled with rocks and stones providing large surfaces to promote evaporation. Then the pit has to be covered except for a rather small opening as inlet for the wastewater and also for the evaporation of collected water. Due to the high temperatures in Konna, which is situated in the Sahel zone, the wastewater volume is constantly reduced due to evaporation. Solid particles in the wastewater (including the agents of diseases—e.g., bacteria and eggs of *Schistosoma haematobium*) remain in the pit, are cut off from the surface water, and thus remain out of reach of people.

In the year 2001 the German association Konna e.V. started the construction of three sewage pits as a pilot project. Due to their high graded efficacy, many more sewage pits have been completed since then in public places in the following years.

All these needed measurements to control (to decrease) the infection rates listed here for these two typical tropical parasitosis must be extended onto the control measurements of other agents of diseases. The Konna e.V. project might be a blueprint for other aid programs initiated by the Europeans.

References

Clements AC, Bosqué-Oliva E, Sacko M, Landouré A, Dembélé R et al (2009) A comparative study of the spatial distribution of schistosomiasis in Mali in 1984–1989 and 2004–2006. PLoS Negl Trop Dis 3:e431

Coulibaly D, Rebaudet S, Travassos M, Tolo Y et al (2013) Spatio-temporal analysis of malaria within a transmission season in Bandiagara, Mali. Malar J 12:82

Mehlhorn H (2016) Human parasites, 7th edn. Springer Spektrum, Heidelberg

Mehlhorn H (ed) (2017) Encyclopedia of parasitology, 3 vols, 4th edn. Springer, New York

Sacko M, Magnussen P, Keita AD, Traoré MS et al (2011) Impact of *Schistosoma haematobium* infection on urinary tract pathology, nutritional status and anaemia in school-aged children in two different endemic areas of the Niger River Basin, Mali. Acta Trop 120 S1:S142–S150

Sissoko MS, van den Hoogen LL, Samake Y, Tapily A et al (2015) Spatial patterns of *Plasmodium falciparum* clinical incidence, asymptomatic parasite carriage and *Anopheles* density in two villages in Mali. Am J Trop Med Hyg 93:790–797

Sissoko MS, Sissoko K, Kamate B, Samake Y et al (2017) Temporal dynamic of malaria in a suburban area along the Niger River. Malar J 16:420

Traoré M, Landouré A, Diarra A, Kanté B, Sacko M, Coulibaly G, Sangho A, Simaga SY (2007) Geographic distribution and epidemiology of urinary schistosomiasis in Mali: implications for a control program. Mali Med 22:22–28

Chapter 13
Leishmaniasis in Israel and Palestinian Territories

Heinz Mehlhorn

Abstract The inhabitants and the pilgrims visiting the "holy region of several religions" are endangered by a malicious disease, the agents of which are transmitted by sandflies attacking in the dark.

Keywords Leishmaniasis · Holy country · *Phlebotomus*

This rather dry desert region, where three world religions (Judaism, Christianism, and Islam) have developed their basics, is also worldwide known as *Promised Country* or *Holy Country* (Fig. 13.1). Furthermore it is unique with respect to the fact that this "holy river" (*Jordan* = Nahar ha-Yardén, Nahr al-Urdunn), which has its source in the Lebanon mountains, ends not in the Mediterranean Sea but just in a desert lake (Dead Sea) and finally disappears completely some miles behind. Starting in early times with Babylonian invaders followed by Roman, Arab, and Christian powers, this region has been unstable until the present times, although the state of Israel (founded in the year 1948) tries to govern this region according to democratic rules. This remains difficult, since the Palestinian Arabic territory is completely surrounded by Israel territory. This situation poses problems due to varying reasons. The political, ethnic, and religious differences make it difficult to control consequently and permanently vector-transmitted agents of diseases in both countries. Therefore especially *cutaneous leishmaniasis* remains a permanent problem due to the fact that the breeding sites (e.g., garbage heaps) of the vectors cannot be reduced in times of intifada (=rebellion of the Arab population against Israel), when removal of disposals is stopped or at least reduced, so that new generations of sandflies can develop therein.

The constant fight against these *Leishmania* species (Tables 13.1, 13.2, and 13.3) remains difficult up to impossible. Furthermore the situation becomes even worse

H. Mehlhorn (✉)
Department of Parasitology, Heinrich Heine University, Duesseldorf, Germany
e-mail: mehlhorn@uni-duesseldorf.de

H. Mehlhorn, S. Klimpel (eds.), *Parasite and Disease Spread by Major Rivers on Earth*, Parasitology Research Monographs 12,
https://doi.org/10.1007/978-3-030-29061-0_13

Fig. 13.1 Map of Israel and Palestinian territory (https://mapcruzin.com/free-maps-west-bank-gaza/israel_sm_2008.gif)

Table 13.1 *Leishmania* species in Israel and Palestinian territory

Leishmania species	Main vector (sandflies)	Reservoir hosts
Leishmania major	*Phlebotomus papatasi*	*Rodents*: *Psammomys obesus, Meriones tristrami, M. crassus, Microtus guentheri, Gerbillus* sp.
Leishmania tropica	*Phlebotomus sergenti*	*Rock hyrax*: mainly *Procavia capensis*
Leishmania infantum (rare cases)	Not determined yet	Not determined yet

Table 13.2 *Leishmania* species in Mediterranean regions, Middle East countries, and Ethiopia

Old world species	Clinical features
L. major	*Severely inflamed multiple ulcers*: may coalesce, heal slowly, induce disfiguring and disabling scars
L. tropica	*Multiple painless dry ulcers*: heal spontaneously over 1 year, but tend to leave disfiguring scars
L. infantum	*Single nodules*: may ulcerate and may resolve spontaneously within 1 year but leave atrophic scars
L. aethiopica	*Localized nodules*: oronasal lesions causing localized anatomical distortion, heal within 2–5 years Causes disseminated *cutaneous leishmaniasis*: mostly on limbs and face, but does not induce ulcerations unless in immunosuppressed persons

Table 13.3 Drugs to treat the various forms of leishmaniasis (Pace 2014)

Drug	Mode of action on the parasite	Route, main indication, and application
Pentavalent antimonials—sodium stibogluconate	Inhibition of glycolysis and fatty acid oxidation Dose-dependent inhibition of ATP and GTP formation	*im/iv*: VL, CL, MCL, PKDL *Intralesional*: CL
Pentamidine isethionate	Inhibition of polyamine biosynthesis and disruption of mitochondrial membrane potential	*im*: CL, MCL *Intralesional*: CL
Amphotericin B and lipid formulations	Inhibition of cell membrane synthesis by binding to ergosterol Pore formation in cell membrane	*iv*: VL, CL, MCL, PKDL
Paromomycin	Possible interference with RNA synthesis and membrane permeability	*im*: VL *Topical*: CL
Allopurinol	Interference with protein synthesis (purine salvage cycle)	*Oral*: VL, CL

VL visceral leishmaniasis, *CL* cutaneous leishmaniasis, *MCL* mucocutaneous leishmaniasis, *PKDL* post-kala-azar dermal leishmaniasis, *im* intramuscular, *iv* intravenous

Fig. 13.2 Light micrograph of a sandfly of the genus *Phlebotomus*

due to the fact that not only humans may become infected but also a broad spectrum of animal hosts (Table 13.1), where the parasites do not introduce such very severe symptoms of disease as in humans. Due to the fact that these desert animals live very close to human dwellings, the human population is highly endangered. Especially the Arab population is at risk, since their houses have not the same standards like those in Israel and due to the fact that there is no permanent regular garbage collection. The increasing number of garbage heaps offers further wide breeding sites for the vectors of the agents of leishmaniasis (*Phlebotomus* species, Table 13.3 and Fig. 13.2). The main hosts of *Leishmania* stages are sand rats (*Psammomys obesus*) and jirds (*Meriones crassus*) (WHO 2010; Ben-Shimol et al. 2016).

In villages in Israel and in the territory of the Palestinian Authority occur two main indigenous *Leishmania* species, *L. major* and *L. tropica*, which both induce the so-called cutaneous leishmaniasis leading to skin scars (Figs. 13.3 and 13.4). In former times these two species appeared in separate regions, but recently they are spread in nearly all regions.

Besides these two *Leishmania* species, a third one is present in the Middle East: *Leishmania donovani infantum* (Azmi et al. 2017). This species is the main cause of the *visceral leishmaniasis*, which leads to life-threatening symptoms. However, the incubation period can vary from a few weeks to several months, so that the sites,

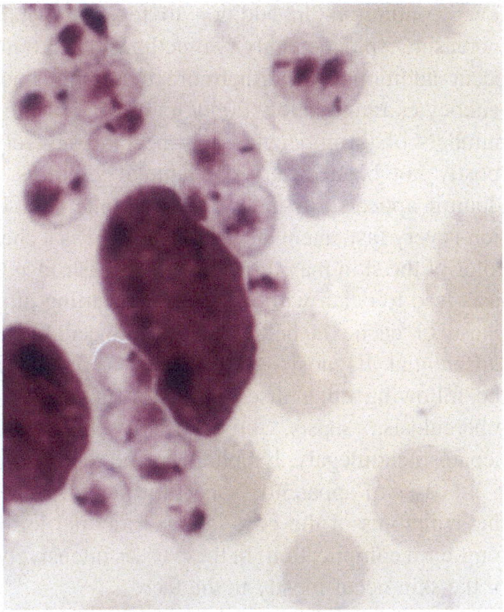

Fig. 13.3 Light micrograph of ruptured host cells releasing stages of *Leishmania major* besides two large host cell nuclei

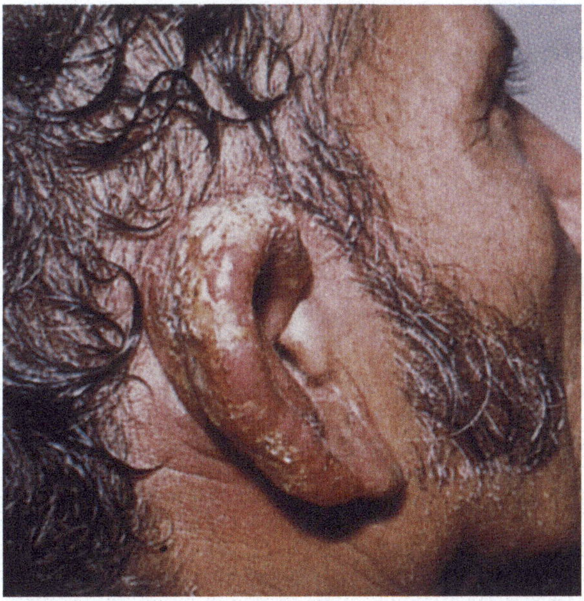

Fig. 13.4 Ear of a patient with *Leishmania tropica* scars

where the infection really did occur, are never completely free from infected persons and infested vectors. Fevers (without shivers) generally occur in cases of *visceral leishmaniasis* and may last several weeks reaching up to 39 and 40 °C. These fevers may be interrupted by fever-free phases and in general finally end in a low-graded fever continuum. In addition to fever, the main symptom of the disease is an increasing splenomegaly. Sometimes hepatomegaly and lymph node swellings occur additionally. Furthermore increasing anemia and leukopenia (numbers of leukocytes range below $2000/m^3$ in 75% of the cases) may occur. In addition the numbers of thrombocytes may become severely reduced, while eosinophilia has mostly not been observed. Gamma globulins become strongly increased, while albumin appears low-graded in electrophoresis. The erythrocyte sedimentation reaction is very fast, anemia occurs, and there is a change in the protein compounds. The color of the skin may turn from pale to ash-faced. These cases of visceral leishmaniasis lead to death without treatment starting after 6 months up to 3 years during a phase of cachexia being mainly induced by secondary bacterial complications. Differential diagnosis at the beginning of the infection must consider (=exclude) the following other diseases leading to splenomegaly: malaria, typhus, (miliary) tuberculosis, sepsis, brucellosis, schistosomiasis in the late phase, and hepatosplenomegaly, lymphomas, and leukemia.

In cases of apparently successful treatment, it may happen that 1 up to 10 years the symptoms of the post-kala-azar dermal leishmaniasis (PKDL) may occur as a long-term complication. In these cases intensively pigmented or erythematous nodes of the skin occur mostly in the face.

Bibliography

Allan MZ et al (2014) Genetic diversity of *Leishmania donovani/infantum* complex in China. Inf Genet Evol 22:112–119

Azmi K, Krayter L, Nasereddin A, Ereqat S, Schnur LF, Al-Jawabreh A, Abdeen Z, Schönian G (2017) Increased prevalence of human cutaneous leishmaniasis in Israel and the Palestinian Authority caused by the recent emergence of a population of genetically similar strains of *Leishmania tropica*. Infect Genet Evol 50:102–109

Ben-Shimol S, Sagi O, Horev A, Avni YS, Ziv M, Riesenberg K (2016) Cutaneous leishmaniasis caused by *Leishmania infantum* in Southern Israel. Acta Parasitol 61:855–858

Cantacessi C et al (2015) The past, present and future of *Leishmania* genomics and transcriptomics. Trends Parasitol 31:100–108

Eroglu P et al (2015) Clinical manifestations and genetic variations of *Leishmania infantum* and *L. tropica* in Southern Turkey. Exp Parasitol 154:67–74

Goto H et al (2010) Current diagnosis and treatment of cutaneous and mucocutaneous leishmaniasis. Expert Rev Anti Infect Ther 8:419–433

Ivens SC et al (2005) The genome of the kinetoplastid parasite *Leishmania major*. Science 309:436–442

Mansueto P et al (2014) Leishmaniasis in travellers. Travel Med Infect Dis 12:563–581

Mehlhorn H (2016a) Human parasites, 4th edn. Springer, New York

Mehlhorn H (2016b) Encyclopedia of parasitology, 4th edn. Springer, New York

Neumayr AL et al (2012) Successful treatment of imported mucosal *Leishmania infantum* leishmaniasis by miltefosine after severe hypokalemia under meglumine antimoniate treatment. J Travel Med 19:124–126

Pace D (2014) Leishmaniasis. BIAM 69:510–518

Peacock CS et al (2007) Comparative analysis of three *Leishmania* species that cause divers human diseases. Nat Genet 39:839–847

Richter J et al (2011) Mucosal *Leishmania infantum*. Parasitol Res 109:959–962
Van der Auwera G, Dujardin JC (2015) Species typing in dermal leishmaniasis. CMR 28:265–293
WHO Library Cataloguing-in-Publication Data (2010) Control of the leishmaniasis: report of a meeting of the WHO Expert Committee on the Control of Leishmaniases, Geneva, 22–26 Mar 2010

Chapter 14
Past and Present of Tropical Diseases in the Amazon River

Felipe Guhl Nannetti and Juan F. Guhl Samudio

Abstract The Amazon River was and is the most important "living vein" in South America that offers life for animals and humans but also endangers their health. Entering this "hell" often led to death in former centuries due to vector-transmitted diseases, which even today have to be constantly controlled.

Keywords Ascariasis · Chagas disease · Fungi · HTLV · Malaria · Syphilis · Toxoplasmosis · Tuberculosis · *Tunga penetrans*

14.1 Introduction

Is the Amazon a *green hell* or a *lost paradise*? To begin the chapter with this question is relevant when attempting to address the historical lack of epidemiological data of this region. There are reports on the diseases in prehistoric America, but most lack scientific evidence. We ignore whether this vast region suffered from pandemics such as the Black Death of Europe. There are records describing a population decline in the settlements of the Amazon River area more than 11,000 years BP, attributable to the environmental and climatic changes that occurred between the Holocene and Pleistocene (8000–6000 BP). However, given the absence of bone material and human or animal remains, these facts are not yet entirely clear (Rapp Py-Daniel 2015), FIAN.

The Amazon was visited by the first European settlers 500 years ago, with more than 9500 years of human presence lacking a written history. The first written chronicles referred to a "biblical paradise" and its people. They were described as

F. Guhl Nannetti (✉)
Faculty of Sciences, Department of Biological Sciences, University of los Andes, Bogotá, Colombia
e-mail: fguhl@uniandes.edu.co

J. F. Guhl Samudio
Universidad Nacional de Colombia, Bogotá, Colombia

H. Mehlhorn, S. Klimpel (eds.), *Parasite and Disease Spread by Major Rivers on Earth*, Parasitology Research Monographs 12,
https://doi.org/10.1007/978-3-030-29061-0_14

having a great facility for cultivation and a high consumption of proteins from animal origins, generated from hunting activities, whose lives were similar to those described in the genesis of the Old Testament, since they had no clothing or evidence of carnal sin. However, the constant harassment from the large number of haematophagous insects (mainly mosquitoes) was crucial to discarding it as a paradise. Given the chronicles written by the first European explorers, we are aware that insect vectors of tropical diseases such as malaria and yellow fever existed.

On 30 January 1500, a Spanish expedition led by Vicente Yañez Pinzón discovered the mouth of the Amazon River on the Brazilian coast, initially called the *Rio Grande*. The explorers were amazed by the swell caused when the river clashed with the ocean tides, whose enormous flow allowed them to fill their containers with fresh water, despite being more than 160 km from the coast in the open sea. For this reason, the Spanish named this area *Santa María de la Mar Dulce* (Pineda 2013).

Years later in 1541, Francisco de Orellana, a Spanish conqueror, was the first European to navigate the river that crosses South America from the west to its mouth on the opposite side. Together with his scribe Friar Gaspar de Carvajal, who lost an eye because of an arrow shot by the Indians during the expedition, upon their return to Spain they described being attacked not only by men but also by numerous warrior women who had mastered archery. Thus, they named it *Rio de las Amazonas*, as a reference to the mythical women who fought Heracles and Achilles (González 2009).

In 1494, the Portuguese Empire, as a result of the Treaty of Tordesillas signed by Pope Alexander VI, divided the domains outside Europe with the Spanish Catholic monarchy. The Lusitanians took under their rule the lands surrounding the mouth of the Amazon in the Atlantic Ocean, leading their raids and conquests upriver against the eastern current. Due to the continuous raids by the Dutch and English who established several *feitories* in the Amazonian estuaries, in 1615 the Fort of Presepio was founded, and several constructions were erected around it to counteract and prevent further inland invasions. Thus, the first European city of the Amazon was born, known as Belem del Gran Pará, which today accounts for more than 1.5 million inhabitants. Lusitanian colonisation began with this settlement, empowering the Portuguese Empire in the mouths of the Amazon (Rosas Moscoso 1986; Riponi 2016; San Roman 1994).

14.1.1 Geographic Context of the Amazon

Located in the centre of South America, the Amazon basin has patterns of precipitation throughout the year based on the intertropical convergence zone, since it is located on the global equatorial line. This hydrographic basin is the largest in the world, covering an area of 6,100,000 km^2 and territories from 10 countries: Bolivia, Brazil, Colombia, Ecuador, Guyana, French Guiana, Peru, Suriname and Venezuela. The rivers in this region represent approximately one fifth of the total fluvial flow worldwide. These waters discharge in two systems: the Amazon River, a superficial fluvial system, and the Hamza River, an underground system which is wider and slower. The Amazon basin covers an area that is nearly the same size as the continental USA and is one of the least populated areas on the planet. With an approximate population of 20 million inhabitants, it has a population density of 4 people per km^2; in contrast, the population densities of Brazil and India are 5 and 75 times higher, respectively (Goulding et al. 2010).

The Amazon River was formed by a large gulf originally open to the Pacific Ocean; its waters had their springs in the Guyanese and Brazilian shield flowing from west to east. Due to the formation of the Andes Mountain Range, the gulf closed to the northwest, creating the huge Pebas Lake located in northern South America. The cumulus of water and sediments descending from the Andean Mountain Range and the separation of the African-South American supercontinent resulted in the current drainage system, shaping its course and outlet in the Atlantic Ocean.

The origin of the Amazon River is located in the heights of the great Andes Mountain Range in Peru, which extends 6750 km. The waters of the Marañón and Ucayali rivers converge in Peru to form the Amazon River, named after the jungle plains. Its extensive water flow shows a small slope, forming a typical plain river descending less than 100 m, allowing navigability in much of its route and tributaries that run through the basin and has historically been the main means of transportation for the inhabitants of this region.

The river width varies from 1.6 km in its high stage to 190 km in the rainy season before reaching its outlet in the Atlantic Ocean where it becomes an estuary that is 240 km wide. The Amazon River crosses three countries, keeping its name throughout (Map 14.1). For the Spanish-speaking countries of the basin, with rivers of Andean origin, the Amazon River begins at the source of the Ucayali River in the glaciers of the mountain range. Others identify the river's origin as the confluence of the Ucayali and Marañón rivers. However, from this point to the confluence with the Negro River, Brazilians refer to it as the Solimões River, and only after these colossal waters converge do they call it the Amazon River.

In this chapter, we use the most common mapping interpretation to define the Amazon River. Over 80% of the riverbed is in Brazil, 18% is exclusively in Peru and approximately 2% is shared between Colombia and Perú.

There are more than 1000 tributaries of the Amazon that flow from the highlands of Guyana, Brazil and the Andes. Six of these tributaries are in Colombia, Caquetá, Japurá, Juruá, Madeira, Negro, Purus and Xingú in Brazil, covering over 1600 km, and the Madeira River exceeds 3200 km from source to mouth. The largest ships currently sailing the ocean can navigate 1609 km upriver to the city of Manaus in Brazil, while at any time of the year, smaller cargo ships can reach Iquitos, Peru, which is 2090 km upriver.

The flood patterns of the Amazon River are complex due to large drainages in the south and north receiving rain peaks at different times of the year. The course and floodplain in the period of highest precipitation cover over 62,121 km^2. Between Santarém and Iquitos, the annual fluctuation average is approximately 4.5–10.0 m. The central Amazon features the highest fluctuations, near Manaus. These values

Map 14.1 Amazon River course from its source in Ucayalí (Perú) to its mouth in Belem do Pará (Brazil). The river's total length is approximately 6750 km

gradually decrease towards the upper part of the river and are more marked at the lower part of the Negro River mouth. Between April and July, the entire Amazon River above the Xingu River is flooded. Only in October and November does the Amazon River have a period of low water (Goulding et al. 2010).

14.1.2 First Human Populations in the Amazon Basin and Their Diseases

Man has migrated throughout history to populate almost every ecosystem on the planet. The Amazon basin is no exception, and the first findings of human presence in Brazil were painted stone dating to over 11,000 years BP (Roosevelt et al. 2002).

The presence and efficacy of shamanism (Fig. 14.1) and its practices regarding health can be verified in the historical descriptions of the expeditions of the first Europeans to the Amazon, demonstrating that ecology has impacts on the well-being of people. Some diseases of the Amazon rainforest can only be explained through their ecology. The knowledge of the natives in curing some of these pathologies is widely recognised, and the plants they used (unknown in Europe) were so effective that were finally adopted by the world pharmacopoeia, as in the case of cinchona

Fig. 14.1 A recognised shaman in the Amazon River

(*Cinchona officinalis*), native to the Amazon rainforest, whose bark (known as *quina* or *quinaquina*) is extracted for its ability to cure malaria.

Archaeological evidence shows that the formative period of the Amazon starts approximately 4000 years BP with the settlement of sedentary communities entering the jungle environment and developing cultivation practices. As a result, there was a considerable increase in the population (Meeggers and Evans 1957; Lathrap 1970; Carneiro 1995; Denevan 1996; Heckenberger 2005; Gnecco and Mora 1997). Although these authors have theoretical disagreements as to the existence of chiefdoms or complex societies in the Amazon, their excavations show that there were populations with a considerable demographic density, giving rise to new societies that of course were exposed to new health problems. Leles and Araújo (2015) show evidence of bacteria, viruses, parasites and fungi in prehistoric America and how the people in the Amazon were exposed to them. Table 14.1 lists the various diseases and pathogens that would have affected the inhabitants of the Amazon. The authors propose that the Amazonian communities were exposed to various infectious agents due to the migration of individuals from the Andean population. Therefore, infections in the jungle were due to intra-Amazonian exchange networks, assuming that these pathogens must have been present prior contact with Europeans.

In turn, Sotomayor (1995) proposes that the population densities of the Amazon were very low. As a consequence, the communities were isolated and therefore healthy, with little probability of disease transmission. This hypothesis is focused in

Table 14.1 Main infectious agents or aetiological agents in the Amazonian context

Group	Main infections aetiological agents	Importance in the Amazonian context
Bacteria	Tuberculosis	The paleoparasitological findings prior to European contact with North America are superior to those found in South America, and this discrepancy is even greater in relation to the Amazonian region
	Syphilis	There is discussion about whether syphilis is American or European in origin, and findings from the Amazon may help to clarify this issue
	Helicobacter pylori	Further studies on these bacteria in the Amazon region may clarify its origin
Virus	HTLV	The different distributions of HTLV1 and HTLV2 in the modern population suggest possible entry routes into America in addition to its origin. The absence and presence of some of these virus groups in the modern American native population contrast with the findings of HTLV1 and HTLV2 in pre-contact material
Fungi	Coccidioidomycosis and paracoccidioidomycosis	These are important as agriculture markers in past societies and represent an important discussion in archaeology
Protozoa	Malaria	Although it is almost entirely accepted that malaria was introduced to America, there is a lack of data to corroborate this
	Toxoplasmosis	The Amazon is one of the areas featuring the greatest diversity of new *T. gondii* genotypes. The study of the archaeological material of this region may contribute empirically to elucidate the origin and dispersion of this protozoan
	Chagas disease	Paleoparasitology may contribute to recent discoveries on the distribution of *Trypanosoma cruzi* in the region
Helminths	Ascariasis	The close relationship between the natives and wild pigs of South America and those introduced by Europeans may contribute to help clarify the controversy of whether there is a unity of species or whether these originated with humans or *sainos* first
	Hepatic capillariasis	The increasing number of cases of false parasitism by *Capillaria hepatica*, specifically in the Amazon, has been attributed to the consumption of game meat. Some authors point to a new transmission mechanism in the modern population, apparently corroborated by paleoparasitology due to the absence of cases of this helminth in archaeological material in Brazil. This scenario is being altered due to recent paleoparasitological data
Ectoparasites	*Tunga penetrans*	Some parasites infect the host's exterior, and tungiasis is a classic example of an ectoparasitosis originating in the Americas that still constitutes a real plague among natives today

Leles and Araújo (2015), Black (1992) and Boletim epidemiologico (2019)

the northwest of the basin, where no record of hierarchical societies exists and where there should be extensive exchange networks on the banks of the Amazon River.

Plenty of other archaeological evidence shows that the Amazon was populated since remote times and has demonstrated that anthropogenic influence was crucial for the structuring of the basin's biota and species distribution. Notably, bioarchaeological investigations of the Amazon River show that human remains are very scarce, since few skeletons have been found in funeral urns, and it has been known that people suffered from tooth decay and fractures. However, there is scarce evidence on the parasitic diseases or viruses that could have affected people in tropical areas. The current lack of knowledge regarding health issues associated with the inhabitants of the Amazon in pre-Columbian times is evident.

14.2 European Incursions into the Amazon

Francisco de Orellana, Pedro de Ursua, Friars Domingo de Brieva and Andrés Toledo were the first European explorers of the Amazon River in the sixteenth century. Their chronicles described great empires and cities populated by natives on the riverside. The arrival of explorers, and later of African slaves, brought new diseases to the region, such as smallpox, measles, influenza, rubella and mumps, which are viral infections transmitted by air; others transmitted by vectors, such as yellow fever, dengue and very likely malaria; and those transmitted by sexual or body contact, such as hepatitis B and varicella, respectively.

According to Crosby (1993), Amerindians knew diseases such as *purú purú* or Pinta disease, whose causal agent is *Treponema carateum*; Bouba or Yaws caused by *Treponema pallidum pertenue*; venereal syphilis caused by *Treponema pallidum pallidum*; polio caused by poliovirus; some varieties of tuberculosis (not those usually associated with lung diseases); and several intestinal parasites. However, the natives never seemed to have had any experience with Old World diseases (Table 14.2).

These diseases, in addition to the wars between the Portuguese and Spanish colonisers, represented by the Carmelite and Jesuit orders, respectively, led to the extinction of large indigenous populations such as the Omaguas, who inhabited an area from the mouth of the Negro River to the Amazonian highlands.

By the late seventeenth and early eighteenth century, Priest Samuel Fritz reported that he had led more than 60 missions with native inhabitants through the Amazon, from the mouth of the Negro River to the vicinity of the Andean Mountain Range, while there were 90 Carmelite missions on the Amazon River and its tributaries (Cabrera 2002).

Francis L. Black, in his work *Why did they die?* states that in the New World, 56 million people died from contact with Europeans, either due to wars and social breakdown or (mostly) because of diseases introduced to the continent. As a consequence, only 10% of the population survived (Black 1992).

Table 14.2 Main diseases introduced to America by European incursions

Agents of disease	Disease
Viruses	Smallpox
	Measles
	Influenza
	Rubella
	Mumps
	Yellow fever
	Dengue
	Hepatitis B
	Varicella
Bacteria	Whooping cough
	Leprosy
	Gonorrhoea
Protozoa	Malaria
Helminthiasis	Onchocerciasis
	Tapeworms
	Schistosomiasis
	Echinococcus
	Uncinariasis
	Ancylostomiasis

Spanish colonisation continued downhill from the Andes, creating settlements, *haciendas* and *factories* in the Amazonian eyebrow,[1] where natives were forced to become the workforce of the colonisers. In addition, under dissimilar organisational systems such as vassalage, *mitayos*[2] and servants, the natives were incorporated and stripped of their beliefs. Between 1711 and 1722, a great epidemic called the black plague of Peru took place, leading to depopulation and increasing native rebellions against the missionaries and settlers. A significant rebellion was carried out by Santos de Atahualpa, who created a messianic movement arguing for the return of the great Inca. As a consequence, the Spaniards were expelled from the central jungle in 1742, which marked the first insurrection against the Spanish empire by South American natives.

In the spiritual government of the first Bishop of Pará from 1721 to 1733, D. Fray Bartholomeu do Pilar described a large smallpox epidemic that took the lives of thousands of people. In addition, he described another epidemic of malignant measles, which occurred in 1749 and devastated the black and native populations (Viana 1906 in Santos 2012).

[1]The Amazonian eyebrow is the high forest, corresponding to the eastern area of the Andean Mountain Range in Perú, where the thickness of its forests and the rivers that make it up are part of the Amazon basin.

[2]The appropriation of the land by the European colonisers generated dissimilar ways of organising the native populations; the mitayos were natives who own lands and had forms of cultivation but had to pay a tax on their production to the Spanish lords.

The mistreatment of the local population continued, as this was a devastated region that was impoverished and afflicted by new diseases. The death of approximately 40,000 individuals due to smallpox was reported between 1743 and 1750 (Pineda 2013).

The delimitation of the Amazon by the two European powers greatly changed the dynamics of their territories. In Brazil, the Negro River captaincy was transferred to Barra, where buildings, shipyards and factories were built and the colonial presence was strengthened in cities such as Tefe, Tabatinga and Javari (now Manaus). Portugal, interested in natural resources, sent the first scientific expedition up the Negro River, departing from Belem do Pará. The expedition was led by Rodriguez Ferreira, who described the hospitalisation of a soldier who suffered from leprosy and another who had smallpox. He also mentioned that smallpox was primarily striking the native population.

Charles Marie de La Condamine toured the Amazonian riverbeds during the same period and described several poisonous snakes, a constant danger to the health of the inhabitants of the great river. This French expeditionary, who travelled from Quito to Belem do Pará, described how in this latter city he spent a long time waiting for the winds to sail back to his country. He realised that many natives were fleeing to the mountains, afraid of the deaths caused by smallpox. In his descriptions, he tells how a missionary inoculated the natives who had not fallen ill and saw that they did not die. This is the first evidence of people's immunity to disease (La Condamine 1993).

Theodor Koch-Grünberg, a German explorer and ethnographer (1909), travelled the Amazon River at the beginning of the twentieth century and estimated that the population of Belém do Pará was 100,000 people and that of Manaus was 40,000. During this tour he described the homes of the natives and claimed that his people were exposed to malaria. He also described pneumonia and flu and the rituals that the shamans used to reduce the mortality of their patients.

14.2.1 Urban Settlements, Commerce and Navigation on the Amazon River

Navigation on the Amazon River increased substantially due to the technological innovations of the time, generating a significant impact on the floodplains of the Amazon River. The introduction of steam ships in 1850 facilitated the journey and transport of people into the Amazon while also increasing the flow of infectious agents and diseases. Syphilis reached many communities and affected their population, as described by the physician Antonio José Moreira (Moreira 1854 in Santos da Silva 2012). Large quantities of wood were required for the operation of these ships, impacting species such as *mulatto* wood (*Calycophyllum spruceanum*), which was apparently almost extinct by the time steam ships were abandoned in the 1930s. During the first and second world wars, the maximum exploitation of rubber took place, granting the Amazon worldwide popularity. To extract the vegetable latex, a large workforce of native people and *caboclos* was required. In the Amazonian

Fig. 14.2 The great theatre of the opera in Manaus built in the rubber bonanza era (1896)

highlands, the Arana House was the main representative of this industry, and the company used debt-bondage, enslavement and torture against the native population. Given its strong penetration into the jungle, this company was responsible for the spread of epidemics and diseases that significantly reduced the communities. This has been called the Amazon holocaust by historians and anthropologists (Biocca 1944; Pineda 2013).

The situation was not very different in the mid- and lower Amazon regions. The main cities grew rapidly, and their demography changed based on the migration from the Brazilian northeast. The rubber boom made it possible for the richest cities in Brazil to gain access to the electricity network before Rio de Janeiro and Sao Paulo. The construction of an opera house and other buildings highlighted the opulence and the excessive wealth of the time. Large groups of migrants arrived in poor health and experienced dismal working conditions. As a consequence, the first medical commissions and care regulations for migrants were organised in Brazil. Celso Furtado (2007) estimated that approximately 260,000 immigrants arrived at the region between 1872 and 1900, and from 1900 to 1910, this figure reached a total of 500,000 people (Fig. 14.2).

14.2.2 Occurrence of Diseases in Boom Times

The presence of multiple diseases and the unfair conditions of the rubber workers is attributed to the massive migration and demographic change. This represented a major health issue for the big river cities such as Manaus, where there were large groups of rubber workers. Health policies increased to the extent that vaccination campaigns were compulsory for all new arrivals. Additionally, clinical examinations were conducted, sending sick individuals to the island of Umirisal for recovery or to die. The commerce boom was so lucrative at the time that transatlantic steam ships demanded a larger island for quarantines, and the islands of Tatuoca and Marapatá served this purpose. To avoid the spread of vectors that could transmit disease, fumigation was implemented for the first time inside the vessels.

The data shown in Graph 14.1 is based on the medical records of the physician Alfredo da Matta from 1895 to 1914 and on the National Department of Health records by the physician Achilles Scorzelli Junior from 1915 to 1920. It shows that the greatest cause of mortality in this period was malaria (64.4% of the total deaths), followed by tuberculosis (12.7%, considered only for pulmonary tuberculosis, with other types of tuberculosis on the list). Yellow fever was in the third place, followed by influenza, with an increased incidence in 1918 when the Spanish flu pandemic occurred (Santos 2012).

The epidemiological data from the late nineteenth century from the city of Manaus is notable. However, records from the river populations are very scarce. This indicates that the malaria epidemic that was besieging the city was expanding to the communities and latex extraction areas. Reports from the northwest of the basin show large epidemics of measles, together with flu and malaria, causing the natives to take refuge in the jungle until the diseases disappeared (Pineda 1985).

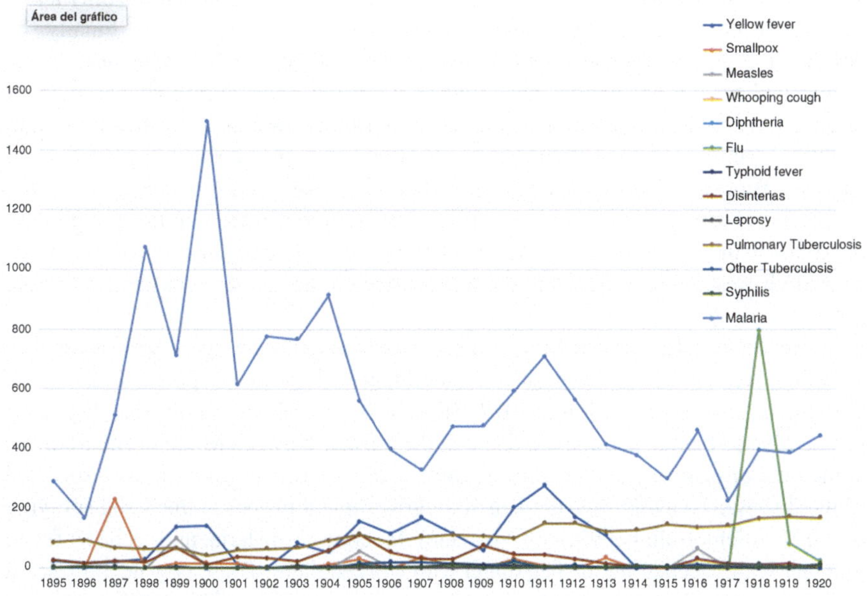

Graph 14.1 Mortality caused by different diseases in Manaus during 1895–1920

The fall of the Arana House was caused by the accusations of R. Casmet, alleging the inhuman treatment of the natives and opening a trial against J.C. Arana in the British House of Lords. This event is recorded in his famous book *The Red Book of Putumayo*. Meanwhile, the borders between Colombia and Perú were defined through warfare, which compelled the Colombian armies to include medical brigades to attend the soldiers, being the first of this type in the Colombian Amazon.

After the rubber boom, the Amazon entered into a sort of lethargy and abandonment; the trade of wild animal skins and some timber species became the main drivers of local economies. The great theatres where Caruzo (the greatest tenor of his time) once performed were left desolate and the buildings were in decline. On the other hand, the native populations remained in relative peace. They reorganised in multi-ethnic communities as a result of the massive displacements due to the rubber industry. The forest returned to tranquillity for a short time.

During this period, the Italian microbiologist Ettore Biocca carried out an expedition to analyse the different pathogens that affected the populations. He identified that in the Negro River area, nearly all of the settlers suffered from malaria. His research on pulmonary tuberculosis showed that changes in the diet of the natives, assimilating the rice consumption of the "civilised" world, caused vitamin deficiency, but could be addressed by eating the forest fruits that they usually consumed. Additionally, the overcrowding of the young population in boarding schools resulted in the spread of tuberculosis.

In relation to gastrointestinal parasites, Biocca reported that one of the most frequent helminthiases was caused by uncinariasis, introduced by black slaves brought by Europeans to the area. He also reported a high prevalence of syphilis and leprosy, which was widespread throughout the Amazon basin, while only a few cases were known in the "civilised" settlers of the Negro River area. Finally, he described a disease typical of the native population, which they call *purú purú*, propagated by the vegetation they used to whip the body during rituals, thereby infecting the epidermis through cuts in the skin. Before his expedition it was thought that this pathology was a dermomycosis. However, upon performing serological analysis, Biocca found that it was a spirochetosis caused by *Treponema carateum*.

The harvesting of cinchona was another important occurrence in the exploitation of Amazonian natural resources. Workers employed to extract the antimalarial drugs carried viruses and parasites with them that affected the native communities (Zarate 2001).

In the 1970s, the construction of the trans-Amazonian highway was planned to exploit the resources of the region and integrate it with the development dynamics of the rest of the country (Smith 1977; Riponi 2016). Many people rejected these projects based on the fateful experience of the railway between Madeira and Mamoré at the beginning of the century, where thousands of workers from the Amazon died of the disease. To avoid this, the Brazilian government proposed the aerial spraying of DDT (dichlorodiphenyltrichloroethane) to control malaria and yellow fever, together with vaccinating the personnel who worked there. These new health-related techniques would prevent the death of thousands of workers. This project included the creation of new population centres and the foundation of *agrovilas* that were planned every 10 km and included 48–60 houses, a school, a health centre, a church and community centres. On the other hand, the *agropolis* were intermediate settlements that concentrated services for the *agrovilas*, with local administrative structures, a secondary school, hospital and ample commerce. The *rurópolis* were cities that already existed along highway BR-230 and had centralised administration, along with all the abovementioned services at a greater scale, including hospitals, health centres, technical schools, banks, hotels and airports. Consequently, they attracted the people living in the *agrovilas* and the *agrópolis* (Smith 1977 in Martins de Souza 2015).

The dictatorship prohibited publicising the epidemics produced by the trans-Amazonian highway construction. However, newspapers reported on new diseases, such as the haemorrhagic syndrome of Altamira, transmitted by the *Simulium sextobecium* insect. This disease was associated with nose and mouth bleeding, which caused childhood deaths and complications in adults. Likewise, inhabitants who witnessed the construction of the highway affirm that deaths were caused by malaria, yellow fever, leishmaniasis and snake bites and that the highway itself became a risk factor due to traffic accidents.

The development continued along the river and continued to strongly impact the ecosystem, disturbing the environment and the health of the communities and native populations, who had historically survived and coexisted in harmony with the environment. Development projects include the construction of the Amazon perimeter highway in Perú, oil exploitation in Loreto in 1971, gas exploitation in Aguaytía (Ucayali) and others in the Amazonian highlands.

There is a clear trend towards urbanisation along the Amazon River. The main cities are Belem do Pará, Manaus and Iquitos (Map 14.1). Together with Leticia, Tabatinga, Tefé, Itacoatiara, Parintis, Óbidos, Santarém, Almeirim and Macapá, they constitute the urban centres that control the economy and host migration from rural areas. Manaus is the main Amazonian city and has a large amount of industry and factories. It has experienced an increase in its population that has led to the establishment of favelas and poor living conditions where its inhabitants are exposed to the pandemics that have devastated the region. These migrations and situations of vulnerability led to the first cases of HIV in the Amazon in the 1980s, which currently is a major health issue.

The population is distributed as follows: the state of Loreto in Perú has almost 1 million inhabitants, of which 157,529 live in the city of Iquitos. Rural migration fell from 67.7% (1940) to 34.6% (2007). This is probably due to the constant internal migration of the population from the countryside to the city. The department of Amazonas in Colombia has a population of approximately 80,000 people, of which 42,280 live in Leticia and almost half of them in rural areas. In turn, the state of Amazonas in Brazil consists of approximately 4 million inhabitants, of which 2,755,490 live in cities and 1,802,014 live in Manaus. The state of Pará consists of 7,581,051 people, of which 5,191,559 live in urban spaces and 1,393,399 live in Belem do Pará.

By the time Manaus and other cities experienced a rapid growth in the 1960s, the forests of the central and western floodplains were the main sources of wood. By 1980, most of the floodplains of the Amazon River had already been exploited. Even the magnificent Kapok tree (*Ceiba pentandra*) was destroyed on a large scale in most of the surrounding river areas. Illegal logging in these plains is a problem even in the Mamiraúa Sustainable Development Reserve (at the confluence of the Solimões and the Caquetá-Japurá Rivers) and at the source of the Amazon River.

Cattle have been raised on the floodplains of the Amazon River for at least 150 years. Livestock destroys the forest due to the increase of grass production and other herbaceous plant species that are needed for livestock feed. Cattle, in turn, destroy floating meadows and other communities of herbaceous plants. Thus, large-scale cattle ranching constitutes a destructive cycle for the floodplain ecosystem. This impact, in addition to other human interventions in the ecosystem such as hydroelectric works, large-scale and illegal mining and palm, soybean and coca crops, among others, have drastically impacted the health of the population. It was demonstrated that the affected areas have much higher rates of malaria and yellow fever than areas with a high degree of environmental conservation. Some diseases that these activities have generated are associated with the consumption of mercury derived from gold mining, affecting many riverine communities.

14.3 Ecosystem Transformation and Its Impact on Health: Chagas Disease and Yellow Fever in the Amazon

One of the main characteristics of the Amazon is its enormous variety of palms. Carl Friedrich P. von Martius and zoologist Johan B. von Spix travelled more than 2200 km (1817–1820) along and across the Amazonian basin. The result of this journey was *The Book of Palms,* a detailed work containing illustrations and a modern palm taxonomy, along with the first maps of the biogeographic distribution of Brazilian palms. Years later, Alfred Russel Wallace, after carrying out multiple expeditions in the Brazilian Amazon (1840–1852), published a detailed account of the distribution of palms and their uses, including detailed drawings and maps (Wallace 1992; Guhl and Ramirez 2013; Ministerio da Saude 2011) (Map 14.2).

Palms constitute natural ecotypes that serve as a refuge for a large variety of animals such as snakes, amphibians, reptiles, birds and many types of mammals, including bats, rodents and small and large marsupials such as *didelphids* and *tamanduas*. Many of these mammals act as reservoirs for different parasites such as leishmaniasis and American trypanosomiasis (Chagas disease) and serve as a food source for multiple bloodsucking insects that transmit them while maintaining the zoonotic cycles of parasites in the environment.

The changes over time generated by both native communities and settlers have led to interventions in the natural ecosystem, such as hunting animals, harvesting palm fruits to feed domestic animals and using palm leaves to build roofs. When humans enter the natural cycle, they become additional actors in the transmission of infectious agents, along with domestic animals, thus beginning an overlap of the sylvatic and domestic transmission cycles, with humans as accidental actors.

Map 14.2 Distribution of palms in the Americas, drawn by AR Wallace (1853). The shaded area refers to the density of palm forests in the Amazon (Wallace 1992)

Fig. 14.3 Model predicting the structure of the transmission cycle of Trypanosoma cruzi genotypes (Guhl and Ramírez 2013)

Figure 14.3 shows how human incursion into the forest leads to a permanent exchange of vectors, reservoirs and transmission of different genotypes of *Trypanosoma* cruzi between the sylvatic and domestic cycles.

Additionally, the Amazon has large areas of fruit palms (Fig. 14.4a, b) that are used by local inhabitants for consumption and commercialisation. The assai palm *Euterpe oleracea* is very important to the locals due to its good tasting fruits that have high vitamin content.

Many Triatominae insects inhabit the palms and are naturally infected with *Trypanosoma cruzi.* When the palms are harvested, the faeces of the insects contaminate the fruits causing frequent outbreaks of oral Chagas disease in the population, many of them with cases of high mortality. *T. cruzi* transmission is through food contamination, particularly in juices from palms such as assai which today is being considered as endemic in the Amazon region with more than 1500 cases notified (Coura and Junqueira 2015).

On the other hand, the expansion of trade and fluvial transport associated with the storage of fresh water in the boats offered the favourable conditions for the breeding and transport of the *Aedes aegypti* mosquito from one port to another. This is why from the first experiences with yellow fever it was thought that it was confined to urban communities along the riverbed.

Today we know that there is yellow fever in the jungle where *Ae. aegypti* is replaced by other mosquitoes of the genus *Haemagogus* and *Sabethes.*

Both urban cycles and the silvatic cycles of yellow fever overlap, increasing their vectorial transmission potential to primates and humans.

Fig. 14.4 (**a**) The assai palm Euterpe oleracea. (**b**) The fruit of the palm constitutes a big industry in the Amazon for human consumption

14.4 Current Status of Epidemiology: The Case of the State of Amazonas in Brazil

In 1979, the Alma-Ata treaty was signed in Russia, where all countries commit themselves to bring health brigades to the peripheral regions of their countries, leading to that in the 1980s the indigenous communities were treated and countries began to know the epidemiological reality of this vast region. There are systematic health records of the most important Amazonian cities, where people are assisted in first-level medical services that show the existence of several pathologies that differ from the underreporting cases in the whole rural indigenous population.

To exemplify the diseases that cause the greatest morbidity and mortality in the Amazon basin, we will rely on the data presented by the Ministry of Health of Brazil in 2010 and on the epidemiological bulletins of the year 2019, for the state of Amazonas, which allows to extrapolate, as is the situation in most of the basin.

As we have discussed throughout the chapter, there are diseases of the American continent and others that were introduced from different latitudes centuries ago and that today continue to be a problem for the health of the inhabitants of this region:

1. *Tuberculosis* 2292 cases are documented and the disease is reported for more than a century. This shows that the migration of people to cities and the changes of settlements in rural communities generate overcrowded conditions and poor

hygiene that allow the bacillus to continue to cause havoc, especially in the cords of misery in cities.

2. *AIDS* had its first notification in 1986 and was registered 7342 cases for 2010. The five municipalities had the highest presence: Manaus 6572 cases, followed by Parintins with 176, Tabatinga with 108, Itacoatiara with 67 and Tefé with 59. Although this is alarming and the finding that the records are made in urban centres, it is probably sure that there is underreporting and HIV is pandemic and is already a problem of large dimensions throughout the Amazon basin, where campaigns for prevention and for use of condoms should extend not only to cities.
3. *Acute diarrheal diseases (ADD)* are a constant problem and have a high incidence in the population. 664,569 cases have been reported during 1999 and 2010. This shows their importance and urgently needed attention, since children younger than 5 years are affected the most, often ending up in fatal cases.
4. *Viral hepatitis (A, B and C)*, which affected 10,525 cases during the same decade, should be addressed in a relevant and effective way, to avoid further progressing along the Amazon.
5. *American tegumentary leishmaniasis* was shown in 1439 persons and remains an indisputable problem in public health in the region.
6. *Malaria*, a disease that has ravaged the region for centuries, is still a problem of great magnitude. For example 74,130 cases had been reported in 2010, concentrating about 80% of the cases in 15 municipalities (Manaus, São Gabriel da Cachoeira, Eirunepé, Atalaia do Norte, Coari, Barcelos, Tabatinga, Manicoré, Tefé, Labrea, Borba, Tapaua, Rio Preto da Eva, Itacoatiara and Guajará), demonstrating that the plans for fumigation and use of mosquito nets and repellents have not been sufficient and that the anthropic intervention on the forest represents an increasingly serious scenario for the propagation of the *Anopheles* mosquitoes that transmit this parasite.
7. Among the diseases that have arrived in recent years are *dengue fever*, of which 3837 cases occurred in 2017 and 2407 in 2018. 247 chikungunya cases have been reported in 2017 and 79 in 2018, and in the same years, 415 and 370 cases occurred due to the *Zika virus*, respectively.

   This shows that the majestic Amazonas River spreads many agents of disease while transporting humans.
8. Other cases of disease such as leptospirosis, rubella and flu occur at low levels of cases; however, this is perhaps only due to underregistration.

14.5 Final Remarks

We can conclude that at present the inhabitants of the big cities along the Amazon receive adequate health care in hospitals and health centres. On the other hand, indigenous settlements along the Amazon River are precarious, since their living conditions away from the big urban cities are poor at a low nutritional status due poverty.

The permanent human intervention in the ecosystem over the years has generated new fronts of colonisation for agricultural development and the exploitation of wood and other natural resources as is the case of illegal mining added to the construction of roads and roads of penetration in the jungle causing profound changes in the different scenarios of transmission of zoonotic diseases involving man in their cycles of transmission including prostitution and alcoholism.

The vertiginous increase in river transport has created a permanent flow of migrants. Some come from the large riverside cities, importing infectious entities unknown to the inhabitants of the jungle and others that travel to the big cities in search of work and a better future. The migratory flow includes not only humans but all kinds of exotic animals that are sold as pets in urban centres and in the other direction domestic animals that are introduced into the forest increasing the dispersion of new infectious agents, some of which have become endemic entities either in the cities or in the indigenous villages.

Natural ecosystems and little intervention have generated throughout history a balance between man and nature that has allowed an ideal coexistence in terms of good health and well-being. The inhabitants that inhabit the course of the largest river in the world that also contains the largest water reserve on the planet deserve that this unique ecosystem must be protected and conserved for their well-being and also the well-being of humanity.

Acknowledgements Financial support was provided by the Research Committee of the Faculty of Sciences, University of Los Andes.

References

Biocca E (1944) Traducción: Alzate B (2007): Viaje entre los indígenas del río en Negro. Instituto Italiano de cultura, Bogotá Colombia

Black L (1992) Why did they die? Science 258(5089):1739–1740

Boletim Epidemiológico (2019) Secretaria de Vigilância em Saúde, Ministério da Saúde, vol 50, Jan 2019

Cabrera G (2002) La Iglesia en la frontera: misiones católicas en el Vaupés 1850–1950. Editorial Universidad Nacional de Colombia

Carneiro L (1995) The history of ecological interpretations of Amazonia: does Roosevelt have it right? In: Sponsel LE, Netting RM (eds) Indigenous peoples and the future of Amazonia: an ecological anthropology of an endangered world. The University of Arizona Press, Tucson, pp 45–69

Coura JR, Junqueira ACV (2015) Ecological diversity of *Trypanosoma cruzi* transmission in the Amazon basin. The main scenarios in the Brazilian Amazon. Acta Trop 151:32–50

Crosby A (1993) Imperialismo Ecologico – A Expansao biologica da Europa: 900–1900. Cia das Letras, Sao Paulo

da Saúde M (2011) Sistema Nacional de Vigilância em Saúde, Relatório de Situação Amazonas, 5a edn, Brasilia

Denevan M (1996) A bluff model of riverine settlement in prehistoric Amazonia. Ann Assoc Am Geogr 86(4):654–681

FIAN *Historia oral y proceso esclavista en el Caquetá* (1985) Bogotá D.C., Fundación de Investigaciones Arqueológicas Nacionales, Banco de La República
Furtado C (2007) Formação econômica do Brasil, 34a edn. Companhia das Letras, São Paulo
Gnecco C, Mora S (1997) Early occupations of the tropical forest of northern South America by hunter-gatherers. Antiquity 71(273):683–690
Gonzales C (2009) Fray Gaspar de Carvajal, cronista de la expedición de Francisco de Orellana en el descubrimiento del río Amazonas. Coloquios históricos de Extremadura, España
Goulding M, Barthem R, Ferreira E (2010) The Smithsonian Atlas of the Amazon. Princeton Editorial Associates, Princeton, NJ, p 232
Guhl F, Ramírez JD (2013) Retrospective molecular integrated epidemiology of Chagas disease in Colombia. Infect Genet Evol 20:148–154
Heckenberger J (2005) The ecology of power: culture, place, and personhood in the Southern Amazon, A.D. 1000–2000. Routledge, New York
Koch-Grünberg T (1909) Zwei Jahre unter den Indianern Reisen in Nordwest-Brasilien 1903–1905. Traducción Dos años entre los indios. (1995) Watzke A, Camacho R. Editorial Universidad Nacional de Colombia, Bogotá
La Condamine CM (1993) Viaje a la América meridional por el río Amazonas, 2nd edn. Quito, Ecuador, Editorial Abya-Yala
Lathrap W (1970) The upper Amazon: ancient peoples and places. Thames & Hudson, Southampton
Leles D, Araújo A (2015) Potencialidades da Amazônia para a paleoparasitologia. Revista de patología tropical 44:229–244
Martins de Souza C (2015) Morte, saúde e ditadura na construção da transamazônica. Tempos Históricos 19:65–91
Meggers B, Evans C (1957) Archeological investigations at the mouth of the Amazon. Bulletin 167. Smithsonian Institution, Washington D.C.
Pineda R (1985) Historia oral y proceso esclavista en el Caquetá. Bogotá D.C., Fundación de Investigaciones Arqueológicas Nacionales (FIAN), Banco de La República
Pineda R (2013) El río Amazonas: un gigante indomado. Una mirada hacia su historia contemporánea (1500–2010) Boletín Cultural y Bibliográfico, vol 47, no 87
Rapp-Py Daniel A (2015) Os contextos funerários na arqueologia da calha do rio amazonas. Tese de Doutorado apresentada ao Programa de Pós-Graduação em Arqueologia do Museu de Arqueologia e Etnologia da Universidade de São Paulo
Riponi L (2016) A única vida possivel. Itinerarios de Ermanno Stradeli na Amazônia. Editora UNESP, Brasil
Roosevelt A, Douglas J, Brown L (2002) The migrations and adaptations of the first Americans: Clovis and Pré-Clovis viewed from South America. In: Jablonski N (ed) The first Americans, the Pleistocene colonization of the new world, vol 27. Memoirs of the California Academy of Sciences, San Francisco, pp 159–235
Rosas Moscoso F (1986) La Amazonia en conflict (Siglos XVI al XVIII). Rev Anthropol 4 (4):39–68
San Roman JV (1994) Perfiles Históricos de la Amazonía Peruana. Centro de Estudios Teológicos de la Amazonía. Centro Amazónico de Antropología y Aplicación Práctica Instituto de Investigaciones de la Amazonía Peruana, Iquitos
Santos Da Silva J (2012) Adoecendo na cidade da borracha: Manaus (1877–1920). Tese de Doutorado apresentada ao Universidade Federal do Amazonas - UFAM Instituto de Ciências Humanas e Letras - ICHL Programa de Pós-graduação em História - PPGH
Smith N (1977) O modelo brasileiro de assentamento na Rodovia Transamazônica: agrovilas, agrópolis e rurópolis. In HEBETTE, Jean et al. Colonização, desenvolvimento e modelos econômicos. Belém, NAEA/UFPA
Sotomayor HA (1995) Una historia epidemiológica de Colombia. Enfoque de las grandes etapas históricas de la salud en Colombia. En Medicina y salud en la historia de Colombia Javier Guerrero (compilador). Colección Memorias de historia. IX Congreso de Historia de Colombia.

Tunja 2,3,4 y5 de mayo de 1995. Editado por la sección de Publicaciones de la Universidad Pedagógica y Tecnológica de Colombia, pp 32–56

Viana A (1975) As Epidemias no Pará. Universidade Federal do Pará. Belem (primeira edição em 1906)

Wallace A (1992) Una Narración de viajes por el Amazonas y el río Negro (1848–1852). Monumenta Amazónica. IIAP, Iquitos

Zárate C (2001) Extracción de quina: la configuración del espacio Andino-Amazónico de fines del siglo XIX. Universidad Nacional de Colombia, Bogotá

Chapter 15
Brazil's Big River Projects and Their Impact on the Occurrence of Tropical Diseases

Big River Projects and Tropical Diseases in Brazil

Alberto Novaes Ramos Jr, Marta Cristhiany Cunha Pinheiro, Sharmênia Araújo Soares Nuto, Edenilo Baltazar Barreira Filho, and Jorg Heukelbach

Abstract Big river projects have been important in achieving sustainable development in Brazil. However, these projects often have considerable impacts on human and social development, with negative effects on biodiversity and environmental preservation, leading to the emergence and re-emergence of tropical diseases.

Keywords Communicable diseases · Schistosomiasis · Malaria · Yellow fever · Dengue · Zika · Chikungunya · West Nile virus · Filariasis

15.1 Background

Water infrastructure projects, which include irrigation systems, large and small dams and reservoirs, may cause profound changes in aquatic environments, as well as in the surrounding ecosystems and communities (Gergel and Turner 2017; Heinrich et al. 2017; Steinmann et al. 2006). Large water projects have been responsible for important environmental, economic, and sociocultural changes, interfering in the daily life of the surrounding populations and their livelihoods,

A. N. Ramos Jr · M. C. C. Pinheiro · J. Heukelbach (✉)
School of Medicine, Department of Community Health, Federal University of Ceará (UFC), Fortaleza, Ceará, Brazil

S. A. S. Nuto
Oswaldo Cruz Foundation (FIOCRUZ), Eusébio, Ceará, Brazil

University of Fortaleza (UNIFOR), Fortaleza, Ceará, Brazil

E. B. Barreira Filho
Faculty of Education and Culture of Ceará (FAECE), Law School, Fortaleza, Ceará, Brazil

H. Mehlhorn, S. Klimpel (eds.), *Parasite and Disease Spread by Major Rivers on Earth*, Parasitology Research Monographs 12,
https://doi.org/10.1007/978-3-030-29061-0_15

especially regarding health conditions and quality of life (Souza et al. 2017; Oliveira et al. 2006, 2007; WHO 2000; Lereer and Scudder 1999).

Although irrigation systems date back to ancient times, large-scale dam construction is a relatively more recent phenomenon (Gergel and Turner 2017). In the past 100 years, water consumption has increased almost sixfold and maintains an annual increase of 1%. The water crisis has emerged to an increasing extent in the twenty-first century, as aggravated by the social, economic, and environmental context (Wallet et al. 2016; Tundisi 2008). At the current rate of population growth, economic development, and climate change, it is estimated that the number of people living in areas with scarcity of water for at least 1 month per year will increase from currently 3.6 billion to 5.7 billion in 2050 (UN WATER 2018). The construction of large dams has become an alternative in the management of water use, to enable economic and social development in many countries (Wallet et al. 2016; Tundisi 2008; WHO 2000). Construction of dams requires considerable financial investment and political will, involving cutting-edge technology for the provision of a renewable source of energy (Wallet et al. 2016). Worldwide there are approximately 60,000 large dams (defined as a dam with a height of 15 m or greater measured from the lowest foundation to the crest or those between 5 and 15 m in height and a reservoir capacity over 3 million cubic meters). Most of them are single-purpose dams (49.5%), with irrigation as the main purpose (50.0%) (ICOLD 2018; UN WATER 2018). Estimates indicate that 32 countries have more than 80% of their electricity needs covered by hydropower generation dams, such as Brazil, Mozambique, Nepal, and Norway (THE WORLD BANK 2014). Brazil, because of its continental dimensions and privileged hydrology, has more than 1300 hydroelectric power plants in operation, 40% located in the Amazon region (ANEEL 2019).

Major ecological changes are observed both upstream and downstream of dams. Floodings upstream of dams may result in the permanent destruction of terrestrial ecosystems through inundation, eventually causing terrestrial plants and animals to disappear in submerged areas (Gyasi et al. 2018; Wallet et al. 2016; Zarfl et al. 2015). Transformations of natural landscapes caused by dams have effects on the soil, water, atmosphere, fauna, flora, biodiversity, and the human population itself, contributing to the development and dissemination of several diseases, with proliferation of disease vectors and geographical spread of emerging and reemerging diseases (Tubaki et al. 2004; Tundisi 2008; Sleigh and Jackson 2001; WHO 2000). Even though substantial advances in biomedical sciences and public health measures have facilitated the control of many infectious diseases, human health and ecosystem issues have become more and more important (Barcellos et al. 2009). There has been an increase in the incidence and prevalence of diseases that mainly affect populations living in poverty, with inadequate sanitation, lacking access to safe drinking water, and with poor access to health services (Gyasi et al. 2018; Sanchez-Ribas et al. 2012; Sleigh and Jackson 2001; WHO 2000). In Brazil, particularly intestinal schistosomiasis, a parasitic waterborne disease that leads to an expressive number of severe forms and deaths annually, is an important public health issue (Bezerra et al. 2018; Brasil 2018; Martins-Melo et al. 2014, 2015, 2018; Steinmann et al. 2006).

Large dams also influence health through changes in the social determinants of water and food security, in addition to the social disturbances caused by construction. In general, communities close to large dams do not directly benefit from revenues associated with water transfer and electricity generation, but experience the negative impacts on their lives (Gyasi et al. 2018; Lereer and Scudder 1999). The Brazilian Movement of People Affected by Dams (*Movimento dos Atingidos por Barragens—MAB*) estimates that 1 million Brazilians have suffered negative impacts from the construction of dams and that 70% of these populations have not received any compensations. MAB, originally formed by people affected by the large Itaipu dam in the 1970s, has grown into a strong national movement of dam-affected communities (MAB 2019).

15.2 Health Impact Assessments of Large Dams

Prior to the implementation of large water projects, a proper assessment of the impacts on the environment and on local communities and their health has to be performed, in addition to standard economic and environmental feasibility studies (Poma 2018; Bacci and Pataca 2008; Zhouri and Oliveira 2007). Despite its importance, the study of dam impacts on health is a process that has been implemented only recently (Wallet et al. 2016; WHO 2000). Health determinants are the factors causally related to the health status of communities. Table 15.1 presents some of the main social determinants of health.

The assessment of human health impacts resulting from large water projects helps to understand the burden of diseases associated with the economic activities already installed and in phase of installation in the territories (Souza et al. 2017; Fase and Ettern 2011). In fact, social impacts of dams include livelihoods, health, social systems, and people's cultures (Wallet et al. 2016). However, the complexity of health aspects requires the integration of multi-professional technical teams, in order to optimize the health of the populations affected by the construction of large dams (Sleigh and Jackson 2001).

Environmental Impact Assessments (EIAs) were adopted and formalized in many countries during the 1980s, although many developing countries adopted EIA legislation only in the 1990s. In many cases, however, these assessments were regarded as a merely bureaucratic element with administrative constraints (Wallet et al. 2016). Health Impact Assessments (HIA) were defined, understood as a set of procedures, methods, and systematic tools (Souza et al. 2017; Bhatia 2011; Quigley et al. 2006). HIA identify unintentional changes in the process of determining health in the communities, resulting from projects with potential territorial damage (Bhatia 2011; WHO 2000). The systematic identification of hazards, risks, and opportunities for health can be used during the planning of projects for implementation of measures of mitigation and health promotion. HIA can also inform the public and subsidize decision-makers on projects with potential impacts on human health (Quigley et al. 2006). HIA should consider the abovementioned determinants of

Table 15.1 Examples of health determinants and their classification

Main categories	Fields	Examples of health determinants
Individual/ family	Biological	Genetics, age, gender, immunity, nutritional status
	Behavior/ lifestyle	Risk of acceptance and behavior, occupation, education
	Circumstantial	Poverty, empowerment, family structure
Environmental	Physical	Air, water sources and soil, infrastructure, disease vectors, housing, power, land use, environmental pollution, agriculture, food, traffic
	Social	Community structure, culture, crime rates, discrimination, social cohesion
	Economic/ financial	Unemployment, investment rate, interest rate, inflation rate
Institutional	Health services	Primary health care, specialized services, access, timely diagnosis, medicines supply
	Other services	Police, transportation, public constructions, municipal authorities, local government, ministry of project sector, local community organizations, NGOs, emergency services, access, education
	Public policies	Regulations, jurisdictions, laws, goals, limits, priorities, standards

Source: Adapted from WHO (2000)

human health and assess the changes in health risks attributable to the project under evaluation. In turn, impacts should not be considered only from a negative point of view: both positive and negative impacts must be determined. The HIA should be properly funded and included at the beginning of the project, having adequate and sound technical basis, but also have clear components of population monitoring in regard to health aspects after construction (Wallet et al. 2016).

In Brazil, the EIA and the Environmental Impact Report (*Relatório de Impacto Ambiental—RIMA*) are the most important documents of the environmental licensing process. Together, they provide support to the licensing of projects whose production process presents health risks, aiming to decide on the viability of their implantation, considering the impacts to the biological, physical, social, and cultural dimensions, as required by law (BRASIL 2014). Although the regulations concerning environmental licensing in Brazil also highlight the need for HIA in EIAs and their respective RIMAs (CONAMA 1986, 1997), these procedures still occur in an insufficient and superficial way. With the aim of integrating critical elements for a correct assessment, the World Commission on Dams recommended the following components to be included in the assessments (Wallet et al. 2016; WHO 2000):

- Determination of the number and location of the populations affected. This includes all people living in vicinity of the reservoir (downstream, upstream, and in catchment areas whose properties, livelihoods, and non-material resources are affected).

- Baseline assessment of the health status of the affected populations, considering also socioeconomic aspects. The health diagnosis comprises questionnaires for specific research, construction of health status indicators (such as incidence and prevalence of diseases, mortality, nutritional, and psychological status, among others), and recognition of risk factors (water sources, water quality, sanitation, pesticide use, hygiene standards, waste disposal, air quality, noise levels, presence of insects or molluscan vectors, biodiversity, access to health services, among others), and should involve local people and resources. Impacts vary according to the population's baseline health status and according to the population's ability to withstand new health concerns induced by this type of enterprise.
- Classification of planned changes (classified as defined, likely and possible) regarding health determinants related to the project. The changes can produce health impacts or alterations in health status, and direct and indirect effects. Comparison with health conditions should be made before the enterprise becomes fully operational, but the project duration is so long that the health impact may occur earlier, as it develops. Furthermore, the time lag between the decision to build a dam and starting the construction may lead to stagnation of local development because governments, entrepreneurs, farmers, and other key players are reluctant to make productive investments in areas that can be flooded later. As all health impacts cannot be anticipated and some expected may not manifest as initially anticipated, HIA should be complemented by systematic long-term monitoring.
- Development of measures to prevent, minimize, and mitigate the effects of the enterprises. It is essential that potentially affected individuals participate in the HIA. Appropriate actions identified by the HIA can be applied as a Public Health Management Plan, estimating the cost of preventing or mitigating potential health impacts in the overall project cost estimation. Determining what constitutes an acceptable level of risk should be undertaken as part of a public policy process with broad democratic participation through multi-criteria analysis.

In accordance with the provisions of the Brazilian legal system, the environmental license is possible, even if the environmental impact study is unfavorable. Such an assertion is based on Articles 170, V and 225 of the Constitution itself, when referring to the existence of sustainable development, with a view to achieving a balance between environmental protection and free competition, guiding the economic development. If EIS/RIMA are unfavorable, "the balance between the ecologically balanced environment and the economic development will be the object of a study by the administration to grant or not the environmental license" (Fiorillo 2018). In case the EIA/RIMA is totally or partially unfavorable, it is up to the public administration, according to the convenience and opportunity to release the license or not. In case of not releasing the license, a justification by the Public Entity is not necessary. However, if management deems the granting of the license to be timely, it is imperative to address each of the points that have been presented as impacting.

Table 15.2 Major categories of health issues and the extent of current knowledge regarding their association with large reservoirs/dam projects

Health issues	Examples	Knowledge base
Communicable diseases	Vector-borne, waterborne, sexually transmitted, zoonoses, other parasitoses	Large, reliable, ecosystem-specific, with some quantification
Noncommunicable diseases	Poisoning by minerals, biological toxins, pesticide residues, industrial effluents	Geographically limited, reliable, generalizable and often well quantified
Injuries/health concerns	Drowning, construction damage, domestic and community violence, catastrophic failures, seismic activity, traffic accident injury	Limited, reliable, transferable, some statistics
Malnutrition	Lack of protein, carbohydrates or essential elements	Limited and controversial, limited transferability, reasonably quantified, limited reliability
Psychosocial disorder	Stress, suicide, substance abuse, social disruption, restlessness, violence, decreased tolerance	Low volume, poor reliability with little quantification and cultural variation
Social welfare	Quality of life, social cohesion and support structures, self-determination, human rights, equity	Low volume, of variable reliability and quantification, with considerable cultural variation

Source: Adapted from WHO (2000)

When the EIA/RIMA is favorable, it conditions, which means that the administration is supposed to grant the environmental license.

15.3 Large Dams and Public Health

In contrast to economic development, large dams often generate negative impacts on human and social development, and on the biodiversity and sustainability of countries that do not adopt strategic policies for proper assessment and implementation (Wallet et al. 2016; Leturcq 2016). Table 15.2 summarizes the main health problems related to large dams. Direct and indirect effects of large dams on the transmission dynamics of parasitic and vector-transmitted diseases are highly relevant to the health of human populations (Oliveira et al. 2006, 2007). Infectious diseases may be introduced and/or disseminated within the area of influence of reservoirs of large dams, particularly in locations with a hot climate and significant human population density (Nava et al. 2017; Heinrich et al. 2017; Agrawal 2017; Steinmann et al. 2006).

In principle, health issues that have water as a possible central element in its transmission can be categorized into four categories (Oliveira et al. 2006; Yamana 2004; WHO 2000):

1. Waterborne diseases, caused by the direct consumption of water contaminated by human, animal, or chemical waste. These diseases are especially prevalent in areas lacking access to adequate sanitary facilities and include diarrhea, dysentery, cholera, hepatitis A, and typhoid fever.
2. Diseases with water-based transmission caused by parasites that spend at least part of their life cycle on water in intermediate hosts (e.g., freshwater snails). These include schistosomiasis, prevalent in many parts of the world.
3. Water-related diseases, transmitted by vectors that rely on water to live and reproduce/propagate. Vectors are insects or animals that carry and transmit parasites between infected persons or animals, as occurs with malaria, filariasis, yellow fever, and other hemorrhagic viral infections transmitted by mosquitoes (dengue fever, Chikungunya, Zika infection, and West Nile fever).
4. Diseases related to water, sanitation, and hygiene, which can be prevented by hand washing, such as trachoma and onchocerciasis, or by means of sanitation, such as intestinal parasitic diseases.

In the case of communicable diseases, the main health issues related to large dams fall mostly into categories 2 and 3 (Yamana 2004). Reservoirs formed by dams provide large areas of stagnant water, ideal for parasites and vectors reproduction, and so do irrigation canals and drainage networks. Dams also provide a steady source of water, supporting the life cycle of various microorganisms and insects in humid environments, which is the case for the mosquito-transmitted parasitic leishmaniases (visceral and tegumentary). In fact, the biggest health threats come from diseases endemic to the affected regions, whether controlled or not, directly or indirectly associated with the increased presence of water in large dams (Kibret 2018; Gyasi et al. 2018; Heinrich et al. 2017; Wallet et al. 2016; WHO 2000; Lereer and Scudder 1999). Water projects have demonstrated a relationship with the emergence of diseases in such areas by providing favorable habitats and increasing varieties of vectors and host species (Patz et al. 2004). Global evidence points to an increase in the transmission dynamics of persistent, emerging or reemerging pathogens with a strong correlation with the construction of large dams, such as schistosomiasis, malaria, and Rift Valley fever (Nava et al. 2017; Heinrich et al. 2017; Steinmann et al. 2006). Table 15.3 presents basic aspects of the main communicable diseases related to water projects in terms of agents, vectors, transmission route, interaction with water, control, and prevention.

Dams have long been associated with high burdens of human schistosomiasis (Sokolow et al. 2017). Different studies have shown an increase in schistosomiasis following the construction of dams, such as the Aswan Dam in Egypt, the Diama Dam at the mouth of the Senegal River, and the Three Gorges Dam in China (Heinrich et al. 2017). Schistosomiasis is recognized as a sensitive indicator disease for monitoring ecological transformations because it is widespread throughout the African continent, and its prevalence may change promptly after environmental impacts by water projects (Gergel and Turner 2017). In this way, its occurrence can be used as a strategic tool for the indirect monitoring of other water-related diseases, such as malaria and onchocerciasis, even with different transmission dynamics.

Table 15.3 General characteristics of communicable diseases potentially related to water projects

Disease	Agent, vector, and transmission route	Interaction with water	Control and prevention
Schistosomiasis mansoni (intestinal schistosomiasis)	*Schistosoma* worm infection; worm larvae develop into freshwater snails and penetrate human skin; infected humans transmit the eggs through the feces or urine, which, in turn, infect the snails	Host snails prefer running or slow water; affected by irrigation projects	Better sanitation, safe access to water; prevention and control of snail vectors
Malaria	*Plasmodium* parasite, transmitted by mosquito bites	Closely related to the presence of stagnant or slow flowing water for breeding sites; affected by irrigation and water storage	Mosquito control measures, including reduction of breeding sites and prevention of bites
Yellow fever	Yellow fever virus, transmitted monkey-to-monkey, monkey-to-human, or human-to-human by mosquitoes	Stagnant waters preferred as breeding sites of mosquitoes	Mosquito control measures, including reduction of breeding sites and prevention of bites
Dengue fever	Dengue virus, transmitted by mosquito bites	Stagnant waters preferred as breeding sites of mosquitoes	Mosquito control measures, including reduction of breeding sites and prevention of bites
Chikungunya	Chikungunya virus, transmitted by mosquito bites	Stagnant waters preferred as breeding sites of mosquitoes	Mosquito control measures, including reduction of breeding sites and prevention of bites
Zika infection	Zika virus, transmitted by mosquito bites	Stagnant waters preferred as breeding sites of mosquitoes	Mosquito control measures, including reduction of breeding sites and prevention of bites
West Nile fever	West Nile fever virus transmitted by mosquito bites	Stagnant waters preferred as breeding sites of mosquitoes	Mosquito control measures, including reduction of breeding sites and prevention of bites
Lymphatic filariasis	Parasite of the *Filariidae* family, transmitted by mosquito bites	Stagnant and polluted waters are preferred breeding sites	Mosquito control measures, sanitation improvements

Source: Adapted from Heinrich et al. (2017)

Broadening the hypothesis of such relationship, it has recently been argued that dams could block the reproduction of migratory shrimp that feed on the snail intermediate host of schistosomiasis. In the Senegal river basin there is evidence that shrimp populations have declined and schistosomiasis has increased after completion of the Diama dam. Restoring shrimps to a water-access site upstream of the dam reduced snail density and reinfection rates in humans, though this effect has not been reproduced in another context. A study addressing large dams around the world identified that the construction of dams was followed by a substantive increase in cases of schistosomiasis in traditional shrimp habitats. It was estimated that one-third to one-half of the global population at risk of schistosomiasis could benefit from restoring native shrimp populations as an ecological solution for reducing human disease (Sokolow et al. 2017).

Malaria is a disease that puts the lives of infected people at risk, caused by parasites that are transmitted to humans through bites of infected female *Anopheles* mosquitoes. Communities living near reservoirs are at a greater risk of contracting malaria than those living farther, since damming rivers slows down the water flow and thus provides breeding habitats for mosquitoes (Kibret 2018; Sanchez-Ribas et al. 2012). There is strong evidence that large dams have an impact on malaria transmission in Africa, with estimates of at least 1 million cases each year from the analysis of 956 large dams (Kibret et al. 2015). With current population growth and climate changes, this number is expected to increase to 26 million by 2050, generating 2.1–2.9 million cases of malaria associated with water reservoirs (Kibret et al. 2016).

In Brazil, *Trypanosoma cruzi*, the etiological agent of the parasitic disease transmitted by bugs, American trypanosomiasis (Chagas' disease), was isolated in bats captured in the vicinities of the Belo Monte power plant in the Brazilian Amazon, indicating the potential risk of zoonotic dissemination to growing cities in the region (Costa et al. 2016).

The ecological imbalance generated by an exploratory and unsustainable nature of economic development (Giongo et al. 2015; Granziera 2011), in association with the emergence of new human and animal populations, can affect the life cycle of disease vectors, shifting from a predominant zoophilic pattern to an anthropophilic one. This means for example, that the mosquitoes will attack more and more human populations. Migration is important in this context, as this implies arrival of individuals with high susceptibility to parasites in the affected regions, bringing customs and habits that may contribute to the proliferation and transmission of autochthonous parasites or by installing new transmission foci of non-endemic diseases (Patz et al. 2004).

In fact, the natural history of tropical diseases and the risks of infectious diseases are affected by many factors, including changes in the habitat structure, biodiversity, host demographics, and parasite evolution (Cable et al. 2017). Habitat loss and fragmentation is considered a major threat to ecosystems, with loss of biodiversity and reduced response to external stressors. Modifications in the biodiversity and river hydrology also take place, in addition to direct and indirect impacts on health due to climate changes (Agrawal 2017). There are changes in water quality and algal proliferation, and variation in the mineral profile of the soils of affected areas, with indirect impacts on the health of human populations (Wallet et al. 2016). Two scientific perspectives diverge about the role of biodiversity: some point out that high biodiversity can dilute the parasites in a large number of vector species and intermediate hosts, most of them being less transmission-competent to humans, thus reducing the overall incidence of parasitic diseases; others point out that high biodiversity implies a greater number of parasites and their natural reservoirs, with high levels of disease transmission (Chivian and Bernstein 2008).

Environmental changes generate the possibility of emergence, persistence, or amplification of the transmission dynamics of communicable diseases, including those transmitted by water, as previously discussed, the vector-borne diseases, such as malaria, schistosomiasis, Japanese encephalitis, diarrhea, among others, and those transmitted by animals, known as zoonoses (Wallet et al. 2016; Guergel 2013; Steinmann et al. 2006; Lereer and Scudder 1999). Both issues related upstream, such as loss of biodiversity, increasing agriculture, sedimentation and flooding, as well as changes in the river flow regime, may impact the transmission dynamics of infectious agents. In terms of the reservoir itself, flooding of land, presence of a large artificial reservoir, pollution, changes in mineral content, and decomposing organic material contribute to the impact on the health of local populations. Downstream, water security can be compromised because of lower water levels, poor quality, lack of seasonal variation, and loss of biodiversity in surrounding areas. Finally, in the areas of irrigation created or enhanced, there is greater availability of water and agriculture, however, with risk of growing aquatic weeds, changes in flow and mineral content, pollution, and contamination/impregnation by pesticides (Lereer and Scudder 1999). As a result of these environmental changes, there is the possibility also of noncommunicable diseases and other health concerns, such as poisoning by minerals, biological toxins, pesticide residues, industrial effluents, mercury intoxication in prospective mining areas, potentially leading to fish contamination, dust-induced respiratory illness, and uncontrolled proliferation of mosquitoes (Wallet et al. 2016). Together with the communicable diseases, they can lead to a rapid epidemiological transition in the region, coexisting with health issues that become more complex when associated. These effects on human health are critical, even before the project completion (Nava et al. 2017). The construction phase itself involves respiratory diseases caused by dust, accidents and impacts on workers' health due to issues related to qualification and working conditions (Lereer and Scudder 1999). The construction process requires the mobilization of a large number of workers from other regions, many remain in the region after construction. This fact not only increases the impact on the environment, with a rapid impact on the transmission of malaria, but also increases the occurrence of other infectious diseases, including HIV/AIDS (Fearnside 2006, 2015; Patz et al. 2004; Randell 2016a, b). The dismantling of the social organization of the impacted territories may also lead to an increase in traffic accidents (increased road traffic) drowning incidents, for example, as well as greater impacts on seismic activity, with the possibility of major tragedies, when dealing with constructions with higher risks or lower technical quality.

As a result of large water projects, there are deep social changes, increasing inequality, particularly affecting the most vulnerable populations (migrants, indigenous people, black populations, women, children, and older adults). Changes in lifestyle, social disruption, and loss of referral in displaced communities generate poverty and psychosocial disorders, including stress, fear, anxiety, depression, suicide, substance abuse, violence, and stigma (Wallet et al. 2016). These changes also increase the exposure to sexually transmitted infections, including HIV/AIDS, changes in employment patterns in the areas (informal work and commercial sex work), changes in agricultural production, livestock, and fisheries, in addition to generating food insecurity and malnutrition. The reduction or loss of subsistence capacity of local populations, coupled with uncertainties and lack of empowerment, renders the health concerns even more critical.

In a country with continental dimensions such as Brazil, which has large regional differences and areas with low coverage of sanitary and health infrastructure, the impacts are extremely severe (Souza et al. 2017; Zhouri and Oliveira 2007). A classic example is the change in the epidemiological profile of some parasitic diseases that affect Amazonian populations in the face of environmental changes caused by expansion of agriculture and livestock farming, construction of roads and dams, and mining (Arrifano et al. 2018; Becker 2012): malaria and leishmaniasis are those that respond more quickly to these changes; toxoplasmosis, schistosomiasis, and soil-transmitted helminthiasis are highly likely to respond rapidly to such alterations (Randell 2016a, b; Galardo et al. 2015; Fearnside 2015; Confalonieri et al. 2014). Schistosomiasis-related aspects are discussed now in greater detail (Wallet et al. 2016), taken as a case study.

15.4 Schistosomiasis as an Example

Schistosomiasis (or bilharzia) is prevalent in tropical and subtropical areas, especially in poor communities without appropriate access to water and inadequate sanitation. It is estimated that at least 91% of those in need of treatment for schistosomiasis live in Africa (WHO 2018). Schistosomiasis is caused by blood flukes (trematode worms) of the genus Schistosoma. People are usually infected during agricultural, domestic, occupational, tourism, and recreational activities in which they are exposed to infested water. Poor hygiene and certain habits of children, such as playing, swimming, or fishing in infested rivers and lakes, make them particularly vulnerable. Transmission of schistosomiasis has been reported in 78 countries.

Figure 15.1 depicts the distribution of schistosomiasis in the world.

People are infected when larval forms of the parasite—released by freshwater snails—penetrate the skin during contact with infested water. Transmission occurs when people with schistosomiasis contaminate freshwater sources with their excreta containing parasite eggs, which hatch in water. In the body, the larvae develop into adult schistosomes. Adult worms live in the blood vessels, where the females release eggs. Some of the eggs are eliminated from the body in the feces or urine to continue the life cycle of the parasite. Others are trapped in the tissues of the human body,

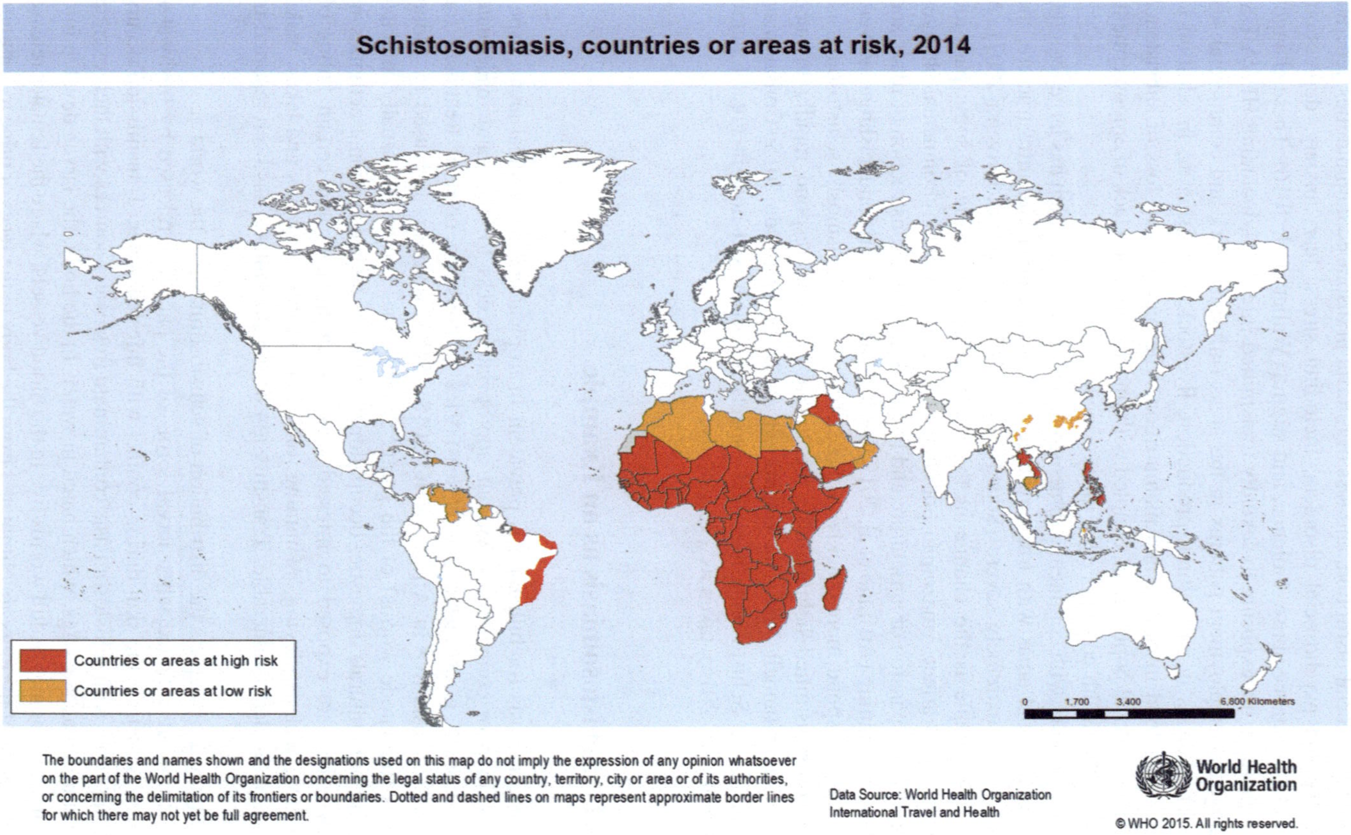

Fig. 15.1 Distribution of schistosomiasis in countries or areas at risk in the world, 2014. World Health Organization (WHO). Source: http://www.travelhealth.ie/images/maps/large/Schistosomiasis2014.png

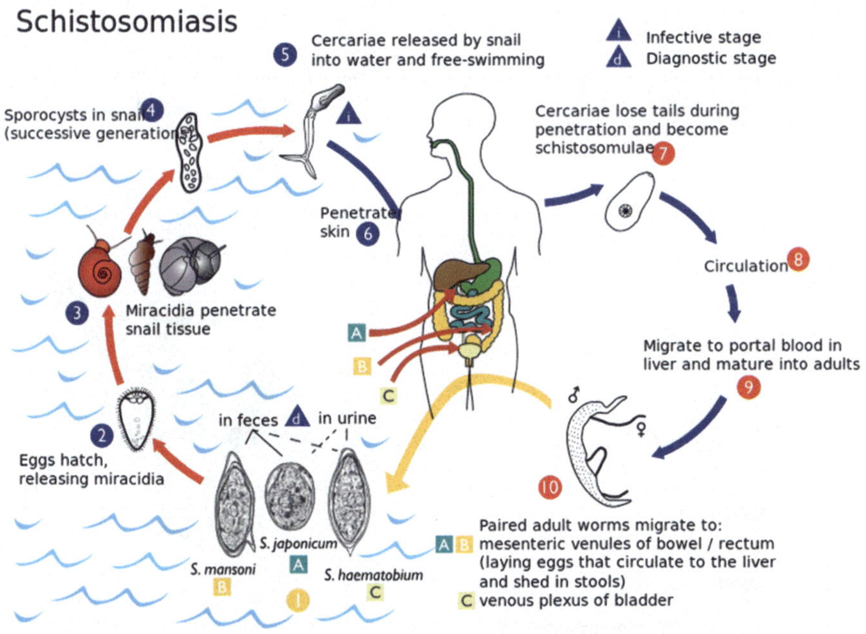

Fig. 15.2 Life cycle of *Schistosoma* spp. Source: Centers for Disease Control and Prevention (CDC)—https://www.cdc.gov/dpdx/schistosomiasis/index.html

causing immune reactions and progressive damage to organs. Figure 15.2 shows the life cycle of *Schistosoma* spp.

Control of schistosomiasis is based on reducing disease burden by means of periodic, large-scale mass treatment with the antiparasitic drug praziquantel. Estimates show that more than 200 million people required preventive treatment in 2016 (WHO 2018). Preventive treatment, which is repeated over several years, has the potential to reduce the population worm burden and prevent severe morbidity. Mass treatment is embedded in integrative control measures, including access to potable water, adequate sanitation, and snail control for transmission reduction (WHO 2018). Since tourism in endemic areas is associated with the risk of occurrence of cases, general preventive measures should be adopted. These include avoidance of swimming, or other contact with freshwater (rivers, lakes) in areas of risk (WHO 2018).

There are two main clinical forms of schistosomiasis—intestinal and urogenital—caused by five major species of Schistosoma, with specific geographic distribution. Intestinal schistosomiasis is caused by four different species: (1) *Schistosoma mansoni* in Africa, the Middle East, the Caribbean, Brazil, Venezuela, and Suriname; (2) *Schistosoma japonicum* in China, Indonesia, the Philippines; (3) *Schistosoma mekongi* in several districts of Cambodia and the Lao People's Democratic Republic; (4) *Schistosoma guineensis* and related *Schistosoma intercalatum* in rainforest areas of central Africa. Urogenital schistosomiasis is

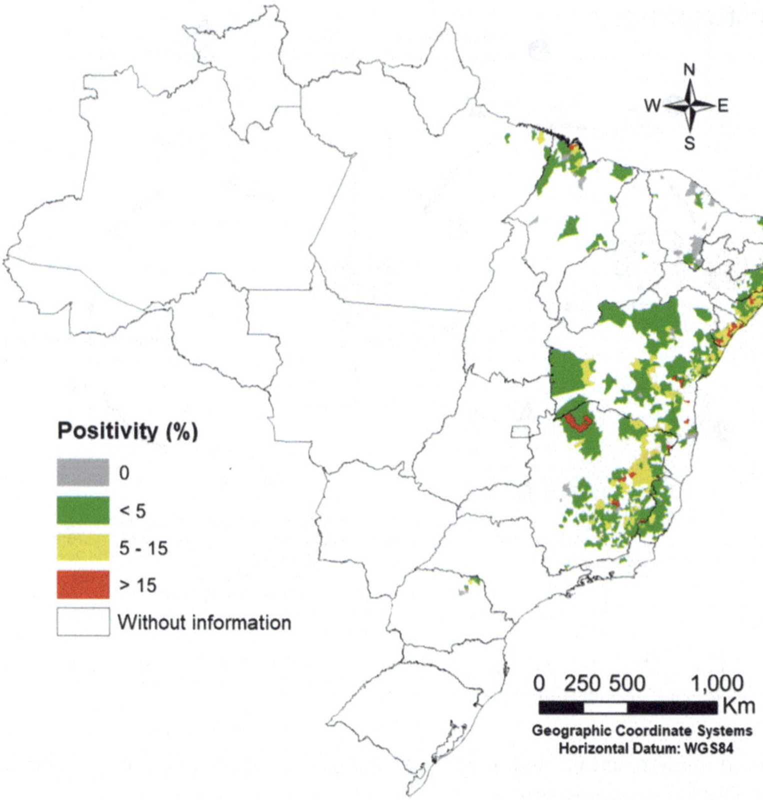

Fig. 15.3 Distribution of schistosomiasis in the endemic area, by positivity range, by municipality in Brazil, 2009–2017. Source: Brazilian Ministry of Health, Secretariat of Health Surveillance, Schistosomiasis Control Program (SISPCE/SVS/MS)—http://portalms.saude.gov.br/images/pdf/2019/janeiro/25/Mapa-distribuicao-esquistossomose-2009-2017.pdf. Endemic area: States of Alagoas, Bahia, Pernambuco, Paraíba, Rio Grande do Norte, Sergipe, Minas Gerais, and Espírito Santo. Focal transmission area: States of Maranhão, Pará, Piauí, Ceará, Rio de Janeiro, São Paulo, Paraná, Santa Catarina, Goiás, Rio Grande do Sul

caused by *Schistosoma haematobium* and occurs in Africa, in the Middle East, and most recently also on the Island of Corsica (France) (WHO 2018). Intestinal schistosomiasis is an important endemic disease in Brazil, with a high morbimortality burden and strong association with environmental changes generated by large dams (Martins-Melo et al. 2018; Steinmann et al. 2006). Official data from the Brazilian Ministry of Health show that the distribution of schistosomiasis coincides with the area of many large-scale water projects in Brazil, such as the transposition of the São Francisco River (Fig. 15.3). The increasing number of large dams in the country generates more and more areas with favorable conditions for the development of snail vectors, an increase in the local population with unfavorable social and tourism conditions, often lacking health risk management.

15.5 Major Water Engineering Projects in Brazil and Their Health Impact

Brazil is the world's fifth-largest country by extension (8,516,000 km^2) and has a vast hydrographic net with different characteristics related to climate, human occupation, and hydrogeological features. Estimates point to an average of 260,000 m^3/s of water flow, with approximately 78,000 m^3/s of surface water availability, but the distribution is not uniform, showing great variability between the Brazilian regions and periods of the year. The North Region (which includes the Amazon) accounts for only 8.6% of the Brazilian population and has about 83% of the country's surface water availability, whereas the most populated urban agglomerations with a high demand for water are located far away, in other regions of Brazil. The mostly semiarid Northeast Region harbors 28% of the Brazilian population, but only 3% of the country's water availability. According to data from the Brazilian National Water Agency (ANA—Agência Nacional das Águas), there was an increase in water demand in Brazil by about 80% in the last two decades. Up to 2030, another increase of 30% is estimated. Population growth, urbanization, and economic development are the main factors. Consequently, Brazil has invested heavily in the construction of large water engineering works for the most diverse purposes, mainly electric power generation, followed by the supply of large urban centers. In 2015, Brazil was the third leading country in energy generation by means of water projects in the world, surpassed only by China and Canada (Table 15.4). Water projects have been responsible for generating 63.1% of the Brazil's total electrical energy (Table 15.5).

Data from the Generation Database (Banco de Informações de Geração—BIG) of the Brazilian Electricity Regulatory Agency (ANEEL—Agência Nacional de

Table 15.4 Hydropower generation in the world—ten leading countries (TWh), 2011–2015

	2011	2012	2013	2014	2015	Δ% (2015/2014)	Part % (2015)
Worldwide	**3468.4**	**3629.2**	**3756.9**	**3855.1**	**3840.5**	**−0.4**	**100.0**
China	681.2	854.2	900.5	1040.6	1103.3	6.0	28.7
Canada	372.0	376.5	387.9	378.7	376.8	−0.5	9.8
Brazil	**428.3**	**415.3**	**391.0**	**373.4**	**359.7**	**−3.7**	**9.4**
United States	319.4	276.2	268.6	259.4	249.1	−4.0	6.5
Russia	164.2	164.2	179.3	173.5	166.3	−4.2	4.3
Norway	119.8	141.2	127.5	134.8	137.3	1.9	3.6
India	129.3	112.7	133.6	128.1	120.3	−6.1	3.1
Japan	82.9	75.1	77.8	81.4	84.8	4.2	2.2
Sweden	66.0	78.3	60.8	63.2	74.7	18.2	1.9
Venezuela	82.8	81.2	82.7	86.3	74.2	−14.1	1.9
Other	1022.6	1054.2	1147.2	1135.6	1093.9	−3.7	28.5

Source: Statistical Yearbook of Electricity 2018, http://www.epe.gov.br/pt/publicacoes-dados-abertos/publicacoes/anuario-estatistico-de-energia-eletrica

Table 15.5 Electricity generation by source in Brazil (GWh), 2013–2017

	2013	2014	2015	2016	2017	Δ% (2017/2016)	Part % (2017)
Total	570,835	590,542	581,228	578,898	587,962	1.6	100.0
Hydraulic[a]	390,992	373,439	359,743	380,911	370,906	−2.6	63.1
Natural gas	69,003	81,073	79,490	56,485	65,593	16.1	11.2
Petroleum products[b]	22,090	31,529	25,657	12,103	12,733	5.2	2.2
Coal	14,801	18,385	18,856	17,001	16,257	−4.4	2.8
Nuclear	15,450	15,378	14,734	15,864	15,739	−0.8	2.7
Biomass[c]	39,679	44,987	47,394	49,236	49,385	0.3	8.4
Wind	6578	12,210	21,626	33,489	42,373	26.5	7.2
Others[d]	12,241	13,540	13,728	13,809	14,976	8.5	2.5

[a]Includes self-production
[b]Petroleum products: diesel oil and fuel oil
[c]Biomass: firewood, sugarcane bagasse, and bleach
[d]Others: coke oven gas, other secondary ones, other non-renewable ones, other renewable and solar ones
Source: Statistical Yearbook of Electricity 2018, http://www.epe.gov.br/pt/publicacoes-dados-abertos/publicacoes/anuario-estatistico-de-energia-eletrica

Energia Elétrica) in Brazil indicate the existence of 1338 operational hydroelectric sources (which use hydroelectric power from rivers in the generation of electricity), generating in total 104,346,169 kW, 126 hydroelectric source projects with construction not yet started (for the generation of 2,268,186 kW) and 40 hydroelectric source projects under construction (for generation of 1,302,101 kW) (ANEEL 2019). Table 15.6 details the installed and operational generation capacity according to rivers and Brazilian Federal Units (ANEEL 2019).

Table 15.7 and Fig. 15.4 show the major water engineering projects in Brazil with a recent high impact (ANA 2017).

15.6 Teles Pires Hydroelectric Power Plant

The Teles Pires Hydroelectric Power Plant is part of the Teles Pires hydroelectric complex and is built in the upper Tapajós basin, on the border of the states of Pará (North) and Mato Grosso (Central-West region), 945 kilometers from Cuiabá (capital of Mato Grosso) (Fig. 15.5). The plant is located between the cities of Paranaíta (State of Mato Grosso) and Jacareacanga (State of Pará), in the area called Cachoeira Sete Quedas, with a reservoir totaling an area of 150 km^2. It has an installed capacity of 1820 MW, being the largest plant of the Teles Pires Complex, tenth place as the largest power generator in the country. The complex includes four large hydroelectric plants, among them Teles Pires, which was financed by more than 80% by public funds of the National Bank for Economic and Social Development (BNDES, the Brazilian Development Bank), approximately 3 billion BRL (Brazilian Real).

Table 15.6 Installed and operational capacity of electricity generation of large water projects in Brazil, according to rivers and Federal Units

River—state	Number of installed plants	Installed capacity (kW)
Tocantins—Pará	1	8,535,000
Xingu—Pará	1	7,566,433
São Francisco—Alagoas	6	7,441,601
Madeira—Rondônia	2	7,318,000
Grande—Minas Gerais	12	7,206,300
Paraná—Paraná	2	7,000,085
Iguaçu—Paraná	6	6,907,467
São Francisco—Bahia	7	6,809,501
Paraná—Mato Grosso do Sul	4	6,544,038
Paraná—São Paulo	3	6,535,200
Paranaíba—Goiás	5	5,642,302
Grande—São Paulo	7	5,220,200
Paranaíba—Minas Gerais	3	4,984,000
São Francisco—Sergipe	1	3,162,000
Tocantins—Tocantins	4	2,731,450
Teles Pires—Pará	2	2,519,800
Paranapanema—São Paulo	11	2,398,156
Uruguai—Santa Catarina	2	2,305,000
Uruguai—Rio Grande do Sul	2	2,305,000
Paranapanema—Paraná	8	2,184,700

Source: Brazilian Electricity Regulatory Agency (ANEEL)—http://www2.aneel.gov.br/aplicacoes/capacidadebrasil/Rio.asp. For rivers that compose the borders of Brazilian states, the capacity is quantified for both states

Table 15.7 Recent major water engineering projects in Brazil

Name	Location (state)	Hydrographic basin	Period of construction	Installed capacity (MWs)	Flooded area (km^2)
Teles Pires Hydroelectric Power Plant	Mato Grosso and Pará	Teles Pires River	2010–2015	1820.0	95.0
Belo Monte Power Plant	Pará	Xingu River	2011–2019	7566.3	478.0
Jirau Hydro-electric Power Plant	Rondônia	Madeira River	2009–2016	3750.0	361.6
Santo Antônio Hydroelectric Power Plant	Rondônia	Madeira River	2008–2016	3568.3	421.56
São Francisco River Transposition	Ceará, Paraíba, Pernambuco e Rio Grande do Norte	São Francisco River	2007–2019	–	–

Source: Brazil National Water Agency (*ANA—Agência Nacional das Águas*)

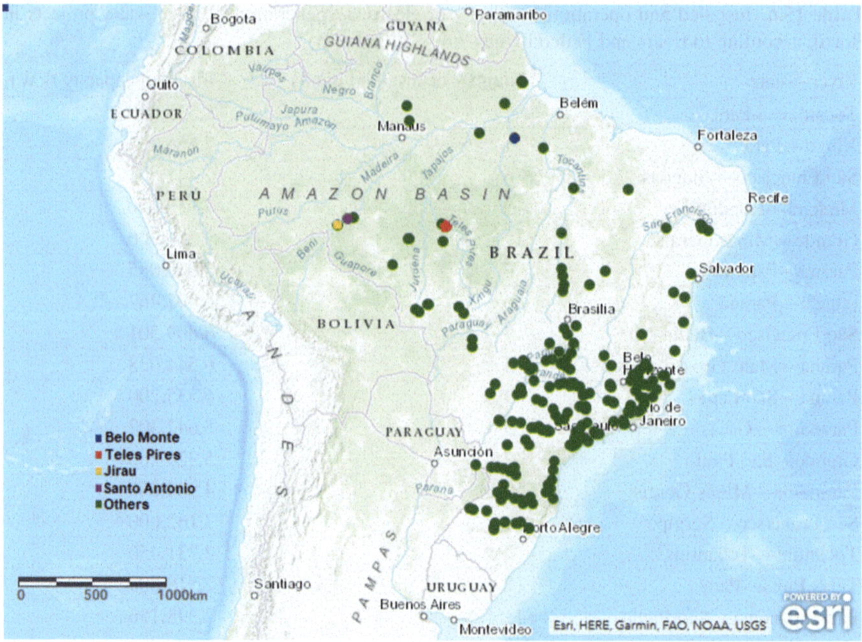

Fig. 15.4 Brazil's major hydropower utilization projects. Source: Brazilian Electricity Regulatory Agency—ANEEL, ANEEL Geoprocessing Service Portal: https://sigel.aneel.gov.br/portal/home/webmap/viewer.html?webmap=947a10caf45249faacd8c45ac74573c7. Highlighted: Teles Pires Hydroelectric Power Plant, Belo Monte Power Plant, Jirau Hydroelectric Power Plant, Santo Antônio Hydroelectric Power Plant, and São Francisco River Transposition

The construction of this plant involves flooding of indigenous lands (example of the Munduruku) and areas of environmental conservation. In addition to energy production, in order to serve the region's agricultural sector (in particular soybean production), waterway plans are designed to make it viable to transport to the Amazon River ports from the Tapajós River and two tributaries (Juruena and Teles Pires). Despite the protections in Brazil's Constitution and Legislation and in International Conventions, the application of "security suspensions" made these ventures viable (Fearnside 2015). The environmental impacts are significant, particularly in the context of the interconnection of the Amazon and the Cerrado in terms of water, fauna, and flora (Fearnside 2015). In association with agricultural activities in Brazilian Amazonia, these impacts may be even more prominent. Considering that these states present high incidence and prevalence of tropical diseases with high morbimortality burden (Martins-Melo et al. 2018), such as malaria and tegumentary leishmaniasis, we anticipate potential direct and indirect impacts on the health of the population.

Fig. 15.5 Teles Pires Hydroelectric Power Plant. Source: Teles Pires Hydroelectric Plant (CHTP), http://www.uhetelespires.com.br/site/wp-content/uploads/2015/05/UHE-Teles-Pires-e1439936818847.jpg

15.7 Belo Monte Hydroelectric Power Plant

The Belo Monte Dam (formerly Kararaô) is located in the Xingu River basin, in the southwestern state of Pará, in the north of Brazil, a region with significant social vulnerability (Franco et al. 2018) (Fig. 15.6). The lake of the plant has an area of 478 km^2 (1/10,000 of the Legal Amazon area). In operation since 2016, the total installed power will be 7566.3 MW on average during the year, which represents approximately 10% of the national consumption. At fully installed capacity, the Belo Monte plant will be the third largest hydroelectric plant in the world, behind only the Chinese Three Gorges (installed capacity of 18,200 MW) and the Itaipu Plant (installed capacity of 14,000 MW), a cooperation project between Brazil and Paraguay. Up to November 2017, R $ 38.6 billion BRL were invested from public and private sources. Its conclusion, with the installation of the last turbine, is scheduled for December 2019. The construction of this hydroelectric dam involves intensive debates on the use of public investments to subsidize the guarantee of electric energy for the smelting of aluminum for export. The Belo Monte case and additional upstream planned dams further indicate the need for adjustments in environmental assessment and licensing to include the impacts of multiple interdependent projects (Fearnside 2006). The definition of the Belo Monte project received strong opposition from Brazilian and international environmentalists, especially from some local indigenous communities, social movements, and members of

Fig. 15.6 Belo Monte Hydroelectric Plant. Source: Public Defender Office of Pará (DPE/PA), http://www2.defensoria.pa.gov.br/Portal/images/justi%C3%A7a/01092016_hidreletricabelomonte_1.jpg

the Catholic Church. The Belo Monte Plant has been studied by different authors, showing indicating impacts of different natures on the communities' lives already marked by great vulnerability (Franco et al. 2018), forced migration (Randell 2016b), welfare reduction with change to urban spaces (Randell 2016a), and environmental conflicts (Fleury and Almeida 2013; Pinto 2012). The construction of the Tucuruí Hydroelectric Power Plant on the Tocantins River (State of Pará), planned and built during the military dictatorship in Brazil (1964–1985), is comparable with that of the Belo Monte power plant on the Xingu River, three decades later. Issues such as lack of transparency, signs of misallocation of public resources, and unmitigated socio-environmental impacts mark these major infrastructure projects in the Amazon (Pinto 2012).

Health impacts have been described in detail, such as the relationship between migration and workers' diseases, syphilis in pregnant women, deaths due to violence (homicides), and traffic accidents (Grisotti 2016). The environmental impacts have also been significant and, as for Teles Pires, this impacted area of Pará state presents a significant incidence and prevalence of tropical diseases (Martins-Melo et al. 2018), such as malaria, tegumentary leishmaniasis, and Chagas disease (Costa et al. 2016; Santos et al. 2014). A study carried out on bats captured in the Belo Monte area recognized the diversity of bat trypanosomes, including some previously unidentified (Costa et al. 2016). The causative agent of the parasitic Chagas disease, *Trypanosoma cruzi*, is a flagellate protozoan capable of infecting many species of mammals and is transmitted by various groups of invertebrates, including the order Chiroptera (Bats). The occurrence of *T. cruzi* in the region of the Belo Monte Hydroelectric Plant in the Amazon Basin was presented in view of the risk of migration of human population required for dam works and new cities that grow in the vicinity of these projects. Chagas disease as a zoonosis is already known in the Amazon region, and the presence of species not previously classified as *Trypanosoma* shows the great unknown parasite biodiversity (Costa et al. 2016).

15.8 Jirau Hydroelectric Power Plant

The Jirau Hydroelectric Power Plant is part of the Madeira River Complex, 120 km from Porto Velho, capital of the State of Rondônia (North region) (Fig. 15.7). It has a re-servicing facility with an area of 362 km^2 when it is at its maximum volume, with an installed capacity of 3750 MW, and entered commercial operation in 2013. Inaugurated in 2016, it represents the third largest hydroelectric plant in Brazil, with an estimated cost of the work in 19 billion reais, within the PAC of the Federal Government. In addition to socio-environmental issues, the impact of this plant on the amplification of dengue virus transmission was documented in Rondônia

Fig. 15.7 Jirau Hydroelectric Power Plant. Source: Jirau Environmental Observatory—http://observatoriojirau.com.br/wp-content/uploads/2017/05/2_USINA_DJI_0024_OPT.jpg

comparing the period before (2000–2008) and after the construction (2009–2013) of the Jirau and Santo Antônio—the latter, described below (Camasmie Abe and Miraglia 2018). There was an association between the artificial flood resulting from the construction of the Jirau hydroelectric plant and the richness and composition of anopheline mosquitoes, transmitting malaria (Rodrigures et al. 2017). The municipality of Porto Velho has been highly endemic to malaria and played an important role in the spread of the parasite to other municipalities in the Amazon and even to non-endemic areas of the country, with migration having a prominent role (Angelo et al. 2017). However, due to recent changes in the human occupation of Brazilian Amazonia, characterized by the intense expansion of transport networks, mobility between urban and rural areas has also become an important factor in the transmission of the disease (Angelo et al. 2017).

15.9 Santo Antônio Hydroelectric Power Plant

The Santo Antônio Hydroelectric Power Plant is also located on the Madeira River, in the middle of the Amazon rainforest, 7 km from the center of the capital Porto Velho, capital of Rondônia (North region) (Fig. 15.8). The total area of the reservoir is 422 km^2. It has capacity for power generation of 3568 MW of installed power. It is the fourth largest hydroelectric plant in operation in Brazil and one of the largest in

Fig. 15.8 Jirau Hydroelectric Power Plant. Source: Jirau Environmental Observatory—http://observatoriojirau.com.br/wp-content/uploads/2017/05/2_USINA_DJI_0024_OPT.jpg

the world, having started operations in March 2012. Its construction cost estimated to be approximately 20 billion BRL. As for the Jirau Hydroelectric Power Plant, the amplification of the risk for transmission of leishmaniasis has been documented in the region (Furtado et al. 2016; Galardo et al. 2015), as well as malaria and arboviruses (Camasmie Abe and Miraglia 2018).

15.10 São Francisco River Transposition

The project for the transposition of the São Francisco river involves the displacement of part of the waters of the São Francisco River in Brazil, called by the Brazilian government the "Integration Project of the São Francisco River with Hydrographic Basins in the Northeast of the North" (PISF) (Figs. 15.9, 15.10, and 15.11). The purpose of the project is to provide water for various purposes (human supply, irrigation, agriculture, animal husbandry, fish and shrimp farming, supply of industrial poles) in an area that currently has about 12 million people (Brazil 2014).

In total, 720 km of canals will be built, nine pumping stations, 27 aqueducts, eight tunnels (including the largest tunnel for water transport in Latin America—CUNCAS I, 15 km long), 35 small reservoirs, two small (with capacity to produce 52 MW), in addition to using pre-existing dams as receivers, serving as water distribution poles. The canals will also send water in the canals of some rivers of

Fig. 15.9 São Francisco River Transposition Project (PISF). Source: Government of Brazil. http://mi.gov.br/web/projeto-sao-francisco

Fig. 15.10 Area of the São Francisco River Transposition Project (PISF). Source: Ministry of National Integration of Brazil. http://mi.gov.br/pt/c/document_library/get_file?uuid=86d666d3-8509-447a-a536-66361d8bce75&groupId=10157

the region, where the water will travel: 60 km in the Salty River; 80 km in the Jaguaribe River, both in the North Axis; and 90 km on the Apodi River; 130 km in the Piranhas-Açu River and 150 km in the Paraíba River, in the East Axis, totaling 510 km of course in other rivers (BRASIL 2004). The PISF crosses the area called Polígono das Secas, located in the Northern Northeast, encompassing part of the semi-arid region north of the São Francisco River (Fig. 15.12). This area and its population suffer intermittently with the effect of prolonged droughts, which compromises survival in dignified conditions, being a factor that generates poverty and misery (BRASIL 2004). Various emergency actions have already been developed in isolation in these areas to alleviate this situation, such as the construction of cisterns (rainwater catchment), wells (groundwater abstraction), and small dams (inefficient strategy due to high levels of evaporation), among others. But none of these actions reached the central point of the problem, which is the scarce and poor distribution of water in the region. The consumption of a resident of the city of São Paulo, on average, 340 liters of water per day; the average in the city of Rio de Janeiro is 220 liters; while in the Northeastern semi-arid the daily consumption is only 50 liters per inhabitant. This figure is far below the 120 liters per day recommended by the

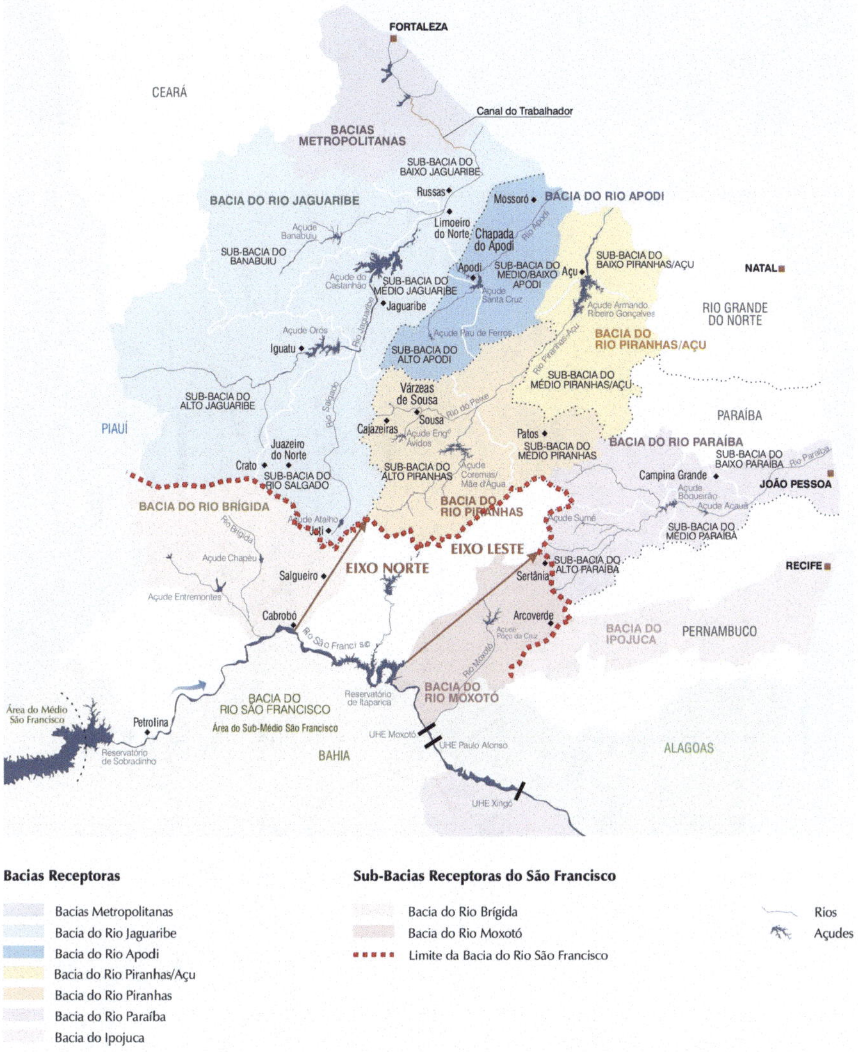

Fig. 15.11 Basins served by the São Francisco River Transposition Project (PISF). Source: National Bank for Economic and Social Development of Brazil—http://www.bndes.gov.br/arquivos/pisf/r32-sintese.pdf

United Nations (UN), considered a critical situation by this institution because it is lower than the low sustainability index (1000 m^3/s per inhabitant per year), jeopardizing employment generation and income for the population of the region (BRAZIL 2016; ISA 2007).

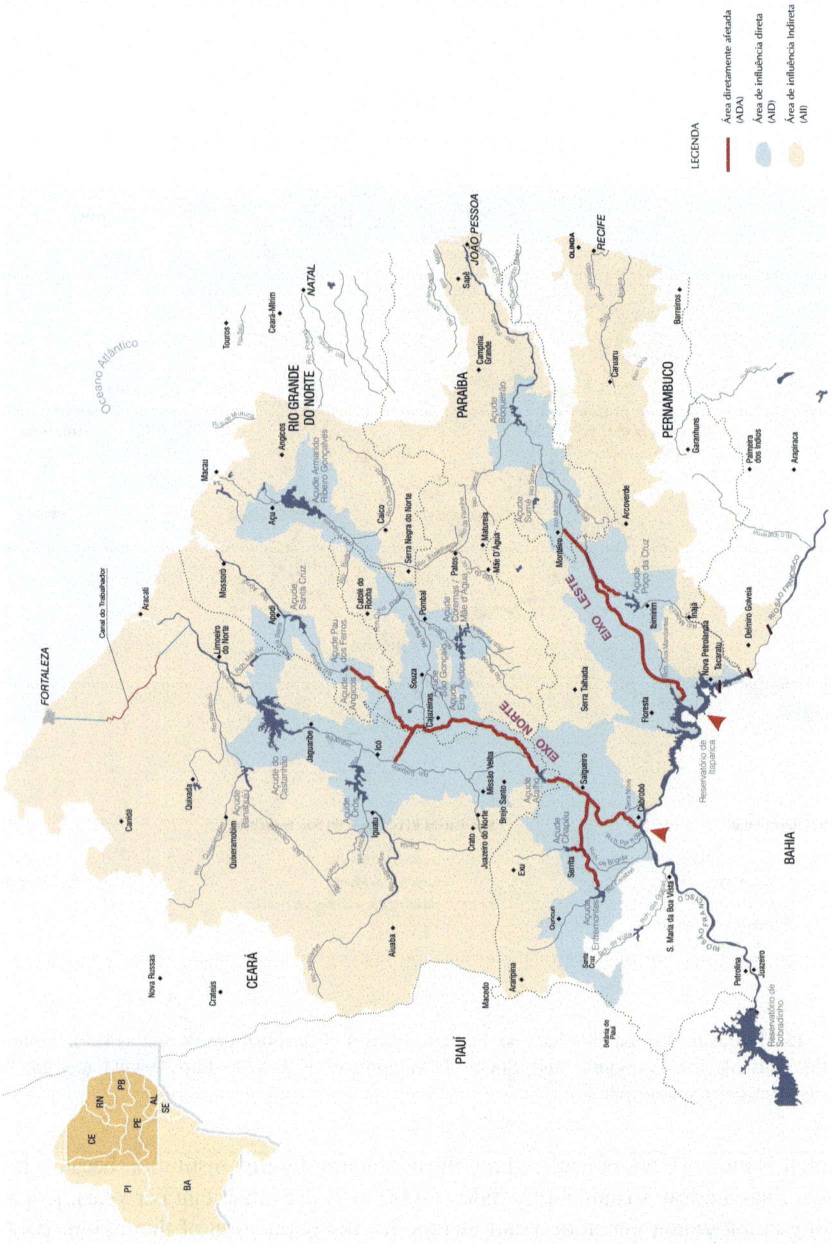

Fig. 15.12 Affected areas of the São Francisco River Transposition Project (PISF). Source: National Bank for Economic and Social Development of Brazil—http://www.bndes.gov.br/arquivos/pisf/r32-sintese.pdf

The East Axis was inaugurated in March 2017, and according to a survey released by the Brazilian government in the first half of 2018, the North Axis had forecast delivery and total operation for the beginning of 2018, and is currently 96.4% of its operation (Brazil 2019). The budget currently estimated is approximately 10 billion BRL, required for irrigation in the Northeastern and semi-arid region of Brazil. However, its approval and implementation have been criticized mainly in terms of the use of water: displacement of areas with a high demand for water for human use and animals, agroindustry, and shrimp farming. In spite of these questions, the purposes are listed as positive in the RIMA due to the consequent generation of employment and income, but with great controversy among social movements and researchers. In addition, the risk of amplification of schistosomiasis transmission has been analyzed by different studies (Bezerra et al. 2018; Silva Filho et al. 2017; Favre et al. 2016). In addition to schistosomiasis, there is concern about the expansion of other tropical diseases, such as Chagas disease, visceral leishmaniasis, and arboviruses.

The transposition of the São Francisco River in the Northeast of Brazil promotes great concerns in the context of public health, since it involves a large territorial portion of four Brazilian states, which have different prevalence and significant parasitic load for schistosomiasis, besides the intermediate hosts of *S. mansoni.* The damming of water and the consequent passage from a lotic to a lentic system, in addition to the environmental changes mentioned above, may contribute to the expansion of habitat suitable for schistosomiasis (*Biomphalaria* spp) transmitting snails. In addition, not less important, it can lead to an increase in human settlements and tourism in these places (Bezerra et al. 2018; Silva Filho et al. 2017; Favre et al. 2016). Figure 15.13 shows the shell morphology of *S. mansoni* intermediate hosts present in Brazil. The population densities of these snails are influenced by a number of different factors (pH, water turbidity and hardness, among others) (Bezerra et al. 2018; Silva Filho et al. 2017). A more recent hypothesis argues that dams block the reproduction of migratory shrimp that feed on the snail host of schistosomiasis (Sokolow et al. 2017). In addition to these investigations, it is necessary to diagnose schistosomiasis in school-age children, which is a sentinel population to monitor the progress of control programs (Bezerra et al. 2018; BRASIL 2014).

The socioeconomic benefits of PISF on the other side are also related to possible negative effects, such as serious environmental changes, including alterations in the composition of the native aquatic biological communities of the catchment areas, with a risk of biodiversity reduction, modification of the fluvial system of receiving drainage, risk of new species introduction (mollusks, plants, fish, among others) potentially harmful to humans and the environment, including influencing the transmission of diseases, such as schistosomiasis. The construction of these canals and new reservoirs affects the ecosystem, as the impoundment of water, and the consequent passage from a lotic to a lentic system, leads to environmental changes in water quality, fauna richness, and species density and in other aspects of biodiversity endows conditions conducive to the establishment of schistosomiasis transmitting molluscs. The increase in the number of dams on a global scale has led to an increase in the prevalence of schistosomiasis; often observed, as in the Aswan dam in Egypt;

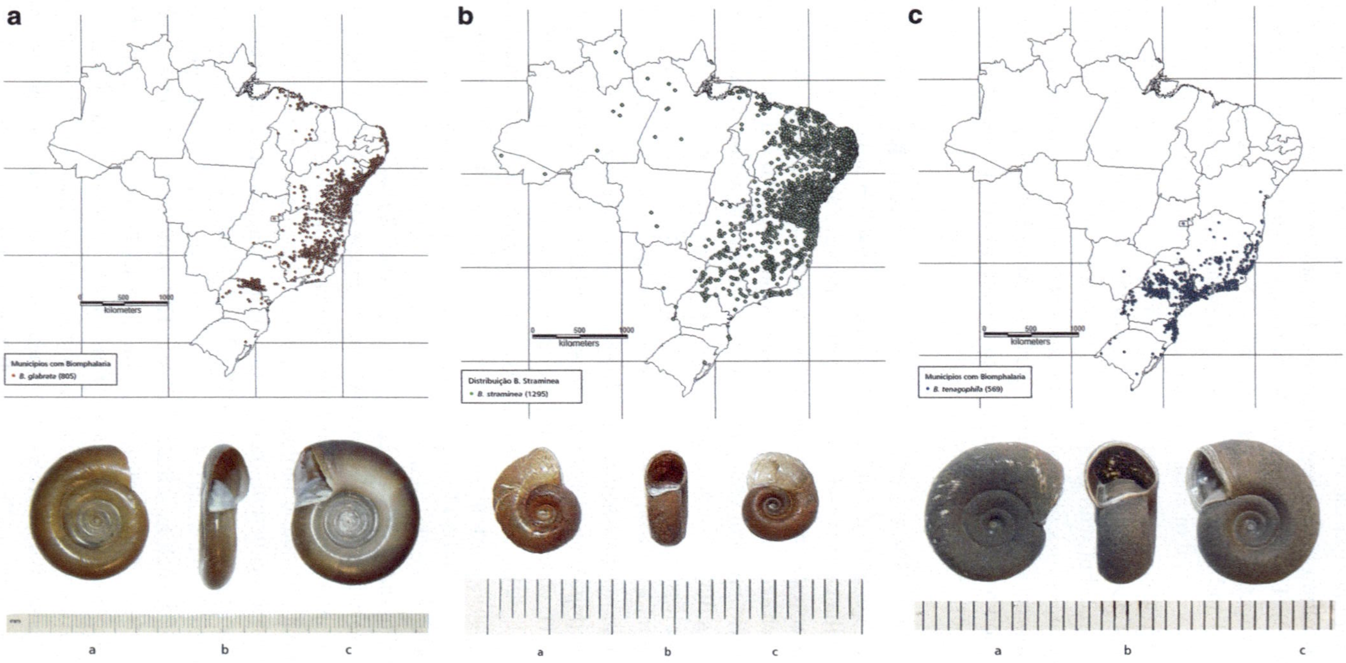

Fig. 15.13 Shell morphology of *S. mansoni* intermediate hosts and distribution in Brazil. *Biomphalaria glabrata* (**a**), *Biomphalaria straminea* (**b**), *Biomphalaria tenagophila* (**c**). Source: Ministry of Health of Brazil—http://bvsms.saude.gov.br/bvs/publicacoes/vigilancia_esquistossome_mansoni_diretrizes_tecnicas.pdf. (BRASIL 2014), Planorbid snails photos originally from René Rachou Institute, Oswaldo Cruz Foundation—Belo Horizonte, Minas Gerais. Laboratory of Intestinal Helminthiasis. "Morphological identification of *Biomphalaria glabrata*, *B. tenagophila* and *B. straminea* intermediate hosts of *Schistosoma mansoni*", 2004. Authors: Carvalho OS, Caldeira RL. Support: Brazil. Ministry of Health. Secretariat of Health Surveillance. Schistosomiasis Series, 6

Fig. 15.14 Area of the Pecém Industrial and Port Complex (CIPP). Rural communities of farmers and fishermen coexist with steel plants, thermoelectric plants, cement plants, among other enterprises. (**a**) Pecém Steel Company; (**b**, **c**) Parada neighborhood in São Gonçalo do Amarante, Ceará State; (**d**) Conveyors of ore and coal to Pecém Steel Company. Source: Authors

Selingue in Mali; Three Gorges in China; and Kossou in Côte d'Ivoire (Barakat 2013; McManus et al. 2010; Steinmann et al. 2006; Traoré 1989; Zheng et al. 2002; N'Goran et al. 1997). It leads to the increase of human settlements in its proximities, often in a precarious way, and tourism in these places (Lima and Batista 2010; Leme 2007). In addition, temporary loss of jobs and income due to expropriations, disruption of socio-community relations and risk of conflicts with riverine communities that needed to be relocated from their lands, pressure on the fragile urban infrastructure of small towns in the affected area by the works (Silva Filho et al. 2017; Favre et al. 2016). The use of the final destination of water to supply the Pecém Industrial and Port Complex (CIPP) in Ceará State has been criticized, where companies with large water consumption are installed, such as two thermoelectric plants and one steel plant (BRASIL 2004) (Fig. 15.14).

Water resources indispensable to agricultural development and other productive processes can contribute to the spread of schistosomiasis in the country, especially when inadequately exploited (Steinmann et al. 2006; Traoré 1989). Therefore, the PISF fosters major concerns in the context of public health, since it involves a large territorial portion of four Brazilian states, which have different prevalence rates and parasitic load of schistosomiasis, as well as a large distribution of the intermediate hosts of *S. mansoni* (Silva Filho et al. 2017).

Surveys carried out by Bezerra et al. (2018) and Favre et al. (2016) have evidenced the wide distribution of the intermediate snail hosts in areas directly affected by the PISF works. As these disease vectors reproduce in slow and quiescent limnic habitats, PISF canals and reservoirs become favorable sites for their development, with the possibility of proliferation and increased contact with populations. In addition, the increase and/or the appearance of new cases of schistosomiasis are related to the greater availability of water (canals and new reservoirs), associated to the lack of adequate treatment before distribution to the population, together with the lack of collection and sewage treatment. This scenario has a positive impact on the transmission of other human diseases of water transmission (Fernandez et al. 2014; Rezende et al. 2009). Favre et al. (2016) also identified schoolchildren parasitized by *S. mansoni* and confirmed spatial coincidence between breeding sites of these planorbids and the homes of infected schoolchildren and inadequate domestic sanitary conditions, indicating active transmission of schistosomiasis even before transposition. Silva Filho et al. (2017) also detected schistosomiasis-positive workers from the sites. This major construction work may also enable the entry of *B. glabrata*, the most effective intermediate host in the transmission of schistosomiasis, because it is highly susceptible to the parasite. Previous studies have already reported that workers were responsible for the introduction and/or dissemination of schistosomiasis in areas affected by water works (Coutinho et al. 1992).

Thus, in spite of the small number of cases detected in the studied PISF areas, the transmission dynamics of schistosomiasis has already been evidenced in this region and may be further aggravated by the arrival of the waters of the São Francisco River. In order to avoid the spread of schistosomiasis, it is necessary to invest in actions to monitor the host of the disease, to map the areas with risk of transmission, and to use more sensitive methods of diagnosis that enable the detection of all cases and the accomplishment of the respective treatments. It is important to train health professionals in these areas to manage positive cases, as well as to plan, implement and execute the actions of control and surveillance of schistosomiasis in relation to this and other water projects (Steinmann et al. 2006). No epidemiological surveillance is currently available in these areas, as disease prevalence is low. Nevertheless, because of the use of less sensitive diagnostic methods, the actual burden of schistosomiasis may still be underestimated (Favre et al. 2016).

15.11 Final Considerations

Projects involving water resources have been important in achieving sustainable development in various parts of the world. However, large water projects may have major effects on biodiversity and environmental preservation, and may lead to the emergence and reemergence of tropical diseases, such as waterborne parasitic diseases. Communities living in areas of large dams generally do not benefit from the revenues generated by the provision of water and the production of electricity. The expanded health component should be integrated with environmental impact and social impact assessments for large dam projects. Brazil is a very significant case, given its territorial dimensions and water potential, but also the social and economic inequalities. There are still important gaps in scientific research to better understand the relationship between the epidemiology of waterborne diseases and different disease vectors. There is an urgent need to mobilize financial resources to support integrated approaches and to carry out environmental studies, where habitat monitoring, species inventories, and assessment of the transmission dynamics of pre-existing diseases in these areas are carried out, as well as risk factors identified. The sustainability of water resources will guarantee global access to safe drinking water, hygiene and sanitation, which are essential factors to combat endemic diseases. The effective cooperation of society with governments and companies, in a synergistic way, extends the possibilities in terms of regulatory, financial, institutional, technological and human resources in order to guarantee benefits of development, reducing social inequalities.

Related Documentary Information

Documentary "O Complexo". https://www.youtube.com/watch?v=Zfvr5mtpOdI. "O Complexo" is a Best Documentary Brazil award-winning documentary at the Brazil International Festival. The short film reports the impact of the construction of four hydroelectric power plants and some Small Hydroelectric Power Plants (SHPP) on the Teles Pires river basin and the lives of fishermen, indigenous people, and settlers living in the vicinity of the construction site. The film shows the discrepancy between the speeches of those responsible for the engineering works and the reports made by those who live near the river and depend on it. Produced by Forest Comunicação. Directed by: Thiago Foresti.

Documentary "Belo Monte—After the Flood". http://www.belomonteaftertheflood.com/https://youtu.be/bw4eHUVlMAk. This documentary presents the current situation in the city of Altamira, in the Amazon. The documentary is a production of the NGO International Rivers, directed by Canadian filmmaker Todd Southgate and narrated by actor Marcos Palmeiras. Link to the original movie site: http://www.belomonteaftertheflood.com/.

Documentary "The Battle of Belo Monte". https://youtu.be/CUqGWNYzSIQ. This report explains the largest and most controversial infrastructure work of Lula and Dilma governments. This video is a compilation in the short-film format of the three blocks of the program: "O Canteiro" (http://youtu.be/sR0rTLLsRtY); "Altamira" (http://youtu.be/oNU0s-DK2Tw); and "The Impactados" (http://youtu.be/9zU5_-HlaBA). Directors: Douglas Lambert, Marcelo Leite and Lalo de Almeida

Documentary Jirau and Santo Antônio: reports of an Amazon war—https://youtu.be/ZFQ11fri3vs. "In the middle of the Amazon there are people and these people talk, and if people do not talk, the river will talk". Child prostitution, water contamination, hundreds of miles of dead trees, tons of fish killed, thousands of families homeless, and unemployed. It seems that the progress promised by the constructions of the Hydroelectric Power Plants on the Madeira River, in Rondônia, went far from that region. This work intends to give voice to the affected and affected who suffer and fight daily for the guarantee of their rights. This is the case of Nilce de Souza Magalhães, a woman, fisherwoman, and fighter, who months after the recording of this film was murdered. For Nicinha, and all the stricken people of the Amazon region, not a minute of silence, but a whole life of struggles.

Documentary "Waters for life! Not to death!". Realization: Movement of People Affected by Dams. Support: Audiovisual Brigade "Eduardo Coutinho" Landless Rural Workers Movement. Additional information http://mabnacional.org.br, https://www.facebook.com/MAB.Brasil. Documentary Chamado do Madeira (Madeira River) ("A sobering thought from Madeira") http://www.mabnacional.org.br/video/chamado-do-madeira. Part 1: https://youtu.be/tKT18fAayII. Part 2: https://youtu.be/xZD6RxqPLE0. Part 3: https://youtu.be/rd7KbDLqGR8. Part 4: https://youtu.be/hSi3QUgI3Tk.

Documentary "Transposição do Rio São Francisco e águas do Ceará: os cursos da privatização". https://www.youtube.com/watch?v=fu7z7Dwbzk8.

References

Agência Nacional de Águas – ANA (2017) Água na indústria: uso e coeficientes técnicos. http://www.snirh.gov.br/portal/snirh/snirh-1/acesso-tematico/usos-da-agua/aguanaindustria_usoecoeficientestecnicos.pdf. Accessed 12 Feb 2019

Agência Nacional de Energia Elétrica – ANEEL (2019) Dados do Banco de Informações de Geração (BIG). http://www2.aneel.gov.br/aplicacoes/capacidadebrasil/capacidadebrasil.cfm. Accessed 06 Mar 2019

Agrawal SK (2017) Impact of climatic variation on parasitic infections - short review. Int J Res Stud Microbiol Biotechnol 3(3):17–21. https://doi.org/10.20431/2454-9428.0303005

Angelo JR, Katsuragawa TH, Sabroza PC et al (2017) The role of spatial mobility in malaria transmission in the Brazilian Amazon: the case of Porto Velho municipality, Rondônia, Brazil (2010–2012). PLoS One 12(2):e0172330. https://doi.org/10.1371/journal.pone.0172330

Arrifano GPF, Martín-Doimeadios RCR, Jiménez-Moreno M et al (2018) Large-scale projects in the amazon and human exposure to mercury: the case-study of the Tucuruí Dam. Ecotoxicol Environ Saf 147:299–305. https://doi.org/10.1016/j.ecoenv.2017.08.048

Bacci DC, Pataca EM (2008) Education for water. EAV 22(63):211–226. http://www.revistas.usp.br/eav/article/view/10302/11958. Accessed 18 Jan 2019

Barakat RM (2013) Epidemiology of schistosomiasis in Egypt: travel through time: review. J Adv Res 4(5):425–432. https://doi.org/10.1016/j.jare.2012.07.003

Barcellos C, Monteiro AMV, Corvalán C et al (2009) Climatic and environmental changes and their effect on infectious diseases: scenarios and uncertainties for Brazil. Epidemiol Serv Saúde 18 (3):285–304. https://doi.org/10.5123/S1679-49742009000300011

Becker BK (2012) Reflections on hydroelectric dams in the Amazon: water, energy and development. Bol Mus Para Emílio Goeldi Ciênc Hum 7(3):783–790. https://doi.org/10.1590/S1981-81222012000300011

Bezerra FSM, Pinheiro MCC, Silva Filho JDD et al (2018) Identification of *Biomphalaria* sp. and other freshwater snails in the large-scale water transposition project in the Northeast of Brazil. Rev Inst Med Trop Sao Paulo 60:e41. https://doi.org/10.1590/s1678-9946201860041

Bhatia R (2011) Health impact assessment: a guide for practice. Human Impact Partners, Oakland, CA. https://humanimpact.org/wp-content/uploads/2011/01/HIA-Guide-for-Practice.pdf. Accessed 8 Feb 2019

Brasil. Ministério da Integração Nacional (2004) Projeto de Integração do Rio São Francisco com Bacias Hidrográficas do Nordeste Setentrional – Relatório de Impacto Ambiental (RIMA). Brasília: MIN. http://www.mi.gov.br/documents/10157/3678963/Rima+-+Relat%C3%B3rio+de+Impacto+Ambiental.pdf/4324863d-cbff-4522-9bd0-eab9d34b8fe2. Accessed 8 Feb 2019

Brasil. Ministério da Integração Nacional (2014) Projeto de Integração do Rio São Francisco. Brasília: MIN. Assessoria de Comunicação Social. http://www.mi.gov.br/c/document_library/get_file?uuid=261be082-5ac5-43b7-8e8b-59bb61b1b108&groupId=2054191. Accessed 8 Jan 2019

Brasil. Ministério da Integração Nacional (2019) Projeto de Integração do Rio São Francisco: entenda os detalhes. http://www.integracao.gov.br/web/projeto-sao-francisco/o-andamento-das-obras. Accessed 8 Jan 2019

Brasil. Ministério da Saúde. Secretaria de Vigilância em Saúde. Departamento de Vigilância de Doenças e Agravos não Transmissíveis e Promoção da Saúde (2018) Health Brazil 2017: an analysis of the health situation and challenges to the achievement of the sustainable development goals. Brasília: Ministério da Saúde. http://bvsms.saude.gov.br/bvs/publicacoes/saude_brasil_2017_analise_situacao_saude_desafios_objetivos_desenvolvimento_sustetantavel.pdf. Accessed 15 Feb 2019

Brasil. Ministério da Saúde. Secretaria de Vigilância em Saúde. Departamento de Vigilância em Saúde Ambiental e Saúde do Trabalhador (2014) Avaliação de Impacto à Saúde – AIS: metodologia adaptada para aplicação no Brasil. Brasília: Ministério da Saúde. http://bvsms.saude.gov.br/bvs/publicacoes/avaliacao_impacto_saude_ais_metodologia.pdf. Accessed 8 Feb 2019

Brasil. Ministério da Saúde. Secretaria de Vigilância em Saúde. Departamento de Vigilância Epidemiológica (2014) Vigilância da Esquistossomose Mansoni: diretrizes técnicas – 4. ed. Brasília: Ministério da Saúde. http://bvsms.saude.gov.br/bvs/publicacoes/vigilancia_esquistossome_mansoni_diretrizes_tecnicas.pdf. Accessed 15 Feb 2019

Brasil. Ministério das Cidades. Secretaria Nacional de Saneamento Ambiental - SNSA (2016) Sistema Nacional de Informações sobre Saneamento: Diagnóstico dos Serviços de Água e Esgotos – 2014. Brasília: SNSA/MCIDADES. http://www.epsjv.fiocruz.br/upload/Diagnostico_AE2014.pdf. Accessed 8 Jan 2019

Cable J, Barber I, Boag B et al (2017) Global change, parasite transmission and disease control: lessons from ecology. Philos Trans R Soc Lond B Biol Sci 372(1719):20160088. https://doi.org/10.1098/rstb.2016.0088

Camasmie Abe K, Miraglia SGEK (2018) Dengue incidence and associated costs in the periods before (2000–2008) and after (2009–2013) the construction of the hydroelectric power plants in Rondônia, Brazil. Epidemiol Serv Saude 27(2):e2017232. https://doi.org/10.5123/S1679-49742018000200012

Chivian E, Bernstein A (2008) Sustaining life: how human health depends on biodiversity. Oxford University Press, New York

Comissão Nacional de Meio Ambiente - CONAMA (1986) Resolução No 1, de 23 de janeiro de 1986. Dispõe sobre critérios básicos e diretrizes gerais para a avaliação de impacto ambiental. http://www.siam.mg.gov.br/sla/download.pdf?idNorma=8902. Accessed 8 Jan 2019

Comissão Nacional de Meio Ambiente - CONAMA (1997) Resolução No 237, de 19 de dezembro de 1997. Dispõe sobre licenciamento ambiental; competência da União, Estados e Municípios; listagem de atividades sujeitas ao licenciamento; Estudos Ambientais, Estudo de Impacto Ambiental e Relatório de Impacto Ambiental. http://extwprlegs1.fao.org/docs/pdf/bra25095.pdf. Accessed 8 Jan 2019

Confalonieri UE, Margonari C, Quintão AF (2014) Environmental change and the dynamics of parasitic diseases in the Amazon. Acta Trop 129:33–41. https://doi.org/10.1016/j.actatropica.2013.09.013

Costa AP, Nunes PH, Leite BHS et al (2016) Diversity of bats trypanosomes in hydroeletric area of Belo Monte in Brazilian Amazonia. Acta Trop 164:185–193. https://doi.org/10.1016/j.actatropica.2016.08.033

Coutinho AD, Silva ML, Gonçalves JF (1992) Epidemiological study on mansoni schistosomiasis in irrigation areas of Northeastern Brazil. Cad Saúde Pública 8(3):302–310. https://doi.org/10.1590/S0102-311X1992000300009

Favre TC, Fernandez MA, Beck LC et al (2016) Assessment of schistosomiasis in the semi-arid Northeast region of Brazil: the São Francisco River large-scale water transposition project. Rev Soc Bras Med Trop 49(2):252–257. https://doi.org/10.1590/0037-8682-0243-2015

Fearnside PM (2006) Dams in the Amazon: Belo Monte and Brazil's hydroelectric development of the Xingu River Basin. Environ Manag 38(1):16–27. https://doi.org/10.1007/s00267-005-0113-6

Fearnside PM (2015) Amazon dams and waterways: Brazil's Tapajós Basin plans. Ambio 44(5): 426–439. https://doi.org/10.1007/s13280-015-0642-z

Fernandez MA, Mattos AC, Silva EF et al (2014) A malacological survey in the Manso Power Plant, State of Mato Grosso, Brazil: new records of freshwater snails, including transmitters of schistosomiasis and exotic species. Rev Soc Bras Med Trop 47(4):498–506. https://doi.org/10.1590/0037-8682-0138-2014

Fiorillo CAP (2018) Curso de Direito Ambiental Brasileiro, 18th edn. Editora Saraiva, São Paulo

Fleury LC, Almeida J (2013) The construction of the Belo Monte hydroelectric power plant: environmental conflict and the development dilemma. Ambient Soc 16(4):141–156. https://doi.org/10.1590/S1414-753X2013000400009

Franco VS, Souza EB, Lima AMM (2018) Floods and social vulnerability: study on the Xingu river in Altamira/PA. Ambient Soc 21:e01573. https://doi.org/10.1590/1809-4422asoc0157r3vu18l1ao

Furtado NV, Galardo AK, Galardo CD et al (2016) Phlebotomines (Diptera: Psychodidae) in a hydroelectric system affected area from Northern Amazoni-an Brazil: further insights into the effects of environmental changes on vector ecology. J Trop Med 2016:9819723

Galardo AKR, Galardo CD, Silveira GA et al (2015) Phlebotominae sand flies (Diptera: Psychodidae): potential vectors of American cutaneous leishmaniasis agents in the area associated with the Santo Antônio Hydroelectric System in Western Amazonian Brazil. Rev Soc Bras Med Trop 48(3):265–271. https://doi.org/10.1590/0037-8682-0088-2015

Gergel SE, Turner MG (2017) Learning landscape ecology, 2nd edn. Springer-Verlag, New York. eBook ISBN 978-1-4939-6374-4

Giongo CR, Mendes JMR, Santos FK (2015) Development, health and environment: contradictions in the construction of dams. Serv Soc Soc 123:501–522. https://doi.org/10.1590/0101-6628.034

Granziera MLM (2011) Direito ambiental, 2nd edn. Editora Atlas, São Paulo

Grisotti M (2016) The construction of health causal relations in the Belo Monte dam context. Ambient Soc 19(2):287–304. https://doi.org/10.1590/1809-4422ASOC0252V1922016

Guergel DR (2013) Water resources development: engineering the future of global health. Boston University Frederick S. Pardee Center for the Study of the Longer-Range Future. https://open.bu.edu/handle/2144/22735. Accessed 10 Feb 2019

Gyasi SF, Boamah B, Awuah E et al (2018) A perspective analysis of dams and water quality: the Bui Power Project on the Black Volta, Ghana. J Environ Public Health 2018:6471525. https://doi.org/10.1155/2018/6471525

Heinrich K, Bach M, Breuer L (2017) Infectious disease research—what role is there for hydrologists? JWARP 9(2):139–162. https://doi.org/10.4236/jwarp.2017.92011

Instituto Social Ambiental – ISA (2007) Abastecimento de água e esgotamento sanitário nas capitais brasileiras, em 2004. ISA, São Paulo. http://www.socioambiental.org/banco_imagens/pdfs/AbastEsgotBrasilNov07.pdf. Accessed 2 Feb 2019

International Commission on Large Dams (2018) World register of dam - general synthesis. ICOLD, Paris. https://www.icold-cigb.org/GB/world_register/general_synthesis.asp. Accessed 2 Feb 2019

Kibret S (2018) Time to revisit how dams are affecting malaria transmission. Lancet Planet Health 2(9):e378–e379. https://doi.org/10.1016/S2542-5196(18)30184-0

Kibret S, Lautze J, McCartney M et al (2015) Malaria impact of large dams in sub-Saharan Africa: maps, estimates and predictions. Malar J 14:339. https://doi.org/10.1186/s12936-015-0873-2

Kibret S, Lautze J, McCartney M et al (2016) Malaria and large dams in sub-Saharan Africa: future impacts in a changing climate. Malar J 15:448. https://doi.org/10.1186/s12936-016-1498-9

Leme FBM (2007) As represas como lugares turísticos: Novas significações e valorizações de uma paisagem sem memória. Cultur: Revista de Cultura e Turismo 1(1):1–21. http://200.128.65.36/revistas/culturaeturismo/edicao1/artigo6.pdf. Accessed 2 Feb 2019

Lereer LB, Scudder T (1999) Health impact of large dams. Environ Impact Assess Rev 19 (2):113–123. https://doi.org/10.1016/S0195-9255(98)00041-9

Leturcq G (2016) Differences and similarities in impacts of hydroelectric dams between north and south of Brazil. Ambient Soc 19(2):265–286. https://doi.org/10.1590/1809-4422ASOC0254R1V1922016

Lima SFS, Batista GT (2010) Impact of the reservoir of Paraibuna Hydroelectric Power Plant, SP, Brazil. Ambiente & Água - Interdiscip J Appl Sci 5(3):208–221

Martins-Melo FR, Pinheiro MC, Ramos AN Jr et al (2014) Trends in schistosomiasis-related mortality in Brazil, 2000–2011. Int J Parasitol 44(14):1055–1062. https://doi.org/10.1016/j.ijpara.2014.07.009

Martins-Melo FR, Pinheiro MC, Ramos AN Jr et al (2015) Spatiotemporal patterns of schistosomiasis-related deaths, Brazil, 2000–2011. Emerg Infect Dis 21(10):1820–1823. https://doi.org/10.3201/eid2110.141438

Martins-Melo FR, Carneiro M, Ramos AN Jr et al (2018) The burden of Neglected Tropical Diseases in Brazil, 1990–2016: a subnational analysis from the Global Burden of Disease Study 2016. PLoS Negl Trop Dis 12(6):e0006559. https://doi.org/10.1371/journal.pntd.0006559

McManus DP, Gray DJ, Li Y et al (2010) Schistosomiasis in the People's Republic of China: the era of the Three Gorges Dam. Clin Microbiol Rev 23(2):442–466. https://doi.org/10.1128/CMR.00044-09

Movimento dos Atingidos por Barragens - MAB (2019) Movimento dos Atingidos por Barragens (MAB) no Brasil - Água e energia não são mercadorias! http://www.mabnacional.org.br/. Accessed 18 Jan 2019

N'Goran EK, Diabate S, Utzinger J et al (1997) Changes in human schistosomiasis levels after the construction of two large hydroelectric dams in central Côte d'Ivoire. Bull World Health Organ 75(6):541–545

Nava A, Shimabukuro JS, Chmura AA et al (2017) The impact of global environmental changes on infectious disease emergence with a focus on risks for Brazil. ILAR J 58(3):393–400. https://doi.org/10.1093/ilar/ilx034

Oliveira FAS, Heukelbach J, Gomide M, Moura RCS (2006) Impact of large dams on public health: I. Upstream effects. Cad Saúde Colet 14(4):575–596

Oliveira FAS, Heukelbach J, Gomide M et al (2007) Impact of large dams on Public Health: II. Dowstream effects. Cad Saúde Colet 15(1):9–26

Patz JA, Daszak P, Tabor GM et al (2004) Unhealthy landscapes: policy recommendations on land use change and infectious disease emergence. Environ Health Perspect 112(10):1092–1098. https://doi.org/10.1289/ehp.6877

Pinto LF (2012) De Tucuruí a Belo Monte: a história avança mesmo? Bol Mus Para Emílio Goeldi. Ciênc Hum 7(3):777–782. https://doi.org/10.1590/S1981-81222012000300010

Poma A (2018) The emotional dimension of resistance movements against dams. Ambient Soc 21: e02070. https://doi.org/10.1590/1809-4422asoc0207vu18l3ao

Quigley R, Den Broeder L, Furu, P et al. (2006) Health impact assessment international best practice principles. Special Publication Series No. 5. International Association for Impact Assessment, Fargo. https://activelivingresearch.org/sites/activelivingresearch.org/files/IAIA_HIABestPractice_0.pdf. Accessed 12 Feb 2019

Randell H (2016a) The short-term impacts of development-induced displacement on wealth and subjective well-being in the Brazilian Amazon. World Dev 87:385–400. https://doi.org/10.1016/j.worlddev.2016.07.005

Randell H (2016b) Structure and agency in development-induced forced migration: the case of Brazil's Belo Monte Dam. Popul Environ 37(3):265–287. https://doi.org/10.1007/s11111-015-0245-4

Rezende HR, Sessa PA, Ferreira AL et al (2009) Effects of the installation of the Rosal hydroelectric powers station, Itabapoana River, States of Espírito Santo and Rio de Janeiro, on anophelinae, planorbidae and phlebotominae. Rev Soc Bras Med Trop 42(2):160–164. https://doi.org/10.1590/S0037-86822009000200013

Rodrigures MS, Batista EP, Silva AA et al (2017) Change in Anopheles richness and composition in response to artificial flooding during the creation of the Jirau hydroelectric dam in Porto Velho, Brazil. Malar J 16(1):87. https://doi.org/10.1186/s12936-017-1738-7

Sanchez-Ribas J, Parra-Henao G, Guimarães AÉ (2012) Impact of dams and irrigation schemes in Anopheline (Diptera: Culicidae) bionomics and malaria epidemiology. Rev Inst Med Trop S Paulo 54(4):179–191. https://doi.org/10.1590/S0036-46652012000400001

Santos TV, Silva FMM, Barata IR et al (2014) A new species of phlebotomine, *Trichophoromyia adelsonsouzai* (Diptera: Psychodidae) of Brazilian Amazonia. Mem Inst Oswaldo Cruz 109(2): 140–147. https://doi.org/10.1590/0074-0276130159

Silva Filho JD, MCC P, Sousa MS et al (2017) Detection of schistosomiasis in an area directly affected by the São Francisco River large-scale water transposition project in the Northeast of Brazil. Rev Soc Bras Med Trop 50(5):658–665. https://doi.org/10.1590/0037-8682-0299-2017

Sleigh AC, Jackson S (2001) Dams, development, and health: a missed opportunity. Lancet 357(9256):570–571. https://doi.org/10.1016/S0140-6736(00)04072-1

Sokolow SH, Jones IJ, Jocque M et al (2017) Nearly 400 million people are at higher risk of schistosomiasis because dams block the migration of snail-eating river prawns. Philos Trans R Soc Lond Ser B Biol Sci 372(1722):20160127. https://doi.org/10.1098/rstb.2016.0127

Solidariedade e Educação - FASE, Laboratório, Estado, Trabalho, Território e Natureza - ETTERN, Instituto de Pesquisa e Planejamento Urbano e Regional - IPPUR (2011) Relatório síntese: Projeto de avaliação de equidade ambiental como instrumento de democratização dos procedimentos de avaliação de impacto de projetos de desenvolvimento. FASE/ETTERN/IPPUR, Rio de Janeiro. https://fase.org.br/wp-content/uploads/2011/07/Relat%C3%B3rio+-+S%C3%ADntese+-+Projeto+Avaliacao+Equidade+Ambiental+final.final_.pdf. Accessed 21 Jan 2019

Souza NCR, Fontes AS, Luz LD et al (2017) Identification of the degree of impact of dams on the hydrologic regime of semi-arid rivers: an evaluation of the DHRAM method. RBRH 22:e13. https://doi.org/10.1590/2318-0331.011716093

Steinmann P, Keiser J, Bos R et al (2006) Schistosomiasis and water resources development: systematic review, meta-analysis, and estimates of people at risk. Lancet Infect Dis 6 (7):411–425

The United Nations World Water Assessment Programme - WWAP, UN-Water (2018) The United Nations world water development report 2018: Nature-Based Solutions for Water. Paris, UNESCO. http://www.unwater.org/publications/world-water-development-report-2018/. Accessed 21 Jan 2019

The World Bank (2014) Database world development indicators - electricity production from renewable sources, excluding hydroelectric (kWh). https://data.worldbank.org/indicator/EG.ELC.RNWX.KH. Accessed 21 Jan 2019

Traoré M (1989) Schistosomiasis in the Sélingué dam area: the integrated approach. Trop Med Parasitol 40(2):228–231

Tubaki RM, Menezes RM, Cardoso RP Jr et al (2004) Studies on entomological monitoring: mosquito species frequency in riverine *habitats* of the Igarapava Dam, Southern Region, Brazil. Rev Inst Med Trop Sao Paulo 46(4):223–229. https://doi.org/10.1590/S0036-46652004000400009

Tundisi JG (2008) Water resources in the future: problems and solutions. Estud Av 22(63):7–16. https://doi.org/10.1590/S0103-40142008000200002

Wallet F, Sicard D, Brey P et al (2016) Health impact of dams: lessons from a case study. Environ Risque Sante 15(5):419–425. https://doi.org/10.1684/ers.2016.0897

World Health Organization - WHO (2000) Sustainable development and healthy environments. Human health and dams: the World Health Organization's submission to the World Commission on Dams (WCD). WHO, Geneva. https://www.who.int/docstore/water_sanitation_health/Documents/Dams/Damsfinal.htm. Accessed 15 Feb 2019

World Health Organization - WHO (2018) Schistosomiasis - key facts, 20 February 2018. https://www.who.int/news-room/fact-sheets/detail/schistosomiasis. Accessed 21 Feb 2019

Yamana T (2004) The impacts of dams and reservoirs on public health. Report no. 1096. http://web.mit.edu/eltahir/www2/Makana/papers/research_healthimpacts.doc. Accessed 15 Feb 2019

Zarfl C, Lumsdon AE, Berlekamp J et al (2015) A global boom in hydropower dam construction. Aquat Sci 77(1):161–170. https://doi.org/10.1007/s00027-014-0377-0

Zheng J et al (2002) Relationships between the transmission of schistosomiasis japonica and the construction of the Three Gorge Reservoir. Acta Trop 82(2):147–156. https://doi.org/10.1016/S0001-706X(02)00046-3

Zhouri A, Oliveira R (2007) Development, social conflicts and violence in rural Brazil: the case of hydroelectric dams. Ambient Soc 10(2):119–135. https://doi.org/10.1590/S1414-753X2007000200008

Chapter 16
The Major Rivers and the Genesis of the Recent Area of Ticks *Ixodes persulcatus* in Western Siberia

Valerie V. Yakimenko

Abstract During the time of the study of carriers of agents of diseases in Western Siberia, which coincides with the period of the study of tick-borne encephalitis (from the 1940s of the twentieth century), a change in the area of common species ticks, including the *Ixodes persulcatus*, is recorded. Important effects of this process are based on the major rivers of the West Siberia, which had become distribution paths for many species in the northern and southern directions starting from the main area. However, the key role in this process belongs to the changing of the landscape as a result of intense human economic activities in times of climate change.

Keywords Ob and Irtysh rivers · Western Siberia · Area · *Ixodes persulcatus* · Vector · Ticks borne encephalitis · Ixodid tick-borne borreliosis

The fauna of ixodid ticks of Russia (family *Ixodidae*) includes five genera and 60 species (Balashov 1998), of which 12 species reside in Western Siberia (genus *Ixodes*, *Dermacentor* (Fig. 16.1a, b), and *Haemaphysalis*). The genus *Ixodes* in Western Siberia is represented by six species:

Ixodes persulcatus P. Sch. 1930 and *I. pavlovskyi* Pom. 1946—with a pasture type of parasitism

I. apronophorus P. Sch., 1924; *I. crenulatus* Koch, 1844; and *I. lividus* Koch, 1894—with a nesting-burrow type of parasitism

I. trianguliceps Bir. 1895 (Fig. 16.2)—with a mixed type of parasitism (Fedorov 1963; Popov 1962; Koklyagina 1963, 1967; Bezzubov 1965; Drozdova 1967; Davydova and Lukin 1969; Malyushina 1966, 1983; Alifanov et al. 1970; Ivanov 1971; Igolkin et al. 1972; Sapegina 1962; Vershinina 1964; Bogdanov 1999, 2000, 2001, 2005; Bogdanov et al. 2010; Romanenko 2004, 2009; Livanova et al. 2011; Malkova et al. 2012; Yakimenko et al. 1991, 2013).

V. V. Yakimenko (✉)
Laboratory of Arbovirus Infections, Research Institute of Nature-Foci Infections, Omsk, Russia
e-mail: yakimenko@oniipi.org

H. Mehlhorn, S. Klimpel (eds.), *Parasite and Disease Spread by Major Rivers on Earth*, Parasitology Research Monographs 12,
https://doi.org/10.1007/978-3-030-29061-0_16

Fig. 16.1 Species of ticks of genus *Ixodes* and *Dermacentor*: (**a**) *I. persulcatus*; (**b**) *D. marginatus*

The most widespread species in the region is *I. persulcatus*, whose current range covers the entire forest zone (with the exception of northern taiga, located in the permafrost zone) and the northern forest-steppe of the flat part of Western Siberia. In the mountainous areas (Salair ridge, Kuznetsk Alatau, Altai), it inhabits the foothill forest-steppe and the entire mountain-forest belt up to the state border of Russia. The species inhabits different forest habitat types, reaching the highest abundance in mixed middle-aged forests that formed in the territory of cut over indigenous taiga (Igolkin 1978; Sapegina 1980; Filippova 1985; Bogdanov 2004).

The range of hosts of the *I. persulcatus* is extensively high with respect to both systematical and ecological aspects, and includes (in the whole area) about 200 species of mammals and more than 120 species of birds. Reptiles and amphibians may serve as further random hosts (Filippova 1977, 1985). In Western Siberia, the species *I. persulcatus* was found on 174 species of hosts including 64 species of mammals, 105 species of birds, 1 species of amphibians, and 4 species of reptiles (Bogdanov 2004).

I. persulcatus in Western Siberia is the main epidemically significant vector of such infectious agents of dangerous human diseases like tick-borne encephalitis and ixodic tick-borne borreliosis. The territories, within which these diseases of people in Western Siberia are recorded, coincide practically with the habitat of these species

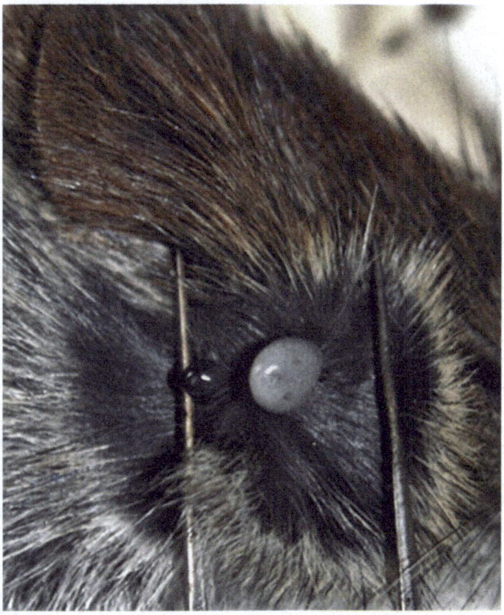

Fig. 16.2 Full-fed nymph and female of *Ixodes trianguliceps* on northern red-backed vole—a tick with a mixed type of parasitism

(http://net-kleshu.ru/o-kleshchah/regiony-zarazheniya/). In addition, this tick species is one of the carriers of the causative agents of tularemia in epidemically active natural foci of this infection. Other natural foci of which are widespread in the north of Western Siberia (Meshcheryakova et al. 2014; Kozlova et al. 2016).

The greatest interest while studying the distribution of ixodic ticks in Western Siberia coincides with the period of studying tick-borne encephalitis. It began at the end of the 1950s and at the beginning of the 1960s of the twentieth century. Until that time information on the distribution of vectors was scarce, mostly only episodic. The period of studying the ranges of ixodid ticks, including the range of the *I. persulcatus*, is not longer than 60 years.

16.1 The Northern Boundary of the *I. persulcatus* Area in Western Siberia (Fig. 16.3)

The first reports on the northern border of the species distribution go back to the 1930s of the twentieth century (60°58′N, village Samarovo, Khanty-Mansi Autonomous Okrug (now within the regional center)—according to Olenev 1934, cit. by Popov 1962). In the middle of 1940s of the twentieth century specimens were recorded on cattle and small mammals south of Khanty-Mansiysk (Repolovo village 60°39′N) in the middle of the 1950s—in the river Ob valley in the Kondinsky district of the Khanty-Mansi Autonomous Okrug (village Anor, 300 km to the north of village Repolovo, about 62°N). According to Popova (1962) the spreading along the river Ob was limited to the north of the village Alexandrovsky Tomsk Oblast

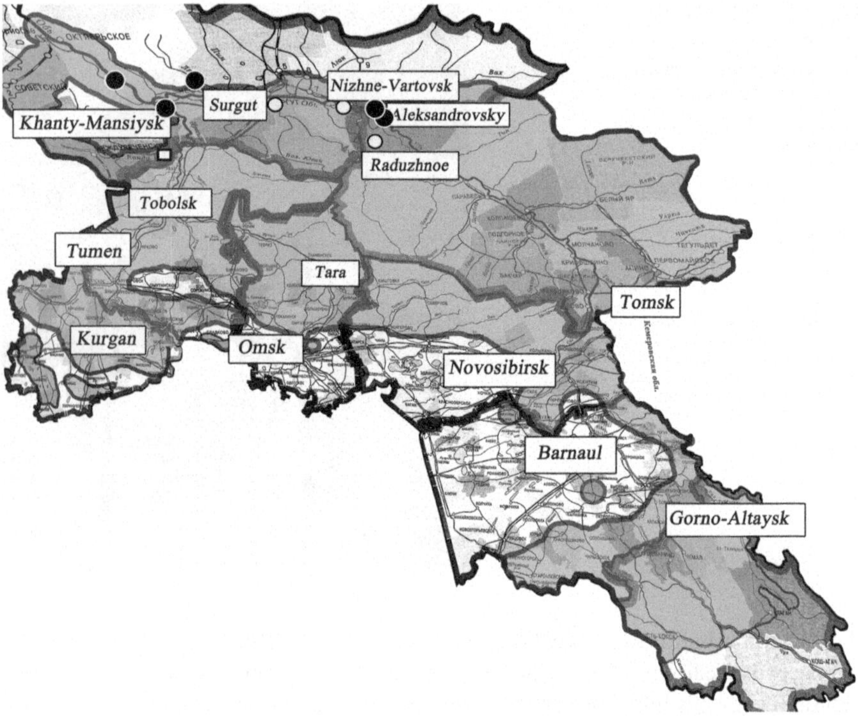

Fig. 16.3 Recent area of the *I. persulcatus* range in Western Siberia. Circles in black are places of findings of *I. persulcatus* ticks from 1950 to 1970; circles with white contents are places of absence of the *I. persulcatus* from 1950 to 1970; the square with a white filling marks the place where *I. persulcatus* was single detected in the period 1950–1970

inclusively (61°10′N). In the downstream region of the river Ob (Nizhnevartovsky district and in the environments of the town Surgut), the ticks were not registered.

Beginning in the first half to the middle of 1960s of the twentieth century, *I. persulcatus* was registered as common species in next the territory of the Khanty-Mansi Autonomous Okrug in the valley of the river Ob at latitude 61°21′–61°37′N (village Elizarovo and Urmanny), in the valley of the right confluent of the river Ob—river Lyamin—61°32′N (village Gornyy), and at river Wah—61°06′N (village Laryak—with mean abundances on rodents—0.57 prevalence (%) on rodents—35.7)—61°15′N (Malushina 1963). In the same period the larvae and nymphs of *I. persulcatus* on animals were recorded sporadic in the valley of the left confluent of river Ob—river Konda—59°41′N (village Chesnok) and in the valley of the river Ob in the neighborhood of the town Surgut (61°14′N). Reports were published from findings of larvae of *I. persulcatus* on small mammals in habitats of the floodplain of the river Ob the valley at 62 N. In the late 1960s and in the beginning of the 1970s of the twentieth century, the species has not yet been recorded in the area of the valley of the river Ob in the east of the town Surgut and in the west of the mouth of the river Wah of the Khanty-Mansi Autonomous Okrug and in the area of the town Strezhevoy of the Tomsk Region (60°42′N).

In the beginning of the 1970s of the last century, *I. persulcatus* was registered as common species in the valley of the Konda river (left confluent of river Ob) before 59°45′N (mean abundances on rodents: 1.15–36.4% prevalence).

In the modern period (the beginning of the XI century), the spread of the *I. persulcatus* in the valley of the river Ob on the territory of the Khanty-Mansi Autonomous Okrug-Ugra generally remains within the boundaries registered previously (Malkova et al. 2012; Yakimenko et al. 2013). In this territory the species is distributed unevenly: its abundance in various types of locations varies from 6 to 80 specimens/km of registered route. During the years of the abundance peaks in the local (generally located in the river valley) territories of the Khanty-Mansiysk, Oktyabrsky and Nefteyugansk regions, it may even exceed 150 specimens/km.

In general, the highest abundances of *I. persulcatus* were observed on the following sections of the floodplain of the river Ob: on the left bank of the river Ob from the lade mouth of the river Wah in the west to the lade mouth of the Irtysh river; in the east in the Khanty-Mansi Autonomous Okrug; in the left and right bank of the river Ob below the mouth of the Irtysh river to the north, to line connecting Yaguryakh (61°14′N) and Kedrovy (61°28N) villages; and in the left and right bank of the river Ob of the Tomsk Region. Northern of the territory the tick was recorded in the valley of the river Ob near the village Bolshaya Atlem Oktyabrsky district—62°16′N). In the Trans-Ural portion of Western Siberia, the tick *I. persulcatus* is regularly observed in forests along the river Konda. In the south-western region of the Kondinsky district (Leushinsk elevation) the abundance of ticks varies from 70 to 150 specimens/km (near the villages Mezhdurechensky, Leushi and Yagodny). To the north, on the territory of the Sovetsky district of the Khanty-Mansi Autonomous Okrug, in the floodplain habitats of rivers and streams and in the forests of the coastal swamps, the abundance of *I. persulcatus* does not exceed 10 specimen/km (Malkova et al. 2012).

The distribution of the species outside the floodplain on the right bank of the river Ob between the mouth of the Irtysh and Wah rivers has apparently two reasons (Yakimenko et al. 2010; Malkova et al. 2012). The first reason is the chance of the microclimatic conditions, which provide better chances for the metamorphosis of ticks in the deepfrozen soil (abandoned roads in the marshes—"plank beds" (Fig. 16.4), canned oil and gas well, dumping of oil and gas main). Here *I. persulcatus* stages are found up to the 62 nd km of the road from Nizhne-Vartovsk to Raduzhnoe (61°27′N), but its abundance is less than 1 specimen/km. North of the region transfer is registered only by birds. The second way of spreading occurs along the floodplains of the right confluent of river Ob (along the river Agan and river Tromyegan rivers, where the species has already been registered within 61°47′ and 62°10′N).

Fig. 16.4 Abandoned roads in the marshes—"plank beds" (42 km of the Nihne-Vartovsk-Raduzhnoe road, Nizhne-Vartovsky district, right bank of the river Ob)

16.2 The Southern Boundary of the *I. persulcatus* Range in Western Siberia

In the 1930s of the twentieth century in the flat part of Western Siberia, tick search was carried out in the regions between 56° and 57° N (Popov 1962). In the Omsk region in the 1950s of the last century, the species was not recorded south of the Bolsherechensky district in the valley of the Irtysh river and in adjacent territories. In the 1960s of the twentieth century, the southern distribution boundary was noted along the northern forest-steppe of the Kurgan, Omsk, and Novosibirsk regions (Alifanov and Netsky 1954; Bezzubova 1965; Loginovsky 1965; Bogdanov 1968; Davydova and Lukin 1969). According to updated data until 1980 in the south-west of Western Siberia, the *I. persulcatus* was widespread south of the main area (Fig. 16.5a), forming a protrusion into the south-east (from the Ural range) direction and reaching in the valley of the Tobol river (approximately 55° N). Autonomous territories of the species existed until the south (in the Kurgan region) and to the east (on the border of the Tyumen and Kurgan regions). In the Tyumen region between the main area of the *I. persulcatus* and its southern fragment there was a gap (Fig. 16.5a) that persists even now. The southern boundary of the main area of these ticks in the Tobol-Ishim and Ishim-Irtysh interfluves in the same period ran along 56°30′N. with a protrusion along the valley river Ishim up to 55°30′N. In the Ob-Irtysh interfluve, the species was distributed until approximately 56° N. with the exception of some territories on the right bank of the Irtysh river (Nizhne-Omsky district of the Omsk region) and of the right bank of the river Ob (Iskitimsky, Moshkovsky, and Maslyaninsky districts of the Novosibirsk region).

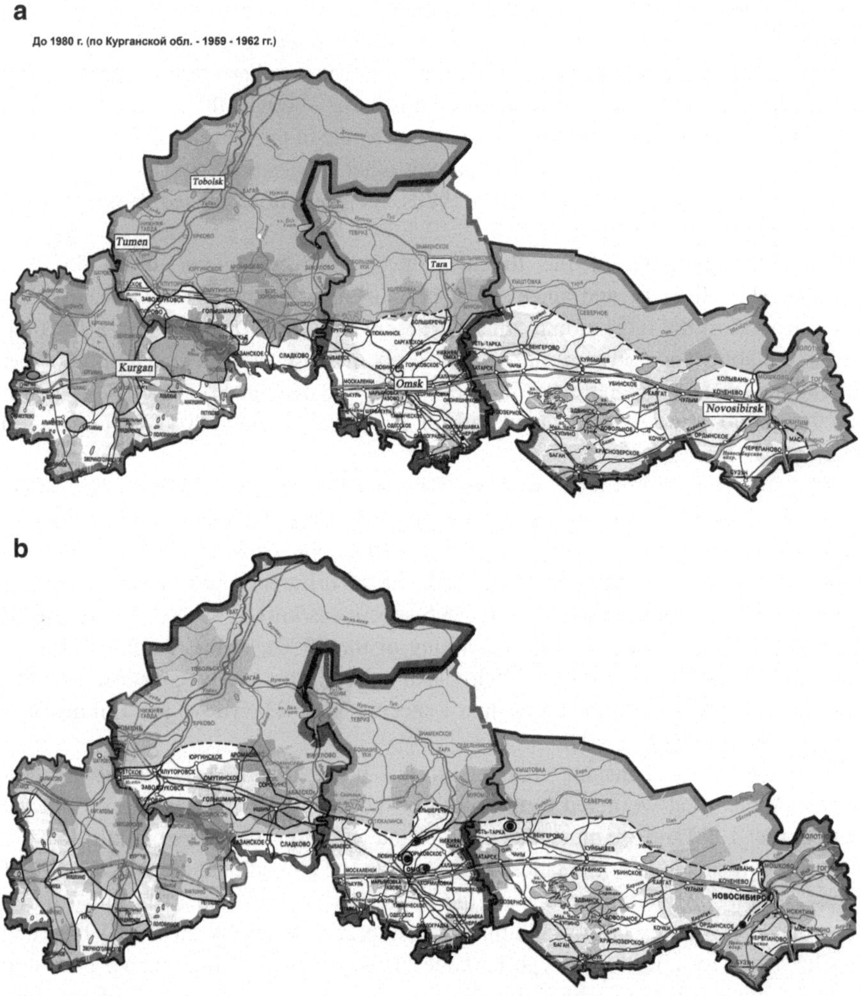

Fig. 16.5 The southern boundary of the *I. persulcatus* range in Western Siberia: (**a**) The distribution of the *I. persulcatus* in the south of Western Siberia until 1980. (Area with a translucent gray filling); (**b**) The distribution of the *I. persulcatus* in the south of Western Siberia after 1990. (Area with a translucent gray filling). Circles with a black gully show the places of detection of the *I. persulcatus* outside the main range; black solid line shows the established boundaries of the area; black dashed line shows the estimated boundaries of the area

An analysis of the generalized data on the distribution of *I. persulcatus* ticks in the flat part of Western Siberia from 1990 to the present allows to note a certain displacement of the southern boundaries of the area of this species (Fig. 16.5b). Thus, in the south-west of the region, the territory of the permanent habitat of the ticks (the southeast protrusion in the Kurgan and Tyumen regions) extended to the border of the Tyumen and Omsk regions. Simultaneously the southern boundary of the main area of this species in the Tyumen region moved to the north (approximately 57°20′N). On the left bank of the Irtysh river in the Omsk region, the *I. persulcatus* wandered to the south up to 55°40′N (southern part of Tyukalinsky and northern part of Sargatsky districts). Thus in the 1960s of the twentieth century, this species of ticks was not numerous in the subzone of the northern forest-steppe of the Omsk region, and its proportion among other *Ixodes* ticks was no higher than 5–6%. However, already in the 1970–1980s, its proportion reached peaks of up to 60% (Bogdanov and Shuteev 1981).

On the right bank of the Irtysh river, it inhabits regions up to 55°25′N (Gor'kovskiy district), and since 1992, permanent groupings of the species have been formed in the recreational zone of the city of Omsk (about 55° N). In some years, the abundance of these ticks reached about 15 specimens/km. Its proportion among the *Ixodes* ticks in this territory in 1988–1992 averaged 0.7 ± 0.2%, whereas in 1996–2000, it increased to an average of 20.9 ± 0.9% and in 2001–2011 even up to 29.6 ± 1.1%.

In the western part of the Novosibirsk region (northern forest-steppe; the bordering territories of the Ust-Tarksky and Vengerovsky districts), the southern boundary of the main area of *I. persulcatus* is currently approximately 55°40′N, and its proportion among the fauna of *Ixodes* ticks in the different types of habitats varies from 7.1% (with an abundance of 1.7–3.5 specimens/km) to 35.1% (10 specimen/km). At the same time, until the beginning of the 1990s, in the Barabinsk forest-steppe (Novosibirsk region), in Ob-Irtysh interfluve, this tick was registered individually occurring in numbers less than 1 specimen/km (Davydova and Lukin 1969; our data from 1989 to 1991).

In general, the constant changes in the boundaries of the distribution of *I. persulcatus* (that we noted to the south of the main range) started mainly in the period after 1990 and are most likely associated with different types of anthropically based transformations of the landscape. Thus, in the recreational zone of the city of Omsk, stable population groups of the species were found in areas of artificial planting of conifers with a predominance of spruce, which by the mid-1990s had reached ages forming forest-type herbage. This provided a more stable and high humidity regime being highly favorable for the existence of all phases of development of the *I. persulcatus* ticks in these moisture-deficient territories. Probably, similar reasons may be associated with changes in the boundaries of the species range in the Kurgan region. In the Novosibirsk region, the change in the nature of the territorial distribution of *I. persulcatus* in the forest-steppe zone is most likely associated with the desistance of forest management measures. This caused the accumulation of dying trees in the surface layer, which could affect the microclimate change in the topsoil, which is favorable for tick metamorphosis. In addition, this species invaded the valley of the river Ob in the Ordynsky and Suzunsky districts of the Novosibirsk region, where it had not been registered before.

The existence of protrusions to the north and south of the main area of the *I. persulcatus* ticks (usually forming along the valleys of large rivers) is determined by the possibility of carrying the species into these territories and the presence of microclimatic conditions for its existence at the place of introduction. The distribution of the *I. persulcatus* stages outside the main area is carried out mainly by migratory birds during seasonal migrations (for this species—mainly spring migrations) that overlap in time with the period its seasonal activity. As mentioned above, in Western Siberia, *I. persulcatus* was found on 105 species of birds (Bogdanov 2004). Many of which became common hosts. In general, the larvae and nymphs of this species show ornitophilicity (Filippova 1985), especially manifested during the periods of depression in the number of main hosts: small mammals such as rodents and insectivores. Within the area of some territories and during the seasons of population depression of small mammals, birds often feed more than the half of the larvae and nymphs of the ticks (e.g. Ravkin and Sapegina 1967; Malkov and Bogdanov 1966; Uspenskaya 1978) and carry out regular transfers within their individual territories during the breeding season.

The nature of the migratory migration of birds has two options: the span "wide front" and the span along ecological channels, which are presented by river valleys, piedmont plains and large lake systems (Dolgushin 1960; Kumari 1972 etc.). For many groups, above all—for species of the order Passeriformes—such channels are the valleys of large rivers. Ecological groups of birds of this order, while feeding on the terrain (mainly species of Emberizidae, Motacillidae, Turdidae, Fringillidae and some others) (Ostapenko 1976), are attacked by adult ticks or their immature developmental phases (larvae and nymphs) that had survived the winter period in a hungry state. The period of activity of larvae of this species in Western Siberia (Davydova and Lukin 1969; Sapegina 1962; Filippova 1977; Bogdanov 2004; Yakimenko et al. 2013) starts from late April (early May) to mid-August (At the maximum—from early May to early June to erly July). Nymphs are active from mid-May to early September (with a maximum of activity from at July). Adults are active from mid-April (rarely at the end of March) to mid-August (with a maximum—in the second—third decade of May in the forest-steppe zone, and until the end of June—in the forest zone). For imago stages of *I. persulcatus*, the period from the onset of their seasonal activity and the duration of this period depends on the geographical (consequently, natural and climatic) habitat characteristics, which vary considerably within areas of this species in Western Siberia. The duration of bloodsucking of larvae takes up to 4 days, that of nymphs up to 7 days, while females feed from 7 to 10 days, rarely longer. Longer feeding periods are occasionally observed at the end of seasonal activity.

It should be noted that the migratory movements of birds (spring or autumn) are long processes, including short-term phases, but long-range movements, alternating with long stops (about 2 weeks in average, sometimes longer) during which the birds resume their energy reserves by feeding, which is necessary for the next migration flight (Gavrilov 1976). In most cases, places of such stops during migration are located in areas that are quite favorable for feeding, before crossing the so-called "environmental barriers"—spaces that are unfavorable for renewing energy resources (Payevsky 1976). Migratory birds crossing such unfavorable territories have been developed in the process of evolution of migrations selecting birds, which are able to cross such a territory (either it is energetically fit and has enough fat or dies during flight).

Fig. 16.6 Optimal habitat of *I. persulcatus* within the main area: (**a**) Mixed forest with a predominance of conifers in the subzone of southern taiga (Omsk region, Znamensky district). (**b**) Mixed forest with a predominance of deciduous forest in the southern taiga (Omsk region, Znamensky district). (**c**) Mixed and coniferous forests of the floodplaces of the right confluent of the Irtysh river (small-leaved forests subzone, Omsk region, Tarsky district)

To successfully transfer ticks over long distances, it is necessary that the time of its attack on the bird coincides with the completion of the next feeding period and the beginning of the migration movement. Given the duration of feeding, a tick individual that had attacked a bird at the time of the start of a migration flight or the day before can be transferred by it over a distance of 200–600 km (depending on the bird species and weather conditions during the migration flight). The further fate of the introduced tick specimens depend on the micro- and mesoclimatic conditions in the introductions territory—e.g. if there are conditions for metamorphosis and subsequent feeding or not. From our point of view, landscape changes under the influence of human economic activity contribute to the creation of favorable conditions for ticks faster and to a greater extent than the observed climate changes (the role of which, however, cannot be denied).

16.3 Landscape Characteristic of the Distribution of *I. persulcatus*

Within the main areas *I. persulcatus* inhabit diverse groups of forest habitats reaching its highest and most stable numbers in mixed forests (Fig. 16.6). At the northern and southern borders of the areas, the distribution and numbers of the tick species are inconstant. In the northern limits of distribution, the biotopes with the highest (sometimes abnormally high) numbers of specimens are located on the upper floodplain terraces of large rivers. As a rule, these regions are mixed (coniferous-deciduous) or dark-coniferous forest types (Fig. 16.7) with well-developed layers of herbs. The numbers of *I. persulcatus* stages on the left bank of the river Ob (40 km to the east from city Khanty-Mansiysk, 2009) range in the years of high tick populations from 25–27 to 40 specimen/km. In a distance of several kilometers from the floodplain habitats, the number of ticks in the same habitats in the same time decreases to 5–15 specimens/km. A similar picture was observed in the left-bank territories of the river Ob in the Nizhne-Vartovsky district (a region with mixed and dark coniferous forests on the terrace above the floodplain), where ranges occur from 19 to 28 specimens/km. At a distance of several kilometers from the floodplain no more than 5 specimens/km occur (e.g. at the left bank of the Ob River, about 50 km to southeast from town Nizhne-Vartovsk). The valleys of large rivers (Fig. 16.8) are unsuitable as *I. persulcatus* habitats due to the hydrological (flood) regime in spring, when a significant part of the valley is flooded.

On the southern border of the *I. persulcatus* area in Western Siberia, the species also becomes more stenotopic, prefers the birch forest in swamp depressions (Fig. 16.9a) and occurs in birch-aspen and birch forests located along the lake shore of large lakes (Fig. 16.9b), floodplains terraces of large rivers, in the coombs of the floodplains of large rivers. Here their numbers in some years increase to up to 19 specimens/km. In other types of forests and artificial forest plantations, its number does not exceed 1 specimen/km. An exception are the artificial spruce forests close to the city of Omsk, where the number of species in different periods of long-term population fluctuations varies from 2 to 5 specimens/km.

Fig. 16.7 Optimal habitat of *I. persulcatus* in the subzone middle taiga: (**a**) Mixed (coniferous-deciduous) forest (left bank of the river Ob, 40 km to east of city Khanty-Mansiysk). (**b**) Dark-coniferous forest types (left bank of the river Ob, Nizhne-Vartovsky district, to the east of town Nizhne-Vartovsk)

Fig. 16.8 Biotopes in the floodplain of the Ob river in the Khanty-Mansi Autonomous Okrug: (**a**) Left bank south-east of town Nizhne-Vartovsk. (**b**) Left bank to the east of the city of Khanty-Mansiysk

Fig. 16.9 Habitat of *I. persulcatus* on the southern border of area: (**a**) Birch forest in swampy depressions (northern forest-steppe, Barabinskaya lowland, Novosibirsk region). (**b**) Birch-aspen and birch forests, which are located along the shores of large lakes (eastern shore of lake Tenis, northern forest-steppe, Tyukalinsky district, Omsk region)

References

Alifanov VI, Netsky GI (1954) Ixodic ticks of the Omsk region. Proc Omsk Sci Res Inst Epidemiol Mikrobiol Hyg 2:53–62

Alifanov VI, Netsky GI, Malkov GB, Bogdanov II (1970) Ecological and faunistic complexes of ixodic ticks in Western Siberia. Issues Infect Pathol 2:82–84

Balashov YS (1998) Ixodid ticks-parasites and carriers of infections. SPb. Science, p. 287
Bezzubov VP (1965) Ixodaceous ticks of the Novosibirsk region. Nat Focal Dis West Siberia 1965:184–192
Bogdanov II (1968) Some features of the biology of the tick *Dermacentor pictus* Herm. Med Parasitol Parasitic Dis 37(4):456–462
Bogdanov II (1999) Ixodic ticks (Acarina, Parasitiformes, Ixodidae) of Western Siberia. Message I. species composition. Nat Sci Ecol 4:161–165
Bogdanov II (2000) *Ixodes* ticks (Acari, Parasitiformes, Ixodidae) of Western Siberia. Message 2. The structure of the areas of numerous and common types of ticks. Nat Sci Ecol 5:200–204
Bogdanov II (2001) Ixodid ticks (Ixodidae, Parasitiformes) of Western Siberia. Communication III. *Dermacentor nuttalli* Ol. Nat Sci Ecol 6:229–231
Bogdanov II (2004) Ixodic ticks of Western Siberia. Report 5. Comparative characteristics of regional biology and ecology of ticks of the genus Ixodes Latr. and their life schemes. Nat Sci Ecol 8:79–88
Bogdanov II (2005) Ixodid ticks (Ixodidae, Parasitiformes) of Western Siberia. Post IV. *Haemophysalis concinna* Koch. Nat Sci Ecol 9:173–175
Bogdanov II, Shuteev MM (1981) Perennial changes in the species composition of ectoparasites and nidicols of small mammals in natural foci of Omsk hemorrhagic fever. Nat Foci Man 1981:203–211
Bogdanov II, Yakimenko VV, Malkova MG (2010) Ixodic ticks (Ixodidae, Parasitiformes, Acarina) of the Omsk region. Nat Sci Ecol 14:173–185
Davydova MS, Lukin AM (1969) Landscape-geographical distribution of ixodic ticks. Biol Zone Novosibirsk Reg 1969:250–264
Dolgushin IA (1960) Birds of Kazakhstan 1:1–469
Drozdova YV (1967) The number and landscape distribution of ixodic ticks of the North-Eastern Altai. Nat Tick-Borne Encephal Foci Altai 1967:21–29
Fedorov YV (1963) Materials for the fauna of ixodic ticks of the Omsk region. Nat Focal Dis 1963:86–88
Filippova NA (1977) Ixodid ticks of subfamilies. Ixodinae 4(4):1–396
Filippova NA (ed) (1985) Taiga tick *Ixodes persulcatus* Schulze (Acarina, Ixodidae): morphology, systematics, ecology, medical importance. In: Species of the fauna of the USSR and adjacent countries
Gavrilov VM (1976) Ecological and physiological bases of migrations of birds. Vert Zool Bird Migr 1976(9):92–133
Igolkin NI (1978) Ectoparasite complexes of small mammals in the south-eastern part of Western Siberia. 1978:1–239
Igolkin NI, Davydova MS, Semenov PV, Popov VV (1972) Ixodes ticks, their location, number and epidemiological significance in the floodplain of the Ob. Biol Resour Ob Floodplain 1972:292–305
Ivanov DI (1971) Geographic distribution of tick *Ixodes apronophorus*. Issues Infect Pathol 1971:258–263
Koklyagina AT (1963) Fauna of *Ixodes* ticks of the Altai Territory. Tick-borne encephalitis in the Altai region 1963:19–25
Koklyagina AT (1967) Geographic distribution of ixodic ticks in the Altai Territory. Materials of the Regional Conference of microbiologists, epidemiologists and infectious disease specialists on natural focal infections in the Altai region 1967:31–37
Kozlova II, Pakhotina VA, Kashapov NG, Minyailo LA, Ostapenko NA, Fayzullina NM, Fedko TF (2016) Epidemiological and epizootological characteristics of tularemia in Khanty-Mansiysk Autonomous Okrug – Ugra. Prophyl Clin Med 3(60):33–42
Kumari EV (1972) Aspects of bird migration in North Asia. Transcontinental connections of migratory birds and their role in the distribution of arboviruses 1972:23–26

Livanova NN, Livanov SG, Panov VV (2011) Features of the spread of ticks *Ixodes persulcatus* and *Ixodes pavlovskyi* at the border of the forest and forest-steppe zones of Priobye. Parasitology 45:94–102

Loginovskyi GE (1965) Distribution and seasonal variation of *Ixodes persulcatus* activity in the Kurgan region. Med Parasitol Parasitic Dis 4:487–488

Malkov GB, Bogdanov II (1966) The composition of the population and tick-borne birds in the foci of tick-borne encephalitis and Omsk hemorrhagic fever in areas of Western Siberia. The role of migratory birds in the distribution of arboviruses 1966:49–51

Malkova MG, Yakimenko VV, Tantsev AK (2012) Changing the boundaries of the ranges of pasture ixodid ticks of the genus *Ixodes* in Western Siberia. Parasitology 5:369–383

Malushina EP (1963) On the northern boundary of the distribution of *Ixodes persulcatus* P. Sch. in the Tyumen region. Nat Focal Dis 1963:54–55

Malushina EP (1966) Spread *Ixodes (Exopalpiger) trianguliceps* Bir. First Acarol Meet 1966:128–129

Malushina EP (1983) Ixodidae of the Tyumen region. Ecol Anim Faun 1983:52–71

Meshcheryakova IS, Dobrovolsky AA, Demidova TN, Kormilitsyna MI, Mikhailova TV (2014) Transmissible epidemic outbreak of tularemia in Khanty-Mansiysk in 2013. Epidemiol Vaccine Prev 78:14–19

Ostapenko VA (1976) Medical aspects of bird migration. Vert Zool Bird Migr 1976(9):253–285

Payevsky VA (1976) Population-demographic aspects of bird migration. Vert Zool Bird Migr 1976 (9):8–60

Popov VM (1962) Ixodic ticks of Western Siberia (taxonomy, characterization, ecology and geographical distribution of certain species, epidemiological and epizootological significance, the fight against ixodic ticks 1962:1–259

Ravkin YS, Sapegina VF (1967) The importance of birds in the feeding of ixodic ticks in the outbreak of tick-borne encephalitis of the north-eastern Altai. The nature of foci of tick-borne encephalitis in Altai 1967:134–147

Regions of tick-borne infections. http://net-kleshu.ru/o-kleshchah/regiony-zarazheniya/

Romanenko VN (2004) Ixodic ticks occurring in the territory of Tomsk and in adjacent forests. Siberian zoological conference: proceeding report 2004:398–399

Romanenko VN (2009) Duration of the season of activity of ticks of the genus *Ixodes* in anthropogenic landscapes of the taiga zone of Western Siberia. Parasitological studies in Siberia and the Far East: materials of the III interregion. Sci Conf 2009:243–245

Sapegina VF (1962) To ecology *Ixodes trianguliceps* Bir. Problems of zoological research in Siberia: materials of the meeting of zoologists of Siberia 1962:210–211

Sapegina VF (1980) Distribution of ixodid ticks in the forest zone of Western and Central Siberia. Problems of zoogeography and history of the fauna. Science 1980:67–76

Uspenskaya IG (1978) On the role of birds in the feeding of preimaginal stages of ixodide in conditions of a reduced number of rodents. Transcontinental connections of migratory birds and their role in the distribution of arboviruses 1978:317

Vershinina GP (1964) Tick *Ixodes apronophorus* P. Sch. 1924 in the Shegarsky District of the Tomsk Region. Zool J 43(12):1873–1874

Yakimenko VV, Bogdanov II, Tagiltsex AA (1991) Arthropods of refuge complex in colonial settlements of the sand-martin in the territory of Western Siberia and Kazakhstan. Parasitology 25(1):39–46

Yakimenko VV, Malkova MG, Bogdanov II, Tantsev AK et al (2010) On trends in the tick-borne encephalitis virus distribution in Western Siberia. J Infect Pathol 17(3):157–159

Yakimenko VV, Malkova MG, Shpynov SN (2013) Ixodic ticks of western Siberia (fauna, ecology, basic research methods). Publishing center "Omsk scientific messenger" 2013:1–240

Chapter 17
The Main River and Main–Danube Canal: A Hub for Ponto-Caspian Parasite Invasion

K. G. Alt, S. Emde, J. Kochmann, D. D. Doerge, and S. Klimpel

Abstract The Main and the Main–Danube Canal create a link between the North Sea and the Black Sea, which facilitates transfers and transportation of goods over long distances. At the same time, they act as a gateway for species, which become undeliberately introduced beyond former ranges, especially from the Ponto-Caspian region. Once species have arrived in their new habitat in Germany, they might invade native freshwater communities. In addition, these non-indigenous species can act as carriers of parasites and pathogens and lead to changes in local food webs. Here, few major examples of recent invasions of species in the Main river system are presented.

Keywords *Neogobius melanostomus* · *Dikerogammarus villosus* · Invasion corridor · River system · Host–parasite interaction

17.1 The Main–Danube Canal

The Main River's catchment crosses Germany from east to west and is part of the Trans-European Inland Waterway network. It is a tributary to the Rhine, which was connected to the Danube via the Main–Danube Canal (MDC) in 1992 (Fig. 17.1). The MDC links the Danube to the Main and, thus, indirectly connects the North Sea

K. G. Alt (✉) · J. Kochmann · D. D. Doerge · S. Klimpel
Institute of Ecology, Evolution and Diversity, Senckenberg Biodiversity and Climate Research Centre (SBiK-F), Senckenberg Gesellschaft für Naturforschung (SGN), Goethe University (GU), Frankfurt am Main, Germany
e-mail: katharina.alt@senckenberg.de

S. Emde
Department 26: Fisheries Ecology, North Rhine Westphalian State Agency for Nature, Environment and Consumer Protection, Kirchhundem, Germany

H. Mehlhorn, S. Klimpel (eds.), *Parasite and Disease Spread by Major Rivers on Earth*, Parasitology Research Monographs 12,
https://doi.org/10.1007/978-3-030-29061-0_17

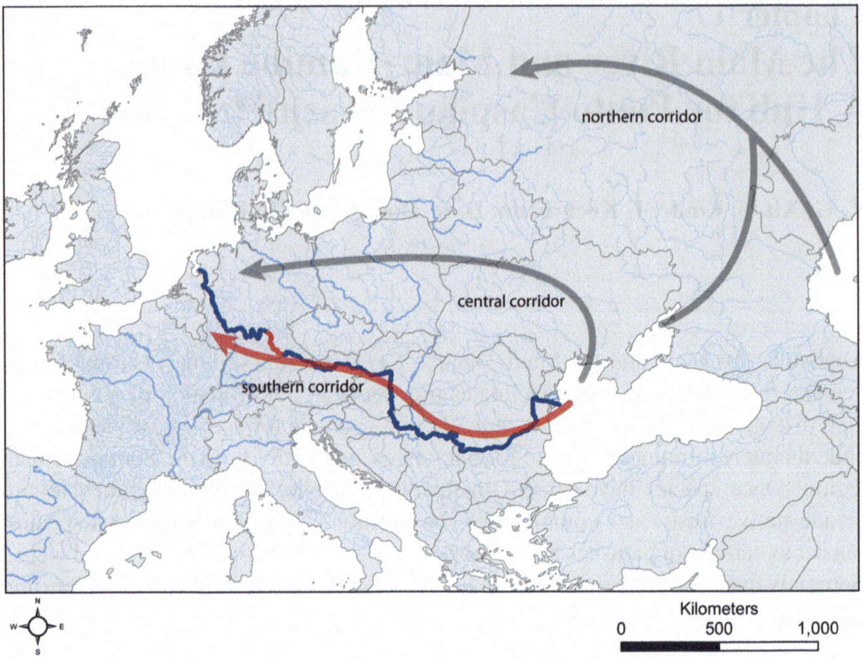

Fig. 17.1 Invasion routes of Ponto-Caspian species along European river systems along the northern, central and southern corridors. The southern corridor (red arrow) is formed by Rhine–Main–Danube river system (dark blue), which is connected by the Main–Danube Canal (red) (Based on Bij de Vaate et al. 2002)

at the Rhine Delta in Rotterdam (Netherlands) to the Black Sea via the Danube Delta (Romania). It is a Waterway Class Vb according to the European Conference of Ministers of Transport (CEMT), allowing vessels up to 6000 t. These modifications and the creation of new links between river systems promote the spread of non-native species, e.g. through vectors like vessels or active migration.

Many invaders originate in the Ponto-Caspian basin and have spread very quickly to European countries via the so-called central and southern corridor, the river Rhine (central) and the Main–Danube Channel (southern) in Germany, respectively (Bij de Vaate et al. 2002; Emde et al. 2012).

17.2 Introduced Agents of Disease: The Crayfish Plague in Central Germany

The causative agent of the crayfish plague, *Aphanomyces astaci*, is a parasitic water mold (Class: Oomycetes) that was introduced to Europe through invasive North American freshwater crayfish species. Known introduced vector species of the crayfish plague in Germany are *Pacifastacus leniusculus*, *Oronectes limosus*,

Procambarus clarkii (Diéguez-Uribeondo 2006) and *Oronectes immunis* (Schrimpf et al. 2013; Filipová et al. 2013). These crustaceans are reservoir hosts of the parasite, releasing a seasonally varying amount of spores (Strand et al. 2014). According to a study by Schrimpf et al. (2013), the main factor preventing coexistence of the endangered European crayfish (*Astacus astacus*) and the introduced North American crayfish species is *A. astaci*. Since the immune system of European crayfish is not adapted to this parasite, the species is vulnerable to infections.

Higher temperatures seem to promote the sporulation of *A. astaci* (Strand et al. 2012; Svoboda et al. 2013; via Wittwer et al. 2018), which could lead to a higher abundance of the pathogen during increasingly hot periods in summer. Wittwer et al. (2018) established a method to monitor the occurrence of the pathogen *A. astaci* without requiring host tissue samples of infected crayfish populations. Thus, the authors were able to detect spores of the oomycete in samples from a tributary river of the Main River through sampling of eDNA. This method is now applied for crayfish plague detection by the responsible governmental authorities (Heinz 2015). Apart from measures preventing the spread of invasive crayfish and the pathogen through targeted fisheries, ongoing research also suggests that European crayfish might adapt to the parasite (Jussila et al. 2011).

17.3 Invaders in Host–Parasite Communities

The host spectrum of parasites with multi-host life cycles typically forms over a long period, and different theories of co-evolution of the parasites and their mandatory hosts exist. Founding populations of introduced host species are often small, and their parasite fauna tends to be less diverse than in their native range (Emde et al. 2012). However, generalistic parasites with low host specificity can become co-introduced and will establish if they are able to complete their life cycle in the new habitat. In the following paragraphs, the amphipod *Dikerogammarus villosus*, the round goby *Neogobius melanostomus*, both originating from the Ponto-Caspian region, and their associated parasites, are presented as an example for a new host–parasite interaction in the Main River (Fig. 17.2).

17.3.1 Dikerogammarus villosus

The invasion and low species' richness of the macrozoobenthos communities along the River Main might be associated to manipulated riverine habitats, which have little structural elements that would otherwise support species with different niches (Kley and Maier 2006). Over the past two decades, the Ponto-Caspian amphipod *Dikerogammarus villosus* has become the predominant gammarid in macrozoobenthos communities of rivers in South Germany. Its history of invasion in Germany is linked to the completion of the MDC, which created a direct connection between the rivers Danube and Main.

Fig. 17.2 Invasive Ponto-Caspian species (*Dikerogammarus villosus*, *Neogobius melanostomus*) serve as intermediate hosts of parasites in their non-native range, transmitting helminths to their definitive hosts, i.e. predatory fishes (*Esox lucius*, *Anguilla anguilla*, *Squalius cephalus*) and birds (*Ardea cinerea*, *Phalacrocorax carbo*)

Inferred invasion patterns from population genetic analyses of *Dikerogammarus* spp. showed that *D. villosus* arrived one year after *D. haemobaphes* (Müller et al. 2002). *Dikerogammarus haemobaphes* was the first species from the genus *Dikerogammarus* found in the Rhine drainage immediately after the opening of the MDC in 1992 (Schleutner et al. 1994; Tittizer 1996). One year later, *D. haemobaphes* was replaced by the larger and more robust *D. villosus* (Bij de Vaate and Klink 1995). These results support the "successive invasion wave" hypothesis for Ponto-Caspian amphipods into the Rhine–Main catchment (Müller et al. 2002).

Generally, gammarid diversity was found to decline with increasing habitat disturbance, where *D. villosus* often remains the most robust species (Pöckl 2007). Its foraging behaviour and reproductive characteristics make *Dikerogammarus* species strong competitors for amphipod niches. *Dikerogammarus* spp. are R-strategists with bigger clutch sizes, a high fecundity and large number of eggs but small egg size (Kley and Maier 2006). They have a much shorter reproductive cycle than indigenous gammarids, e.g., the reproduction of *Gammarus fossarum* and *G. roeseli* takes 1.8–3.5 times longer than in *Dikerogammarus* spp., depending on temperature (Pöckl 2007). The foraging behaviour of *D. villosus* seems to be more opportunistic in comparison to native gammarid species (*G. pulex* and *G. roeseli*). Thus, *Dikerogammarus villosus* has been found to consume up to 25 mg of macroinvertebrate biomass per day, exhibited a higher predation rate than the smaller native species and showed a less selective foraging behaviour than its native competitors in mixed prey experiments (Krisp and Maier 2005).

Between 2002 and 2004 Kley and Maier (2006) observed different amphipod communities at different sampling sites along the upper Rhine and Danube, the Main River and the Main–Danube Canal and detected almost no native species. The native amphipods *Gammarus roeseli* and *G. pulex* were only found in the Danube upstream of the Main–Danube Canal and in a Rhine tributary south of Karlsruhe. The composition of species in the MDC and in the Main River was dominated by the invasive *D. villosus* and *Echinogammarus ischnus*, with *D. villosus* being the only species found at downstream sampling sites. *Echinogammarus ischnus*, another amphipod from the Ponto-Caspian region, and *D. villosus* can coexist, if appropriate substrates are available (boulder for *E. ischnus*, gravel for *D. villosus*). In a study by Emde et al. (2014a), amphipod communities in the Main were dominated by *D. villosus*, whereas *E. trichiatus* was dominant in the Rhine. Apart from these two species, other introduced amphipod species recorded were *E. ischnus*, *Chelicorophium robustum* and *Cryptorchestia cavimana*.

17.3.2 Neogobius melanostomus

Neogobius melanostomus is a Ponto-Caspian goby species now found abundantly in central European rivers (Borcherding et al. 2013; Bij de Vaate et al. 2002). There are at least two possible invasion pathways of this species into the Rhine, one from the Baltic Sea (where the species had established earlier via the "northern corridor") to Duisburg (inland harbour) and another one through the MDC—the "southern corridor" (Roche et al. 2013). To date, neither of both could be confirmed due to reporting gaps along the MDC (Roche et al. 2013). First reports of *N. melanostomus* in the River Main were made in 2007 (Borcherding et al. 2011; Emde et al. 2014a). Since then, high abundances have been found by fishermen and during samplings by the HLUG (Hessian Agency for Nature Conservation, Environment and Geology, Schlag 2014).

The round goby has typical characteristics found in invasive species, such as high temperature tolerance (4–30 °C), high biological plasticity and acclimatisation to conditions in highly manipulated environments. It thrives in areas with rockfill embankments but seems to be obstructed by faster currents (Harka and Biró 2007). As rockfills are common in waterways such as the River Main, the round goby profits from the shelter of rock interspaces and finds a suitable nursing habitat. Round gobies intensely care for their fry, which gives their offspring an advantage, and a short generation time (1–4 years) and several spawnings per year support a rapid population growth. Amphipods constitute a main diet component of the round goby in Germany, although diet analyses suggest seasonal variations in feeding patterns, depending on the availability of certain food items, e.g. insect larvae in summer and crustaceans in spring and autumn (Emde et al. 2014a).

17.3.3 Associated Parasites

Compared to at least ten parasite species in its native range (Kvach 2005), *N. melanostomus* has been reported as a host to 94 parasite species worldwide (Kvach and Stepien 2008). In its introduced range in Europe, 35 parasite species have so far been identified (Ondračková et al. 2005; Kvach and Skóra 2007; Kvach and Stepien 2008; Francová et al. 2011; Emde et al. 2014a).

Whereas 10–16 parasite species per individual host are known from the native range (Machkevsky et al. 1990; Kvach 2005), studies comparing infection levels of round gobies between the rivers Main and Rhine showed a lower parasite diversity in *N. melanostomus* than in its native range (Emde et al. 2012, 2014a). Round gobies were parasitised by the Nematoda *Paracuaria adunca* (prevalence P = 2.5%), *Raphidascaris acus* (P = 2.5%), the cystacanth stage of the Acanthocephala *Pomphorhynchus tereticollis* (P = 90%) and *Pomphorhynchus* spp. in the River Rhine near Düsseldorf and in the River Main in Frankfurt. Acanthella larvae of *P. tereticollis* (P = 0.04%), a parasite suggested to use *N. melanostomus* as intermediate host, and barbel, chub, trout and catfish as definitive hosts, were found at the same site in the River Rhine in the intermediate host *Dikerogammarus villosus* (Emde et al. 2012). *Pomphorhynchus* sp. was not found in amphipod samples from the River Main, although *D. villosus* was more abundant than in the Rhine (Emde et al. 2014a). Potential explanations of the authors included site-effects, seasonality, missing hosts and the low infection rate of *Pomphorhynchus* spp. in gammarids.

Since a steady increase in the number of recorded parasites in round gobies was observed in the Baltic Sea at the Gulf of Dansk (Kvach and Skóra 2007), parasite diversity in the River Main might be similarly expected to increase over time (Emde et al. 2014a).

In contrast to the Main River, *N. melanostomus* is found in sympatry with related Ponto-Caspian gobies (*N. fluviatilis*, *Ponticola kessleri* and *P. semilunaris*) in the upper Rhine, and its parasites include *Gyrodactylus protororhini*, larval cestodes, metacercariae of Echinostomatidae, the nematodes *Pseudocapillaria tomentosa*, *Raphidascaris acus* ($P > 50\%$), *Anguillicoloides crassus*, *Streptocara crassicauda* and the acanthocephalan *Pomphorhynchus laevis* (Ondračková et al. 2015). Although a much higher prevalence of *R. acus* (P = 56.4%) was reported by Ondračková et al. (2015) than by Emde et al. (2012) (P = 2.5%), infections with *Pomphorhynchus* spp. appear to be similar in both studies ($P \sim 90\%$) (Emde et al. 2014a). According to Emde et al. (2014a), the parasite fauna of *N. melanostomus* appears to be more diverse in the Rhine than in the Main River.

While the goby does not yet serve as a definitive host of parasites in the German sections of the Main and Rhine rivers, its relevance as intermediate host for the invasive nematode *Anguillicoloides crassus* has been observed by Emde et al. (2014b). This parasite seems to be one relevant factor contributing to the decline of European eel (*Anguilla anguilla*) as it infects the eel's swim bladder and thus interferes with the long-distance migration of the fish during its life cycle. Another possible explanation might be an impairment of the immune system of European eel, resulting in an increase of its susceptibility towards pathogens (Terech-Majewska et al. 2015). The natural definitive host of *A. crassus* is the Japanese eel, which seems to be more robust towards infections (Knopf and Mahnke 2004; Knopf 2006; Münderle et al. 2006; Heitlinger et al. 2009). Its first obligatory host are copepods (Lefebvre et al. 2012). Since the round goby has become a relevant food component for eels in European rivers, this new intermediate host seems to be a key element in the life cycle and transmission of *A. crassus* in German rivers (Fig. 17.2). A possible mechanism relies on *A. crassus* hyperparasitising the cysts of *Pomphorhynchus* spp., thus evading an immune-response of the goby (Emde et al. 2014b; Hohenadler et al. 2018).

17.4 Risk Factors for Human Health in the Main River: Drug Resistance and Antibiotics

While drinking water quality underlies strict controls in Germany, standards of wastewater treatment still require improvement, e.g. the EU Water Framework Directive does not include the monitoring of microbial contamination or measures to reduce potential pathogens. To implement a standard for this issue, the BMBF (German Federal Ministry of Education and Research) funded a joint research project (RiSKWa/HyReKA February 2016–January 2019) to evaluate the microbial burden and risk through the dissemination of drug-resistant bacteria into the environment. This is especially relevant for rivers in densely populated areas such as the Rhine–Main region.

In areas with high population densities, a corresponding number of hospitals are an important part of the health infrastructure. Hereby, a high volume and treatment of clinic wastewater influence the potential burden of multidrug-resistant (MDR) bacteria in a water body. Multidrug-resistant bacteria are resistant to several groups of antibiotics like penicillin, cephalosporin, fluorchinolone and reserve antibiotics like carbapenems (Hembach et al. 2017; Heudorf and Götz 2018) and are causing nosocomial infections—hospital-acquired infections like pneumonia, sepsis, meningitis and urinary tract infections. For example, carbapenem-resistant Gram-negative bacteria (MDR/XDR) *Pseudomonas aeruginosa* and *Acinetobacter baumanii* have been found in a German catchment system (Müller et al. 2018), and antibiotic resistance patterns of *Pseudomonas* spp. were detected in samples from a Danube River survey along 2500 river kilometres (Kittinger et al. 2016). These bacteria form a reservoir for antibiotic resistance genes, which could be spread to other species by horizontal gene transfer and thus promote the increase of resistance in potential or facultative pathogens (Kittinger et al. 2016).

A 2017 study of the Main River and its tributaries by the health department in Frankfurt (Heudorf and Götz 2018) identified MDR bacteria and antibiotics in water and sediment samples. Only some of the bacteria found were characterised as pathogenic to humans; however, they contained carbapenem-degrading proteins (subtype 2) that can spread among bacteria via plasmid exchange. Antibiotics are usually detected downstream of wastewater treatment plants as they normally exit an organism without being degraded and therefore become part of the communal wastewater. However, the environmental stability of different groups of antibiotics varies (Kümmerer 2009a, b) and could explain why some groups of antibiotics were more frequently found in water samples in the 2017 study.

In a similar manner, antibiotics applied in agriculture (colistin) are probably linked to the presence of respective resistance genes (Hembach et al. 2017). Thus, some carbapenem-resistant genes were found upstream of sewage treatment plants (Heudorf and Götz 2018), which would suggest different input pathways, such as agriculture and wild birds (Dolejska et al. 2016; Vittecoq et al. 2017; Vergara et al. 2017; Heudorf and Götz 2018).

References

Bij de Vaate A, Klink AG (1995) *Dikerogammarus villosus* Sowinsky (Crustacea: Gammaridae) a new immigrant in the Dutch part of the lower Rhine. Lauterbornia 20:51–54

Bij de Vaate A, Jazdzewski K, Ketelaars HA et al (2002) Geographical patterns in range extension of Ponto-Caspian macroinvertebrate species in Europe. Can J Fish Aquat Sci 59:1159–1174. https://doi.org/10.1139/f02-098

Borcherding J, Staas S, Krüger S et al (2011) Non-native Gobiid species in the lower River Rhine (Germany): recent range extensions and densities. J Appl Ichthyol 27:153–155. https://doi.org/10.1111/j.1439-0426.2010.01662.x

Borcherding J, Dolina M, Heermann L et al (2013) Feeding and niche differentiation in three invasive gobies in the lower Rhine, Germany. Limnologica 43:49–58. https://doi.org/10.1016/j.limno.2012.08.003

Diéguez-Uribeondo J (2006) The dispersion of *Aphanomyces astaci*-carrier *Pacifastacus lenuisculus* by humans represents the main cause of disappearance of the indigenous crayfish *Austropotamobius pallipes* in Navarra. Bull Fr Pêche Piscic 380–381:1303–1312. https://doi.org/10.1051/kmae:2006036

Dolejska M, Masarikova M, Dobiasova H et al (2016) High prevalence of *Salmonella* and IMP-4-producing Enterobacteriaceae in the silver gull on Five Islands, Australia. J Antimicrob Chemother 71:63–70. https://doi.org/10.1093/jac/dkv306

Emde S, Rueckert S, Palm HW, Klimpel S (2012) Invasive Ponto-Caspian amphipods and fish increase the distribution range of the acanthocephalan *Pomphorhynchus tereticollis* in the River Rhine. PLoS One 7:e53218. https://doi.org/10.1371/journal.pone.0053218

Emde S, Kochmann J, Kuhn T et al (2014a) Getting what is served? Feeding ecology influencing parasite-host interactions in invasive round goby *Neogobius melanostomus*. PLoS One 9: e109971. https://doi.org/10.1371/journal.pone.0109971

Emde S, Rueckert S, Kochmann J et al (2014b) Nematode eel parasite found inside acanthocephalan cysts - a "Trojan horse" strategy? Parasit Vectors 7:504. https://doi.org/10.1186/s13071-014-0504-8

Filipová L, Petrusek A, Matasová K et al (2013) Prevalence of the crayfish plague pathogen *Aphanomyces astaci* in populations of the signal crayfish *Pacifastacus leniusculus* in France: evaluating the threat to native crayfish. PLoS One 8:e70157. https://doi.org/10.1371/journal.pone.0070157

Francová K, Ondračková M, Polačik M, Jurajda P (2011) Parasite fauna of native and non-native populations of *Neogobius melanostomus* (Pallas, 1814) (Gobiidae) in the longitudinal profile of the Danube River. J Appl Ichthyol 27:879–886. https://doi.org/10.1111/j.1439-0426.2010.01582.x

Harka Á, Biró P (2007) New patterns in Danubian distribution of Ponto-Caspian gobies–a result of global climatic change and/or canalization. Electron J Ichthyol 1:1–14

Heinz P (2015) Was sind gebietsfremde Krebsarten und welche Gefahren gehen von ihnen in unseren Gewässern aus? Regierungspräsidium Darmstadt, Dezernat V 51.1 - Landwirtschaft, Fischerei und internationaler Artenschutz. https://rp-darmstadt.hessen.de/sites/rp-darmstadt.hessen.de/files/content-downloads/Flyer%20Krebspest.pdf. Accessed 15 Mar 2019

Heitlinger EG, Laetsch DR, Weclawski U et al (2009) Massive encapsulation of larval *Anguillicoloides crassus* in the intestinal wall of Japanese eels. Parasit Vectors 2:48. https://doi.org/10.1186/1756-3305-2-48

Hembach N, Schmid F, Alexander J et al (2017) Occurrence of the mcr-1 Colistin resistance gene and other clinically relevant antibiotic resistance genes in microbial populations at different municipal wastewater treatment plants in Germany. Front Microbiol 8:1282. https://doi.org/10.3389/fmicb.2017.01282

Heudorf U, Götz E (2018) Oberflächengewässer in Frankfurt am Main 1996–2017 – Bericht zur hygienischen Qualität. Gesundheitsamt, Frankfurt am Main

Hohenadler MA, Honka KI, Emde S et al (2018) First evidence for a possible invasional meltdown among invasive fish parasites. Sci Rep 8:15085. https://doi.org/10.1038/s41598-018-33445-4

Jussila J, Makkonen J, Vainikka A et al (2011) Latent crayfish plague (*Aphanomyces astaci*) infection in a robust wild noble crayfish (*Astacus astacus*) population. Aquaculture 321:17–20. https://doi.org/10.1016/j.aquaculture.2011.08.026

Kittinger C, Lipp M, Baumert R et al (2016) Antibiotic resistance patterns of *Pseudomonas* spp. isolated from the River Danube. Front Microbiol 7:586. https://doi.org/10.3389/fmicb.2016.00586

Kley A, Maier G (2006) Reproductive characteristics of invasive gammarids in the Rhine-Main-Danube catchment, South Germany. Limnologica 36:79–90. https://doi.org/10.1016/j.limno.2006.01.002

Knopf K (2006) The swimbladder nematode *Anguillicola crassus* in the European eel *Anguilla anguilla* and the Japanese eel *Anguilla japonica*: differences in susceptibility and immunity between a recently colonized host and the original host. J Helminthol 80:129–136. https://doi.org/10.1079/JOH2006353

Knopf K, Mahnke M (2004) Differences in susceptibility of the European eel (*Anguilla anguilla*) and the Japanese eel (*Anguilla japonica*) to the swim-bladder nematode *Anguillicola crassus*. Parasitology 129:491–496. https://doi.org/10.1017/S0031182004005864

Krisp H, Maier G (2005) Consumption of macroinvertebrates by invasive and native gammarids: a comparison. J Limnol 64:55–59. https://doi.org/10.4081/jlimnol.2005.55

Kümmerer K (2009a) Antibiotics in the aquatic environment – a review – part I. Chemosphere 75:417–434. https://doi.org/10.1016/j.chemosphere.2008.11.086

Kümmerer K (2009b) Antibiotics in the aquatic environment – a review – part II. Chemosphere 75:435–441. https://doi.org/10.1016/j.chemosphere.2008.12.006

Kvach Y (2005) A comparative analysis of helminth faunas and infection parameters of ten species of gobiid fishes [Actinopterygii: Gobiidae] from the North-Western Black Sea. Acta Ichthyol Piscat 2:103–110

Kvach Y, Skóra KE (2007) Metazoa parasites of the invasive round goby *Apollonia melanostoma* (*Neogobius melanostomus*) (Pallas) (Gobiidae: Osteichthyes) in the Gulf of Gdańsk, Baltic Sea, Poland: a comparison with the Black Sea. Parasitol Res 100:767–774. https://doi.org/10.1007/s00436-006-0311-z

Kvach Y, Stepien CA (2008) Metazoan Parasites of introduced round and tubenose gobies in the Great Lakes: support for the "enemy release hypothesis". J Great Lakes Res 34:23–35. https://doi.org/10.3394/0380-1330(2008)34[23:MPOIRA]2.0.CO;2

Lefebvre F, Wielgoss S, Nagasawa K, Moravec F (2012) On the origin of *Anguillicoloides crassus*, the invasive nematode of anguillid eels. Aquat Invasions 7:443–453. https://doi.org/10.3391/ai.2012.7.4.001

Machkevsky VK, Mordinova TN, Parukhin AM (1990) Gelmintofauna bychkov Yegorlytskogo zaliva i oz. Ekol Morya 36:69–75

Müller JC, Schramm S, Seitz A (2002) Genetic and morphological differentiation of *Dikerogammarus* invaders and their invasion history in Central Europe. Freshw Biol 47:2039–2048. https://doi.org/10.1046/j.1365-2427.2002.00944.x

Müller H, Sib E, Gajdiss M et al (2018) Dissemination of multi-resistant Gram-negative bacteria into German wastewater and surface waters. FEMS Microbiol Ecol 94:fiy057. https://doi.org/10.1093/femsec/fiy057

Münderle M, Taraschewski H, Klar B et al (2006) Occurrence of *Anguillicola crassus* (Nematoda: Dracunculoidea) in Japanese eels *Anguilla japonica* from a river and an aquaculture unit in SW Taiwan. Dis Aquat Org 71:101–108. https://doi.org/10.3354/dao071101

Ondračková M, Dávidová M, Pečínková M et al (2005) Metazoan parasites of Neogobius fishes in the Slovak section of the River Danube. J Appl Ichthyol 21:345–349. https://doi.org/10.1111/j.1439-0426.2005.00682.x

Ondračková M, Valová Z, Hudcová I et al (2015) Temporal effects on host-parasite associations in four naturalized goby species living in sympatry. Hydrobiologia 746:233–243. https://doi.org/10.1007/s10750-014-1967-5

Pöckl M (2007) Strategies of a successful new invader in European fresh waters: fecundity and reproductive potential of the Ponto-Caspian amphipod *Dikerogammarus villosus* in the Austrian Danube, compared with the indigenous *Gammarus fossarum* and *G. roeseli*. Freshw Biol 52:50–63. https://doi.org/10.1111/j.1365-2427.2006.01671.x

Roche KF, Janač M, Jurajda P (2013) A review of Gobiid expansion along the Danube-Rhine corridor – geopolitical change as a driver for invasion. Knowl Manag Aquat Ecosyst 411:01. https://doi.org/10.1051/kmae/2013066

Schlag E (2014) Ergebnisse der Fischbestandserhebung in Hessen. In: Hessisches Landesamt für Umwelt und Geologie (ed) Hessisches Landesamt für Umwelt und Geologie – Jahresbericht 2013. Wiesbaden, pp 27–38

Schleutner M, Schleutner A, Potel S, Banning M (1994) *Dikerogammarus haemobaphes* (Eichwald 1841) (Gammaridae) aus der Donau erreicht über den Main-Donau-Kanal den Main [*Dikerogammarus haemobaphes* (Eichwald 1841) a crustacean of Danube origin proved as newcomer in the River Main]. Lauterbornia 19:155–159

Schrimpf A, Maiwald T, Vrålstad T et al (2013) Absence of the crayfish plague pathogen (*Aphanomyces astaci*) facilitates coexistence of European and American crayfish in Central Europe. Freshw Biol 58:1116–1125. https://doi.org/10.1111/fwb.12112

Strand DA, Jussila J, Viljamaa-Dirks S et al (2012) Monitoring the spore dynamics of *Aphanomyces astaci* in the ambient water of latent carrier crayfish. Vet Microbiol 160:99–107. https://doi.org/10.1016/j.vetmic.2012.05.008

Strand DA, Jussila J, Johnsen SI et al (2014) Detection of crayfish plague spores in large freshwater systems. J Appl Ecol 51:544–553. https://doi.org/10.1111/1365-2664.12218

Svoboda J, Kozubíková-Balcarová E, Kouba A et al (2013) Temporal dynamics of spore release of the crayfish plague pathogen from its natural host, American spiny-cheek crayfish (*Orconectes limosus*), evaluated by transmission experiments. Parasitology 140:792–801. https://doi.org/10.1017/S0031182012002223

Terech-Majewska E, Schulz P, Siwicki AK (2015) Influence of nematode *Anguillicoloides crassus* infestation on the cellular and humoral innate immunity in European eel (*Anguilla anguilla* L.). Cent Eur J Immunol 40:127–131. https://doi.org/10.5114/ceji.2015.52824

Tittizer T (1996) Vorkommen und Ausbreitung aquatischer Neozoen (Makrozoobenthos) in den Bundeswasserstraßen. In: Gebhart H, Kinzelbach R, Schmidt-Fischer S (eds) Gebietsfremde Tierarten. Ecomed, Fellbach, pp 49–86

Vergara A, Pitart C, Montalvo T et al (2017) Prevalence of extended-spectrum-β-lactamase- and/or carbapenemase-producing *Escherichia coli* isolated from yellow-legged gulls from Barcelona, Spain. Antimicrob Agents Chemother 61:e02071-16. https://doi.org/10.1128/AAC.02071-16
Vittecoq M, Laurens C, Brazier L et al (2017) VIM-1 carbapenemase-producing *Escherichia coli* in gulls from southern France. Ecol Evol 7:1224–1232. https://doi.org/10.1002/ece3.2707
Wittwer C, Stoll S, Strand D et al (2018) eDNA-based crayfish plague monitoring is superior to conventional trap-based assessments in year-round detection probability. Hydrobiologia 807:87–97. https://doi.org/10.1007/s10750-017-3408-8

Chapter 18
"Father Rhine": From the Alps to the North Sea

Monika Niehaus

Abstract This river passing 6 recent countries is one of the most important life veins of Europe in the past and today being the border between Roman culture and Asian invaders in former times. Even early hominids settled there. However, the Rhine also helped to spread agents of diseases in former times.

Keywords History · Rhine gold · Malaria · Romans · Neanderthal man · Cholera · Plague · Tourism

18.1 Introduction

This river has always impressed the people who lived on its banks and used it in many ways, and numerous myths, fairy tales, and legends have been entwined around it since antiquity. The Celtic population called "him" Rênos, which probably derives from their term for "run, flow," and the Romans called him Rhenus Pater, father Rhine, and worshiped him as river god. Like all major waterways worldwide, the Rhine has always strongly influenced the fate of people in its catchment area: It brought them prosperity because it favored the exchange of goods and knowledge but unfortunately also paved the way for misery and epidemics such as the plague and cholera and thus helped to spread these both deadly diseases easily and quickly (Fig. 18.1).

M. Niehaus (✉)
Department of Biology, Düsseldorf University, Düsseldorf, Germany
e-mail: drmonika.niehaus@googlemail.de

H. Mehlhorn, S. Klimpel (eds.), *Parasite and Disease Spread by Major Rivers on Earth*, Parasitology Research Monographs 12,
https://doi.org/10.1007/978-3-030-29061-0_18

Fig. 18.1 Father Rhine as baroque sculpture in the park of the castle at Schwetzingen (Germany) created by the artist Peter Anton von Verschaffelt, eighteenth century. https://commons.wikimedia.org/wiki/File:Schwetzingen_Schlosspark_Skulpturen_26.jpg

18.2 The Rhine: A Short Geographical Journey from Its Source to Its Estuary

The age of the original Rhine is estimated at 14–15 million years based on fossil finds in its sediments—including one of the first ancestors of deer and a dinotherium ("tusk elephant")—but since then it has changed its course many times (Böhme et al. 2012; Breier and Stiefenhofer 2016).

Today, the Rhine is the only river to flow from the Alps to the North Sea. With a total length of around 1230 km, it is one of the ten longest rivers in Europe (see Table in Chap. 1). Among those that cross the region with a German-speaking population, it occupies the second place after the almost twice as long river Danube before the river Elbe. The Rhine covers about 370 km in Switzerland, 700 km in Germany, and 160 km in the Netherlands.

It rises in the Swiss Tomasee at an altitude of 2300 m as the "Vorderrhein" (Anterior Rhine); joins its second source river, the "Hinterrhein" (Posterior Rhine), to form together the "Alpenrhein" (Alpine Rhine); and flows into Lake Constance. At Constance (Germany) it leaves Lake Constance and becomes the "Hochrhein" (High Rhine). Via the Rhine Falls of Schaffhausen, it reaches the Swiss town of Basel; from here, the river, now called the Upper Rhine, is navigable, and the international count of its length begins. From the confluence point joining the Swiss Aare onward, the Rhine is declared as a stream (Fig. 18.2).

Near Bingen (Germany) it enters the narrow and rugged Rhine Valley, which leads through the Rhenish Slate Mountains, as the so-called Middle Rhine. After about 130 km, the riverbed widens again, and the river becomes the Lower Rhine,

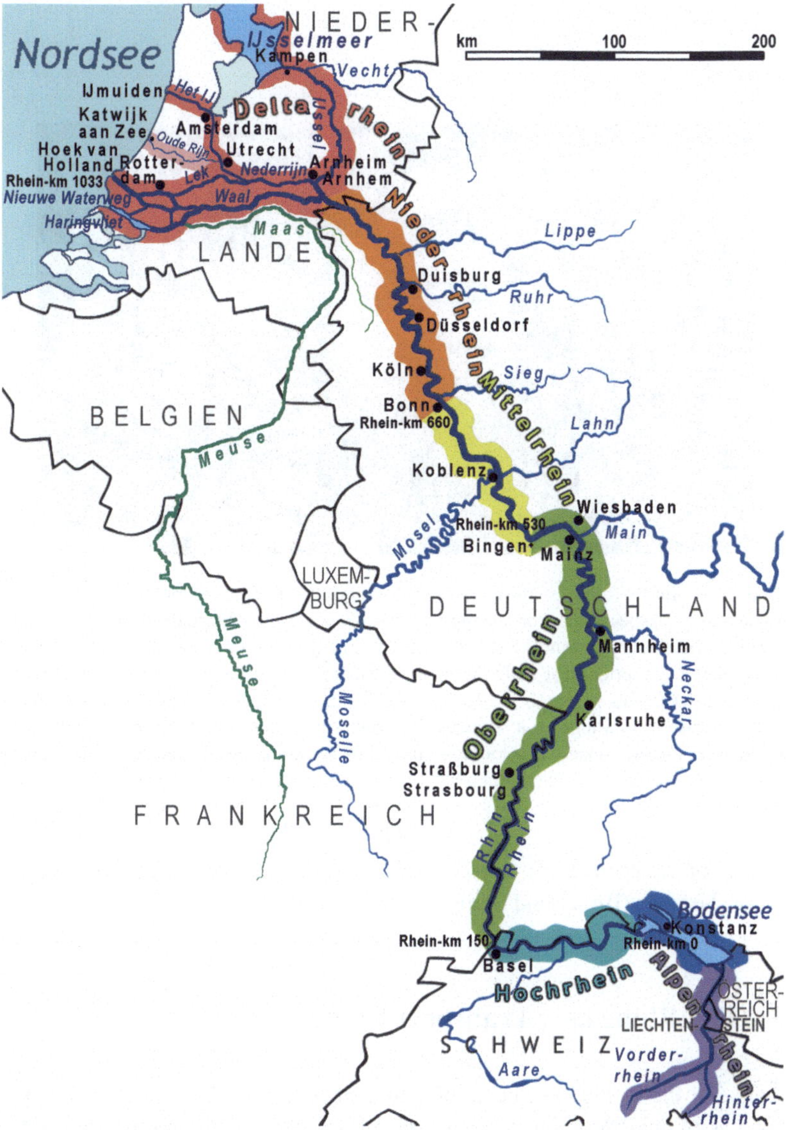

Fig. 18.2 On its way from the Alps to the North Sea, the Rhine links six European countries: Switzerland, Liechtenstein, Austria, France, Germany, and the Netherlands. https://upload.wikimedia.org/wikipedia/commons/2/29/Rhein-Karte2.png

where the Sieg river opposite to the former capital of the Federal Republic, Bonn, flows into the Rhine. After having passed the Roman-founded town Cologne, the largest city along the Rhine, the river crosses the border to the Netherlands at the town Emmerich, spreads out to the Rhine-Meuse delta—known as the Delta Rhine for short—and flows into the North Sea at Rotterdam, Europe's largest seaport.

Fig. 18.3 The Middle Rhine is considered to be the most "romantic" section of this river with the highest density of castles in the world along its banks. This was less due to the beautiful view than to the lucrative opportunity to squeeze off customs from passing sailors and merchants—the lords of the castle had been often just little more than robber knights and highwaymen (Breier and Stiefenhofer 2016). This figure shows Katz Castle. In the background rises the Loreley rock into the river. According to a legend, a sorceress had lured the Rhine sailors to their deaths with her singing and her charms since when looking upward to her combing the hair they did not note the rocks that broke their ships. https://upload.wikimedia.org/wikipedia/commons/1/12/Mittelrhein_Burg_Katz.jpg

A drop of water takes 31 days to travel the full length from the source to the mouth of the river (Breier and Stiefenhofer 2016) (Fig. 18.3).

18.3 The Rhine as a Transport Route

The banks of the Rhine have been inhabited since the Ice Age about 600,000 years ago; people of the Paleolithic were the first Rhinelanders. At that time, the landscape there must have been reminiscent of present-day Scandinavia: moose in the woods and salmons in the river. In other ancient times, Celts and Germanic tribes lived at the Rhine until the Romans came, and the Rhine became the border between Roman high culture and Germans, who were considered as "barbarians." Thus the Rhine was a border between civilization and "wilderness." The Romans built the first bridges across the river and constructed fortified camps, which later developed into large cities such as Cologne. Trade flourished at the time of the Romans, and one of the most curious "goods" the Germanic Teutons had to offer was blond hair, made into wigs, which was highly coveted by the noble Roman women (Breier and

Stiefenhofer 2016). Thus many Germans head lice might have found a new home on Roman heads this way.

In the Middle Ages (starting 476 at the end of the West Roman Empire lasting until around 1453, when the Turks captured Constantinople), the Rhine became one of the most important European transport and trade routes, on which people, cattle, and goods of all kinds were transported on rafts, barges, and ships of various sizes from Lake Constance downstream to the North Sea. Upstream, it was only possible with the muscle power of man, ox, or horse (Wegener 2003); for this purpose, so-called towpaths were created to pull the ships against the current. The trade and customs duties levied on goods were such a profitable business for the Rhine residents, that numerous castles were built in the narrow Middle Rhine Valley to "skim off profits" (see above). Today their ruins, as a symbol of the "romantic Rhine," are a no less lucrative tourist attraction for the residents.

Rhine Gold The Rhine is rich in legends and myths. Probably the most famous legend on the Rhine is the *Nibelungen Saga*, which dates back to lyrics in the early thirteenth century (von Plessen 2016). It tells the story of Siegfried the Dragon Slayer, who fought for King Nibelung's enormous treasure and shows his love for Kriemhild (the King's daughter), and his death by the hands of Hagen, henchman of Kriemhild's rival Brunhild. Killer Hagen murdered the hero from behind with a spear and sank the Nibelungen treasure into the Rhine. Kriemhild's terrible revenge followed. She married the King of the invading huns, invited the Nibelungs to the court and killed them by activation of hun warriors. This story reflects the events during the period of migration of nations. The gold of the Nibelungs was never found.

In the nineteenth century the German composer Richard Wagner created the four-part opera cycle "the Ring of the Nibelungs" from the legend, including "Das Rheingold." By the way, the Rhine was already famous in ancient times for its gold deposits; the Greek historian Strabo called it "the gold-rich." Gold was washed on its banks for many centuries. The Grand Duke Leopold of Baden, for example, had minted some 40,000 ducats made of Rhine gold during his reign (Tümmers 1999; von Plessen 2016).

While wine from the Rhine and Moselle, wood from the Black Forest, and many other goods were transported down the Rhine to Holland (the Netherlands of present days) (see below), salted fish, especially herring, was one of the most important goods shipped upstream. At that time Christianity had to hold about 100 days of fasting a year when meat and poultry were taboo, but fish was allowed to be eaten. Salted herring was so important for the Dutch economy that it was said that Amsterdam was "built on herring bones."

The Dutch Rafts: Giants of Rhine Navigation Rafts were already used by the Roman army commander Julius Caesar to cross the Rhine. In the course of time, these rafts became ever larger finally reaching truly gigantic proportions: In the seventeenth/eighteenth century, the Netherlands had become a global maritime and trading power and thus needed enormous quantities of wood to build ships; the necessary wood—mainly firs and oaks—originated from the Black Forest in Southern German regions and was assembled into rafts: up to 400 m long and up to 80 m wide, these giant rafts had the size of a modern aircraft carrier. A crew of about 500 men was needed to maneuver such a raft, and a corresponding amount of food (or livestock) had to be carried along, including about 20,000 liters of beer—5 liters per rafter and day; additionally there were accommodations and facilities such as laundry, kitchen, bakery, slaughter facilities, etc. on the giant raft.

The timber merchants sent in general giant rafts to Holland, which were much more difficult to maneuver instead of several smaller ones. But this was needed due to the numerous customs stations along the Rhine, since it was quicker and cheaper to bribe a customs officer with a larger sum than to haggle over many smaller rafts each time.

By the way, rafting on the Rhine was only stopped in the years 1967/1968, just about 50 years ago (Breier and Stiefenhofer 2016; Neumann 2018; Tümmers 1999; Wegener 2003; Weikinn 2000).

18.4 A Tropical Disease at the Rhine: Malaria

As an important transport route, father Rhine brought prosperity to his children, but this prosperity had its price. From antiquity to modern times, thousands of people living along the Rhine died of a deadly disease, which we now generally locate in the tropics: malaria. This disease—also known as swamp fever—is one of the best-known human parasitic diseases and still claims more victims than all other parasitic diseases combined (Piperaki 2018; Mitchell 2016).

Malaria, which is associated with high fever attacks, ague, sweating, as well as headache and pain in the limbs, was probably introduced to Europe from the Southeast as early as the Neolithic Age (4000–3000 BC). The pathogen—*Plasmodium vivax*, *P. malariae*, and *P. falciparum*—settled down in such places where it was climatically favorable and where sufficiently large populations of humans and biting mosquitoes of the genus *Anopheles* occurred. In Western and Northern Europe especially female *Anopheles atroparvus* served as vectors. Humid river valleys and floodplains provided good conditions for development for the mosquito larvae.

Remnants of proteins of *P. falciparum* were already found in Egyptian mummies of 3200 BC. Since the Egyptians proceeded a busy trading with populations of Greece and other Mediterranean countries, the parasites might have reached Europe in this way. Since ancient times, the Pontine swamps near Rome have been a "paradise" for *Anopheles* mosquitoes for almost two millennia. The Romans considered the malaria fever as a consequence of bad evaporations of the swamp and thus called it "mal' aria" = bad air. Thus they had no idea of the transmission of the dangerous unicellular passengers by mosquito bites. The medieval warm period between the middle of the tenth and thirteenth centuries favored the spreading of malaria to the north. Thus *Plasmodium malariae* was probably also endemic everywhere on the banks of the Rhine, Rhône, and Danube in the thirteenth century (Piperaki 2018).

In the Netherlands, the Rhine-Meuse Delta was a malaria endemic area in the sixteenth century, and it has been handed down that the great German Renaissance painter Albrecht Dürer also had become infected with the fever there while waiting at the Emperor's Court to be supported by a grant. In Germany, Mannheim, situated at the mouth of the Neckar river into the Rhine, was hit by a severe malaria epidemic in the year 1761 (Winkle 2005). Especially along the Upper Rhine, the climate in early modern times could be tropical and humid in summer. Around human settlements there were extensive swamps, which offered an ideal breeding ground for mosquitoes, which could transmit agents of diseases such as malaria. In these areas, floods and high tides repeatedly transformed vast areas near rivers into swamplands. Therefore malaria epidemics occurred again and again, which, according to records of the Catholic church, had costed the lives of more than 100 people in the small town of Burkheim on the Upper Rhine just in 1720. The same occurred in the neighboring communities (Blackbourn 2008; Breier and Stiefenhofer 2016).

At this time, malaria and other diseases such as cholera, typhoid fever, dysentery, and other infectious diseases were the worst scourges of the population. In the eighteenth century, malaria on the Rhine claimed more victims than all warlike conflicts (Breier and Stiefenhofer 2016).

18.5 The Straightening of the Rhine in the Nineteenth Century

With its high water, the Rhine had frightened the people living along its banks again and again. Probably the most devastating flood in historical times was the Magdalene flood of 1342, named after the saint's name day on 22 July: At that time, the Rhine and its tributaries flooded the entire Middle Rhine Valley. One eyewitness reported that "the walls of the city of Cologne were crossed with barges" (Weikinn 2000). Long-lasting, torrential rains, but also the rigorous deforestation at the beginning of the century, were to blame.

The constant flooding, which was able to wash away entire villages, the recurrent flare-up of malaria, and the increased requirements for shipping traffic finally led to a mammoth project being tackled in the nineteenth century: the straightening of the Rhine. Hardly any river voluntarily flows straight ahead, but they form loops and meanders. The engineer Johann Gottfried Tulla from the Baden region began in

1817 to force the "multi-armed" Upper Rhine into a single riverbed by cutting off river loops (Breier and Stiefenhofer 2016). This construction work—the most extensive ever carried out in Germany—was completed in the year 1876. Since then the Rhine is navigable over 850 km from Basel to the North Sea. Its length between Basel and Worms had been shortened by about 80 km. Thus it flows since then much faster. Thus along both sides large swamps were cut off from the river and silted up. However, as a result of the straightening of the Rhine, the chance for gold panning also declined more and more, and thus the last professional gold washer on the Upper Rhine gave up in 1874 (von Plessen 2016; Tümmers 1999).

The diminution of the mosquito breeding areas, but also the separation of human and animal dwellings (people did no longer sleep with the mosquito-plagued cattle under the same roof), led to a strong decrease of malaria, but it took further 100 years until the disease was largely defeated in Europe. Since 1975, Germany has been regarded as "malaria-free," even if the disease has flared up locally and is introduced still today by tourists or refugees (Piperaki 2018).

18.6 King Cholera Travels Down the Rhine and Elbe

Like other major rivers, the Rhine, as a busy waterway, was always a gateway for epidemics that spread upstream or downstream in the cities along its banks. Cholera was an often fatal disease that still claimed many victims in Germany in the nineteenth century.

The name of this infectious disease, which is associated with vomiting and severe, watery diarrhea, derives from Greek *chole* "bile" and *rhein* (sic!) "flow"—"bile" meaning here "bodily fluids" that are expelled from the body vehemently. This loss of fluid quickly leads to a fatal dehydration of the body. The symptoms were commonly known, but there was neither an effective remedy nor the cause of the infection or way of propagation was known.

When the cholera epidemic reached Germany in 1831, it had already come a long way. In 1817 it broke out in Southeast Asia, and it took several years until it reached the West via Russia. The predication was made by the physician Friedrich Schnurrer from Tübingen in his "Chronicle of Diseases" in 1824 (Schnurrer 1824) "that it [the disease] could arrive on the Rhine in four years." The first wave of cholera did not reach the northeast of Germany until 1831 but spread across rivers such as the Rhine and Elbe and reached Hamburg in the same year. Between 1831 and 1892, there occurred epidemics of cholera during 16 years, which was mainly due to the poor quality of the drinking water (see below). The authorities were completely helpless, since they had no idea what caused the epidemic. A few "experts" even identified the stars as cause of the epidemic, as did also an expert from the University of Munich (Vasold 2003). And the doctors recommended what they always recommended as an effective remedy: phlebotomy. This "treatment" gave some of the already dehydrated patients the rest (Fig. 18.4).

A COURT FOR KING CHOLERA.

Fig. 18.4 King Cholera holds court in nineteenth-century London as shown on a drawing in the journal Punch in 1852. The spreading was favored by the coexistence of many people of a very small space, poor living conditions, and inadequate hygiene: There were no sewage systems, no clean drinking water, and contaminated food. https://commons.wikimedia.org/wiki/File:Punch-A_Court_for_King_Cholera.png

The medical fraternity was no great help either: there had been people called miasmatics, who believed in dangerous soil effluvia as they did with malaria, and others called contactgionists, who suspected some kind of infections (although they did not know how), were irreconcilably opposed to each other. However, the idea of an infection promoted by mobility and trade gave the authorities in Koblenz and Düsseldorf the idea of strictly controlling shipping on the Rhine and taking strict quarantine measures. This showed a certain success. While almost half a million people died in the Kingdom of Prussia in the course of the waves of cholera between 1831 and 1873, Koblenz remained disease-free and Düsseldorf got off lightly.

About a decade later (1884), the German physician and microbiologist Robert Koch discovered the pathogen of the deadly infectious cholera disease: a short, rod-shaped bacterium called *Vibrio cholerae*, which later turned out to be spread by contaminated drinking water or contaminated food. Although the young doctor Robert Koch was initially unable to prevail over the followers of the miasma theory, time was on his side. The cities began to construct canals, separated drinking and wastewater, and generally payed more attention to hygiene. Only in Hamburg cholera flared up again in 1892: There the water was still taken directly from the Elbe without investing in effective purification plants—a deadly thrift.

Thanks to sanitation and improved hygiene, cholera disappeared from Europe in the twentieth century to a large extend and retreated to Southern Asia, where its mortality was significantly reduced due to saline infusions and antibiotics. But we still should not forget: In a world which is growing ever closer together due to our increased mobility, cholera, which took more than 10 years in the nineteenth century to travel from South Asia to the Rhine and Elbe, is always just a few flying hours away in the twenty-first century.

18.7 The Plague Moves Up the Rhine

Malaria and cholera are without any doubt scourges of humanity, but their horror cannot be compared to the panic fear of the "Black Death," *the plague*, which killed more than a third of the European population in the fourteenth century. The cause of this terrible disease is the bacterium *Yersinia pestis*, which is spread by rat fleas. According to recent studies (see below), human ectoparasites—human fleas and head lice—could have played a crucial role in the fourteenth century in addition to rat fleas: The cramped, unhygienic living and sleeping conditions, in which entire families spent the night together on a straw bed, offered these bloodsuckers the best opportunity to switch from one host to another (Dean et al. 2018). The *bubonic plague*, which accounts for 90% of all plague cases, killed about one third of the infected persons.

After the extinction of the Black Death in 1352, Europe was able to catch its breath briefly, but the plague returned again and again until the eighteenth century. Already in 1356 the plague broke out again in the Rhineland and brought a "second great misery and a second great dying," according to the *Limburg Chronicle*. Further "Jammer" (German term for misery) followed in two waves between 1372 and 1387 (Winkle 2005), during which the epidemic frequently originated at port cities and was spread by the ship rats infested with fleas—as well as by human fleas and lice after human infections had taken place (Fig. 18.5).

In the middle of the fifteenth century, there occurred another plague epidemic, which spread from the Rhineland to France and Spain. At the turn of the sixteenth century, the epidemic raged in London and then again around 1563. The first exact data from this time are available now: Within a half year 14,086 people died in the British capital.

Afterward the London plague epidemic moved upstream the Rhine. A Cologne patrician, Hermann von Weinsberg, wrote at the time: "In 1564 the plague raged in a terrible way from the beginning of August. All shared flats ("Bursen") were closed, all students—spiritual and secular, poor and rich—left the city." It is reported that between August and October 10,000–12,000 people died in Cologne and that the disease did not stop (Rath 1955). Even further upstream the Rhine, in Strasbourg and Freiburg, people died "like flies," and the authorities, churches, or doctors had not

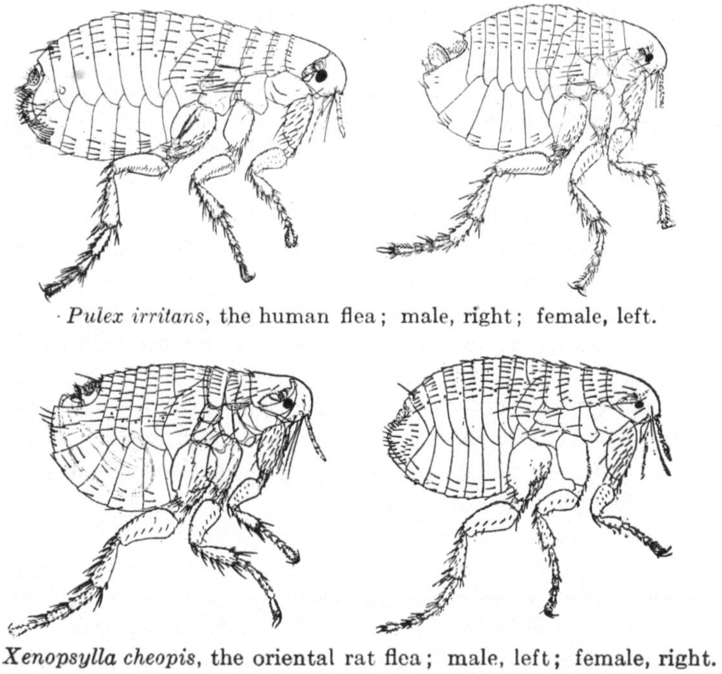

Fig. 18.5 At least during the plague epidemics in the fourteenth century—according to the latest model calculations (Dean et al. 2018)—human fleas (*Pulex irritans*, above) and body lice (*Pediculus humanus corporis*) besides rat fleas (*Xenopsylla cheopis*, below) seem to have ensured rapid spread of the agents of disease. Although human and rat fleas are two different genera, the parasites can hardly be distinguished externally by naked eyes, as this figure shows. https://en.wikipedia.org/wiki/Oriental_rat_flea. https://upload.wikimedia.org/wikipedia/commons/c/cd/PulexIrritans.jpg

been able to help them. Like cholera and malaria, the plague was generally attributed to "miasmas" or bad celestial bodies. The famous Eau de Cologne (Cologne water) was originally used to control plague miasmas.

The last major plague epidemic in Europe broke out again in London in 1663 claiming nearly 70,000 lives within 1 year. And again, starting from the German North Sea coast, the plague went up the great waterway, the Rhine. Repeatedly the plague flared up again and again along the Rhine in the seventeenth century for shorter or longer periods. It passed through Cologne, moved southward along the Rhine, and raged in Bingen, Mainz, and Frankfurt. The Alsace was not spared either, nor was the Rhine city of Basel. The last epidemic broke out in the French port of Marseille in 1720/1721. Since then there have been no more plague epidemics in Europe.

18.8 An Old Pest: The Fish Tapeworm

However, the Rhine not only served as a waterway for major epidemics upstream and downstream but also harbors its very own parasitic fauna, from which the largest representative is selected here: We know that the Romans were already infected by the fish tapeworm (*Diphyllobothrium latum*), which can reach a length of 20 m. Its eggs had already been found in the petrified excrements of Roman hosts. The pathogens apparently came from Germania and Gaul, where they can be traced back to the Iron and Bronze Ages. It is controversial whether the tapeworm eggs were exported to Rome with the famous fish sauce "Garum" as was recently suspected or with the soldiers and slaves returning from northern regions (Tümmers 1999).

Certain is, however, that the tapeworm existed in fish in the Rhine and Elbe rivers until modern times. A textbook on intestinal worms at the end of the eighteenth century, for example, states that tapeworms are frequently found in many different fish, including Rhine and Elbe salmons.

It is also certain that human excrements were disposed untreated into the Rhine for a long time. For example the Dutch rafts looked like floating villages with a certain "infrastructure", however, sanitary facilities are not mentioned anywhere. Apparently the crew probably placed their feces directly into the river.

Today, the fish tapeworms no longer threaten the inhabitants of the Rhine area, unless these are "raw" sushi lovers (Düsseldorf is home to the largest Japanese colony in Europe). In meat of raw salmon and other raw fish, which is processed to these Japanese and other delicacies, larval stages of this "long guy" can be found and might infect humans even today (Craig 2012; Goeze 1782).

18.9 The Rhine Today

Even in the twenty-first century, the Rhine still brings floods and mosquito plagues to people along its borders, because mosquitoes (such as *Aedes vexans*, sometimes also called "Rheinschnake," and *A. sticticus*) find optimal breeding conditions in the old branches of the Rhine. Larvae and pupae are mainly controlled biologically with *Bacillus thuringiensis* (Becker 1997; Breier and Stiefenhofer 2016). However, local mosquito species are epidemiologically irrelevant as vectors along the Rhine so far. However, the possible spread of the Asian tiger mosquito (*A. albopictus*), a vector for the Zika, Chikungunya, and Dengue viruses, is a cause of concern. In 2007, an adult female was found near Weil am Rhein (Werner et al. 2012); now further larger populations in the Upper Rhine plain have been found.

The plague disease has been fully stopped and cholera as well as malaria seem to be largely defeated in Europe, but what is the risk of the return of malaria? Higher temperatures, as predicted by climatologists for the future, could support the recurrence of malaria. There are many *Anopheles* species suitable as vectors in Germany/

Europe; more recently, *A. plumbeus* has caused some concern. This originally forest-dwelling species has adapted its life to human dwellings and can apparently serve as a vector for *Plasmodium vivax* and *P. falciparum*, as cases in Germany and England have shown (Piperaki 2018). In any case, it is crucial to ensure that malaria cases are diagnosed and treated quickly to avoid to create a human reservoir that can pass on the gametocytes of the parasites to the mosquito population (Piperaki 2018). The risk seems manageable—as long as infrastructure and sanitation are working accordingly (Mehlhorn 2016a, b).

And what's about the Rhine itself? After being used for a long time as a sewer and wastewater disposal of the chemical and heavy industry along its banks, the Rhine is now—thanks to the expansion of sewage treatment plants and a growing environmental awareness—clean enough to be home of salmons and trouts again.

Even a tropical fish species has settled in a Rhine tributary meanwhile. The 30 °C warm cooling water of the lignite-fired power plant at Niederaussem near Cologne ensures that the water temperature in the Gillbach does not drop below 19 °C even in winter. This comes just handy for an ornamental fish from Guatemala, the Zebra cichlid (*Amatitlania nigrofasciata*). This fish, which was apparently accidentally released there, also brought along a tropical roommate, the parasitic threadworm (*Camallanus cotti*) from Asia, which is common in aquaristics. A study showed that many local fish species in the warm stream were also infested by non-native, subtropical, and tropical fish parasites. The Gillbach flows into the Rhine via the Erft. Whether the nematode can survive in the cooler Rhine water remains to be seen.

"Panta rhei" ... everything flows. For more than 1000 years, the Rhine has been one of the continent's most important waterways, and thus it is only logical that the European Parliament should have its seat in Strasbourg. The Rhine is a truly European river.

∗ Comment: During my study of biology at Düsseldorf University more than 40 years ago, I remember buying fish regularly from the Rhine boatmen for preparation courses: The fish, mostly roaches (*Rutilus rutilus*) and red feathers (*Scardinius erythrophthalmus*), often had ulcers on their head and hull, and their liver didn't look good at all. They certainly wouldn't have gotten on my plate back then....

References

Becker N (1997) Microbial control of mosquitoes: management of the upper Rhine mosquito population as a model programme. Parasitol Today 13:485–487

Blackbourn D (2008) Die Eroberung der Natur: Eine Geschichte der deutschen Landschaft. Pantheon, München

Böhme M et al (2012) The antiquity of the Rhine River: stratigraphic coverage of the Dinotheriensande (Eppelsheim formation) of the Mainz Basin (Germany). PLoS One 7: e36817. https://doi.org/10.1371/journal.pone.0036817

Breier F, Stiefenhofer C (2016) Der Rhein: (1) Von Vulkanen und Riesenflößen; (2) Von Malaria und der Jahrtausendflut. https://www.zdf.de/dokumentation/terra-x/der-geschichte-des-rheins-teil-1-von-vulkanen-und-100.html.bzw. https://www.zdf.de/dokumentation/terra-x/die-geschichte-des-rheins-teil-2-von-malaria-und-100.html. Accessed 10/17 July 2016

Craig N (2012) Fish tapeworm and sushi. Can Fam Physician 58:654–658

Dean KR et al (2018) Human ectoparasites and the spread of plague in Europe during the second pandemic. PNAS 115(6):1304–1309. https://doi.org/10.1073/pnas.1715640115

Goeze JAE (1782) Versuch einer Naturgeschichte der Eingeweidewürmer thierischer Körper Blankenburg, 1782. http://mdz-nbn-resolving.de/urn:nbn:de:bvb:12-bsb10231405-6

Koch R (1884) An address on cholera and its bacillus. Br Med J 2:453–459

Mehlhorn H (ed) (2016a) Encyclopedia of parasitology, 4th edn. Springer, Berlin

Mehlhorn H (2016b) Human parasites, 8th edn. Springer, Cham

Mitchell PD (2016) Human parasites in the Roman world: health consequences of conquering an empire. Parasitology 144:48–58

Neumann U (2018) Geschichte der Floßschifffahrt. https://www.planet-wissen.de/technik/schifffahrt/vom_floss_zur_kogge/pwiegeschichtederflossschiffahrt100.html. Accessed 18 Oct 2018

Piperaki ET (2018) Malaria eradication in the European world: historical perspective and imminent threats. *IntechOpen*. https://doi.org/10.5772/intechopen.76435. https://www.intechopen.com/books/towards-malaria-elimination-a-leap-forward/malaria-eradication-in-the-european-world-historical-perspective-and-imminent-threats

Rath G (1955) Pestepidemien des ausgehenden Mittelalters und der Neuzeit. Ciba 7:2411–2413

Schnurrer F (1823–1824) Chronik der Seuchen in Verbindung mit den gleichzeitigen Vorgängen in der physischen Welt und in der Geschichte des Menschen. 2 Bde, Tübingen

Tümmers HJ (1999) Der Rhein: Ein europäischer Fluss und seine Geschichte. C.H. Beck, München

Vasold M (2003) Epidemie: “Es ist eine böse Zeit”. Zeit online – Die Zeit vom. Accessed 30 April 2003

von Plessen ML (2016) Der Rhein: Eine europäische Flussbiografie. Prestel, München

Wegener U (2003) Die Schwarzwald-Flößer. Internet Archive, SWR2 Wissen. Accessed 16 Dec 2003

Weikinn C (2000) In: Börngen M, Tetzlaff G (eds) Quellentexte zur Witterungsgeschichte Europas von der Zeitwende bis zum Jahr 1850. Gebrüder Borntäger, Berlin

Werner D, Kronefeld M, Schaffner F, Kampen H (2012) Two invasive mosquito species, *Aedes albopictus* and *Aedes japonicus japonicus*, trapped in south-west Germany, July to August 2011. Euro Surveill 17(4):20067

Winkle S (2005) Geißel der Menschheit: Kulturgeschichte der Seuchen. Artemis und Winkler, Düsseldorf

Chapter 19
The Rhine as Hotspot of Parasite Invasions

Bernd Sures, Milen Nachev, and Daniel Grabner

Abstract The Rhine has been severely impacted by non-indigenous free-living and parasite species, largely triggered by the connection of the Ponto-Caspian and the Central European area due to the inauguration of the Main-Danube Canal in 1992. In the present review, we give a brief summary of species invasions in the Rhine with a focus on non-indigenous free-living and parasite species and related changes in species communities that have been recorded in the last decades. Among the most relevant invaders, we find the "killer shrimp," *Dikerogammarus villosus*, and various species of gobiids. Together with non-indigenous hosts, several parasite species were introduced and stayed mostly with their original host, e.g., the monogenean *Gyrodactylus proterorhini* on gobiids and the acanthocephalan *Paratenuisentis ambiguus* in its specific non-indigenous intermediate host *Gammarus tigrinus*. Other parasites colonized indigenous host species after their introduction (parasite spillover), e.g., the monogeneans *Pseudodactylogyrus bini* and *Pseudodactylogyrus anguillae*, the nematode *Anguillicola crassus*, the tapeworm *Schyzocotyle acheilognathi*, and acanthocephalans of the genus *Pomphorhynchus*. As both *A. crassus* and *Pomphorhynchus* sp. use non-indigenous gobies as paratenic hosts, we also find an example of parasite spillover in the Rhine. Furthermore, in cases of co-infection of the same host, *Pomphorhynchus* sp. cysts in the gobies serve as a "refuge" for the larvae of *A. crassus*, which probably increases the efficiency of transmission to the eel final host. Therefore, this combination of an invasive host species and the two parasites is a unique example of an invasional meltdown. Finally, we illustrate the drastic changes in parasite communities caused by

---

B. Sures (✉)
Aquatic Ecology and Centre for Water and Environmental Research, University of Duisburg-Essen, Essen, Germany

Department of Zoology, University of Johannesburg, Johannesburg, South Africa
e-mail: bernd.sures@uni-due.de

M. Nachev · D. Grabner
Aquatic Ecology and Centre for Water and Environmental Research, University of Duisburg-Essen, Essen, Germany

H. Mehlhorn, S. Klimpel (eds.), *Parasite and Disease Spread by Major Rivers on Earth*, Parasitology Research Monographs 12,
https://doi.org/10.1007/978-3-030-29061-0_19

interactions of indigenous and non-indigenous species and their parasites, using the examples of eel and gobies as host species.

Keywords Rhine history · Fish parasites · Invading fish species · Richness in species

19.1 The Rhine: Recent History

The Rhine originates in the Swiss Alps, flows mostly through Western Germany and the Netherlands, and finally enters the North Sea. With a length of 1233 km, it is the largest river in Western Europe (Tittizer and Krebs 1996). Its natural structure was largely changed into a channel and industrial discharge heavily polluted the Rhine until the 1970s (Villeneuve 1996; Wilken 2006; Lelek and Köhler 1990). The following efforts to eliminate pollution and to improve water quality were successful until 1986 when the Sandoz accident near Basel in Switzerland caused a spill of organophosphate insecticides, organochlorines, mercury compounds, and other toxic chemicals into the river (Giger 2007). As a result, a mass mortality of fish, mainly eel, and invertebrates occurred with toxic effects visible also downstream in the Netherlands (Giger 2007). Even though the water quality improved quickly and populations of native species partly recovered (Lelek and Köhler 1990) another threat became obvious in the following years: more and more non-indigenous species were introduced (e.g., with ballast water of ships) and established themselves in the Rhine system (Giger 2007). The major event that opened the way for a massive migration of non-native invertebrate and vertebrate species was the opening of the Main-Danube Canal in 1992 that connects the Danube and the Rhine system via the Main River. Some of the invaders from the Ponto-Caspian area became dominant in the new distribution range. Currently, it can be estimated that the total number of invertebrate species for the whole Rhine from the Alps to the North Sea sums up to 500 (IKSR 2015). With respect to local monitoring sites, the number of invertebrate species established at around 150 species per site, of which approximately 20% are non-indigenous species (IKSR 2015; Nehring 2003). Together with non-indigenous species from other areas, these neozoans cause a major shift in the communities of free-living, but also of indigenous and non-indigenous parasite species that utilized the invaders as hosts (Sures and Streit 2001; Thielen et al. 2007; Hohenadler et al. 2019).

19.2 Major Invaders in the Rhine

Several non-native invertebrate species are relevant in terms of their behavior as highly invasive species such as mussel species, e.g., the zebra mussel (*Dreissena polymorpha*), the quagga mussel (*Dreissena rostriformis bugensis*), and the Asian clam (*Corbicula* sp.) as well as crustaceans such as the killer shrimp (*Dikerogammarus villosus*), the amphipod *Chelicorophium robustum,* and the isopod *Jaera sarsi*. Some of these non-indigenous species turned out to be involved as intermediate hosts for the transmission of heteroxenic fish parasites. Particularly some amphipod species play major roles for the introduction and transmission of fish parasites such as acanthocephalans (e.g., species of the genus *Pomphorhynchus* or *Paratenuisentis ambiguus*). The amphipod species that probably received most attention is *D. villosus*, which arrived in the Rhine in the 1990s and successfully established stable populations along the river (Bij de Vaate 2002; Rewicz et al. 2014). Due to its comparatively large size and its predatory behavior, *D. villosus* quickly dominated macroinvertebrate communities (Rewicz et al. 2014). It also replaced other non-indigenous amphipod species such as *Gammarus tigrinus*, which was introduced intentionally from North America (Haas et al. 2002). Other non-indigenous amphipod species such as the Ponto-Caspian *Dikerogammarus haemobaphes*, which can be locally abundant, had less substantial impact on local invertebrate communities (Bij de Vaate 2002). Furthermore, two non-indigenous species of the genus *Echinogammarus* are found frequently: firstly, *Echinogammarus berilloni*, an early invader from South-West Europe (France and Spain) that was recorded in Germany already in 1924 (Tittizer et al. 2000), and, secondly, *Echinogammarus ischnus*, another Ponto-Caspian species that was found in high abundance in the Rhine, as it is able to co-exist with *D. villosus* (Kley and Maier 2006; Thielen et al. 2007). In addition to invertebrates, also non-indigenous vertebrate species, i.e., fish, were able to successfully invade the Rhine. Among fish, the Ponto-Caspian gobiids are the most successful invaders as they were able to spread into different regions of Europe such as the Baltic, Aegean, and North Sea basins (Skóra and Stolarski 1993; Kakareko et al. 2009). They also established successfully in the North American Great Lakes (Jude et al. 1992; Corkum et al. 2004). In newly invaded ecosystems, the gobiids often have a strong impact on local species assemblages as they reproduce quickly and gain high population densities (Jurajda et al. 2005; Adámek et al. 2007; Lavrinčíková and Kováč 2007; Kováč et al. 2009).

After the inauguration of Main-Danube Canal, five Ponto-Caspian gobiid species, the tubenose goby *Proterorhinus semilunaris*, the bighead goby *Neogobius kessleri*, the monkey goby *Neogobius fluviatilis*, the racer goby *Neogobius gymnotrachelus,* and the round goby *Neogobius melanostomus,* invaded the Rhine system (Freyhof and Huckstorf 2006; Stemmer 2008; Borcherding et al. 2011a, b). Historically, the tubenose goby was the first goby species, which was recorded in the Lower Rhine in 1999 (Borcherding et al. 2011a). Five years later, in 2004 *N. melanostomus* was found for the first time in the Dutch Rhine Delta (van Beek 2006), followed in 2006 by *N. kessleri*, which was registered in the German Lower Rhine (Borcherding et al. 2011a). Subsequently, in autumn 2008, *N. fluviatilis* was found in the German part of the Rhine, and in spring 2009 further specimens were recovered further downstream in the Dutch part (van Kessel et al. 2009). In 2010, Borcherding et al. (2011b) found the racer goby, *N. gymnotrachelus*, at two locations along the Rhine in North Rhine-Westphalia. According to Borcherding et al. (2011a), the most abundant goby species in the Lower Rhine is *N. kessleri* followed by *N. melanostomus* and *N. fluviatilis*.

19.3 Parasite Invasions: General Concepts

Besides direct ecological effects, non-indigenous hosts also influence the parasite fauna in the new habitats not only by co-introducing their own non-indigenous parasites but in multiple ways. These introductions can have significant effects on the recipient communities and ecosystems (Sures 2011; Sures et al. 2017; Taraschewski 2006; Thieltges et al. 2008). In the following chapters, we will describe the major invasive parasites of the Rhine and illustrate the effects of non-indigenous free-living and parasitic species on animal communities in the Rhine. At first, we introduce the major concepts of parasitism for invasion biology and the associated effects on hosts and parasites (Fig. 19.1). Theory predicts that some non-indigenous host species will arrive in their new distribution range after invasion with a lower number of parasites compared to their native range ("Enemy release hypothesis," Torchin et al. 2003). Such a loss of parasites—or enemies in general—would be an ecological advantage for the invaders contributing to the invasion success, as these species are not forced to invest in parasite defense measures. However, there are also examples showing that non-indigenous host species introduce their own parasites, which then can "spillover" to the naïve native host species, leading to high parasite abundances in the new host species of the invaded area (Kelly et al. 2009; Fig. 19.1). If the non-indigenous hosts lose some of their parasites, these might be replaced by native parasites of the invaded area. An abundant non-indigenous host, where the parasite can multiply efficiently, can cause an increase of parasite intensities and prevalence in indigenous host species. This phenomenon is called "parasite spill-back" (Daszak 2000; Kelly et al. 2009, Fig. 1). Of course, the non-indigenous host might also be unsuitable for native parasites to develop and therefore act as a dead end for the parasite. Infection of these unsuitable hosts will reduce the infection pressure for the indigenous host individuals, which is called "dilution effect" (Telfer et al. 2005; Kopp and Jokela 2007; Thieltges et al. 2008; Johnson and Thieltges 2010). As in the fifth scenario displayed in Fig. 19.1, two non-indigenous parasites might co-occur in a non-indigenous free-living species, both being able to establish themselves successfully in native hosts of the invaded area. This so-called invasional meltdown hypothesis has recently been demonstrated for host-parasite associations in the Rhine (Hohenadler et al. 2018a).

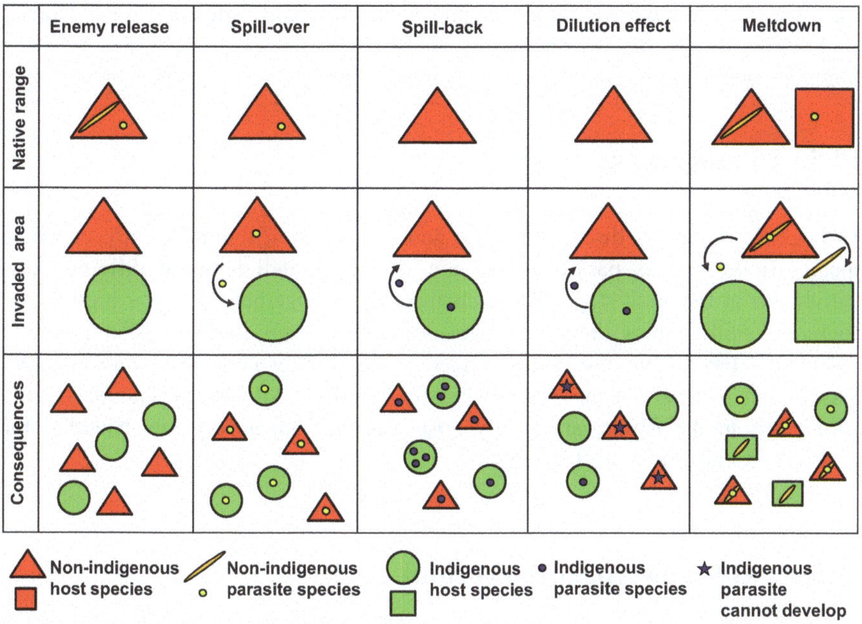

Fig. 19.1 Schematic diagram of the general concepts in parasite invasion biology. See text for detailed explanations

Generally, parasites are assumed to co-invade together with their host and are expected to establish themselves in the new range if the life cycle can perpetuate. Non-indigenous parasites, requiring only a single host species in their life cycle (monoxenous parasites), are more likely to establish after invasion compared to heteroxenous parasites that require one or several intermediate hosts (Sures 2011). In the latter case, the intermediate hosts must also invade, or suitable indigenous species must be available to be used as alternative intermediate hosts (parasite spillover, Daszak 2000). In this case, generalist parasites have an advantage over host specialists (Sures 2011).

19.4 Examples of Selected Non-indigenous Parasites in the Rhine

In the following, we briefly describe the most prominent non-indigenous helminth found in the Rhine including their origin, invasion history (if known), life cycle, and effects on the host. Since the 1980s various free-living neozoans were introduced intentionally or unintentionally into the Rhine system, together with several new parasite species that were co-introduced. Most of the parasites are endohelminths; however, there are also a couple of ectoparasites such as monogeneans, which were introduced and act as important pathogens of fish (Bakke et al. 2007).

19.5 Monogeneans

Monogeneans have a direct life cycle and transmission requiring only one host species (monoxenous parasites) and are able to switch between different host individuals at the adult stage. These characteristics, together with their high reproductive potential, turn them into parasites with a very high invasion potential (Kennedy 1994). In the Rhine, species of the genera *Gyrodactylus* and *Pseudodactylogyrus* are the most successful invaders. All species, known to have established in the Rhine, are highly host specific with a very low potential for spillover to indigenous host species.

19.5.1 Gyrodactylus proterorhini

The monogenean *Gyrodactylus proterorhini* was originally described from tubenose goby (*Proterorhinus marmoratus*) in the Danube in Slovakia (Ergens 1967) and is a specialist parasite of gobiids. Its distribution in the non-native range throughout Europe has been well documented (Gaevskaya and Dmitrieva 1997; Mierzejewska et al. 2011; Ondračková et al. 2012, 2015, 2016; Huyse et al. 2015). Recently, the parasite has also been found in gobiids in the Rhine (Ondračková 2016), with the tubenose goby having the highest infection rates (76% prevalence). Other goby species such as *N. kessleri, N. fluviatilis,* and *N. melanostomus* were also infected, but with much lower prevalences (45.9%, 30.5%, and 3.9%, respectively). In contrast to newly invaded regions, the prevalence and abundance levels of *G. proterorhini* in the native range are relatively low (see Naidenova 1974; Kvach et al. 2015), whereas many different goby species from freshwater, brackish, and marine habitats can serve as appropriate host. For example, the prevalence reported from the Ponto-Caspian region rarely exceeded 40% (Naidenova 1974; Krasnovyd et al. 2012; Kvach et al. 2015). The higher prevalence levels of *G. proterorhini* on gobiids in the non-native range of distribution were suggested to be a result of the overall lower parasite load and thus more free niches on the host (Ondračková 2016).

Studies on the ability of *G. proterorhini* to switch to non-gobiid fish showed that the parasite can occasionally infect other fish species such as the Chinese sleeper, *Perccottus glenii* (Mierzejewska et al. 2012), and the European perch, *Perca fluviatilis* (Ondračková 2016), under natural conditions. The latter author suggested that high prevalence and abundance levels of the parasite (100% and 39%, respectively, in *Neogobius gymnotrachelus*) observed in some non-native areas (River Vistula) in combination with high fish density can favor such a host switch. As host-switch events tend to occur rarely, for now *G. proterorhini* does not represent a threat to native fish species. However, a potential of spillover still exists and might be intensified through environmental factors, further species adaptation, or differentiation that may facilitate the host switch to indigenous species.

19.5.2 Pseudodactylogyrus anguillae *and* Pseudodactylogyrus bini

Two eel-specific monogeneans, *Pseudodactylogyrus anguillae* and *Pseudodactylogyrus bini*, were introduced to Europe as a consequence of international trade with live eel. These parasites were recorded for the first time in Japan (Kikuchi 1929) and are native to East Asia, even though their native range is not fully known (Christison and Baker 2007; McHugh et al. 2017). Particularly *P. anguillae* might actually be native to Europe and Africa (Kennedy 1993; Cone and Marcogliese 1995). The natural host is the Japanese eel (*Anguilla japonica*), and the first record of European eel (*Anguilla anguilla*) originates from an eel farm in the former Soviet Union in 1977 (Golovin 1977). Today, these monogeneans are found in most European countries and North America (Hayward et al. 2001; Marcogliese and Cone 1993). *P. anguillae* was also found in South Africa (Parker et al. 2011). Both species parasitize the gills and can cause severe symptoms in *A. anguilla*, which is highly susceptible to the parasite (Buchmann et al. 1987). Harmful effects, especially under aquaculture conditions (Buchmann 1988), include hyperemia, hemorrhages, hyperplasia of gill tissue, and destruction of the gill filament structure (Buchmann et al. 1987). In the Rhine, *Pseudodactylogyrus* spp. are commonly found on the gills of European eels, mostly with high prevalence. In eel samples taken in 1995, Sures et al. (1999) detected prevalence levels of *Pseudodactylogyrus* sp. of 45.0% and 45.9% at two sampling sites. Four years later, Sures and Streit (2001) found a prevalence around 90% for *P. anguillae* and *P. bini* (species not distinguished) at two other sites along the Rhine. Based on seasonal eel samplings at different sites in the Rhine in 2003, Thielen et al. (2007) found prevalences for *Pseudodactylogyrus* sp. between 20.9% and 80.0%. Accordingly, these monogeneans can be rated as established non-indigenous parasites attaining high prevalence and intensity in eels in the Rhine.

19.6 Acanthocephala

Several acanthocephalan species were co-introduced with their hosts into the Rhine. While the invasion history of some species such as *Paratenuisentis ambiguus* was well documented, some others like members of the genus *Pomphorhynchus* (e.g., *P. laevis*, *P. bosniacus*) appear to be less understood, mainly due to misidentifications of species during the last decades (Hohenadler et al. 2018b; Reier et al. 2019). However, *P. laevis* and *P. bosniacus* appear to be introduced with free-living Ponto-Caspian non-indigenous species after the opening of the Main-Danube Canal (Hohenadler et al. 2018b; Reier et al. 2019). The origin of some other species such as *Acanthocephalus rhinensis*, which was newly found and described in eels from the Rhine, was not elucidated, but some evidence indicates that it might be a neozoan species.

19.6.1 Paratenuisentis ambiguus

The acanthocephalan *Paratenuisentis ambiguus* is native for the east coast of North America (Bullock and Samuel 1975) where the American eel (*Anguilla rostrata*) represents the main definitive host (see Taraschewski et al. 1987). It is highly host specific regarding the intermediate host, being able to infect only the North American amphipod *Gammarus tigrinus* (Taraschewski et al. 1987). The latter is a native species of the coastal waters between the St. Lawrence estuary and Florida (Steele and Steele 1972) and prefers mixohaline coastal habitats. Therefore, the distribution of *P. ambiguus* is restricted mainly to habitats with increased salinity (brackish water), where also the intermediate host is abundant. *G. tigrinus* was intentionally introduced to the rivers Weser and Werra in 1957 as the local amphipod community disappeared mainly due to increased salt concentrations (Pinkster et al. 1977). Following its introduction this species successfully extended its home range into many rivers of Europe including the Rhine (Pinkster et al. 1977). The establishment of stable populations of *G. tigrinus* was the prerequisite for the occurrence of *P. ambiguus*, which was first documented by Taraschewski et al. (1987) and thereafter became one of the most dominant parasites in the intestine of European eels from the Rhine occurring with mean intensities of up to 42 worms (Sures et al. 1999; Sures and Streit 2001). However, due to the massive invasion of Ponto-Caspian amphipods into the Rhine system (Dick and Platvoet 2000; Haas et al. 2002; Krisp and Maier 2005), as well as due to improved water quality and low salinity, *P. ambiguus* almost disappeared together with its intermediate host.

19.6.2 Acanthocephalus rhinensis

The first description of the *Acanthocephalus rhinensis* dates back to 2008 and was based on several sexually mature individuals found in three eels during the period between 2003 and 2004 (Amin et al. 2008). It was suggested to be an eel-specific parasite, as it was found only in the European eel (*A. anguilla*). Dezfuli et al. (2011) confirmed this assumption after finding *A. rhinensis* in an eel population from Lake Piediluco in Italy. This acanthocephalan species was the most prevalent (86.49%) parasite in that lake, occurring with intensities of up to 350 worms per eel (mean 67.38 ± 65.16). A parasitological survey on amphipods found in the stomach of eels and at the location, where the eels were sampled, showed that the amphipod *Echinogammarus tibaldii* serves as intermediate host for *A. rhinensis*.

In the original description, Amin et al. (2008) suggested two scenarios for the occurrence of *A. rhinensis*. According to the first scenario, *A. rhinensis* might be an indigenous species in the Rhine, but is driven close to extinction due to invasion of Ponto-Caspian species. The second scenario states that *A. rhinensis* is a non-indigenous species in the Rhine and was introduced after the inauguration of the Main-Danube Canal in 1992. The findings of Dezfuli et al. (2011) together with the fact that the acanthocephalan was not previously found in any parasitological investigation of eels from Central and Western Europe support the assumption of its foreign origin. As this acanthocephalan seems to be transmitted by amphipods of the genus *Echinogammarus*, it is likely that the transmission of *A. rhinensis* in the Rhine is mainly sustained by the invasive amphipod *Echinogammarus ischnus* or possibly other species of the genus *Echinogammarus*.

19.6.3 **Pomphorhynchus** *spp.*

The acanthocephalan genus *Pomphorhynchus* has caused some confusion due to the presence of cryptic species, and there is an ongoing effort to disentangle the identity and origin of the species *P. laevis*, *P. bosniacus*, and *P. tereticollis* (Špakulová et al. 2011; Hohenadler et al. 2018b; Reier et al. 2019). Generally, the life cycle of *Pomphorhynchus* spp. includes a fish as final host and an amphipod as intermediate host (Kennedy 2006). The latter orally takes up the eggs that are released from the adult worms living in the intestine of the final fish host. After oral uptake, the acanthor larvae hatch in the intestine of the amphipod host, migrate to the body cavity, and develop to the cystacanth stage, which is infective for the final host. Barbel (*Barbus* spp.), chub (*Squalius cephalus*), and brown trout (*Salmo trutta*) are reported among others as adequate final hosts (Hine and Kennedy 1974; Kennedy 1978) for different *Pomphorhynchus* species. Recently, Perrot-Minnot et al. (2019) elucidated that *P. laevis* has a higher preference for chub, while *P. tereticollis* seems to develop more efficiently in barbel. Other fish species can also become infected, but the worms will not become mature (Hine and Kennedy 1974). Most parasitological publications considered those *Pomphorhynchus* sp. found in Central Europe as *P. laevis*. Recently, molecular and morphological analyses lead to the redescription of *P. tereticollis*, another species found in Central Europe (Špakulová et al. 2011). Using an established PCR method to distinguish *P. laevis* and *P. tereticollis*, Hohenadler et al. (2018b) tested the distribution of these two species in the Rhine over a period of 12 years. Their results indicate that *P. laevis* is co-invasive from the Ponto-Caspian area together with *Dikerogammarus villosus* as intermediate and invasive gobies as paratenic hosts and that *P. laevis* replaced *P. tereticollis* in the Rhine. This idea is supported also by other studies, but the identity of the *Pomphorhynchus* sp. is sometimes discussed controversially (Emde et al. 2012, 2014a). However, another recent study added a new dimension of complexity to the *Pomphorhynchus* issue. Reier et al. (2019) detected the species *Pomphorhynchus bosniacus* in the Danube River in Austria and found that this species is by far the most dominant in the Danube near Vienna. *P. bosniacus* is closely related to *P. laevis* and was originally described and reported from the Balkans (Kiskároly and Čanković 1967). Prior to the study of Reier et al. (2019), the presence of *P. bosniacus* was not expected in Central Europe. Interestingly, *Pomphorhynchus* isolates from the Upper Rhine that were originally classified as an exotic *P. laevis* haplotype (David et al. 2018) actually cluster with sequences of *P. bosniacus* (Reier et al. 2019). Therefore, it can be assumed that either *P. laevis* or *P. bosniacus* co-exists in the Rhine or that *P. bosniacus* is dominant at least in some section of the Rhine, similar to the Danube. The haplotype diversity of *P. bosniacus* from the Danube River did not suggest a recent invasion (Reier et al. 2019), but the status of the specimens from the Rhine has to be re-analyzed by sequencing of samples that were so far only distinguished based on the size of PCR bands, which is obviously not sufficient to distinguish between all three species of *Pomphorhynchus*. Also, the specificity of different (indigenous and non-indigenous) intermediate and final hosts for the three species of *Pomphorhynchus* has to be tested, as intermediate host specificity was found in cryptic lineages of *Polymorphus minutus* (Zittel et al. 2018). Prevalence and intensities are not reported from target fish hosts in the Rhine, but, for example, *B. barbus* sampled in the Danube River showed a prevalence of 100% with mean intensities of more than 100 individuals per intestine (Nachev and Sures 2009). Even in fish hosts in which no further development of *Pomphorhynchus* sp. is possible (so-called dead-end hosts), the parasite occurs in high prevalence. For example, in eel from the Rhine, a prevalence of up to 80.6% was described by Thielen et al. (2007). In invasive gobies that can serve as paratenic host for the parasite (Hohenadler et al. 2018a), the prevalence of *Pomphorhynchus* cysts in mesenteries and liver ranged between 25% and 90% with up to 69 individuals per host (Emde et al. 2012, 2014a, b; Hohenadler et al. 2019). In conclusion, even after several years of research on the occurrence of *Pomphorhynchus* sp. in the Rhine, no clear picture emerged mainly due to uncertainties associated with the species' identification, which was often unreliable during recent years.

19.7 *Anguillicola crassus*

In addition to some of the acanthocephalans mentioned above, also the nematode *Anguillicola crassus* is an eel-specific parasite. It originates from East Asia, where it parasitized the Japanese eel, *Anguilla japonica*, without severe pathological effects (see Knopf 2006 and references therein). *A. crassus* was introduced by eel trade to Europe, Africa, and America during the last 39 years where it turned out to be highly invasive and thus established stable populations in different eel species (for summaries see Kirk 2003; Taraschewski 2006; Knopf 2006; Sasal et al. 2008; Jakob et al. 2016). Only one other species of the genus *Anguillicola*, *A. novaezelandiae* from New Zealand, was introduced to Europe, but did not spread and finally disappeared again (Dangel et al. 2015).

Eel species are the only final hosts in which the parasite can develop to the adult stage in the swim bladder. There, the worms feed on host blood and reproduce sexually. Second stage larvae (L2) migrate to the intestine via the *Ductus pneumaticus* and are released to the environment. The life cycle of this parasite further involves a copepod intermediate host that takes up the larvae. In the copepod, the parasite develops to the L3 stage, which is infective for the final host (De Charleroy et al. 1990). In addition, several vertebrate and invertebrate species can serve as paratenic host for the parasite (see Knopf 2006 and references therein). The infection of European eel with *A. crassus* did not affect the condition index of the host (Würtz et al. 1998), but increased HSP 70 and cortisol levels in *A. anguilla* infected with *A. crassus* indicate a stress response of the host, mainly during the early stage of infection when L3 develops to adults (Sures et al. 2001; Dangel et al. 2014; Keppel et al. 2016). Accordingly, the antibody and cellular immune response was mainly targeting the migrating larval stages (Knopf 2006). The major concern about this parasite is the pathological modification of the swim bladder wall due to the feeding activity of the adult worms (Würtz and Taraschewski 2000), leading to changes in gas composition inside the swim bladder (Würtz et al. 1996) and morphological changes of the swim bladder wall (Würtz and Taraschewski 2000). These changes reduce the normal function of the swim bladder (Pelster 2015), which might impair the spawning migration of the eel to the Sargasso Sea (Sures and Knopf 2004; Palstra 2007).

In eel samples from six different sites in the Rhine that were taken between June 1994 and November 1995, a mean prevalence for *A. crassus* of about 80% and a mean intensity of around 5 were found. During the time of investigation, no seasonal trend of prevalence and intensity was observed (Würtz et al. 1998; Sures et al. 1999). Sures and Streit (2001) found higher values compared to the previous studies with a prevalence range of 89.5–94.3% and mean intensities of 8.6–9.1 for *A. crassus* in eel collected at two sites in the Rhine in 1999. In samples from 2003, the prevalence of *A. crassus* in the Rhine ranged between 63.6% and 95% (mean intensity not shown) (Thielen et al. 2007). Considering recent data (Hohenadler et al. 2019), it emerges that not much variation occurs through time. Between 2014 and 2106, the prevalence of *A. crassus* was 73.5% with a mean intensity of 4.2 individuals. Interestingly, recently larval stages of *A. crassus* were discovered in cysts of *Pomphorhynchus* sp. found in the body cavity of invasive gobies (see below for further details) suggesting that the invasional meltdown hypothesis also applies to parasites (Emde et al. 2014b; Hohenadler et al. 2018a).

19.7.1 Schyzocotyle acheilognathi

To date, only a single non-indigenous cestode was found in the Rhine. The introduction of further cestode species was probably prohibited by the lack of suitable hosts in the new range, as the life cycles of this parasite group are complex and involve at least two different host species. The Asian fish tapeworm, *Schyzocotyle acheilognathi*, formerly described as *Bothriocephalus acheilognathi* was considered to be one of the parasites with the highest invasion potential worldwide. It was introduced to Europe from Asia during the 1950s and 1960s mainly with imported cyprinids such as common carp (*Cyprinus carpio*) and grass carp (*Ctenopharyngodon idella*). At the beginning, it was found only in farmed fish; however, due to intensive and uncontrolled translocation of different cyprinids (e.g., guppies, mosquito fish, etc.) and its low host specificity, *S. acheilognathi* could spread also in wild fish populations on all continents except Antarctica (summarized by Kuchta et al. 2018). In Germany *S. acheilognathi* was found mainly in farmed fish (Kuchta et al. 2018); however, Emde et al. (2016) reported infections in the neozoic cichlid fish *Amatitlania nigrofasciata* inhabiting a tributary of the Rhine, which is thermally impacted by cooling water of a power plant. The parasite was also found during routine parasitological surveys in young carp in the Lower Rhine (river kilometer 843, near town Rees) in the years 2014 and 2015 with a prevalence around 10% and a mean intensity between 1 and 2 worms (Pitronaci 2016). However, the distribution of *S. acheilognathi* along the basin of Rhine still remains uninvestigated. *S. acheilognathi* is highly pathogenic to newly colonized host species, as it was previously described for various fish species from different regions of the world (Kuchta et al. 2018). Therefore, it has a high potential to spread and to colonize the populations of some native fish species in the Rhine.

19.8 Eel Parasite Community Changes Associated with Invasion Processes in the Rhine

Among the fish species investigated so far, the European eel has achieved most attention mainly due to the fact that it is considered a critically endangered species. In own studies, the parasite communities of European eel have been investigated occasionally during the last 25 years in the Rhine close to the city of Karlsruhe. Parasite communities were found to fluctuate dramatically which was—at least partly—influenced by the immigration of non-indigenous free-living and parasite species during the last decades. A particularly illustrative example is the acanthocephalan *Paratenuisentis ambiguus* whose population dynamics indicates the ongoing invasion into the Rhine. As described above, this parasite requires the amphipod *Gammarus tigrinus* as a specific intermediate host. Therefore, after *G. tigrinus* could establish stable populations in the Rhine, *P. ambiguus* turned out to become the dominant eel-specific intestinal helminth (Sures et al. 1999). This situation was stable until the massive migration of Ponto-Caspian species occurred. During these years, beginning shortly after the inauguration of the Main-Danube Canal, the diversity of the intestinal helminth community fluctuated considerably (Fig. 19.2). Compared to the situation in 1995, the Shannon-Wiener index, as a measure of

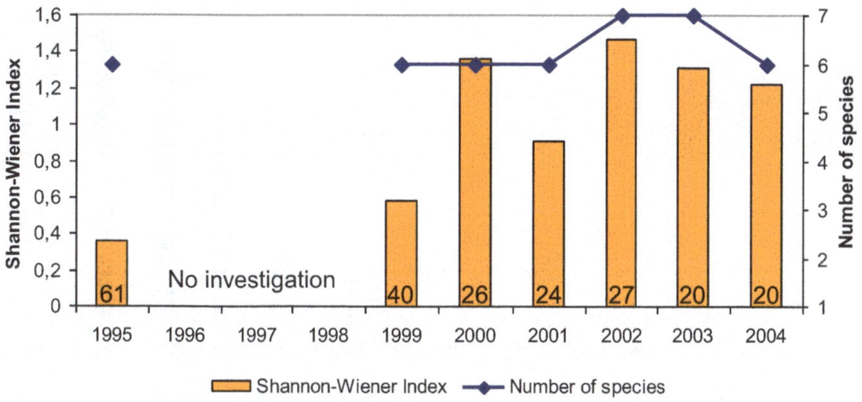

Fig. 19.2 Parasite diversity in eel from year 1995 to 2004. Numbers in bars show the number of dissected eels

diversity, increased at least threefold during the years 1999–2004 although the species richness was rather stable and fluctuated only between six to seven species (Fig. 19.2). These parasite community changes can clearly be attributed to changes in prevalence and abundance of *P. ambiguus*. Sures and Streit (2001) showed that the migration and subsequent dominance of invasive amphipods, mainly *D. villosus*, changed the whole amphipod community. Due to their importance as intermediate hosts for several parasite species, these changes in the amphipod community were reflected by altered helminth communities in eels. For example, *D. villosus* outcompeted many of the native and former invasive amphipod species, whereas none of the newly arrived Ponto-Caspian amphipods were a suitable host for *P. ambiguus*. With the decline of *G. tigrinus*, also its associated acanthocephalan parasite disappeared (Fig. 19.3) and was replaced by the eel-specific nematode *Paraquimperia tenerrima*. The life cycle of this parasite still remains unclear, at least for the Rhine (Thielen et al. 2007). According to Shears and Kennedy (2005), *P. tenerrima* uses minnows (*Phoxinus phoxinus*) as an intermediate host. However, this fish species was not recorded in studies on the fish fauna in the area of Karlsruhe (summarized in Thielen et al. 2008). Therefore, neither the origin nor the associated hosts are known for this nematode. Moreover, it is also possible that *P. tenerrima*, which is a rather small nematode living in the mucosa of the intestine, has been overlooked in those investigations which used frozen viscera. Accordingly, special emphasis should be placed on this nematode in future studies. Even if not all relationships between species can be explained, the described community changes of eel parasites unravel possible effects of invasive free-living species on host-parasite associations.

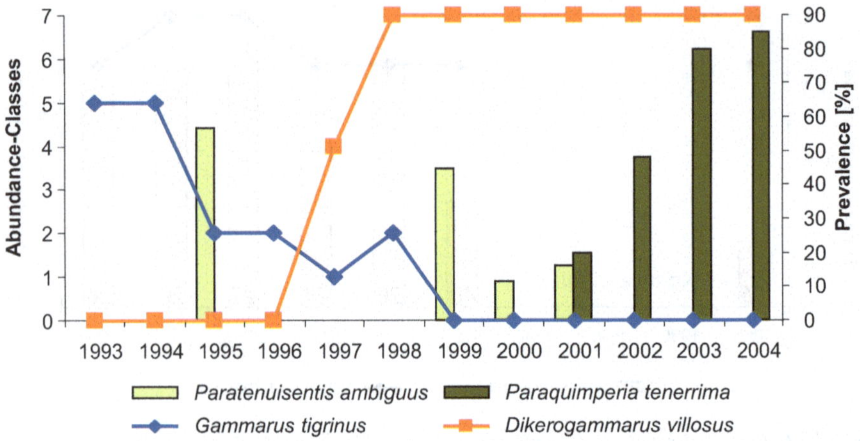

Fig. 19.3 Abundance of *Gammarus tigrinus* and *Dikerogammarus villosus* (abundance classes: 1 = single finding, 7 = mass occurrence) and the prevalence of the eel-specific acanthocephalan *Paratenuisentis ambiguus* and the nematode *Paraquimperia tenerrima* in eel from the Rhine. Samples from 1993 to 2004

19.9 The Influence of Invasive Gobiids on Parasite Communities of Native Fish Species in Rhine

In the native range of gobies (Lower Danube River and Black Sea area), far more than 20 different parasite species were reported (e.g., Kvach 2002, 2004, 2005; Ondračková et al. 2006; Özer 2007; Krasnovyd et al. 2012). This variety exceeds the number of goby parasites in the invaded areas, e.g., the Rhine, where a total of eight different species for *N. melanostomus* (Emde et al. 2012, 2014a; Ondračková et al. 2015) and seven for *N. kessleri* were recorded (Ondračková et al. 2015). According to the enemy release hypothesis, the gobies benefit from lower parasite loads in the invaded areas and are therefore able to establish themselves with high population densities (see Fig. 19.1). Generally, Ponto-Caspian gobies show high infestation rates with acanthocephalans from the genus *Pomphorhynchus* in both the native and the non-native range of distribution, with prevalence levels often exceeding 90% and high infection intensities (Ondračková et al. 2006, 2015; Kvach and Skóra 2007; Emde et al. 2014a). Interestingly, these acanthocephalans cannot complete their life cycle in the gobiids and remain encysted in their abdominal cavity as cystacanths or pre-adult stages (Emde et al. 2014a). Whether the gobies in the invaded areas can serve as paratenic hosts (*spill-back*) which can transmit the parasite to appropriate definitive hosts preying on the gobies or if they are dead-end hosts for the parasite (*dilution effect*) is not elucidated until now. Acanthocephalans of the genus *Pomphorhynchus* require mainly cyprinids as definitive hosts, with barbel (*Barbus barbus*), chub (*Squalius cephalus*), or idle (*Leuciscus idus*) representing the most important fish (Hine and Kennedy 1974). Following experimental infections using *P. laevis* cystacanths dissected from gobies, the acanthocephalans can further develop in chub (unpublished data). Based on these results it is likely that gobies can serve as paratenic hosts and can potentially multiply the occurrence of *Pomphorhynchus* sp. in the local fish fauna, if they become part of the diet of appropriate definitive hosts. It is already known that larger cyprinid specimens become partly piscivorous, and the diet of large chub and barbel can comprise more fish than benthic arthropods (Bašić et al. 2014; Kottelat and Freyhof 2007; own observation). Furthermore, *P. laevis* seems to rely on transmission via paratenic hosts also in non-invaded systems (Médoc et al. 2011). Therefore, it can be assumed that in newly invaded habitats the gobies have become an essential part of the diet of piscivorous and some omnivorous fish species, as reported for pike perch from the newly colonized Kiel Canal (Thiel et al. 2014) or for pike perch and perch from the western Baltic Sea (Oesterwind et al. 2017). Thus, in the Rhine the gobiids can also be favored as food source by different cyprinids and thereby contribute significantly to the transmission of *Pomphorhynchus* species to the definitive host. Due to the high prevalences and intensities found in the different goby species, they might increase the acanthocephalan loads in the local fish fauna (spill-back in the case of *P. tereticollis* and spillover in the case of *P. laevis* or *P. bosniacus*).

Hohenadler et al. (2018b) suggested that *P. laevis* (which now probably is considered as *P. bosniacus* according to Reier et al. 2019) is a neozoic acanthocephalan species in the Rhine being co-introduced with the Ponto-Caspian amphipods and gobies. The authors found that the co-introduced *Pomphorhynchus* sp. probably outcompeted the local *P. tereticollis* which seems to be endemic for Central and Western Europe (Hohenadler et al. 2018a). Today, *P. tereticollis* is found only in isolated tributaries of the Rhine (e.g., River Wupper), which are free of Ponto-Caspian neozoans. In its native range (Ponto-Caspian region), the co-introduced *Pomphorhynchus* sp. commonly uses *D. villosus* as intermediate host (Djikanovic et al. 2010). Investigations on amphipods from the Rhine and adjacent rivers support such host specificity/preference of different *Pomphorhynchus* species. For example, *P. tereticollis* is only found in native amphipods such as *Gammarus pulex,* whereas *P. laevis*/*P. bosniacus* only occur in the invasive *D. villosus* (Hohenadler et al. 2018b). This host specificity together with the fact that the native amphipods have a low abundance or have been completely replaced by invasive species in the Rhine led to changes in the parasite communities as already described for eel. The migrating goby species might have accelerated this process as paratenic hosts, when acting synergistically to *D. villosus* in the transmission of acanthocephalans. As outlined above, recent findings question the identity of *P. laevis* which might in fact be *P. bosniacus* (Reier et al. 2019); therefore further studies are required to clarify the status of *Pomphorhynchus* spp. as indigenous or non-indigenous.

Another interesting phenomenon associated with *Pomphorhynchus* sp. cysts was recently reported by Emde et al. (2014b) who showed that the cysts of *Pomphorhynchus* sp. in gobies from the Rhine harbor larval *A. crassus* with relatively high prevalences. Hohenadler et al. (2018a) successfully proofed the infectivity of the nematode larvae from the cysts for eels and confirmed the role of gobies as potential paratenic hosts in the life cycle of *A. crassus*. The authors demonstrated for first time an example for invasional meltdown of two invasive parasite species, with the *Pomphorhynchus* sp. cysts being able to protect the nematodes from the host immune system. Thus, if gobiids contribute markedly to eel diet, as described above for other fish species, this might lead to a further increase in infection rates and intensities of *A. crassus* in European eel (spill-back).

References

Adámek Z, Andreji J, Gallardo JM (2007) Food habits of four bottom-dwelling gobiid species at the confluence of the Danube and Hron rivers (South Slovakia). Int Rev Hydrobiol 92:554–563

Amin OM, Thielen F, Münderle M, Taraschewski H, Sures B (2008) Description of a new echinorhynchid species (Acanthocephala) from the European eel, *Anguilla anguilla*, in Germany, with a key to species of *Acanthocephalus* in Europe. J Parasitol 94:1299–1304

Bakke TA, Harris PD, Cable J (2007) The biology of gyrodactylid monogeneans: the "Russian-Doll Killers". Adv Parasitol 64:161–376

Bašić T, Britton JR, Jackson MC, Reading P, Grey J (2014) Angling baits and invasive crayfish as important trophic subsidies for a large cyprinid fish. Aquat Sci 77:153–160

Bij de Vaate A, Jazdzewski K, Ketelaars HAM, Gollasch S, Van der Velde G (2002) Geographical patterns in range extension of Ponto-Caspian macroinvertebrate species in Europe. Can J Fish Aquat Sci 59:1159–1174. https://doi.org/10.1139/F02-098

Borcherding J, Staas S, Krüger S, Ondračková M, Šlapanský L, Jurajda P (2011a) Non-native Gobiid species in the lower River Rhine (Germany): recent range extensions and densities. J Appl Ichthyol 27:153–155

Borcherding J, Gertzen S, Staas S (2011b) First record of Pontian racer goby, *Babka gymnotrachelus* (Gobiidae: Teleostei), in the River Rhine, Germany. J Appl Ichtyol 27:1399–1400

Buchmann K (1988) Epidemiology of pseudodactylogyrosis in an intensive eel-culture system. Dis Aquat Org 5:81–85

Buchmann K, Mellergaard S, Køie M (1987) *Pseudodactylogyrus* infections in eel: a review. Dis Aquat Org 3:51–57

Bullock WL, Samuel G (1975) *Paratenuisentis* gen. n. for *Tanaorhamphus ambiguus* Van Cleave 1921 (Acanthocephala), with reconsideration of the Tenuisentidae. J Parasitol 61:105–109

Christison KW, Baker GC (2007) First record of *Pseudodactylogyrus anguillae* (Yin & Sproston, 1948) (Monogenea) from South Africa. Afr Zool 42(2):279–285

Cone DK, Marcogliese DJ (1995) *Pseudodactylogyrus anguillae* on *Anguilla rostrata* in Nova Scotia: an endemic or an introduction? J Fish Biol 47:177–178

Corkum LD, Sapota MR, Skora KE (2004) The round goby, *Neogobius melanostomus*, a fish invader on both sides of the Atlantic Ocean. Biol Invasions 6:173–181

Dangel KC, Keppel M, Tabujew K, Sures B (2014) Effects of *Anguillicola novaezelandiae* on the levels of cortisol and hsp70 in the European eel. Parasitol Res 113:3817–3822. https://doi.org/10.1007/s00436-014-4049-8

Dangel KC, Keppel M, Le TTY, Grabner D, Sures B (2015) Competing invaders: performance of two *Anguillicola* species in Lake Bracciano. Int J Parasitol Parasit Wildl 4:119–124

Daszak P, Cunningham AA, Hyatt AD (2000) Emerging infectious diseases of wildlife - threats to biodiversity and human health. Science 287:443–449

David GM, Staentzel C, Schlumberger O, Perrot-Minnot M-J, Beisel J, Hardion L (2018) A minimalist macroparasite diversity in the round goby of the upper Rhine reduced to an exotic acanthocephalan lineage. Parasitology 145:1020–1026

De Charleroy D, Grisez L, Thomas K, Belpaire C, Ollevier F (1990) The life cycle of *Anguillicola crassus*. Dis Aquat Org 8:77–84

Dezfuli BS, Lui A, Squerzanti S, Lorenzoni M, Shinn AP (2011) Confirmation of the hosts involved in the life cycle of an acanthocephalan parasite of *Anguilla anguilla* (L.) from Lake Piediluco and its effect on the reproductive potential of its amphipod intermediate host. Parasitol Res 110:2137–2143

Dick JTA, Platvoet D (2000) Invading predatory crustacean *Dikerogammarus villosus* eliminates both native and exotic species. Proc R Soc Lond B 267:977–983

Djikanovic V, Gacic Z, Cakic P (2010) Endohelminth fauna of barbel *Barbus barbus* (L. 1758) in the Serbian section of the Danube River, with dominance of acanthocephalan *Pomphorhynchus laeavis*. Bull Eur Assoc Fish Pathol 30(6):229–236

Emde S, Rueckert S, Palm HW, Klimpel S (2012) Invasive ponto-caspian amphipods and fish increase the distribution range of the acanthocephalan *Pomphorhynchus tereticollis* in the river Rhine. PLoS One 7:e53218

Emde S, Kochmann J, Kuhn T, Plath M, Klimpel S (2014a) Getting what is served? Feeding ecology influencing parasite-host interactions in invasive round goby *Neogobius melanostomus*. PLoS One 9:e109971

Emde S, Rueckert S, Kochmann J, Knopf K, Sures B, Klimpel S (2014b) Nematode eel parasite found inside acanthocephalan cysts - a “Trojan horse” strategy? Parasit Vectors 7:504

Emde S, Kochmann J, Kuhn T, Dörge DD, Plath M, Miesen FW, Klimpel S (2016) Cooling water of power plant creates “hot spots” for tropical fishes and parasites. Parasitol Res 115:85–98

Ergens R (1967) New species of the genus *Gyrodactylus* (Monogenoidea) from the Danube basin. Folia Parasitol 14:377–379

Freyhof J, Huckstorf V (2006) Conservation and management of aquatic genetic resources: a critical checklist of German freshwater fishes, pp 113–126

Gaevskaya AV, Dmitrieva EV (1997) Overview of Black Sea monogenean fauna. Ekologija morja 6:7–17

Giger W (2007) Brandkatastrophe in Schweizerhalle 1986 - Rückblick und Bilanz. UWSF - Z Umweltchem Ökotox 19(1):11–23

Golovin PP (1977) Monogeneans of eel during its culture using heated water. Investigation of Monogenoidea in U.S.S.R. Zoological Institute, U.S.S.R. Academy of Sciences, Leningrad, pp 144–150

Haas G, Brunke M, Streit B (2002) Fast turnover in dominance of exotic species in the Rhine River determines biodiversity and ecosystem function: an affair between amphipods and mussels. In: Leppäkoski E, Gollasch S, Olenin S (eds) Invasive aquatic species of Europe: distribution, impacts and management. Kluwer, Dordrecht, pp 426–432

Hayward CJ, Iwashita M, Crane JS, Ogawa K (2001) First report of the invasive eel pest *Pseudodactylogyrus bini* in North American eels. Dis Aquat Org 44:53–60

Hine PM, Kennedy CR (1974) Observations on the distribution, specificity and pathogenicity of the acanthocephalan *Pomphorhynchus laevis* (Müller). J Fish Biol 6:521–535. https://doi.org/10.1111/j.1095-8649.1974.tb04569.x

Hohenadler MAA, Honka KI, Emde S, Klimpel S, Sures B (2018a) First evidence for a possible invasional meltdown among invasive fish parasites. Sci Rep 8:15085

Hohenadler MAA, Nachev M, Thielen F, Taraschewski H, Grabner D, Sures B (2018b) *Pomphorhynchus laevis*: an invasive species in the river Rhine? Biol Invasions 20:207–217

Hohenadler MAA, Nachev M, Freese M, Pohlmann JD, Hanel R, Sures B (2019) How Ponto-Caspian invaders affect local parasite communities of native fish. Parasitol Res:1–13

Huyse T, Vanhove MPM, Mombaerts M, Volckaert FAM, Verreycken H (2015) Parasite introduction with an invasive goby in Belgium: double trouble? Parasitol Res 114:2789–2793

Internationale Kommission zum Schutz des Rheins (IKSR, 2015) Die Biologie des Rheins Synthesebericht zum Rhein-Messprogramm Biologie 2012/2013 und nationale Bewertungen gemäß WRRL. Bericht Nr. 232, Koblenz

Jakob E, Walter T, Hanel R (2016) A checklist of the protozoan and metazoan parasites of European eel (*Anguilla anguilla*): checklist of *Anguilla anguilla* parasites. J Appl Ichthyol 32:757–804

Johnson PTJ, Thieltges DW (2010) Diversity, decoys and the dilution effect: how ecological communities affect disease risk. J Exp Biol 213:961–970. https://doi.org/10.1242/jeb.037721

Jude DJ, Reider RH, Smith GR (1992) Establishment of Gobiidae in the Great Lakes Basin. Can J Fish Aquat Sci 49:416–421

Jurajda P, Černý J, Polačik M, Valová Z, Janáč M, Blažek R, Ondračková M (2005) The recent distribution and abundance of non-native *Neogobius* fishes in the Slovak section of the River Danube. J Appl Ichthyol 21:319–323

Kakareko T, Plachocki D, Kobak J (2009) Relative abundance of Ponto-Caspian gobiids in the lower Vistula River (Poland) 3- to 4 years after first appearance. J Appl Ichthyol 25:647–651

Kelly DW, Paterson RA, Townsend CR, Poulin R, Tompkins DM (2009) Parasite spillback: a neglected concept in invasion ecology? Ecology 90:2047–2056

Kennedy CR (1993) Introductions, spread and colonization of new localities by fish helminth and crustacean parasites in the British Isles: a perspective and appraisal. J Fish Biol 43:287–301

Kennedy CR (1994) The ecology of introductions. In: Pike AW, Lewis JW (eds) Parasitic diseases of fish. Samara, Tresaith, pp 189–208

Kennedy CR (2006) Ecology of the Acanthocephala. Cambridge University Press, Cambridge

Kennedy CR, Broughton PF, Hine PM (1978) The status of brown and rainbow trout, *Salmo trutta* and *S. gairdneri* as hosts of the acanthocephalan, *Pomphorhynchus laevis*. J Fish Biol 13:265–275. https://doi.org/10.1111/j.1095-8649.1978.tb03434.x

Keppel M, Dangel KC, Sures B (2016) The Hsp70 response of *Anguillicola* species to host-specific stressors. Parasitol Res 115:2149–2154

Kikuchi H (1929) Two new species of Japanese trematodes belonging to Gyrodactylidae. Annot Zool Jpn 12:175–186

Kirk RS (2003) The impact of *Anguillicola crassus* on European eels. Fish Manag Ecol 10:385–394

Kiskároly M, Čanković M (1967) *Pomphorhynchus bosniacus* nov. sp. aus Barben *Barbus barbus* (L.) des Save-Gebietes. Zool Anz 182:69–74

Kley A, Maier G (2006) Reproductive characteristics of invasive gammarids in the Rhine-Main-Danube catchment, South Germany. Limnologica 36:79–90. https://doi.org/10.1016/j.limno.2006.01.002

Knopf K (2006) The swimbladder nematode *Anguillicola crassus* in the European eel *Anguilla anguilla* and the Japanese eel *Anguilla japonica*: differences in susceptibility and immunity between a recently colonized host and the original host. J Helminthol 80:129–136. https://doi.org/10.1079/JOH2006353

Kopp K, Jokela J (2007) Resistant invaders can convey benefits to native species. Oikos 116:295–301. https://doi.org/10.1111/j.0030-1299.2007.15290.x

Kottelat M, Freyhof J (2007) Handbook of European freshwater fishes. Kottelat, Cornol and Freyhof, Berlin, p 646

Kováč V, Copp GH, Sousa RP (2009) Life-history traits of invasive bighead goby *Neogobius kessleri* from the middle Danube with a prediction of who will win the goby competition. J Appl Ichthyol 25:33–37

Krasnovyd V, Kvach Y, Drobiniak O (2012) The parasite fauna of the gobiid fish (Actinopterygii, Gobiidae) in the Sukhyi Lyman, Black Sea. Vestnik Zoologii 46:1–8

Krisp H, Maier G (2005) Consumption of macroinvertebrates by invasive and native gammarids: a comparison. J Limnol 64(1):55–59

Kuchta R, Choudhury A, Scholz T (2018) Asian fish tapeworm: the most successful invasive parasite in freshwaters. Trends Parasitol 34:511–523

Kvach Y (2002) The round goby's parasites in native habitats and in a place of invasion. Oceanol Hydrobiol St 31:51–57

Kvach Y (2004) The metazoan parasites of gobiids in the Dniester Estuary (Black Sea) depending on water salinity. Oceanol Hydrobiol Stud 33:47–56

Kvach Y (2005) A comparative analysis of helminth faunas and infection of ten species of gobiid fishes (Actinopterigii: Gobiidae) from the North-Western Black Sea. Acta Ichtyol Piscat 35 (2):103–110

Kvach Y, Skóra KE (2007) Metazoa parasites of the invasive round goby *Apollonia melanostoma* (*Neogobius melanostomus*) (Pallas) (Gobiidae: Osteichthyes) in the Gulf of Gdańsk, Baltic Sea, Poland: a comparison with the Black Sea. Parasitol Res 100(4):767–774

Kvach Y, Boldryev V, Lohner R, Stepien CA (2015) The parasite community of gobiid fishes (Actinopterigyii: Gobiidae) from the lower Volga River region. Biologia 70:948–957

Lavrinčíková M, Kováč V (2007) Invasive round goby *Neogobius melanostomus* from the Danube mature at small size. J Appl Ichthyol 23:276–278

Lelek A, Köhler C (1990) Restoration of fish communities of the Rhine River two years after a heavy pollution wave. Regul Rivers Res Manag 5:57–66

Marcogliese DJ, Cone DK (1993) What metazoan parasites tell us about the evolution of American and European eels. Evolution 47:1632–1635

McHugh KJ, Weyl OLF, Smit NJ (2017) Parasite diversity of African longfin eel *Anguilla mossambica* Peters with comments on host response to the monogenean *Pseudodactylogyrus anguillae* (Yin and Sproston). J Fish Dis 40:959–961

Médoc V, Rigaud T, Motreuil S, Perrot-Minnot M-J, Bollache L (2011) Paratenic hosts as regular transmission route in the acanthocephalan *Pomphorhynchus laevis*: potential implications for food webs. Naturwissenschaften 98:825–835. https://doi.org/10.1007/s00114-011-0831-y

Mierzejewska K, Martyniak A, Kakareko T, Dzika E, Stańczak K, Hliwa P (2011) *Gyrodactylus proterorhini* Ergens, 1967 (Monogenoidea, Gyrodactylidae) in gobiids from the Vistula River - the first record of the parasite in Poland. Parasitol Res 108:1147–1151

Mierzejewska K, Kvach Y, Woźniak M, Kosowska A, Dziekońska-Rynko J (2012) Parasites of an Asian Fish, the Chinese sleeper *Perccottus glenii*, in the Wloclawek reservoir on the lower Vistula River, Poland: in search of the key species in the host expansion process. Comp Parasitol 79:23–29

Nachev M, Sures B (2009) The endohelminth fauna of barbel (*Barbus barbus*) correlates with water quality of the Danube River in Bulgaria. Parasitology 136:545–552

Naidenova NN (1974) Parasite fauna of gobiid fishes of the Black Sea and Sea of Azov. Naukova Dumka, Kiev

Nehring S (2003) Gebietsfremde Arten aus in den deutschen Gewässern – ein Risiko für die Biodiversität – Schriftenreihe des Bundesministeriums für Verbraucherschutz, Ernährung und Landwirtschaft. Reihe A: Angewandte Wissenschaft 498:40–52

Oesterwind D, Bock C, Förster A, Gabel M, Henseler C, Kotterba P, Menge M, Myts D, Winkler HM (2017) Predator and prey: the role of the round goby *Neogobius melanostomus* in the western Baltic. Mar Biol Res 13:188–197

Ondračková M (2016) *Gyrodactylus proterorhini* in its non-native range: distribution and ability to host-switch in freshwaters. Parasitol Res 115(8):3153–3162

Ondračková M, Trichkova T, Jurajda P (2006) Present and historical occurrence of metazoan parasites in *Neogobius kessleri* (Pisces: Gobiidae) in the Bulgarian section of the Danube River. Acta Zool Bulg 58(3):399–406

Ondračková M, Šimková A, Civáňová K, Vyskočilová M, Jurajda P (2012) Parasite diversity and microsatellite variability in native and introduced populations of four *Neogobius* species (Gobiidae). Parasitology 139:1493–1505

Ondračková M, Valová Z, Hudcová I, Michálková V, Šimková A, Borcherding J, Jurajda P (2015) Temporal effects on host-parasite associations in four naturalized goby species living in sympatry. Hydrobiologia 746:233–243

Özer A (2007) Metazoan parasite fauna of the round goby *Neogobius melanostomus* Pallas, 1811 (Perciformes: Gobiidae) collected from the Black Sea coast at Sinop, Turkey. J Nat His 41 (9–12):483–492

Palstra AP, Heppener DFM, van Ginneken VJT, Székely C, van Den Thillart GEEJ (2007) Swimming performance of silver eels is severely impaired by the swim-bladder parasite *Anguillicola crassus*. J Exp Mar Bio Ecol 352:244–256

Parker D, Weyl OLF, Taraschewski H (2011) Invasion of a South African *Anguilla mossambica* (Anguillidae) population by the alien gill worm *Pseudodactylogyrus anguillae* (Monogenea). Afr Zool 46:371–377

Pelster B (2015) Swimbladder function and the spawning migration of the European eel *Anguilla anguilla*. Front Physiol 6:1–10

Perrot-Minnot M-J, Guyonnet E, Bollache L, Lagrue C (2019) Differential patterns of definitive host use by two fish acanthocephalans occurring in sympatry: *Pomphorhynchus laevis* and *Pomphorhynchus tereticollis*. Int J Parasitol Parasit Wildl 8:135–144. https://doi.org/10.1016/j.ijppaw.2019.01.007

Pinkster S, Smit H, Brandse-de Jong N (1977) The introduction of the alien amphipod *Gammarus tigrinus* Sexton, 1939, in the Netherlands and its competition with indigenous species. Crustaceana Suppl 4:91–105

Pitronaci S (2016) Parasitengemeinschaften des Karpfens aus dem Rhein. Staatsexamensarbeit, Universität Duisburg-Essen

Reier S, Sattmann H, Schwaha T, Harl J, Konecny R, Haring E (2019) An integrative taxonomic approach to reveal the status of the genus *Pomphorhynchus* Monticelli, 1905 (Acanthocephala: Pomphorhynchidae) in Austria. Int J Parasitol Parasit Wildl 8:145–155

Rewicz T, Grabowski M, MacNeil C, Bacela-Spychalska K (2014) The profile of a 'perfect' invader – the case of killer shrimp, *Dikerogammarus villosus*. Aquat Invasions 9:267–288. https://doi.org/10.3391/ai.2014.9.3.04

Sasal P, Taraschewski H, Valade P, Grondin H, Wielgoss S, Moravec F (2008) Parasite communities in eels of the Island of Reunion (Indian Ocean): a lesson in parasite introduction. Parasitol Res 102:1343–1350

Shears JA, Kennedy CR (2005) The life cycle of *Paraquimperia tenerrima*: a parasite of the European eel *Anguilla anguilla*. J Helminthol 79:169–176

Skóra KE, Stolarski J (1993) New fish species in the Gulf of Gdansk, *Neogobius* sp. [cf. *Neogobius melanostomus* (Pallas 1811)]. Bull Sea Fish Inst 1(128):83–84

Špakulová M, Perrot-Minnot M-J, Neuhaus B (2011) Resurrection of *Pomphorhynchus tereticollis* (Rudolphi, 1809) (Acanthocephala: Pomphorhynchidae) based on new morphological and molecular data. Helminthologia 48:268–277

Steele DH, Steele VJ (1972) The biology of *Gammarus* (Crustacea, Amphipoda) in the northwestern Atlantic. VI. *Gammarus tigrinus* Sexton. Can J Zool 50(8):1063–1068. https://doi.org/10.1139/z72-144

Stemmer B (2008) Flussgrundel im Rhein-Gewässersystem. Natur in NRW 4/08:57–60

Sures B (2011) Parasites of animals. In: Simberloff D, Rejmánek M (eds) Encyclopedia of biological invasions. University of California Press, Berkeley, pp 500–503

Sures B, Knopf K (2004) Parasites as a threat to freshwater eels? Science 304:208–209

Sures B, Streit B (2001) Eel parasite diversity and intermediate host abundance in the River Rhine, Germany. Parasitology 123:185–191

Sures B, Knopf K, Wurtz J, Hirt J (1999) Richness and diversity of parasite communities in European eels *Anguilla anguilla* of the River Rhine, Germany, with special reference to helminth parasites. Parasitology 119:323–330

Sures B, Knopf K, Kloas W (2001) Induction of stress by the swimbladder nematode *Anguillicola crassus* in European eels, *Anguilla anguilla*, after repeated experimental infection. Parasitology 123:179–184. https://doi.org/10.1017/S003118200100823X

Sures B, Nachev M, Pahl M, Grabner D, Selbach C (2017) Parasites as drivers of key processes in aquatic ecosystems: facts and future directions. Exp Parasitol 180:141–147. https://doi.org/10.1016/j.exppara.2017.03.011

Taraschewski H (2006) Hosts and parasites as aliens. J Helminthol 80:99–128

Taraschewski H, Moravec F, Lamah T, Anders K (1987) Distribution and morphology of two helminths recently introduced into European eel populations: *Anguillicola crassus* (Nematoda Drancunculoidea) and *Paratenuisentis ambiguus* (Acanthocephala, Tenuisentidae). Dis Aquat Org 3:167–176

Telfer S, Bown KJ, Sekules R, Begon M, Hayden T, Birtles R (2005) Disruption of a host-parasite system following the introduction of an exotic host species. Parasitology 130:661–668. https://doi.org/10.1017/S0031182005007250

Thiel R, Horn L, Knörr C, Tonn M (2014) Analyse der ökologischen Einnischung der invasiven Schwarzmundgrundel in relevanten Brack und Süßgewässerhabitaten Schleswig Holsteins. Final Report. Landesamt für Landwirtschaft, Umwelt und ländliche Räume Schleswig Holstein, p 199

Thielen F, Münderle M, Taraschewski H, Sures B (2007) Do eel parasites reflect the local crustacean community? A case study from the Rhine River system. J Helminthol 81:179–189

Thielen F, Weibel U, Hirt J, Münderle M, Marten M, Taraschewski H, Sures B (2008) Ichthyofauna in the upper Rhine River close to the city of Karlsruhe as determined by the analysis of fish impingement by cooling-water intakes of a power plant. Limnologica 38:76–85

Thieltges DW, Reise K, Prinz K, Jensen KT (2008) Invaders interfere with native parasite–host interactions. Biol Invasions 11:1421–1429

Tittizer T, Krebs F (1996) Ökosystem-Forschung: Der Rhein und seine Auen. Springer, Berlin, p 468

Tittizer T, Schöll F, Banning M, Haybach A, Schleuter M (2000) Aquatic neozoan invertebrates in the inland water ways of the Federal Republic of Germany. Lauterbornia 39:1–72

Torchin M, Lafferty K, Dobson A (2003) Introduced species and their missing parasites. Nature 421:628–630

van Beek GCW (2006) The round goby *Neogobius melanostomus* first recorded in Netherlands. Aquat Invasions 1:42–43

van Kessel N, Dorenbosch M, Spikmans F (2009) First record of Pontian monkey goby, *Neogobius fluviatilis* (Pallas, 1814), in the Dutch Rhine. Aquat Invasions 4:421–424

Villeneuve C De HV (1996) Western Europe's artery: the Rhine. Nat Resour J 36:441–454

Wilken RD (2006) The recovered Rhine and its history. In: Knepper TP (ed) The Rhine: the handbook of environmental chemistry, vol 5L. Springer, Berlin

Würtz J, Taraschewski H (2000) Histopathological changes in the swimbladder wall of the European eel *Anguilla anguilla* due to infections with *Anguillicola crassus*. Dis Aquat Org 39:121–134

Würtz J, Taraschewski H, Pelster B (1996) Changes in gas composition in the swimbladder of the European eel (*Anguilla anguilla*) infected with *Anguillicola crassus* (Nematoda). Parasitology 112:233–238

Würtz J, Knopf K, Taraschewski H (1998) Distribution and prevalence of *Anguillicola crassus* (Nematoda) in eels *Anguilla anguilla* of the rivers Rhine and Naab, Germany. Dis Aquat Org 32:137–143

Zittel M, Grabner D, Wlecklik A, Sures B, Leese F, Taraschewski H, Weigand AM (2018) Cryptic species and their utilization of indigenous and non-indigenous intermediate hosts in the acanthocephalan *Polymorphus minutus* sensu lato (Polymorphidae). Parasitology 145:1421–1429

Chapter 20
Free-Living Amoebae in Rivers and Ponds and Their Multiple Role in Environmental Health

Patrick Scheid

Abstract Free-living amoebae (FLA) such as Acanthamoebae, Naegleriae and Vermamoebae have been identified from natural environments. Among the protozoa living in rivers and ponds, the FLA may also be parasites for humans and animals. These potentially pathogenic FLA are widely distributed depending on the environmental and climatic conditions. The most abundant, potentially parasitic FLA in rivers and ponds are the Acanthamoebae. Other highly pathogenic FLA, such as the "brain-eating amoebae" (*Naegleria fowleri*), require higher water temperatures and are therefore to be found only in warmer climates.

As well as being parasites for humans and animals, FLA are known as reservoirs of and vectors for different microorganisms. In some cases, these intracellular microorganisms proliferate inside their host amoebae without being digested. Those microorganisms being able to survive the intracellular conditions in the amoebal trophozoites and cysts benefit from their intracelluar life because they are protected to a certain degree from any hostile environment. Furthermore they are dispersed by the FLA within their environment. These intracellular organisms are called "endocytobionts". Among those organisms resisting the lysis within their amoebic host, there are obligate and facultative pathogenic microorganisms affecting the health of humans or animals. Especially if FLA are present in biofilms in rivers and ponds, e.g. in the littoral zone, there are many possibilities for the FLA to get in contact with other microorganisms. The resulting relationship of "host and endocytobiont" may lead to the development of pathogenicity or to the enhancement of virulence of the endocytobionts. This passage through FLA together with survival of the phagocytosis mechanisms and even the subsequent proliferation may be an essential factor for the development of virulence factors. Thus the relationship may contribute to evolutionary processes. The contact of humans with parasitic FLA or FLA harbouring pathogenic endocytobionts may even increase in the future as a result of human activities in recreational waters due to global warming. At the same time, a higher abundance of the FLA is to be expected following global warming. In

P. Scheid (✉)
Central Institute of the Bundeswehr Medical Service Koblenz, Koblenz, Germany

University Koblenz-Landau, WG Parasitology and Infection Biology, Koblenz, Germany

H. Mehlhorn, S. Klimpel (eds.), *Parasite and Disease Spread by Major Rivers on Earth*, Parasitology Research Monographs 12,
https://doi.org/10.1007/978-3-030-29061-0_20

general, changes of the climate and the environment may increase the importance of many infectious diseases or pathogenic microorganisms affecting people, domestic animals and wildlife. FLA may contribute to this prospective development either as parasitic FLA or as vectors of pathogenic endocytobionts. Therefore they have to be considered in terms of "public health" and "environmental health".

Keywords Free-living amoebae, FLA · Endocytobionts · Vectors · Rivers · Ponds · Global warming

20.1 Introduction

The term "free-living amoebae" (FLA) confines a heterogeneous assemblage of aquatic dwelling amoebae, e.g. in contrast to the parasitic intestinal amoebae. However the term "FLA" has no phylogenetic, systematic or taxonomic relevance (Walochnik and Aspöck 2007). The typical morphology of different FLA as trophozoites and/or cysts is shown in Figs. 20.1, 20.2, 20.3, 20.4, 20.5, 20.6 and 20.7 (*Saccamoeba* sp.; *Willaertia* sp.; *Vannella* sp.; *Acanthamoeba* sp.). The globally distributed FLA are to be found in diverse environments. Their environmental niches include environmental habitats like air, dust, soil, rivers, lakes and ponds, surface water, springs or wells but also engineered habitats like tap water systems (even in hospitals), cooling towers, water tanks, wastewater treatment plants, pharmaceutical factory drains, swimming pools, sewage, contact lens cases, air-conditioning equipment, dialysis units, dental treatment units, etc. (for overview, see Scheid 2018a). The dispersal within the natural environments of most FLA, e.g. in the cyst stage (if available), results in a worldwide distribution, due to passive transfer in water or active transport, e.g. by waterfowl (Scheid 2018a, b). FLA live in biofilms, adherent

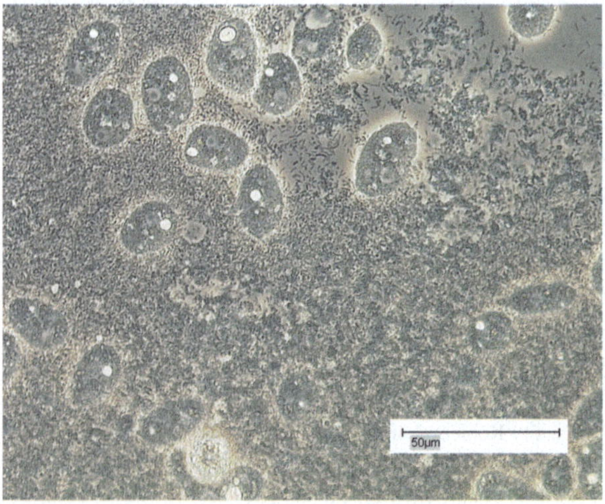

Fig. 20.1 *Acanthamoeba* sp. trophozoites isolated from a pond in Germany; bar 50 µm; phase contrast

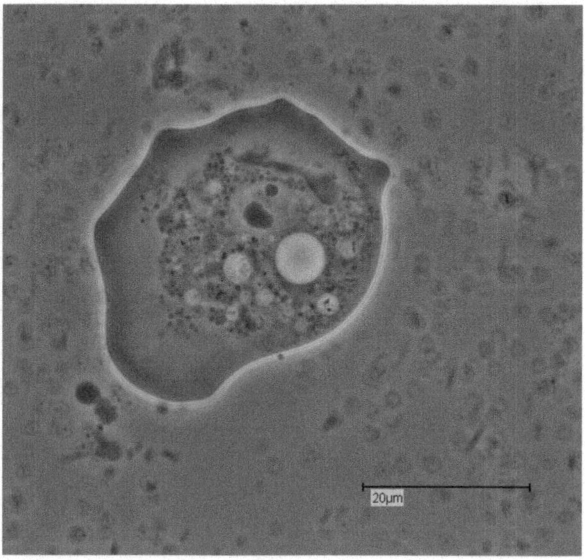

Fig. 20.2 *Vannella* sp. trophozoite; bar 20 μm; phase contrast

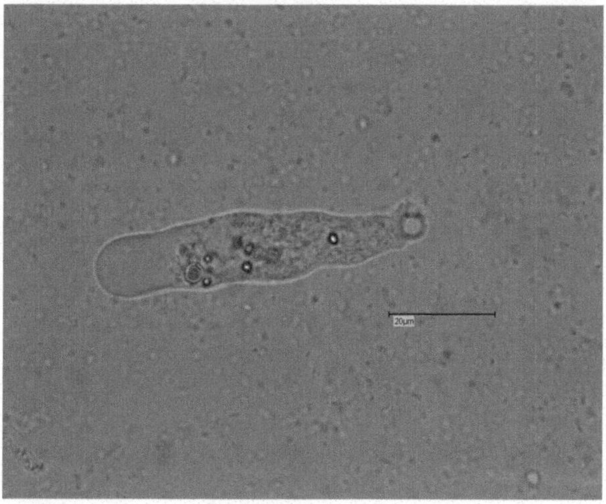

Fig. 20.3 *Saccamoeba limax* trophozoite as an example for a trophozoite with uroid; bar 20 μm

on various surfaces and substrates, and are part of the biocoenosis herein. As heterotrophic microorganisms, the FLA excrete materials (e.g. minerals), which were bound in the biomass, so that these are available within the environment. This is why FLA are initially involved in biofilm formation and subsequently in biofilm succession. FLA are a part of the ecological biofilm community (Khan and Panjwani 2000). At the same time the FLA are provided with nutrition and protection within the biofilms. FLA feed in the trophozoite stage (Figs. 20.1, 20.2, 20.3, 20.4 and 20.5) as predators actively on bacteria, algae, cyanobacteria, smaller

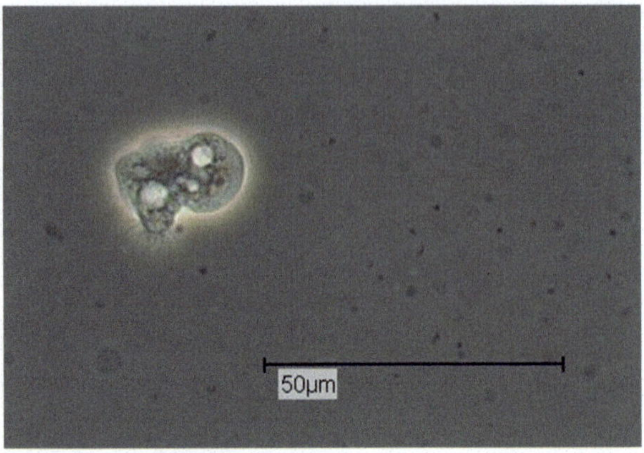

Fig. 20.4 *Willaertia magna* trophozoite as an example for a trophozoite with uroid; bar 50 µm

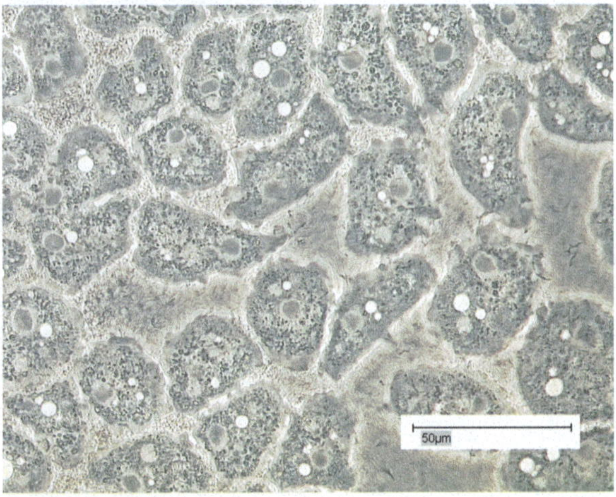

Fig. 20.5 *Vannella* sp. trophozoites; bar 50 µm; phase contrast

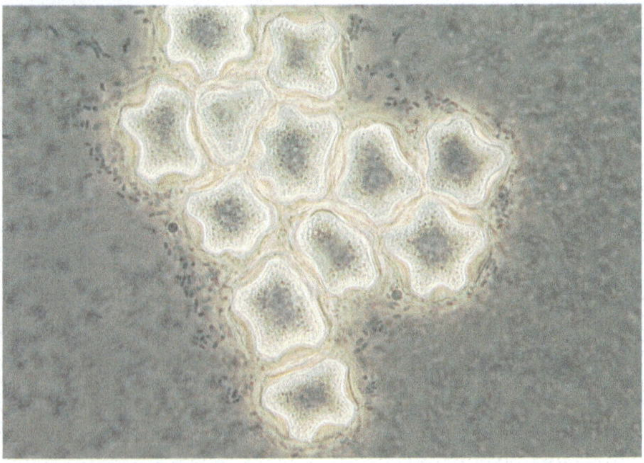

Fig. 20.6 *Acanthamoeba* sp. cysts; phase contrast

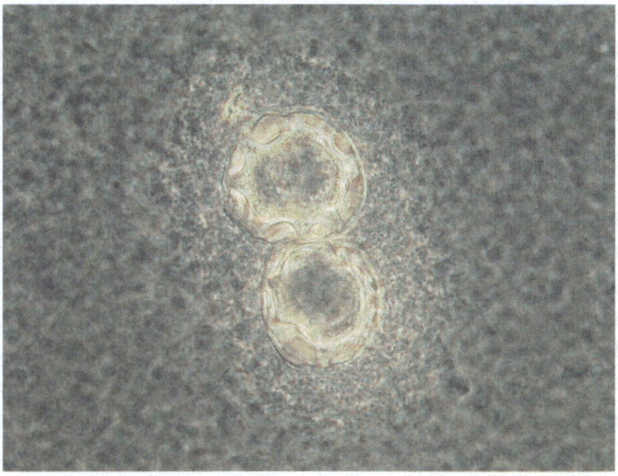

Fig. 20.7 Two *Acanthamoeba* sp. cysts; phase contrast

protozoa and fungi. But at the same time FLA serve as prey for other predatory microorganisms, such as several amoebophagous fungi (Kurek et al. 2010; Scheid 2018a).

Certain FLA form a cyst as a dormant stage with a high tenacity. Within this cyst stage they are protected against adverse environmental conditions (Figs. 20.6 and 20.7).

20.2 FLA as Parasites for Humans and Animals

FLA are no obligate parasites but they are able to survive endozoically. Certain facultative parasitically living species or strains can therefore affect humans and animals (Barker et al. 1993; Scheid 2018a, b). Several genera or species are described to be pathogenic for humans: *Acanthamoeba* strains (Figs. 20.1, 20.6, and 20.7), *Balamuthia mandrillaris* and *Naegleria fowleri*, and some other FLA species are shown to be of (potential) medical relevance. A human infection may result especially when there is an intensive and close contact between humans and FLA, as happens (for example) through the use of contaminated contact lenses when placed on the cornea. The pathogenic FLA can cause severe diseases including amoebic keratitis, amoebic meningoencephalitis or cutaneous amoebiasis in human and animals. Swimming, diving and bathing in rivers or ponds and in unchlorinated pools with these contact lenses, with the uptake of contaminated water via the nose, are examples for risk factors to acquire a parasitic FLA.

Naegleria fowleri, the "brain-eating amoeba," which prefers higher water temperatures and may cause an acute, progressive encephalitis in humans, has made headlines in the USA due to several human fatalities over the years. This typical encephalitis was reported in previously healthy children and young adults with a history of swimming and other recreational activities (e.g. diving, water-skiing) in warm freshwater lakes and ponds (Scheid 2018a). *Naegleria fowleri* prefers elevated temperatures (it tolerates up to 45 °C) and can be found in warm artificial fresh water bodies or warm lakes and ponds (see case reports from Asia or the USA). This may be one of the reasons why (to the best of our knowledge) *Naegleria fowleri* has not been found in German rivers or ponds so far.

Balamuthia mandrillaris, another parasitic amoeba, expresses two stages, a motile, pleomorphic trophozoite and a resistant, dormant cyst stage. The human infection leads to a granulomatous amoebic encephalitis, known as granulomatous *Balamuthia* amebic encephalitis (BAE). The *Balamuthia mandrillaris* trophozoites inhabit water bodies, dust and soil, where they feed from other (smaller) FLA. Predominantly BAE cases are reported from warm regions like Mexico. To the best of our knowledge, *Balamuthia mandrillaris* has not been detected in German rivers, ponds or soil so far, but DNA of *Balamuthia mandrillaris* has been found in a contact lens case during routine screening for AK in a German patient (Scheid and Balczun 2017).

In contrast, *Acanthamoeba* are widely distributed all over the world in rivers and ponds and humid soils including Germany. The most prominent association of *Acanthamoeba* and humans is with wearing contact lenses (risk factors!). The pathogenesis of *Acanthamoeba* and the role of *Acanthamoeba* as parasites for humans and animals are described sufficiently in the literature (see Khan 2009; Scheid 2018a). FLA can be even more harmful and lead to life-threatening conditions, especially for vulnerable persons, such as immunodeficient patients (Scheid 2018a). Acanthamoebiasis can show various clinical signs and pathogenesis, e.g. a cutaneous form, a multifocal encephalitis, a granulomatous amebic encephalitis (GAE) or the *Acanthamoeba* keratitis (AK).

Members of the genus *Acanthamoeba* have a life cycle with a trophozoite as the motile trophic stage and a cyst form (Figs. 20.1, 20.6 and 20.7). *Acanthamoeba* shows the vegetative form when the environment is humid, favourable and nutrition-rich. When exposed to adverse conditions, *Acanthamoeba* are able to form dormant cyst within their biocoenosis. In the cyst form the *Acanthamoeba* (as well as other cyst-forming FLA) show a high resistance. Their high tenacity enables them to be resistant to a wide range of adverse environmental conditions (temperature, pH, salinity, etc.). As indicated for the most FLA, *Acanthamoeba* may also benefit from a global warming effect (as a result of climate change) in natural or man-made water bodies. Their high abundances in rivers and ponds will even arise, resulting in a high level rate of contact with humans. Some *Acanthamoeba* strains are able to stand 32–35 °C (corneal temperature) or even 37 °C (body temperature) which makes them capable of infecting humans (Scheid 2018a). The pathogenicity varies among the *Acanthamoeba* species, strains or sequence types (genotypes; T1–21, based on the nuclear small subunit ribosomal RNA (rRNA) gene). The most prevalent genotype in our region is the T4 genotype, while other genotypes have also been reported as parasites of humans and animals.

There are also other FLA detected as potential parasites in recent years. While Acanthamoebae are described very frequently as human parasites, there have been published several case reports of other FLA as being aetiological/pathogenic agents.

Sappiniae have been proved to be the aetiological agent of the so-called *Sappinia* amebic encephalitis (SAE). Vahlkampfiae, Hartmannellae, *Paravahlkampfia francinae* (n. sp.), *Allovahlkampfia spelaea* and *Dictyostelium polycephalum* are other examples of FLA as parasites (for overview, see Scheid 2018a, b). Several recent cases (two of them in Germany) with the involvement of *Vermamoeba vermiformis* support its potential role as direct human parasite, either as the only aetiological agent or in combination with other pathogens. As more FLA turned out to be opportunistic human parasites, this could indicate that we may have detected only the tip of the iceberg so far. Differences in pathogenicity, susceptibility or climatic preferences will be responsible in the future to show which FLA species, strains or genotypes will profit most from global warming.

20.3 FLA as Reservoirs and Vectors of Other Microorganisms

In their biocoenoses inside biofilms in waters, the FLA come in close contact with many other water-borne microorganisms (bacteria, viruses, fungi, protozoa). FLA feed from bacteria via phagocytosis. As predatory microorganisms, FLA usually ingest other microorganisms by food cup building and lyse them subsequently in their food vacuoles. But even after the uptake of prey microorganisms by FLA, some of these microorganisms avoid being digested. They resist or escape the fate that otherwise awaits them through exposure to oxygen radicals, the acidic environment and the lysosomal enzymes in the phagosomes. Among those intracellularly surviving microorganisms, there are also microorganisms with hygienic/pathogenic relevance. As some of the ingested microorganisms are adapted to an intracellular life within FLA, they benefit from the suitable growth conditions and the protection against adverse environmental conditions. Another advantage for those intracellular resisting microorganisms is the dispersal in the environment with the help of their hosts or vehicles. This phenomenon of FLA serving as hosts has led to the term "FLA as the Trojan horse of the microbial world" years ago (Barker et al. 1993; Khan 2009). These intracellular microorganisms, which are associated with FLA and proliferate intracellularly, are called "endocytobionts" (see Scheid et al. 2008, 2010).

Examples for endocytobionts in FLA are given in Figs. 20.8 and 20.9 and Table 20.1. The described "FLA-endocytobiont" relationships may result in parasitism, phoresy, zoochory or mutualism. The specific expression of these relationships varies depending on temperature (Legionellae and Acanthamoebae as an example) or other different environmental conditions.

As some of these endocytobionts, developing within FLA, are of pathogenic relevance, the FLA can be called therefore "hosts, vehicles or vectors", besides their role as potential human parasites (Scheid 2007; Scheid et al. 2008).

In 1956, *Sarcobium lyticum* (*Legionella lytica*) was detected in an *Acanthamoeba* strain isolated from soil. This bacterium was not culturable on any culture media

Fig. 20.8 Two *Vannella* sp. trophozoites with endocytobionts (here: with intranuclear *Paramicrosporidium* sp. stages); light microscopy; phase contrast

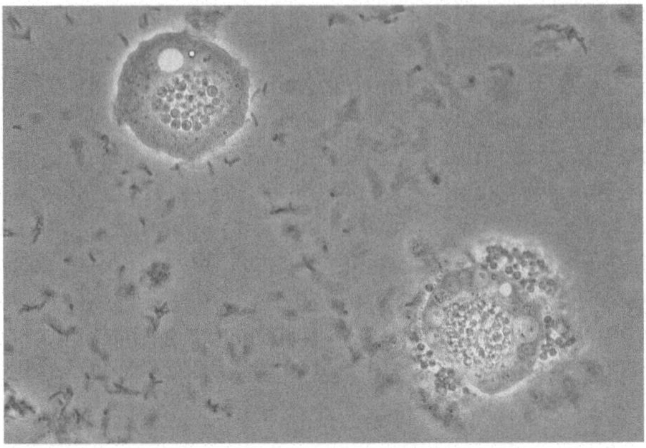

Fig. 20.9 Several *Acanthamoeba* sp. trophozoites with endocytobionts (here: with "giant viruses", *Pandoravirus inopinatum*); light microscopy; phase contrast

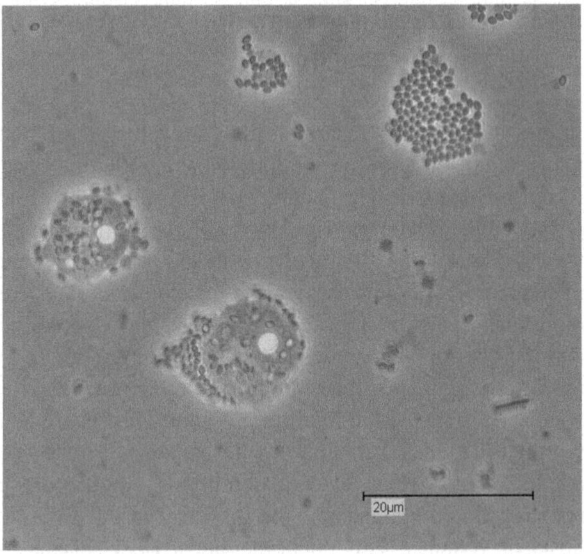

available and therefore proved to be obligate intracellularly (Drozanski 1956, 1991). The most recognized endocytobionts of FLA are the intracellularly proliferating *Legionella* bacteria in *Acanthamoeba*. Rowbotham successfully carried out cocultivation assays of *Acanthamoeba* and Legionellae in 1980. The Legionellae were located intracellularly within the trophozoites, and they proliferated at 35 °C in the FLA. Since the time of this discovery, FLA proved intensively to serve as hosts, reservoirs, vectors and vehicles of Legionellae protecting them from the environmental conditions (Rowbotham 1980). Chlamydiae, important human obligate intracellular, pathogenic microorganisms, were detected as endocytobionts of FLA a few years later (Michel et al. 2004, 2005). The diversity of bacteria, viruses, fungi and protozoa within FLA is remarkably high (for overview, see Table 20.1; taken from Balczun and Scheid 2017). Acanthamoebae are the predominant FLA discovered as hosts of these

Table 20.1 Selection of endocytobionts with public health importance, associated with FLA

Species	Host amoeba
Bacteria	
In situ interaction	
Legionella pneumophila	Several FLA, e.g. Acanthamoebae
Listeria monocytogenes	Acanthamoebae
Mycobacterium avium	Acanthamoebae
Afipia felis	Acanthamoebae
Pseudomonas saccharophila	Acanthamoebae
Pseudomonas aeruginosa	Echinamoebae; Acanthamoebae; Hartmannellae
Ralstonia (*Burkholderia*) *picketti*	Acanthamoebae
Stenotrophomonas maltophilia-Komplex	Acanthamoebae; Naegleriae
Rickettsia-like	Acanthamoebae
Neochlamydia hartmannellae	Hartmannellae; *Hartmannella* (*Vermamoeba*) *vermiformis*
Procabacter acanthamoebae (Betaproteobacterium)	Acanthamoebae
Amoebophilus asiaticus	Acanthamoebae
In vitro interaction	
Salmonella enterica, *S. typhimurium* (and other Salmonellae)	Acanthamoebae
Yersinia enterocolitica	Acanthamoebae
Campylobacter jejuni	Acanthamoebae (*Acanthamoeba polyphaga*)
Arcobacter butzleri	Acanthamoebae (*Acanthamoeba castellanii*)
Mycobacterium spp., *M. leprae*	Acanthamoebae
Staphylococcus aureus (incl. MRSA)	Acanthamoebae
Mobiluncus curtisii	Acanthamoebae
Mycobacterium leprae	Acanthamoebae
Mycobacterium xenopi	Acanthamoebae
Mycobacterium spp., *Mycobacterium avium*	Acanthamoebae
Bacillus anthracis	*Acanthamoeba castellanii*, *Hartmannella* (*Vermamoeba*) *vermiformis*
Acidovorax temperans	Naegleriae
Burkholderia cepacia and *Burkholderia pickettii*	Acanthamoebae
Caulobacter vibrioides	Echinamoebae
Flavobacterium johnsoniae	Naegleriae
Escherichia coli O157	Acanthamoebae
Helicobacter pylori	Acanthamoebae
Porphyromonas gingivalis	Acanthamoebae
Prevotella intermedia	Acanthamoebae
Chlamydophila pneumoniae	Acanthamoebae
Parachlamydia acanthamoebae	Acanthamoebae

(continued)

Table 20.1 (continued)

Species	Host amoeba
Waddlia sp., *Waddlia chondrophila*, Other *Chlamydia*-like endocytobionts	Acanthamoebae, Naegleriae, Hartmannellae, Hyperamoebae, Vahlkampfiae, *Dictyostelium discoideum*
Coxiella burnetii	Acanthamoebae
Francisella tularensis	Acanthamoebae
Protists and fungi	
In situ interaction	
Paramicrosporidium gen. nov	*Vannella* sp.
Cochlonema euryblastum	Thecamoebae; *Thecamoeba quadrilineata*
In vitro interaction	
Cryptosporidium parvum	Acanthamoebae
Cryptococcus neoformans	Acanthamoebae
Viruses	
In situ interaction	
Adenoviruses	Acanthamoebae
Acanthamoeba polyphaga Mimivirus	Acanthamoebae
Pandoravirus inopinatum	Acanthamoebae
Pithoviruses	Acanthamoebae
In vitro interaction	
Adenoviruses	Acanthamoebae
Coxsackie virus	Acanthamoebae
Marseillevirus	Acanthamoebae
Megavirus chilensis	Acanthamoebae
Pandoravirus salinus, *Pandoravirus dulcis*	Acanthamoebae
Pithovirus sibericum	Acanthamoebae
Faustovirus	*Vermamoeba vermiformis*
Mollivirus sibericum	Acanthamoebae

Source: Balczun, C.; Scheid, P.: Free-Living Amoebae as Hosts for and Vectors of Intracellular Microorganisms with Public Health Significance; Viruses 9(4):65 April 2017. https://doi.org/10.3390/v9040065; see also for selected references

endocytobionts, being the most abundant FLA in the environment at the same time. During the last 15 years, especially the research of the association of FLA and viral endocytobionts has increased significantly. Some of these viruses were called "giant viruses", mainly due to their unexpected size and/or their enormous gene repertoire.

An overview of microorganisms, confirmed as endocytobionts within FLA as hosts or vehicles, is presented in Table 20.1, providing additional informations about their host amoeba affiliation. There is a significant qualitative difference as to whether the endocytobionts could be isolated from real specimens in situ (e.g. in the context of environmental studies or from patient specimens) or only within the framework of cocultivation assays (under laboratory conditions). Several recently discovered viral endocytobionts (including some "giant viruses") are included in the list (Table 20.1). The relationship between FLA and fungi or protozoa is not examined that extensively. Endozoochory and parasitism have been described in these associations (Michel et al. 1995; Scheid and Schwarzenberger 2011; Scheid 2014; Scheid 2018a).

The fact that FLA also serve as hosts of and vehicles (vectors) for microorganisms within biofilms has an impact on public health and environmental health, affecting the quality of water as well as the range of measures against pathogenic microorganisms. The preferred habitats of the FLA and their endocytobionts are the biofilms, whether in natural habitats (rivers, ponds, lakes) or artificial water systems in industry, private houses and even in healthcare areas. There are reports of FLA and other potentially human pathogenic microorganisms from tap water, dialysis centres, hospitals and insufficiently disinfected contact lens containers. These findings underline the importance of biofilms, where interesting and relevant communities of microorganisms (co-)exist.

In the extracellular, fibrous, polymer, exopolysaccharide biofilm matrix of acellular substances in the littoral zone of rivers and lakes, microorganisms are protected significantly against exogenous influences such as UV radiation, fluctuations in pH, salts, hydraulic loads, toxic heavy metals, antimicrobial agents and disinfectants (Borde et al. 2013). The consequence of this "lifestyle" within biofilms is an increase in the resistance to adverse environmental conditions or counter-measures. In contrast to suspended and planctonic bacteria, it is a major challenge to control microorganisms (hiding) in biofilms. The existing diffusion and reaction barriers contribute to the resistance properties. Several microorganisms (e.g. most of the bacteria) switch to a different physiological state when growing within biofilms with the effect that they may tolerate higher biocide concentrations and a longer exposure time compared to the suspended form (Flemming 1994). While the microorganisms are thus protected in biofilms, they are additionally protected when living as endocytobionts within FLA. An even higher protection is provided for endocytobionts in cyst-forming FLA. In synopsis, it becomes clear that there is a double or triple protection against unfavourable environmental conditions and counter-measures for endocytobionts surviving inside FLA in biofilms compared to their free-floating forms (Greub and Raoult 2004; Wieser and Schubert 2011; Scheid 2014).

Such adaptations to intracellular life within predatory protozoa or other "phagocytosis-active cells" document the evolutionary potential of these relationships. In consequence these host-endocytobiont associations (whether observed in situ or in vitro) were called "intracellular biofilms". The adhesion of a biofilm on the surface within a river or pond is a first step in the biofilm generation. Likewise is the adhesion of microorganisms to FLA the initial step of the intracellular biofilm generation (Wieser and Schubert 2011). This close relationship between FLA and endocytobionts within biofilms may facilitate the development of human pathogenic properties. The interaction of intracellular endocytobionts surviving the host defense has an impact on the virulence and the tenacity of the involved microorganisms. In an evolutionary sense, FLA may be "test organisms" or "training grounds" for evolutionary "experiments". Following the survival within FLA, these endocytobionts (e.g. *Legionella pneumophila* or *Mycobacterium avium*) may survive as well intracellularly in phagocytes (phagolysosomal cells like macrophages) of humans (Steinert et al. 1998; Fritsche et al. 1998; Molmeret et al. 2005; Cirillo et al. 1997; Steenbergen et al. 2001; Greub and Raoult 2004). This implies that the amoebic passage is a prerequisite for the development and the selection of virulence factors of microorganisms—and thus for the development of human pathogenic microorganisms (see also Greub and Raoult 2004). The uptake processes into both phagocytosis-active "cell lines" (macrophages and amoebic trophozoites) are very similar. The same counts for the replication sites, the release of microorganisms after proliferation (often by lysis, degradation or rupture) and the evasion mechanisms used to escape phagocytosis (e.g. avoiding phagosome-lysosome fusion) (Greub and Raoult 2004). The adaptation of microorganisms to life in a eukaryotic host is in evolutionary terms a preadaptation for the ability to infest or infect metazoa cells, i.e. as human pathogenic microorganisms. Additionally virulence properties and tenacity to antibiotics are known to be associated with in bacteria (Cirillo et al. 1997; Steenbergen et al. 2001). Consequently, FLA can be viewed as evolutionary triggers, incubators or sources for the evolutionary development of a range of microorganisms.

20.4 Conclusion

Global warming as a result of climate change will have an impact on ecological processes like dispersal of individual organisms and population dynamics. In general it is expected that several effects including global warming could favour suitable conditions for parasites in rivers and ponds. When ecological barriers are opened by climatic change, a new distribution of FLA may result. Thermotolerant parasitic FLA (e.g. *Naegleria fowleri*) may appear in regions where they haven't been endemic before. The same counts for thermophilic endocytobionts replicating intracellularly within these FLA (Scheid 2018a). The fact that FLA also serve as hosts of and vehicles (vectors) for microorganisms within biofilms has an impact on the environmental health and public health sector, affecting the quality of water as well as measures against certain microbes. Since many of the pathogenic endocytobionts are water-borne, there is a risk to quality in water networks (Scheid 2018a). To sum up, the possibility of these endocytobionts being able to adapt to human cells by the FLA (or other protozoan) passage is one possibility to explain the development of human pathogenicity of microorganisms. The interaction of endocytobiontic microorganisms with their host FLA leads to the evolution of mechanisms that are found later in the human host as pathogenic mechanisms. The intracellular "lifestyle" of microorganisms over a long period without any damage to the endocytobiont may provoke changes (also at the genome level) in the microorganisms involved. These co-evolved microorganisms have significantly different ecological properties which may serve as evolutionary driving force. Conclusively it can be stated that FLA-endocytobiont relationships are of relevance for the origin of pathogens and evolutionary processes. Therefore, FLA should be considered in the context of environmental health and public health, no matter whether focusing on their role as parasites or vectors or whether focusing on their distribution within natural habitats in rivers, ponds or within man-made environments.

Acknowledgements The author would like to thank Dr. David Lam (MD, MPH, Shaman Medical Consulting) for review and English language editing of the article.

References

Balczun C, Scheid P (2017) Free-living amoebae as hosts for and vectors of intracellular microorganisms with public health significance. Viruses 9:65. https://doi.org/10.3390/v9040065

Barker J, Lambert PA, Brown M (1993) Influence of intra-amoebic and other growth conditions on the surface properties of *Legionella pneumophila*. Infect Immun 61(8):3503–3510

Borde J, Helwig P, Hauschild O (2013) Gelenkprothesen-Infektionen. Krankenhaushygiene up2date 8:89–97

Cirillo J, Falkow S, Tompkins L, Bermudez L (1997) Interaction of *Mycobacterium avium* with environmental amoebae enhances virulence. Infect Immun 65:3759–3767

Drozanski W (1956) Fatal bacterial infection in soil amoebae. Acta Microbiol Pol 5(3–4):315–317

Drozanski WJ (1991) *Sarcobium lyticum* gen. nov., sp. nov., an obligate intracellular bacterial parasite of small free-living amoebae. Int J Syst Bacteriol 41:82–87. https://doi.org/10.1099/00207713-41-1-82

Flemming H (1994) Biofilme, Biofouling und mikrobielle Materialschädigung. Stuttgarter Siedlungswasserwirtschaftliche Berichte 129. Oldenburg, München

Fritsche T, Sobek D, Gautom R (1998) Enhancement of in vitro cytopathogenicity by *Acanthamoeba* spp. following acquisition of bacterial endosymbionts. FEMS Microbiol Lett 166:231–236

Greub G, Raoult D (2004) Microorganisms resistant to free-living amoebae. Clin Microbiol Rev 17:413–433

Khan N (2009) *Acanthamoeba*, biology and pathogenesis. Caister, Norfolk

Khan N, Panjwani N (eds) (2000) Pathogenesis of Acanthamoeba infections, VIII edn. European Multicolloquium of Parasitology, Poznan

Kurek R, Scheid P, Michel R (2010) Darstellung von pilzartigen Endoparasiten bei freilebenden Amöben nach spezifischer Fluoreszenzanfärbung. Mikrokosmos 99(6):327–330

Michel R, Müller K-D, Schmid E (1995) *Ehrlichia*-like organismus (KSL1) observed as obligate intracellular parasites of *Saccamoeba* species. Endocytobio Cell Res 11:69–80

Michel R, Steinert M, Zöller L, Hauröder B, Henning K (2004) Cocultivation of protozoa and the Chlamydia-like bacterium *Waddlia chondrophila* isolated from an aborted bovine foetus in Germany. Acta Protozool 43:37–42

Michel R, Müller K-D, Zöller L, Walochnik J, Hartmann M, Schmid E (2005) Free living amoebae serve as host for the Chlamydia-like bacterium *Simkania nevegensis*. Acta Protozool 44:113–121

Molmeret M, Horn M, Wagner M, Santic M, Abu Kwaik Y (2005) Amoebae as training grounds for intracellular bacterial pathogens. Appl Environ Microbiol 71:20–28. https://doi.org/10.1128/AEM.71.1.20-28.2005

Rowbotham T (1980) Preliminary report on the pathogenicity of *Legionella pneumophila* for freshwater and soil amoebae. J Clin Pathol 33:1179–1183

Scheid P (2007) Mechanism of intrusion of a microspordian-like organism into the nucleus of host amoebae (*Vannella* sp.) isolated from a keratitis patient. Parasitol Res 101(4):1097–1102. https://doi.org/10.1007/s00436-007-0590-z

Scheid P (2014) Relevance of free-living amoebae as hosts for phylogenetically diverse microorganisms. Parasitol Res 113:2407–2414. https://doi.org/10.1007/s00436-014-3932-7

Scheid P (2018a) Amoebophagous fungi as predators and parasites of potentially pathogenic free-living amoebae. Open Parasitol J 6:75–86. https://doi.org/10.2174/1874421401806010075

Scheid P (2018b) Free-living amoebae and their multiple impacts on environmental health. Encyclopedia of environmental health, 2nd edn. https://doi.org/10.1016/B978-0-12-409548-9.10969-8

Scheid P, Balczun C (2017) Failure of molecular diagnostics of a keratitis-inducing *Acanthamoeba* strain. Exp Parasitol 183:236–239

Scheid P, Schwarzenberger R (2011) Free-living amoebae as vectors of cryptosporidia. Parasitol Res 109:499–504. https://doi.org/10.1007/s00436-011-2287-6

Scheid P, Hauroeder B, Michel R (2010) Investigations of an extraordinary endocytobiont in *Acanthamoeba* sp.: development and replication. Parasitol Res 106(6):1371–1377. https://doi.org/10.1007/s00436-010-1811-4

Scheid P, Zoller L, Pressmar S, Richard G, Michel R (2008) An extraordinary endocytobiont in *Acanthamoeba* sp. isolated from a patient with keratitis. Parasitol Res 102:945–950. https://doi.org/10.1007/s00436-007-0858-3

Steenbergen J, Shuman H, Casadevall A (2001) *Cryptococcus neoformans* interactions with amoebae suggest an explanation for its virulence and intracellular pathogenic strategy in macrophages. Proc Natl Acad Sci U S A 98:15245–15250. https://doi.org/10.1073/pnas.261418798

Steinert M, Birkness K, White E, Fields B, Quinn F (1998) *Mycobacterium avium* bacilli grow saprozoically in coculture with *Acanthamoeba polyphaga* and survive within cyst walls. Appl Environ Microbiol 64:2256–2261

Walochnik J, Aspöck H (2007) Amöben: Paradebeispiele für Probleme der Phylogenetik, Klassifikation und Nomenklatur. Denisia 20:323–350

Wieser A, Schubert S (2011) Intra-und extrazelluläre Biofilme uropathogener *E. coli*. Chemotherapie J 20:181–185

Index

H. Mehlhorn, S. Klimpel (eds.), *Parasite and Disease Spread by Major Rivers on Earth*, Parasitology Research Monographs 12,
https://doi.org/10.1007/978-3-030-29061-0

T

U

V

The manufacturer's authorised representative in the EU is Springer Nature Customer Service Centre GmbH, Europaplatz 3, 69115 Heidelberg, Germany. If you have any concerns regarding our products, please contact ProductSafety@springernature.com

Printed and bound by CPI Group (UK) Ltd, Croydon, CR0 4YY

05/10/2026

02228123-0013